Nanotechnology in the Life Sciences

Series Editor

Ram Prasad, Department of Botany, Mahatma Gandhi Central University, Motihari, India

Nano and biotechnology are two of the 21st century's most promising technologies. Nanotechnology is demarcated as the design, development, and application of materials and devices whose least functional make up is on a nanometer scale (1 to 100 nm). Meanwhile, biotechnology deals with metabolic and other physiological developments of biological subjects including microorganisms. These microbial processes have opened up new opportunities to explore novel applications, for example, the biosynthesis of metal nanomaterials, with the implication that these two technologies (i.e., thus nanobiotechnology) can play a vital role in developing and executing many valuable tools in the study of life. Nanotechnology is very diverse, ranging from extensions of conventional device physics to completely new approaches based upon molecular self-assembly, from developing new materials with dimensions on the nanoscale, to investigating whether we can directly control matters on/in the atomic scale level. This idea entails its application to diverse fields of science such as plant biology, organic chemistry, agriculture, the food industry, and more.

Nanobiotechnology offers a wide range of uses in medicine, agriculture, and the environment. Many diseases that do not have cures today may be cured by nanotechnology in the future. Use of nanotechnology in medical therapeutics needs adequate evaluation of its risk and safety factors. Scientists who are against the use of nanotechnology also agree that advancement in nanotechnology should continue because this field promises great benefits, but testing should be carried out to ensure its safety in people. It is possible that nanomedicine in the future will play a crucial role in the treatment of human and plant diseases, and also in the enhancement of normal human physiology and plant systems, respectively. If everything proceeds as expected, nanobiotechnology will, one day, become an inevitable part of our everyday life and will help save many lives.

Karen Esquivel Escalante
Ana Angélica Feregrino Pérez
Alicia Rosales-Pérez • Ram Prasad
Editors

Green Nanomaterials

An Agriculture and Circular Economy Alternative

Springer

Editors
Karen Esquivel Escalante
División de Investigación y Posgrado
Facultad de Ingeniería
Universidad Autónoma de Querétaro,
Cerro de las Campanas S/N Santiago de Querétaro
Querétaro, Mexico

Ana Angélica Feregrino Pérez
Department of Engineering
Autonomous University of Queretaro
Queretaro, Querétaro, Mexico

Alicia Rosales-Pérez
Universidad Tecnologica Metropolitana de Aguascalientes
Aguascalientes, Mexico

Ram Prasad
Department of Botany
Mahatma Gandhi Central University
Motihari, Bihar, India

ISSN 2523-8027 ISSN 2523-8035 (electronic)
Nanotechnology in the Life Sciences
ISBN 978-3-032-04491-4 ISBN 978-3-032-04492-1 (eBook)
https://doi.org/10.1007/978-3-032-04492-1

This Springer imprint is published by the registered company Springer Nature Switzerland AG
The registered company address is: Gewerbestrasse 11, 6330 Cham, Switzerland

This book is dedicated to the researchers and innovators, whose tireless pursuit of knowledge continually drives scientific progress, creating valuable contributions for society and leaving a lasting impact on our planet.

Preface

The rapid evolution of nanotechnology has brought extraordinary opportunities to transform traditional practices across different industries. One example is agriculture, where innovative technologies are needed to address environmental and toxic challenges effectively. Among the most promising developments is the emergence of green nanomaterials, which offer sustainable, environmentally responsible alternatives to conventional agricultural inputs and practices. This book, *Green Nanomaterials: An Agriculture and Circular Economy Alternative*, explores the intersection between nanoscience, sustainability, and circular economy models in the context of modern agriculture.

In response to growing environmental challenges, resource depletion, and the urgent need to reduce chemical usage, new methodologies have been developed for obtaining nanostructured materials through environmentally friendly synthesis methods aimed at minimizing or eliminating environmental impacts. Grounded in the principles of green chemistry, these methods offer attractive alternatives by harnessing natural biological processes and utilizing waste materials to produce nanometric materials that are effective, safe, and aligned with ecological and circular economy principles. Additionally, these nanomaterials are carefully designed and applied according to specific requirements. Their integration into agriculture presents enormous potential for enhancing plant growth and fruit production, improving soil health, managing pests and diseases effectively, and reducing post-harvest losses—all within a framework that promotes resource efficiency and nanomaterial circularity.

This book is structured to guide the reader from the fundamentals of nanotechnology through advanced applications and implications. It begins with Chap. 1, introducing the basic principles of nanotechnology and nanomaterials, followed by Chap. 2, which details the core principles and green synthesis methods used in the field. Chapters 3, 4, and 5 examine the types of green-synthesized nanomaterials used in agriculture and their mechanisms of interaction within plant systems, including plant stress responses. Moving forward, Chaps. 6, 7, 8, 9, and 10 focus on soil health, nanofertilizers, pest and disease management, and post-harvest applications, offering insight into practical and innovative agricultural uses.

The next section, spanning Chaps. 11, 12, and 13, addresses the environmental, toxicological, biosafety, and economic aspects, providing a holistic view of the opportunities and responsibilities that come with deploying these technologies. Finally, Chaps. 14, 15, 16, and 17 explore circular economy principles and their relevance to sustainable agriculture. These chapters underscore how green nanomaterials can align with regenerative models that minimize waste, optimize resource use, and create long-term value across the agricultural value chain.

This book is intended for researchers, students, practitioners, and policymakers who are looking to understand and implement sustainable nanotechnological solutions in agriculture. By merging the disciplines of green chemistry, nanotechnology, and circular economy, we hope this work contributes to a future of farming that is productive, responsible, and resilient.

Querétaro, Mexico — Karen Esquivel Escalante
Queretaro, Querétaro, Mexico — Ana Angélica Feregrino Pérez
Aguascalientes, Mexico — Alicia Rosales-Pérez
Motihari, Bihar, India — Ram Prasad

Acknowledgments

The editors thank each and every one of the authors for their participation, commitment, and scientific contributions.

We also thank the publisher for the opportunity to create this compilation of information, which provides novel and up-to-date information in the field of nanotechnology applications.

KEE and AAFP would especially like to thank the FONFIVE projects (FIN 202413 and FIN202420, respectively) for their support.

Contents

1 **Introduction to Nanotechnology and Nanomaterials** 1
Devanampriyan Rajan, Kalaivani Sundaravadivelu,
Sankari Rajan, and Rakesh Kumar Gupta

2 **Principles and Methods of Green Synthesis** 27
M. L. Jiménez-Gonzalez, Alicia Rosales-Pérez,
Adriana Hernández-Rangel, and Rafael Hernández-Rangel

3 **Green Nanomaterials: An Agriculture and Circular Economy Alternative** 55
Tharindu D. Nuwarapaksha, Shashi S. Udumann,
Nuwandhya S. Dissanayaka, and Anjana J. Atapattu

4 **Green Nanomaterials: Mechanisms of Action in Plant Systems** 79
Yolanda González-García and Antonio Juárez-Maldonado

5 **Plant Stress** 101
Daniel Mendoza Jiménez, Edgardo Ulises Esquivel Naranjo,
and José Antonio Cervantes Chávez

6 **Soil Interaction and Health** 113
Alejandra Jimenez-Hernandez, Valeria Aldaz-Medina,
Jesús Antonio Angole-Tierrablanca, Graciela Avila-Quezada,
and Ramon Gerardo Guevara-Gonzalez

7 **Green Synthesized Nanofertilizers and Plant Growth Promoters** 129
Songita Sonowal and Ram Prasad

8 **Green Synthesized Nanotechnology in Pest and Disease Management** 147
M. Y. Hernández-Ramírez, E. Aguirre-García, and L. A. Martínez-Chávez

9 **Sustainable Nanofertilizers and Plant Nutrition** 189
Sara Khalaji, Sima Panahirad, and Gholamreza Gohari

10 **Post-Harvest Applications of Green-Synthesized Nanomaterials** 207
Deepoo Meena, Rakesh Kumar Gupta, and Prem Prakash Srivastav

11 **Environmental and Ecotoxicological Impact** 241
Mohammad Harun-Ur-Rashid, Abu Bin Imran, and Md. Abu Bin Hasan Susan

12 **Ensuring Safe Nanotechnology: A Biosafety and Regulatory Perspective** 257
Himani Pandey and Rakesh K. Pandey

13 **Economic and Sustainability Aspects** 273
Adriana Hernández-Rangel, Enrique Samaniego-Benítez, Edgar Iván Hernández-Carbajal, and Alicia Rosales-Pérez

14 **Circular Economy Principles** 303
Deepshikha Gupta, Pratibha Sharma, and Tejendra Kumar Gupta

15 **Circular Economy and Sustainability** 333
José L. Navarro, Gustavo Acosta-Santoyo, Irma Robles, Josué D. García Espinoza, and Luis A. Godínez

16 **Sustainability and Economics** 369
Jean Vinícius Moreira

17 **Innovative Applications in Sustainable Agriculture** 393
Aldo Cordoba Guerrero, Diego F. Monroy-Macias, Sergio Ontiveros Seoane, Arturo M. Castillo-Rico, and Karen Esquivel Escalante

Index 415

Abbreviations

Chapter 1

1-D	One-dimensional
2-D	Two-dimensional
3-D	Three-dimensional
Ag	Silver
AI	Artificial intelligence
Au	Gold
BET	Brunauer, Emmett, and Teller
CAN	Calcium ammonium nitrate
CBNs	Carbon-based nanomaterials
CNDs	Carbon nanodots
CNTs	Carbon nanotubes
Cu	Copper
CuO	Copper oxide
EC	Electro-coagulation
Fe_3O_4	Iron oxide
GO	Graphene oxide
ICT	Information communication technology
mi-RNA	Micro RNA
Nano-BC	Nano-biochar
NBS-LRR	Nucleotide binding site
NCs	Nanocomposites
NFC	Near field communication method
NP	Nanoparticles
PEG	Ethylene glycol
PEGS	Paper-based electric-gas sensors
PLGA	Poly(lactic-*co*-glycolic acid)
QD	Quantum dots
RNAi	RNA-interference
SAP	Sub-absorbent polymers

Si-NP	Silicon nanoparticles
TiO_2	Titanium dioxide
TTI	Time-temperature indicator sensors
UHF	Ultra-high frequency
VIGS	Virus-induced gene silencing
ZnNPO	Zinc oxide nanoparticles
ZnO	Zinc oxide
ZnO-NPsZC	Composite with zinc chloride

Chapter 2

AgNPs	Silver nanoparticles
AuNPs	Gold nanoparticles
E-Factor	Environmental Impact Factor
GNPs	Green nanoparticles
$HauCl_4$	Chloroauric acid
IUPAC	International Union of Pure and Applied Chemistry
NPs	Nanoparticles
OECD	Organization for Economic Cooperation and Development

Chapter 3

3D	Three-dimensional
Ag	Silver
Au	Gold
CBNs	Carbon-based nanomaterials
CDs	Carbon dots
CNTs	Carbon nanotubes
Fe	Iron
nZVI	Nanoscale zero-valent iron
PHAs	Poly hydroxyalkanoates
PLA	Polymer polylactic acid
PLGA	Poly lactic-*co*-glycolic acid
UV	Ultraviolet light
Zn	Zinc

Chapter 4

BG	Butylene glycol
BGL	26-beta-glucosidase
CA	Caffeinic acid
CNMs	Carbon nanomaterials
CYC	Cycloartenol synthase
EG	Ethylene glycol
GS	Geissoschizine synthase
LOX	Lipoxygenase
MAPK3	Mitogen-activated protein kinase

NiO NPs	Nickel oxide nanoparticles
NMs	Nanomaterials
NPs	Nanoparticles
ORCA3	Octadecanoid-responsive AP2
PAL	L-phenylalanine ammonia-lyase
PEG	Polyethylene glycol
PG	Propylene glycol
POD	Peroxidase
PPO	Polyphenoloxidase
PUFA	Polyunsaturated fatty acids
PUFA-OOH	Polyunsaturated fatty acid hydroperoxide
RA	Rosmarinic acid
RUBISCO	Ribulose bisphosphate carboxylase oxygenase
Se NPs	Selenium nanoparticles
SiO2 NPs	Silicon dioxide nanoparticles
SOD	Superoxide dismutase
SQS	Squalene monooxygenase
TEG	Tetraethylene glycol
TrEG	Triethylene glycol

Chapter 5

ACC	Acetyl-CoA carboxylase
ACC	Aminocyclopropanecarboxylate
CAT	Catalase
CEC	Cationic exchange capacity
Cr	Chromium
CuNPs	Copper nanoparticles
H_2O_2	Hydrogen peroxide
HS	Heat stress
JA	Jasmonic acid
MeJA	Methyl jasmonate
NMs	Nanomaterials
NO	Nitric oxide
O^{2-}	Superoxide anion
OH^-	Hydroxyl radical
PMs	Primary metabolites
RNS	Reactive nitrogen species
ROS	Reactive oxygen species
SA	Salicylic acid
SMs	Secondary metabolites
SOD	Superoxide dismutase
SS	Salinity stress
TiO_2	Titanium dioxide
UV	Ultraviolet light

ZnO Zinc oxide

Chapter 6

AM	Arbuscular mycorrhizal
CEC	Cation exchange capacity
CNTs	Carbon nanotubes
MONPs	Metallic oxide nanoparticles
NFs	Nanofertilizers
NM	Nanomaterials
NPs	Nanoparticles
OH^-	Hydroxyl ions
PHA	Polyhydroxyalkanoate
PLA	Polylactic acid
ROS	Reactive oxygen species
SEA	Soil enzymatic activity

Chapter 7

AMF	Arbuscular mycorrhizal fungi
NMs	Nanomaterials

Chapter 8

Ag NPs	Silver nanoparticles
AuNPs	Gold-based nanoparticles
CNTs	Carbon nanotubes
GO	Graphene oxide
NMs	Nanomaterials
NPs	Nanoparticles
ROS	Reactive oxygen species
R-SH	Thiol groups
SODs	Superoxide dismutases
ZnO NPs	Zinc oxide nanoparticles

Chapter 9

AgNPs	Silver nanoparticles
ECV	Ecological and economic viability
ZnO NPs	Zinc oxide nanoparticles

Chapter 10

AgNPs	Silver nanoparticles
BIS	Bureau of Indian Standards
CMC	Carboxymethylcellulose sodium
CP	Carambola extract
CS	Chitosan
Cu NPs	Copper nanoparticles

CznO-NPs Colloidal chemistry synthesis zinc oxide nanoparticles
DBT Department of Biotechnology
EFSA European Food Safety Authority
FDA Food and Drug Administration
Fe_3O_4 Iron oxide
FSANZ Food standards Australia New Zealand
FTIR Fourier-transform infrared spectroscopic
GSNP Green-synthesized nanoparticles
GSNPs Green synthesis nanoparticles
GZnO-NPs Green synthesis zinc oxide nanoparticles
MBC Minimum bactericidal concentration
NPs Nanoparticles
ZnO NPs Zinc oxide nanoparticles

Chapter 11

ENMs Engineered nanomaterials
REACH Registration, Evaluation, Authorization, and Restriction of Chemicals
ROS Reactive oxygen species
ZnO Zinc oxide

Chapter 12

ACCC Australian Competition and Consumer Commission
BIS Bureau of Indian Standards
CLP Classification, Labeling, and Packaging
CMP Chemical Management Plan
CSCL Chemical Substances Control Law
ECHA European Chemicals Agency
EPA Environmental Protection Agency
FDA Food and Drug Administration
IEC International Electrotechnical Commission
ISO International Organization for Standardization
LAL Limulus amebocyte lysate
MTS Micronucleus test
NCSN New Chemical Substance Notification
OECD Organization for Economic Co-operation and Development
PLGA Poly (lactide-*co*-glycolide)
PPE Personal protective equipment
REACH Registration, Evaluation, Authorization, and Restriction of Chemicals
TiO_2 Titanium dioxide
TSCA Toxic Substances Control Act
ZnO Zinc oxide

Chapter 13

Ag GNPs	Silver Green nanoparticles
$C_8H_{14}ClN_5$	Atrazine
GCPs	Green Chemistry Principles
GNPs	Green nanoparticles
LCA	Life cycle assessment
$M(NO_3)_3$	Metal nitrates
MnO	Manganese oxide
$NaBH_4$	Sodium borohydride
NMs	Nanomaterials
NPs	Nanoparticles
PLA	Polylactic acid
QDs	Quantum dots
ROS	Reactive oxygen species
SDGs	Sustainable development goals
TEA	Technoeconomic analysis

Chapter 14

AI	Artificial intelligence
CBG	Cane bagasse
CE	Circular economy
COP21	United Nations Climate Change Conference
GHG	Greenhouse gas
IPM	Integrated Pest Management
SDG 2	Zero Hunger
SDG 3	Good Health and Well-being
SDG 4	Quality Education
SDG 5	Gender equality
SDG 6	Public health
SDG 13	Climate action
SDGs	Sustainable Development Goals
SDGs 7–9	Clean energy, economic development and innovation
SDGs 10–12	Reduced inequalities, sustainable cities, and responsible consumption
SDGs 14 & 15	Ecosystems
SDGs 16 & 17	Peace, justice and global partnerships

Chapter 15

AOP	Advanced oxidation processes
BOD/COD	Biological oxygen demand/chemical oxygen demand
CNCs	Cellulose nanocrystals
CNFs	Cellulose nanofibers
COP	Conference of the Parties
COP21	United Nations Climate Change Conference

DO Dissolved oxygen
EC Electrical conductivity
FE Freshwater eutrophication
FRS Fossil resource scarcity
GWP Global warming potential
IPCC Intergovernmental Panel on Climate Change
LCA Life cycle assessment
MAB Man and the Biosphere Program
MDGs Millennium Development Goals
ME Marine eutrophication
MEC Microbial electrolysis cell
MFC Microbial fuel cell
MOFs Metal organic frameworks
NDCs Nationally Determined Contributions
ROS Reactive oxygen species
RSC Residual sodium carbonate index
SAR Sodium adsorption ratio
SDGs Sustainable Development Goals
SS Suspended solids
TA Terrestrial acidification
TET Terrestrial ecotoxicity
TN Total nitrogen

Chapter 16

AI Artificial intelligence
CAP Common Agricultural Policy
CE Circular economy
EGD European Green Deal
FAO Food and Agriculture Organization
GHG Greenhouse gas
NPs Nanoparticles
ROI Return on investment
SDGs Sustainable Development Goals
SiO_2 Silicon dioxide
TiO_2 Titanium dioxide

Chapter 17

Ag Silver
AgNPs Silver nanoparticles
AuNPS Gold nanoparticles
CuO Copper oxide
FeNPs Iron nanoparticles
HAP Hydroxyapatite

HER	Hydrogen evolution reaction
IoT	Internet of Things
NMs	Nanomaterials
NPs	Nanoparticles
SAPs	Super absorbent polymers

List of Figures

Fig. 1.1	Synthesis methods of nanoparticles, such as approaches and methods for synthesis. (Constructed by Sankari Rajan)	11
Fig. 1.2	Green synthesis of silver nanoparticles has antimicrobial activities. (Constructed by Sankari Rajan)	13
Fig. 2.1	Twelve principles of green chemistry	29
Fig. 2.2	Schematic of biosynthesis of nanoparticles	33
Fig. 3.1	Types of nanomaterials used in agriculture. (Constructed by authors)	56
Fig. 3.2	Comparison of conventional and nano fertilizers: impact on plant growth and soil health (constructed by authors)	69
Fig. 4.1	Factors affecting the interplay of NPs with plants. (Constructed by Y. González-García)	85
Fig. 6.1	Positive and negative biochemical interactions of NMs in soils. (Constructed by J.A.A-T)	115
Fig. 6.2	Some beneficial effects of nanomaterials in their interaction with microorganisms. (Constructed by V.A-M and G.A-Q)	121
Fig. 7.1	Physiological functions of plant growth promoters	137
Fig. 7.2	Mechanism of nanofertilizers in plant growth and development	139
Fig. 8.1	Nanopesticides as protectors and delivery systems for agrochemical ingredients in crop production. (Constructed by M.Y. Hernández-Ramírez)	150
Fig. 8.2	Classification of nanopesticides according to the type of nanomaterials used and the specific formulation developed. (Constructed by M.Y. Hernández-Ramírez)	151
Fig. 8.3	Nanocarriers used in nanopesticides formulations. (Constructed by M.Y. Hernández-Ramírez)	156

Fig. 8.4 Advantages of nanopesticides in controlled release of active ingredients (AI). (Constructed by M.Y. Hernández-Ramírez) 162
Fig. 8.5 Principal pathways of nanopesticides in plant uptake and physicochemical properties that influence the cellular uptake of nanopesticides through cell membranes. (Constructed by M.Y. Hernández-Ramírez) 162
Fig. 8.6 Environmental impact of nanopesticides. (Constructed by M.Y. Hernández-Ramírez) 165
Fig. 8.7 Schematic representation of proposed mechanisms of nanomaterial toxicity against bacteria and fungi. (Constructed by E. Aguirre-García) 169

Fig. 9.1 World population growth estimation by 2100 (https://ourworldindata.org/grapher/population-long-run-with-projections) 190
Fig. 9.2 An introduction to nano fertilizers 192

Fig. 10.1 Production and application of green-synthesized nanoparticles 214

Fig. 12.1 (**a**) A schematic figure showing the applications and hazards associated with nanotechnology. (**b**) Articles appearing in PubMed search for the keyword "nanotechnology safety" during 2000–2025. (Constructed by Rakesh K. Pandey) 260

Fig. 13.1 Characteristics and differences between traditional processes and green routes for the synthesis of NPs. (Constructed by E. Samaniego-Benítez and A. Rosales-Pérez) 276
Fig. 13.2 Phases of the synthesis of GNPs. (Constructed by A. Rosales-Pérez) 279
Fig. 13.3 GNPs green synthesis approach summary. (Constructed by A. Rosales-Pérez and E. Hernández-Carbajal) 284
Fig. 13.4 End of life of the GNPs. (Constructed by A. Hernández-Rangel and A. Rosales-Pérez) 290

Fig. 14.1 A generic representation of circular economic framework starting from agriculture waste to value added products 305
Fig. 14.2 17 UN Sustainable Development Goals that can be enabled by implementing circular economy either partially or strongly. (Reproduced from (Velenturf & Purnell, 2021, Henrik Hvid Jensen; Chief Technology Strategist NEE Jan 27, 2022)) 306
Fig. 14.3 (**a**) Usage of wind and solar as renewable energy sources to run different agricultural machines. (**b**) Efficient use of hydro and biomass as renewable energy and recycling of agricultural waste/produce. (Reproduced from (Majeed et al., 2023) under Creative Commons license) 312

Fig. 14.4 Schematic representation of different biofuels and their major applications. (Reproduced from Nair et al. (2022) under Creative Commons license) 316
Fig. 14.5 A block chain process in agriculture 319
Fig. 14.6 CE loop in sustainable model of Araku Coffee Project 321
Fig. 14.7 Conversion of banana waste into useful products via extraction of fires from pseudostem 324
Fig. 14.8 The production process of the Banana peels-polystyrene composites. (Reproduced from (Mohamed et al., 2023) under a common creative license) 325
Fig. 14.9 Sugarcane bagasse valorization through an integrated process for single cell oil, sulfonated carbon-based catalyst, and biodiesel co-production. (Reproduced from (Hassa et al., 2024) under a common creative licence) 327

Fig. 15.1 Pathway to Sustainability: Global Events and Agreements. (**a**) Notion of the Interrelationship between environment and development. (**b**) Creation of UNEP. (**c**) Computational models to predict consequences of unlimited economic and population growth. (**d**) Introduced the concept of sustainable development, linking natural resource conservation with economic and social development. (**e**) UN General Assembly established this commission to study the interactions between economic development and the environment. (**f**) Introduced the concept of sustainable development. (**g**) Agreements like Agenda 21, the Climate Change Convention, and the Convention on Biological Diversity. (**h**) The first binding agreement requiring industrialized countries to reduce their greenhouse gas emissions. (**i**) Also known as Rio + 10, it reviewed progress since 1992 and emphasized integrating social, economic, and environmental goals. (**j**) Achieved during COP21, it marked a milestone in climate action, setting global goals to limit temperature increases below 2 °C. (**k**) A global framework to address social, economic, and environmental challenges through 17 interconnected goals. (Constructed by J.L. Navarro) 337
Fig. 15.2 Wastewater alternative applications toward an agricultural circular economy. (Adapted from (Mannina et al., 2022) 339

Fig. 17.1 (**a**) Different biomass extract precursors, (**b**) bio-synthesis method, (**c**) metal ions are reduced to metal NPs by the active molecules through an intracellular enzymatic redox process. (**d**) Nanoparticle stabilization. (Constructed by Diego F. Monroy-Macias) 395
Fig. 17.2 Operation of IoT technology. (Constructed by S. O. Seoane) 398
Fig. 17.3 Schematic representation of the green NPs production and large-scale application in agriculture. (Constructed by A. Cordoba) 402

List of Tables

Table 1.1 Elemental composition and physical properties of different nanoparticles 9
Table 1.2 Characterization techniques and applications of nanoparticles 14

Table 2.1 Nanostructured systems synthesized by bacterial strains 38
Table 2.2 Nanostructured systems synthesized by fungal strains 40
Table 2.3 Nanostructured systems synthesized by plant extracts 43

Table 6.1 Effects of NMs on physical properties of soils. (Constructed by A.J-H, J.A.A-T, and R.G.G-G) 115

Table 7.1 Some examples of metal based nanofertilizers with their function in plants 133
Table 7.2 Some examples of carbon-based nanofertilizers with roles in selected plants 135

Table 8.1 Advances in the development of nanopesticides based on inorganic, organic, and hybrid NMs for pest control 152
Table 8.2 Advances in the development of nanoformulations for pest control 158
Table 8.3 Classification of nanomaterials by type, their key properties, antimicrobial mechanisms, and agricultural applications 167
Table 8.4 Applications of different nanomaterials in agriculture 173

Table 10.1 Application of GSNP in post-harvest processing 215

Table 11.1 Nanomaterial transformations in different environmental compartments 246
Table 11.2 Summary of nanomaterial-induced toxicity across ecosystems 249
Table 11.3 Comparison of risk assessment frameworks for nanomaterials 251

Table 12.1 Classification of various types of nanomaterials, their applications, and potential risks 259

Table 13.1 Raw materials and synthesis temperature of different metallic GNPs and metallic oxides by green synthesis 277
Table 13.2 Summary of disadvantages of biological/green bottom-up methods for synthesizing NPs 282

Table 15.1 Various impact of nanotechnology on the quantity indicators of crops 344
Table 15.2 Comparison of LCA approaches applied to advanced wastewater treatment 351

Table 17.1 Green-synthesized nanoparticles by microorganism extracts 397
Table 17.2 Antimicrobial activity of biosynthesized Ag-Au NPs 397
Table 17.3 Different types of agricultural solid waste, their by-products, and their possible uses (Ginni et al., 2021) 401
Table 17.4 Provides various proposals for large-scale green synthesis of the most extensively studied NPs for agricultural applications 405

Chapter 1
Introduction to Nanotechnology and Nanomaterials

Devanampriyan Rajan, Kalaivani Sundaravadivelu, Sankari Rajan, and Rakesh Kumar Gupta

1.1 Introduction

Nanotechnology is the field of science and engineering that utilizes the unique properties of materials at the nanoscale to develop new technologies. The nanoscale refers to particles one billion times smaller than a meter. Nanotechnology involves the design, creation, and application of systems at the nanoscale, manipulating matter from individual atoms and molecules up to submicron dimensions and further integrating these nanoscale materials into larger, functional systems (Nasrollahzadeh et al., 2019; Gupta et al., 2022a, b). The nanoscale ranges from 1 to 100 nm particles which fall under this size range are called particles or nanomaterials (McNeil, 2005; Gupta et al., 2024c). During fourth-century C.E., Lycurgus cup showcases the use of gold and silver nanoparticles in glass, resulting in colour changes based on light angle, revealing an early appreciation for the unique properties of nanomaterials. Richard Feynman's lectures were widely considered the catalyst for modern nanotechnology, as he introduced the revolutionary concept of manipulating individual atoms and molecules, which would lay a foundational principles of nanoscience (Sarma et al., 2023). In early twentieth century, scientists began to unravel the distinct properties exhibited by nanosized particles, properties of bulk materials, which establish the fundamental scientific principles in nanotechnology. Furthermore, nanotechnology expanded into diverse sectors, including construction, artificial intelligence, agriculture, sports, and environmental science, which enhances the product performance. Concurrently, the evolution of nanoparticle synthesis methodologies like "top-down" and "bottom-up" approaches offers distinct advantages

D. Rajan · K. Sundaravadivelu · S. Rajan
Department of Food Technology, Kongu Engineering College, Erode, Tamil Nadu, India

R. K. Gupta (✉)
Agricultural and Food Engineering Department, Indian Institute of Technology Kharagpur, Kharagpur, West Bengal, India

K. Esquivel Escalante et al. (eds.), *Green Nanomaterials*, Nanotechnology in the Life Sciences, https://doi.org/10.1007/978-3-032-04492-1_1

and limitations on specific application requirements. Crucial advancement tools in characterization, enabling researchers to analyse the morphologies, structures, compositions, phases, and surface chemistries, which transform the theoretical understanding into practical applications. While nanotechnology has achieved remarkable advancements remains largely untapped, with ongoing research consistently revealing novel applications and possibilities, underscoring the dynamic nature and promising future with transformative innovations (Malik et al., 2024).

Nanotechnology is evolving in agriculture by offering targeted solutions for enhanced productivity and sustainability. Precise plant disease management using nano-capsulated pesticide reduces overall chemical usage, which penetrates plant tissues effectively. Nano fertilizers, incorporating various metal oxide nanoparticles, improve nutrient absorption and increase stress tolerance against environmental challenges like drought and salinity. Furthermore, nanotechnology enables the purification of water and soil, crucial for sustainable agriculture, by removing contaminants. Controlled herbicide release through nano capsules minimizes waste and impact on non-target organisms (Rajan & Mathew, 2025).

This chapter provides a comprehensive overview of nanotechnology, its fundamental principles, diverse classifications, and transformative applications in the agricultural sector. By exploring the fundamentals of nanomaterials, exploring the properties such as physical, chemical, optical, and mechanical, which distinguish them from bulk materials. Subsequently, this chapter explores the nanomaterials relevant to agriculture, highlighting their specific attributes and functionalities. An overview of the synthesis and characterization techniques, crucial for the controlled production. Finally, concludes by exploring the emerging trends, challenges, and future directions of nanotechnology in agriculture, emphasizing its potential to revolutionize farming and irrigation practices.

1.2 Fundamentals of Nanomaterials

Nanomaterials (NM) can be classified as one-dimensional (1-D), two-dimensional (2-D), three-dimensional (3-D), nanocomposite and nanomaterials. Based on the characteristics and implementation, nanomaterials can also be classified as nanoscale devices, nanoscale magnetic, nanostructured biomaterials, and nanopharmaceuticals (Zhang, 2018; Gupta et al., 2024a). Main attributes of the nanomaterials are the size and dimensions will be around 1 and 100 nm. Based on the size, they have been classified as the 0-D, 1-D, 2-D, and 3-D as the nanospheres and nano fibres (Castillo, 2022; Gupta et al., 2024b). Nanomaterials with optical and electrical properties exhibiting a 3D pattern of dimension less than 100 nm are called quantum dots (QD). Quantum confinement has different dimensional structures or films (Awan et al., 2020). Nanomaterials with a size within 100 nm in length, breadth, and thickness are called '0' dimensional nanomaterials. Dimensions of the 0-D material are spherical or a cluster of nanoparticles. Nanoparticles have a unique property, like antibacterial, detectors, electrical NP, such as metals, oxides, hydroxides, sulphides, silica, and inorganic materials, with the breadth and dimension which are in the range of the nm but the length in the range of mm then it is called a 1D nanomaterial. NM, which obtained in the different sizes such as, tubes, wires, fibres, needles, cellulose fibres, carbon fibres, with the length and breadth with a range of micrometre and depth is around a manometer, are classified as the 2D nanomaterials, which are the lightest materials. Nanomaterials can be classified as organic, inorganic, or organic–inorganic materials. 2D nanomaterials are in the form of sheets, films, or layers. The categorization of nanomaterials is determined by their dispersion characteristics in polymer-based systems. In polymer, nanomaterials inherent crystalline structure but exhibit expanded interlayer spacing due to intercalation by polymer chains, resulting in a layered, sandwich-like configuration. Polymer nanomaterials have also been classified as straight-chain polymers and branched polymers. Cellulose nanofibres are a straight-chain polymer that shows more strength and H–H bonding (Karak, 2019).

1.3 Types of Nanomaterials in Agriculture

1.3.1 Carbon-Based Nanomaterials

Carbon-based nanomaterials (CBNs) are atomic structures that have reactive properties that improve plant growth and crop protection. CBNs include fullerene, graphene, carbon nanodots (CNDs), and carbon nanotubes (CNTs) and have numerous applications in agriculture, such as photosynthesis, delayed-release fertilizer, soil decontamination, and plant protection. CNTs improve seed germination by increasing moisture uptake through seed coat penetration in turnip, tomato, onion, and radish at concentrations ranging from 0 to 40 μg/mL. They inhibit replication of the tobacco mosaic virus at a concentration of 200 mg/L, acting as an antimicrobial agent (Safdar et al., 2022). Both fullerene and graphene aid in plant growth by decreasing the effects of drought stress on bell pepper plants. Graphene (1000 mg/L) enhances fruit production (76.2%), fruit dry weight (73.2%), and electrolyte leakage index (83.6%) under 50% irrigation intensity of field capacity and at the same time, the relative water content of the leaves dropped up to 65.9%. The fullerene at 100 and 1000 mg/L raised the flavonoid content to 90. 9% and a total phenol content of 63.1%, respectively (Ahmadi et al., 2024). The hydrophilic two-dimensional material known as graphene oxide (GO) facilitates the delayed release of fertilizers. The coating of KNO_3 fertilizer with GO (0.18%, 0.72%, 1.43%, and 2.86%) greatly extends the release period from 24 to 38 days and boosts nutrient utilization efficiency (An et al., 2017). CNDs are nanoparticles which is <10 nm in size and have vigorous fluorescent effects. They increased the mung bean shoots and roots length by inducing the activity of the Rubisco enzyme at 100 mg/mL, and the growth was reduced at 1000 mg/mL. It aids in moisture retention and carbohydrate production in plants (Chandel et al., 2022).

1.3.2 Metal and Metal Oxide Nanoparticles

Metal and metal oxide nanoparticles have been attracting agricultural studies due to their immense use in nutrient supply, seed germination, and disease detection. These include gold (Au), silver (Ag), titanium dioxide (TiO_2), zinc oxide (ZnO), iron oxide (Fe_3O_4), copper oxide (CuO), and other metal nanoparticles. Nanoparticles like TiO_2 (10 ppm) have been utilized in plant protection from drought stress by increasing the activity of antioxidant enzymes to reduce free radical damage, malondialdehyde, and hydrogen peroxide contents. It also improves shoot growth and essential oil contents in the plant (Mohammadi et al., 2016). Nanoparticles like Ag, Cu, Au, ZnO, and TiO_2 have antimicrobial properties against plant pathogens like *Botrytis cinerea* in tomato, *Fusarium oxysporum* in pepper, *Rhizoctonia solani* in rice, *Alternaria solani* in potato, and *Dickeyadadanti* in sweet potato, respectively. FeO and ZnO at 0–100 mg/L significantly increase the chlorophyll content and reduce the cadmium toxin concentrations while increasing Zn and Fe concentrations in wheat plants. Metal nanoparticles like Ag, Au, and Cu have a notable impact on plant gene expression. Au activates specific pathways for plant signalling to modulate gene expression, plant growth, and development. Copper increases the production of defence signalling molecules in plants, which act against pathogens and pests (Francis et al., 2024). Studies have shown that salinity stress can be overcome by using Fe_3O_4 (40 ppm) in *Moringa oleifera* plants by enhancing plant growth along with photosynthetic pigments. It decreases the hydrogen peroxide and lipid peroxidation contents of *Moringa oleifera* leaves (Tawfik et al., 2021). Despite their benefits, some nanoparticles cause adverse effects on aquatic and land plants. Studies show that Ag, Au, and Cu have toxic effects on the ecosystem. To reduce the risks, quantitative studies for concentrations of metal and metal oxide nanoparticles are much needed.

1.3.3 Polymeric and Bio-Based Nanomaterials

Polymeric nanomaterials promote sustainable agriculture by effectively delivering agrochemicals, moisture retention, controlled nutrient release, and chemical loss reduction and are cost-effective. Functional polymers like poly (ethylene glycol) (PEG), poly (lactic-*co*-glycolic acid) (PLGA), polyvinyl alcohol, and acrylate-based polymers are used as nanocarriers for controlled agrochemical delivery without unwanted leaching to the surroundings. PEG and PLGA are amphiphilic polymers that encapsulate herbicide metolachlor in the ratio 4:1, improving solubility and penetration into root cells (Sikder et al., 2021). The use of sub-absorbent polymers (SAP) in drylands facilitates constant irrigation and preserves soil moisture for extended durations by holding the moisture content. The polymers within the soil, for longer periods, can cause negative effects on the ecological systems (Milani et al., 2017). Whereas bio-based nanomaterials are derived from biological sources like microbes and plants. They are an eco-friendly, biodegradable, and biocompatible nanomaterial. It includes plant-based (cellulose, chitin/chitosan) and biogenic-based (bacteria, algae) nanomaterials. They act against abiotic stress, heavy metal toxicity in plants, viral and fungal infection (Saraswat et al., 2023). Chitosan (10 ppm) promotes 30–50% chlorophyll content, 30–60% photosynthesis intensity, and nutrient uptake (9.8–27.4% N, 17.3–30.4% P, 30–45% K) of coffee plants. In a study, silver nanoparticles are synthesized using marine algae *Padina pavonica* (Linn.) and tested against two important pathogens of cotton, which retard their growth. The biosynthesized nanoparticle inhibited the growth to 12.33 ± 0.33 mm and 10.33 ± 0.33 mm for *F. oxysporum* and *X. campestris* pathogens, respectively (Kokabi & Yousefzadi, 2022).

1.3.4 Nano-composites for Enhanced Agricultural Application

Nanocomposites (NCs) are materials with nanoscale (1–100 nm) components having barrier properties, mechanical and thermal stability. They offer enhanced agriculture practices like pest control, targeted nutrient delivery, crop protection, and soil remediation. It has distinct properties like biodegradability, inexpensive, and availability. Nanocomposites like graphene oxide (GO) and silver (Ag) nanomaterials are used for the removal of pesticides like organochlorine chlorpyrifos, endosulfan, and dichloro-diphenyl-dichloroethylene with 95% degradation efficiency. In another study, NCs with b-cyclodextrin/iron oxide for the remediation of organochlorine pesticides from honey were made using reduced GO (Rawtani et al., 2018). Calcium carbonate NCs are used against the brown rot of stems disease caused by the *R. solani* pathogen, which inhibits rot growth and improves the sucrose level. Similarly, Chitosan NCs are used against leaf spot plant disease and inhibit spore germination caused by *A. alternata*. Some NCs, like silver-doped TiO_2 and gold-doped TiO_2, are completely used against lethal agrochemicals (Kumar et al., 2021). Chitosan–copper NCs possess antifungal activity at higher concentrations (100 mg/L) against *Sclerotium rolfsii* (*S. rolfsii*) and *Rhizoctonia solani* (*R. solani*) prepared using acetone solvent (Rubina et al., 2017). NCs like castor oil-derived polyurethane (PU) release nutrients gradually by utilizing an ion exchange mechanism. A cation exchange material (<5% montmorillonite) is coated with PU to successfully inhibit nitrogen discharge from urea granules (Bortoletto-Santos et al., 2020).

1.4 Properties of Nanomaterials

1.4.1 Physical Properties

The physical properties of nanomaterials include size, shape, crystallinity, etc., smaller particle size directly correlates with enhanced nutrient use efficiency and increased crop yields by facilitating better plant metabolic reactions. Nanoparticle shape also plays a role in plant interaction and nutrient uptake, though specific shape-related effects require further elaboration. Crystallinity influences the nutrient sustained release and amorphous forms, boosting nutrient efficiency and yield. Conversely, agglomeration reduces the effective size, hindering nutrient absorption and diminishing the overall effectiveness of the fertilizers in farming. Optimizing the physical properties like size, shape, crystallinity, and agglomeration is crucial for maximizing nano-fertilizer performance in agriculture, offering a potential benefit to improve nutrient uptake and crop yields (Madlala et al., 2024). Due to their extremely small size, ranging from 1 to 100 nm, nanomaterials exhibit a significantly elevated surface area to volume ratio, leading to distinct physical properties and behaviours (ÇIFTÇI, 2024).

1.4.2 Mechanical Properties

Generally, nanomaterials have superior mechanical properties with enhanced strength and toughness. Carbon nanotubes (CNTs) possess good tensile strength, which is suitable for reinforcing materials across multiple engineering applications (ÇIFTÇI, 2024). Reduced size allows for substantial weight reduction compared to bulk materials, while simultaneously enhancing strength. This property is particularly valuable in low-dusting fertilizers, where lighter components translate to improved crop efficiency and overall performance. Furthermore, certain nanomaterials have enhanced flexibility and ductility, enabling them to deform under stress without fracturing and resilience. Additionally, nanomaterials often exhibit increased toughness, tend to absorb energy and deform plastically before fracturing, a feature crucial for withstanding impact and sudden loads. A defining aspect of these materials is their size-dependent mechanical properties; as dimensions decrease, the surface-to-volume ratio increases, leading to significant changes in mechanical behaviour (Kumar et al., 2024).

1.4.3 Optical and Photo-catalytic (Light Absorption, Plasmonic)

Nanomaterials display optical properties exemplified by quantum dots. These nanoscale particles exhibit size-dependent fluorescence, where their emission colour is directly determined by their dimensions. This unique characteristic enables their application in medical imaging and display systems, where precise control over light emission is crucial (ÇIFTÇI, 2024). Nanomaterials present a diverse array of dynamic optical properties, including tunable absorption, transmission, and reflection, all influenced by their size, shape, and surface characteristics. A particular phenomenon is surface plasmon resonance, where the nanomaterial surface electrons resonate with specific light wavelengths, producing distinct colours and enhanced optical effects. The dimensionality of nanomaterials plays a crucial role in shaping their optical behaviour; as size decreases, the electronic structure shifts, altering the energy levels of HOMO and LUMO, thereby impacting optical properties. Spectroscopic methods are essential tools for characterizing these optical behaviours, providing detailed insights into light-nanomaterial interactions. These unique optical attributes extend to applications like photo-catalysis and photoconductivity, where nanomaterials enhance energy conversion and environmental remediation. Furthermore, nanofillers can be seamlessly integrated with conventional fillers to synergistically enhance various composite material properties. This integration allows for the modification and improvement of optical, electrical, mechanical, thermal, and fire-retardant properties. By leveraging the optical properties and their interactions with other materials, nanomaterials open up possibilities for advancements in agricultural fields (Sharma et al., 2024).

1.4.4 Chemical Reactivity

The high surface area characteristic of nanomaterials results in significantly increased chemical reactivity, which has advantages in catalysis and environmental remediation. Moreover, the enhanced reactivity allows nanomaterials to facilitate chemical reactions far more efficiently than larger particles, leading to improved performance and effectiveness (ÇIFTÇI, 2024). Nanomaterials present a diverse range of chemical properties, such as heat of combustion, oxidation tendency, and partition coefficient are altered which impacting energy applications, environmental behaviour, and biological interactions. Molecular weight, boiling and melting points, and solubility also exhibit size-dependent variations, influencing reactivity, processing, and drug delivery. Stability, flammability, and zeta potential are crucial for long-term applications and suspension stability with surface modifications. Furthermore, corrosion rates, reactivity, and chemical composition directly correlate with durability and performance. Radioactivity, when present, necessitates careful handling but offers potential in soil applications. These varied chemical properties, finely tuneable through synthesis, enable nanomaterials to be tailored for specific applications, ranging from catalysis to crop irrigation management (Gupta et al., 2022a, b).

1.4.5 *Bio-compatibility and Biodegradability*

Biocompatibility is the ability of nanomaterials to interact safely with living organisms in soil and plants. Green synthesis, eco-friendly nanomaterial production without hazardous byproducts, has become popular to overcome issues of old, toxic synthesis processes. Furthermore, biogenic nanomaterials derived from biological sources are gaining prominence due to their potential for enhanced biocompatibility. Their natural origins often result in improved acceptance by biological systems, minimizing adverse reactions and paving the way for safer and more effective irrigation systems (Tonelli et al., 2024).

Biodegradability, the biological breakdown of materials into harmless substances, is crucial for minimizing environmental impact. Bio-based nanomaterials, derived from renewable sources, offer a sustainable alternative, with their biodegradability being a key indicator of environmental friendliness. Developing and promoting these materials is vital for reducing pollution and waste, thus contributing to a sustainable future. Further research is essential to enhance the biodegradability of bio-based nanomaterials by exploring new materials, improving existing ones, and understanding the environmental factors that influence their degradation (Arora et al., 2024). Table 1.1 illustrates the different nanomaterials and its sources with physical properties

1.5 Synthesis and Characterization of Green Nanomaterials

A nanomaterial synthesis through green materials is a single-step methodology where they are eco-friendly with improved durability, exceptional characteristics, and precise dimensions. Synthesis of nanomaterials can be achieved by two different methods: one is by the top-down approach and bottom-up approach.

1.5.1 *Top-Down vs. Bottom-Up Approaches*

Top-down approach splits up bulk particles by the process of micro-fabrication techniques such as wet etching, ejection, laser etching, and mechanical pulverization. In this process, the suitable nanomaterial dimension has been decreased due to the chemical and mechanical methods. Major limitations of this method are deformation in the structure and irregularities in the bottom (Thakkar et al., 2010).

Table 1.1 Elemental composition and physical properties of different nanoparticles

No	Nano material	Source	Elemental composition	Physical properties	Reference
1.	RS-CD	Rice straw	Carbon—(63.39%), oxygen—(20.48%), sodium—(9.19%) chloride—(6.66%)	Size: 2.2–5.5 nm	El-Shaheny et al. (2025)
2.	ZnNPO Zinc oxide nanoparticles	Duckweed extract	NA	Size: 18–38 nm	Shukla et al. (2025)
3.	ZnNPO Zinc oxide nanoparticles	Commercial paste	Zinc—76.81% Oxygen—23.19%	Size: 30–100 nm Crystalline size: 43.25 nm	Nabil et al. (2021)
4.	Zn NP Zinc Nanoparticles	Zinc granules (99.99%)	Zinc—99.9% Impurities—not exceed 1000 ppm	Size: 350 nm Crystalline size: 21 nm	Markov et al. (2024)
5.	Silica nanoparticles	Tetra-ethyl-ortho-silica	Silicon—61.48 wt% Oxygen—23.48 wt%	Particle size: 95 ± 5.59 nm—280 ± nm Band energy gap: 3.92 eV—3.79 eV	Yeasmin et al. (2022)
6.	Silica nanoparticles	*Phyllanthus emblica*	Carbon—27.03% Oxygen—53.43% Silica—19.54%	Size: 1.2 nm Shape: spherical Zeta potential—40.5 mV	Sankareswaran et al. (2022)
7.	Cuprous oxide nanoparticles	Copper sulphate pentahydrate salt	Copper—60.93% Oxygen—20.34% Carbon—5.68% Platinum—4.59% Trace elements—aluminium, silicon, sulphur, zinc, zirconium	Particle size: 4.77 nm (0.001 M $CuSO_4$) 0.82 nm (0.1 M $CuSO_4$)	Regmi et al. (2019)
8.	Carbon dots/iron oxide nanoparticles	Iron (III) acetylacetonate	C/17Fe NPs: Carbon—36.3% Iron—17% C/31Fe NPs: Carbon—26.9% Iron—31%	Quantum yield: C/17Fe NPs—9% C/31Fe NPs—6% Magnetization:	Stachowska et al. (2022)

(continued)

Table 1.1 (continued)

No	Nano material	Source	Elemental composition	Physical properties	Reference
				C/17Fe NPs—4.1 emu/g C/31Fe NPs—17.7 emu/g Size: C/17Fe NPs—6.4 nm C/31Fe NPs—19 nm	
9.	Carbon dots	Citric acid and ethylenediamine	Carbon—48.81 % Hydrogen—5.14% Oxygen—33.68% Nitrogen—12.37%	Size: 10 nm	Zhao et al. (2015)
10.	N-CDs	Citric acid and guanidine carbonate	Carbon—68.44% Nitrogen—15.84% Oxygen—15.72%	Quantum yield: 52% Size: 6.69 nm—15.33 nm Average size: 10.30 nm	Chen et al. (2019)

Bottom-up approach involves the formation of nanoparticles that occurs through the spontaneous organization of atoms into novel nuclei, which subsequently mature into particles with dimensions on the nanoscale, leveraging a range of chemical and biological methodologies. Bottom-up approach is formed by arranging the nanoparticles in the form of layer-by-layer or by organizing the atom-by-atom methods or by molecule-by-molecule layering. No waste materials are obtained in the bottom-up approach (Meyers et al., 2006). Major benefits of bottom-up approach are the improved nanoparticles composition with reduced defects in the synthesized products (Thakkar et al., 2010). These defects have a substantial effect on the morphology and surface reactivity. Many techniques, such as abrasion and thermal decomposition, are among the few methods that are used for the synthesis of the nanoparticles. In the abrasion method, the larger and smaller particles are broken down into small particles by the size pulverization methods. Main parameters that influence the nanoparticles are the residence time of pulverization and the type of components which is used for the synthesis of nanoparticles. There is a growing need for eco-friendly, cost-effective, and non-toxic methods to synthesize nanoparticles, making biological resources, including plants, algae, fungi, agricultural waste, and microorganisms, that can be harnessed for the synthesis of nanoparticles (Thakkar et al., 2010). A variety of nanoparticles, including silver, gold, fluorinated compounds, and titanium dioxide, have been successfully synthesized through advanced

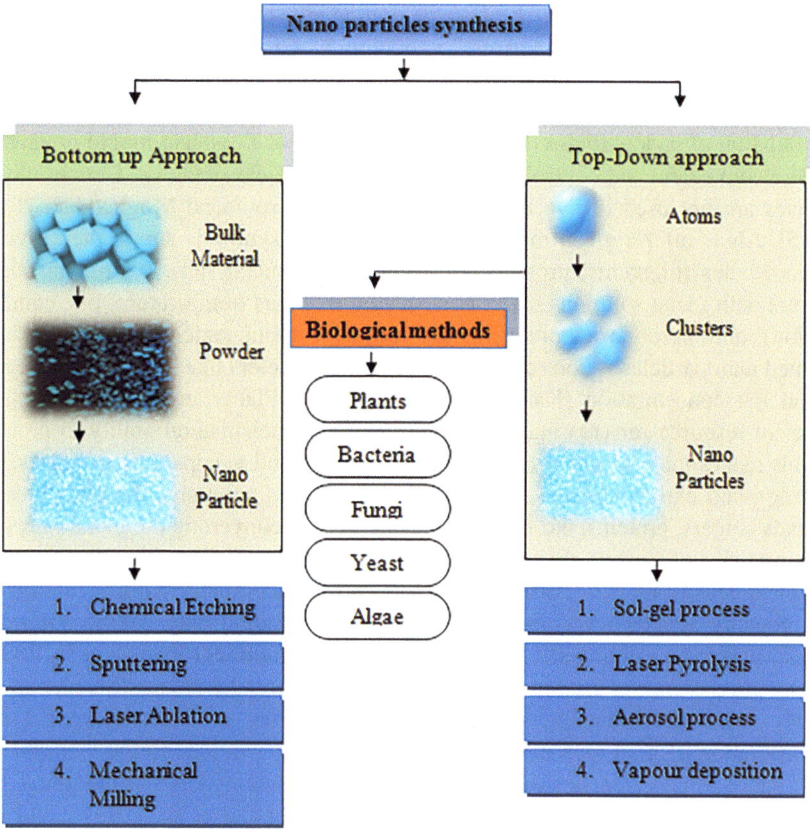

Fig. 1.1 Synthesis methods of nanoparticles, such as approaches and methods for synthesis. (Constructed by Sankari Rajan)

operational and condensation methods, showcasing significant breakthroughs in nanomaterials development. Chemical synthesis involves the use of solvents and chemicals, which are toxic and causes ecological issues. Due to this, biological synthesis is preferred, and green nanoparticle synthesis includes the synthesis of particles from plants and plant extracts (Shukla & Iravani, 2018). Figure 1.1 shows the synthesis and methods of nanoparticles.

1.5.2 Green Synthesis Methods

Biological nanoparticles synthesis offers a green alternative to traditional chemical methods because of a cost-effective, non-toxic approach using organisms like bacteria, algae, and plants, or their extracts. These biological components, such as proteins and enzymes, naturally reduce metal ions, forming nanoparticles in a single step. Due to the possessed risk of stabilizing agents, this method employs natural agents from biological organisms, producing biocompatible nanoparticles with reduced toxicity (Geonmonond et al., 2018; Hussain et al., 2016). Due to the reduction in the cost and ease of developing, bacteria are widely used for the formulation of nanoparticles (Korbekandi et al., 2009). Yeast and mould organisms offer a convenient and controlled method for nanoparticle synthesis. Certain mould species are not used due to their potential risks (Boroumand Moghaddam et al., 2015). Algae offer a green method for nanoparticles synthesis due to their diverse biomolecules (pigments, proteins, etc.) that reduce metal salts. By mixing algae extract with metal solutions under controlled conditions (temperature, pH, concentration), antimicrobial nanoparticles are formed without toxicity. The size of the formed nanoparticles is controllable by varying parameters like incubation time and metal ion concentration (Karaduman et al., 2017). Plants are a sustainable and efficient resource for creating metal nanoparticles. Their natural ability to process metals makes them ideal for both pollution removal and nanoparticle synthesis. By adding plant extracts (leaves, roots, etc.) to metal ion solutions, the plant's compounds (sugars, proteins, etc.) act as reducing agents, converting the metal ions into nanoparticles of various shapes and sizes. This process can occur inside or outside the plant cells (Park et al., 2016). Geranium leaf extract rapidly synthesizes AgNPs by reducing aqueous Ag+ ions. This plant-based method outpaces bacterial and fungal synthesis, yielding stable, crystalline Ag Nanoparticles (Shankar et al., 2003). Lemongrass extract is used for the synthesis of the triangular prisms (Shankar et al., 2004). Researchers have reported that the networking of DNA hydro gel and incorporation of gold ions into the DNA biopolymers can lead to the synthesis of gold nanoparticles (Zinchenko et al., 2014). By incorporating DNA as the source, we can able to synthesize the silver nanoparticles (Majumdar et al., 2013). Gold NP have also been synthesized by using the *Hevea brasiliensis*, which facilitates the reduction of gold ions (Au^{3+}) in solution, even at an elevated temperature 80 °C with dimensions of 40–70 nm with high-purity metal (Cabrera et al., 2013). Figure 1.2 illustrates the green synthesis of silver nanoparticles.

1.5.3 Characterization Techniques

Surface plasmon resonance peaks of spectra analysed at the particular wavelength of 200–800 nm show the presence of copper NPs. UV–vis can be used to analyse the rigidity of the nanoparticles. UV–vis can also be used to analyse the dimension, rigidity, and clustering state of the nanoparticles. The dimension of the nanoparticles can be analysed by SEM and TEM. The crystalline structure of the nanoparticles can be analysed by XRD. The presence of different functional groups of elements, with the type of compounds that are present on the surface of the nanomaterials, in addition to proteins and C–H bond linkages, can be analysed by FTIR. Characterization of nanoparticles shows a different type of nanoparticles with different dimensions (Santhoshkumar et al., 2017). Brunauer, Emmett, and Teller (BET) is

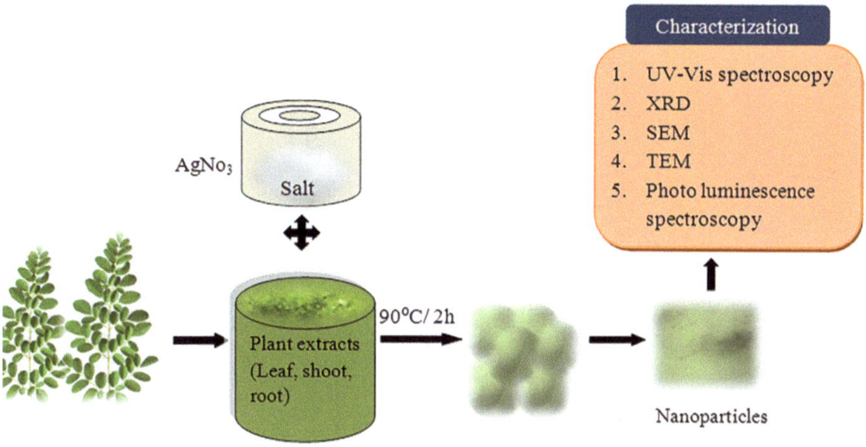

Fig. 1.2 Green synthesis of silver nanoparticles has antimicrobial activities. (Constructed by Sankari Rajan)

used to analyse the total surface area of the nanoparticles, where the average particle diameter is represented in nm (Akbari et al., 2011). Table 1.2 shows the nanoparticles and its wider applications in agriculture, food, and water treatment and the various characterization techniques used to identify the nanoparticles' dimensions and its applications in agriculture.

1.6 Applications of Nanotechnology in Agriculture and Farming

Nanotechnology in farming practices would be a transformative approach in agriculture, enhancing the efficiency, sustainability, and productivity of farmers in terms of pest control and controlling post-harvest loss. By integrating nanomaterials and nano-based tools, farmers can optimize the resources, reduce environmental impact, and improve crop yield. In farming, nano-porous membranes improve desalination and photo-catalytic water purification of the irrigation water. Moreover, potential benefits of nanotechnology in agriculture seem to be significant; the concerns regarding environmental safety, toxicity, and long-term effects need to be addressed for the implementation. Balancing innovation with safety measures is essential for sustainable agricultural practices.

Table 1.2 Characterization techniques and applications of nanoparticles

Source	Nano materials type	Approach method	Size of nanoparticles	Identification techniques	Applications	Reference
Reetha and Shikakai plant extract	Ag-NPs	Chemical-aqueous extraction	20–40 nm and 50 nm	UV-visible, TEM, XRD	Detection of bacteria like *Mycobacterium tuberculosis*	Sur et al. (2018)
Eucalyptus leaf extract	Fe-NPs	Chemical method	95 ± 5 nm	TEM, XRD, FTIR, GCMS.	Water treatment	Liu et al. (2018)
Coccinia grandis	Ag-NPs	Chemical method	40–50 nm	UV–vis, SEM, FTIR	Antibacterial, towards pathogen and toxicity removal	Devi and Ahmaruzzaman (2018)
Eucalyptus leaf extracts	Fe-NPs	Chemical method	70 ± 20 nm	EDS, FTIR, SEM and XRD	Toxicity removal in plants and aquatic systems	Jin et al. (2018)
Aspergillus species	Zn- NPs	Aerosol technique	10–50 nm	TEM	Used as a spray fertilizer, it penetrates inside the plant and improves seed fertility, weed control, and plant nutrition	Tarafdar et al. (2012)
Aspergillus species	Ag -NPs	Fungus-mediated synthesis	60–80 nm	SEM	Antimicrobial and cytotoxic agents, fungicidal agents in agriculture	Rai et al. (2022)
Soil fungi	Zn, Mg, Ti-NPs	Biological method (fungi)	51 ± 05 nm, 49 ± 08 nm, 43 ± 03 nm	TEM, SDS-PAGE	Controls the microbial contamination in the plants	Raliya et al. (2013)
Pathogenic fungi	Ag, Zn, Ti-NPs	Biological method (microbes, reducing agents)	2–20 nm, 4–8 nm, 3 nm	TEM	Antifungal activity for food crops, 20% microbial disease reduction in crops	Sehsahb and Salama (2023)
Fungal species (*Aspergillus*, *Fusarium* and *Penicillium*)	Zn-NPs	Biological method	30–100 nm	UV–vis, SPR, TEM	Antibacterial property against bacterial infection	Hefny et al. (2019)
	Phosphorus-NPs	Biological method	94.7 to 45.1 nm		Growth rate of the plants has increased, used as a nano fertilizer	Yaseen and Amin (2021)

Fungal species (*Aspergillus fumigates*)				PSA, UV-vis, TEM, EDX, and FT-IR		
Rice husk	Silica-NPs	Bottom-up approach (Sol-Gel)	15 nm, 25 nm, and 8.5 nm	XRD, FESEM, EDAX, FTIR, BET	Improve soil fertility, plant growth, increase plant yield	Dorairaj et al. (2022)
Plant metabolites	Gold, silver-NPs	Bottom-up approach	50 nm	UV–vis	Used as nano fertilizers, nano pesticides, and nano herbicides for crops	Castillo-Henríquez et al. (2020)
Metal powders	Alumina-NPs	Plasma reactor technique	27 nm, 5–95 nm	BET, XRD, TEM	Forming agglomerated particles	Akbari et al. (2011)

1.6.1 Nano-fertilizers and Nutrient Delivery Systems

Generally, commercial fertilizers like nitrogen fertilizers such as ammonium sulphate, urea, calcium ammonium nitrate (CAN), and organic fertilizers such as manure, by-products and compost, potash fertilizers, and complex fertilizers such as nitro and ammonium phosphate often result in nutrient losses through leaching in soil, volatilization, or runoff. To minimize the negative effects of conventional agrochemicals, nano-fertilizers are used to improve nutrient efficiency by controlling the release and targeted delivery of essential nutrients. Nano-encapsulated fertilizers like nano-urea (20–50 nm), nano-DAP (<100 nm), nano-apatite, nano-potash, nano-iron, and nano-chitosan were mainly used for controlled and targeted nutrient delivery. Liquid organic fertilizers like Ferbanat and Nanonat enhance cucumber yield. One of the key benefits of nano fertilizers is that they can be utilized for the controlled release of nutrients over an extended time, which maximizes absorption. Chelated metal nanoparticles enhance nutrient uptake, solubility, and controlled release (Dutta et al., 2022). Silicon nanoparticles (Si-NP) may enhance plant resistance to pathogens. Polymeric nanoparticles like dendrimers and nano-hydrogels were employed for long-term release and pesticide encapsulation, thus ensuring sustainable agriculture (Dalwadi & Patel, 2015).

1.6.2 Nano-pesticides for Precision Pest Control

Nano-pesticides enhance potency, improve nutrient efficiency, and are eco-friendly. It provides early spotting and targeted application in precision pest control, addressing the challenges of conventional practices while potentially reducing pesticide usage and minimizing environmental impact. Due to their nano size, they provide greater insect-body surface area coverage. They are also claimed to be safer for supporting organisms, making them natural and effective alternatives that prevent insect damage in stored goods. Controlled release of active ingredients that can act effectively against pests and insects by using a smart delivery system. Silica, alumina, and montmorillonite-based nanocomposites are solid nanoparticles used for the controlled release of nutrients. Composite with zinc oxide nanoparticles had strong antibacterial activity against specific pathogens in the plant stems, like *Escherichia coli* O157:H7 and *Listeria monocytogenes* (Hossain et al., 2023). The nano pesticide made with ZnO-NPsZC (composite with zinc chloride) showed the highest antibacterial activity, proving that adding zinc oxide nanoparticles enhances the antibacterial activity of the plants. Silver nanoparticles (AgNPs) can exhibit antimicrobial properties against plant pathogens. Copper nanoparticles (CuNPs) can act as effective fungicides. Nano-emulsions of essential oils can act as natural pest control agents. Furthermore, a study by Roy and Rhim (2021) shows that the addition of tea essential oil with zinc oxide with the Pickering emulsion process can improve the mechanical property by 30% and the presence of oil emulsion hydrophobicity increases by 10% which have good antioxidant properties due to the presence of catechins and epicatechins group in the tea extract more potent towards the antioxidant properties.

1.6.3 Nanomaterials in Soil and Water Remediation

Generally, nanomaterials focussed on soil and water conditions due to the contamination from heavy metals, pesticides, and pollutants reduces agricultural productivity. Nanomaterials help in remediation through adsorption, degradation, and soil structure improvement. Furthermore, the effective method should be Biochar-based nanomaterials can improve carbon sequestration and nutrient retention. Agricultural waste-derived nano-biochar (nano-BC) offers enhanced specific surface area, adsorption capacity, and mobility, such as higher surface area and porosity compared to traditional biomass-derived biochar. It effectively enhances substrate-induced respiration (Sani et al., 2023). This results in improved soil properties, crop growth, and environmental remediation. Nano-biochar can sequester carbon and reduce greenhouse gas emissions, making it more effective by contributing to soil health and sustainability. Recent studies were majorly focusing in its nanosized form which improve crop productivity through optimizing the physical properties which results in enhancing the soil pH (6–10), pH (0.5–2.0), nutrient availability (P, Ca, Mg), electrical conductivity has 10–30% reduction in soil salinity, cation exchange capacity (CEC) with 20–50% increase in nutrient availability (e.g., K^+, Ca^{2+}, Mg^{2+}), WHC increase with porosity 15–40% increase, lead to enhancement of drought resilience (Dar et al., 2020). Furthermore, magnetic nanoparticles (Fe_3O_4 NPs) can remove heavy metals from soil and water, which offers improved pollutant absorption and mobility in soil, making it a promising tool for waste management and soil remediation with better nutrient retention and microbial activity (Hemmati et al., 2018).

1.6.4 Nano-sensors for Smart Farming and Precision Agriculture

Sensor integration is a novel method to introduce into agricultural farming to ensure the safety and food control to deliver safe food. It enables the real-time monitoring process integrated with spoilage indicators to improve sustainable practices. Nano-sensors can optimize the use of fertilizers and pesticides, water quality monitoring, ensuring safe irrigation practices, and preventing nutrient runoff into water bodies. Innovations like nano-priming enhance seed germination and drought resilience, further supporting sustainable agriculture. Generally, graphene-based sensors can detect plant pathogens like fungi and protozoan species. The advancement in gas sensors, such as paper-based electric-gas sensors (PEGS) used in high-value products such as spinach. PEGS can be linked through smartphones through the near field communication method (NFC) and can be an in-line monitoring method to detect the freshness of the spinach (Naik et al., 2024). Time-temperature indicator sensors (TTI) were used for freshness indicators for radish and beetroot colour change and pigment loss over a time period of 6 days of storage. Carbon dioxide and oxygen gas sensors, along with nano biosensors, were used to monitor the gas production during transportation and storage of the fruits and vegetables. Optical nano sensors can be used to detect plant stress by monitoring changes in chlorophyll levels. RFID sensors with the chipless type can be used to monitor gas levels, such as ethylene, humidity and temperature, as well as bio-monitoring to assess food quality. Ultra-high frequency (UHF) sensors integrated with RFID with coupled-coil resonator sensors can detect the irrigation systems and prevent overwatering or underwatering of crops. Magnetic nano sensors detect specific pathogens and microorganisms, controlling mycotoxigenic fungi and detecting their mycotoxins, which can inhibit microbial growth by 2 log reductions with high accuracy. Nano-enabled drones can effectively spray the fertilizers and reduce the chemical usage. Moreover, nano robots detect soil nutrient deficiencies and disease outbreaks (Zhang et al., 2022)

1.6.5 Nanotechnology in Post-harvest Management and Food Preservation

Nanotechnology is emerging in every stage of crop production, from seed storage to post-harvest management; it efficiently enhances food security and safety by reducing agricultural waste. Typical drawbacks in traditional farming can be restrictions in the utilization of farmlands and the spoilage of fresh produce. Techniques like nano coatings of preservatives, antimicrobial agents, and intelligent and active packaging help to maintain the quality of fresh produce. A study by Rukmanikrishnan et al. (2021) utilized a nanoparticle-based composite film made with the combination of kappa carrageenan, silicon dioxide, hydroxylethyl cellulose, and silver nanoparticles, which are suggested for use in packaging and biomedical applications due to their improved properties. This includes transparency, enhanced thermal (TGA T5% 115.7 and T10% value 150.4) and mechanical ability, shear thinning effect, UV inhibition, and improved storage modulus. This film also showed zone of inhibition (antimicrobial activity) against *Staphylococcus aureus* (34 mm), *Cronobactersakazakii* (25.7 mm), *Salmonella typhi* (22.3 mm), *Bacillus subtilis* (19 mm), *Listeria monocytogenes* (19.7 mm), and *Bacillus cereus* (15.3 mm). The other films without the incorporation of the silver nanoparticle showed no inhibition. A study by Mohamed et al. (2020) shows that used to pack dry fruits like figs and roasted almonds, Walnut film consists of orange essential oil, Trehalose, Tween 20 or Tween 80 series, and 5% (W/v) of glycerol and carrageenan (2%) combination. This is due to trehalose use to prevent aroma loss and colour loss in the product, while the essential oil can act as antibacterial or antimicrobial properties against gram-positive bacteria like *Staphylococcus aureus*, thereby improving the food safety by preventing microbial contamination.

1.7 Emerging Innovations and Future Perspectives

Nanotechnology concepts are creating a drastic growth and innovations to improve the farming practices to focus more on soil health and water quality, pest control by minimizing the drawbacks in traditional farming practices. Some emerging methods like smart sprayers, nano-enabled drones, smart nano fertilizers and coatings, RNA-interference (RNAi) nanoparticles, gene-silencing technology, agri nanobots, gene-editing via nano-carriers, AI or machine learning-based nano sensor networks can collectively address the critical challenges such as nutrient inefficiency, pest resistance, and environmental degradation and utilize the resource effectively.

Nanopriming study was conducted using the silver nanoparticles (AgNPs), which were synthesized using the kefir lime extracts to enhance the germination of our aged rice seeds. Results show that silver nanoparticles around 5–10 ppm entrapped the phytochemicals present in the kefir leaf, which improves the germination efficiency compared to the unprime sample. This is due to the penetration of nanoparticles, which interact with the α-amylase enzyme in the seeds and can act as a nano catalyst, resulting in starch degradation and seed growth (Mahakham et al., 2017).

Advanced gene technology was used for agricultural produce like maize, potato, rice, and tomato to prevent the resistance against biotic and abiotic stress and plant pathogens. Micro RNA (miRNA) was used in maize to prevent the *Fusarium rot.* miRNA 160 was used for potatoes to produce resistance against *Phytophthora infestans* infections. miR482/2118 family was used for tomato (*Solanum lycopersicum*) for the enhancement of internal immunity in plants by regulating the nucleotide binding site (NBS-LRR) receptors. RNA interference was used in barley, wheat, pepper, and plums to protect against the virus pathogen. Barley and wheat utilized the virus-induced gene silencing (VIGS) to reduce the fungal disease caused by the TaMlo virus. Moreover, the fungal infections will be resisted through Fusarium head blight, and Fusarium seed blight can be reduced through the fungal chitin gene mechanism to improve the crop quality (Mujtaba et al., 2021).

Genome surgery is a new emerging trend that is used for creating new generation therapeutic foods through genome editing, which can be done using the smart nanoparticles CRISPR-Cas technology. It is to target the specified cells for curing long-term diseases like gastrointestinal epithelial cancer (CISH gene), sickle cell disease (BCL11A gene) through selective organ-targeting (SORT) nanoparticles, which can reach a new milestone in clinical trials without undesirable effects (Naeem et al., 2021).

Nano-desalination water for smart irrigation, which purifies brackish water, making it usable for farming. It can be done by zero valent iron can be transformed into pellet forms that can be used for agricultural irrigation water. Study by Antia et al. (2022) utilized the graphite oxide nano filter and achieves the 30% to 60% desalination for saline feed water within the salinity range of 4 to 10 g L^{-1} which can address the concerns faced by other methods such as reverse osmosis (RO), thermal desalination methods and electro-coagulation (EC) profound to be a cost-effective solution. Nano-membranes based on carbon nanotubes (CNT) and graphene oxide have good water permeability. By coupling the techniques like nano adsorbents such as zeolite, metal organic frameworks with 2-D nanomaterials have higher salt rejection capabilities. Integration of nanotechnology with artificial intelligence (AI)-based tools to validate the real-time data set with the machine learning algorithms can be an effective method to optimize the water levels and monitor the moisture content and relative humidity in the soil and helps in crop management. Information communication technology (ICT) conjugated with nanosensors and nanoactuators, which employs a machine learning algorithm, can monitor the pest attack and crop colour changes from green to pale yellow, which helps in precision agriculture (Dar et al., 2020).

1.8 Summary

Nanotechnology has emerged as a transformative field that significantly impacts the various sectors, including agriculture. This chapter provided a comprehensive overview of nanotechnology and nanomaterials, exploring their fundamental principles, properties, synthesis, and characterization methods. The classification of nanomaterials, including carbon-based, metal and metal oxide nanoparticles, polymeric, and bio-based nanomaterials, was discussed in detail. Furthermore, the chapter highlighted the nanomaterials' unique physical, chemical, mechanical, and optical properties, which distinguish them from conventional materials. The applications of nanotechnology in agriculture were extensively reviewed, emphasizing nano-fertilizers, nano-pesticides, nano-sensors, and nanomaterials for soil and water remediation. Post-harvest management and food preservation techniques incorporating nanotechnology were also explored. Furthermore, emerging innovations such as RNA-interference nanoparticles, gene-editing nanocarriers, and AI-integrated nano-sensors would showcase the future prospects of nanotechnology in sustainable agricultural practices.

1.9 Conclusion and Prospects

Nanotechnology emerges a shift in modern agriculture by enhancing efficiency, sustainability, and resource optimization. Its wide application in targeted nutrient delivery, precision pest control, soil remediation, and smart farming demonstrates the potential to revolutionize agricultural practices. The integration of green synthesis methods ensures eco-friendly production, reducing environmental concerns associated with conventional nanoparticle fabrication. However, there are still challenges, like ensuring safety and setting proper regulations. Future advancements in nanotechnology, integrated with artificial intelligence and biotechnology, could have a better breakthrough for further innovations in food security, sustainable farming, and environmental conservation. Continued research and implementation could explore the full potential of nanotechnology for agricultural advancements.

Acknowledgments The authors would like to thank the staff of the Department of Food Technology, Kongu Engineering College, for their extensive technical support and also thank all co-authors for their assistance with this chapter. The authors also thank the Ministry of Education, Government of India, for an Institute Research Assistantship, and thank IIT Kharagpur for their assistance in this research.

Conflict of Interest The authors declare no conflicts of interest that are relevant to the content of this chapter.

References

Ahmadi, S. Z., Zahedi, B., Ghorbanpour, M., & Mumivand, H. (2024). Comparative morpho-physiological and biochemical responses of Capsicum annuum L. plants to multi-walled carbon nanotubes, fullerene C60 and graphene nanoplatelets exposure under water deficit stress. *BMC Plant Biology, 24*(1), 116. https://doi.org/10.1186/s12870-024-04798-y

Akbari, B., Tavandashti, M. P., & Zandrahimi, M. (2011). Particle size characterization of nanoparticles – A practical approach. *Iranian Journal of Materials Science and Engineering, 8*(2), 48–56.

An, D., Liu, B., Yang, L., Wang, T.-J., & Kan, C. (2017). Fabrication of graphene oxide/polymer latex composite film coated on KNO3 fertilizer to extend its release duration. *Chemical Engineering Journal, 311*, 318–325. https://doi.org/10.1016/j.cej.2016.11.109

Antia, D. D. J. (2022). Purification of saline water using desalination pellets. *Water, 14*(17), 2639. https://doi.org/10.3390/w14172639

Arora, D., Kumar, M., Bhatt, S., Gautam, R. K., & Taneja, Y. (2024). Biodegradability and sustainability of biobased nanomaterials. In *Biobased nanomaterials: Applications in biomedicine, food industry, agriculture, and environmental sustainability* (pp. 509–535). Springer. https://doi.org/10.1007/978-981-97-0542-9_1

Awan, T. I., Bashir, A., & Tehseen, A. (2020). *Chemistry of nanomaterials: Fundamentals and applications*. Elsevier.

Boroumand Moghaddam, A., Namvar, F., Moniri, M., Md. Tahir, P., Azizi, S., & Mohamad, R. (2015). Nanoparticles biosynthesized by fungi and yeast: A review of their preparation, properties, and medical applications. *Molecules, 20*(9), 16540–16565. https://doi.org/10.3390/molecules200916540

Bortoletto-Santos, R., Plotegher, F., Majaron, V. F., da Silva, M. G., Polito, W. L., & Ribeiro, C. (2020). Polyurethane nanocomposites can increase the release control in granulated fertilizers by controlling nutrient diffusion. *Applied Clay Science, 199*, 105874. https://doi.org/10.1016/j.clay.2020.105874

Cabrera, F. C., Mohan, H., Dos Santos, R. J., Agostini, D. L., Aroca, R. F., Rodríguez-Pérez, M. A., & Job, A. E. (2013). Green synthesis of gold nanoparticles with self-sustained natural rubber membranes. *Journal of Nanomaterials, 2013*(1), 710902. https://doi.org/10.1155/2013/710902

Castillo, Y. H. (2022). Fundamentals of Nanomaterials. In *Applications of nanomaterials for energy storage devices* (pp. 57–75). CRC Press.

Castillo-Henríquez, L., Alfaro-Aguilar, K., Ugalde-Álvarez, J., Vega-Fernández, L., Montes de Oca-Vásquez, G., & Vega-Baudrit, J. R. (2020). Green synthesis of gold and silver nanoparticles from plant extracts and their possible applications as antimicrobial agents in the agricultural area. *Nanomaterials, 10*(9), 1763. https://doi.org/10.3390/nano10091763

Chandel, M., Kaur, K., Sahu, B. K., Sharma, S., Panneerselvam, R., & Shanmugam, V. (2022). Promise of nano-carbon to the next generation sustainable agriculture. *Carbon, 188*, 461–481. https://doi.org/10.1016/j.carbon.2021.11.060

Chen, Y.-J., Liu, Z.-E., Yang, Q., Wang, C.-F., & Zhuo, K.-L. (2019). Nitrogen-doped highly photoluminescent carbon dots derived from citric acid and guanidine carbonate. *Journal of Nanoscience and Nanotechnology, 19*(7), 4058–4063. https://doi.org/10.1166/jnn.2019.16332

ÇIFTÇI, V. (2024). Güncel Pedodonti Çalişmalari IV.

Dalwadi, C., & Patel, G. (2015). Application of nanohydrogels in drug delivery systems: Recent patents review. *Recent Patents on Nanotechnology, 9*(1), 17–25. https://doi.org/10.2174/1872210509666150101151521

Dar, F. A., Qazi, G., & Pirzadah, T. B. (2020). Nano-biosensors: NextGen diagnostic tools in agriculture. In *Nanobiotechnology in agriculture: An approach towards sustainability* (pp. 129–144). Springer. https://doi.org/10.1007/978-3-030-39978-8_7

Devi, T. B., & Ahmaruzzaman, M. (2018). Facile preparation of copper nanaoparticles using Coccinia grandis fruit extract and its application towards the reduction of toxic nitro compound. *Materials Today: Proceedings, 5*(1), 2098–2104. https://doi.org/10.1016/j.matpr.2017.09.206

Dorairaj, D., Govender, N., Zakaria, S., & Wickneswari, R. (2022). Green synthesis and characterization of UKMRC-8 rice husk-derived mesoporous silica nanoparticle for agricultural application. *Scientific Reports, 12*(1), 20162. https://doi.org/10.1038/s41598-022-24484-z

Dutta, S., Pal, S., Panwar, P., Sharma, R. K., & Bhutia, P. L. (2022). Biopolymeric nanocarriers for nutrient delivery and crop biofortification. *ACS Omega, 7*(30), 25909–25920. https://doi.org/10.1021/acsomega.2c02494

El-Shaheny, R., Al-Khateeb, L., El-Maghrabey, M., Magdy, G., & Hashem, H. M. (2025). Double-reverse-signal ratiometric fluorescence probe for residual determination of chlorpyrifos pesticide in surface water based on scalable upcycling of rice straw to luminescent CQDs via energy-efficient alkali treatment. *Talanta Open, 100427*. https://doi.org/10.1016/j.talo.2025.100427

Francis, D. V., Abdalla, A. K., Mahakham, W., Sarmah, A. K., & Ahmed, Z. F. R. (2024). Interaction of plants and metal nanoparticles: Exploring its molecular mechanisms for sustainable agriculture and crop improvement. *Environment International*, 108859. https://doi.org/10.1016/j.envint.2024.108859

Geonmonond, R. S., Silva, A. G. D., & Camargo, P. H. (2018). Controlled synthesis of noble metal nanomaterials: Motivation, principles, and opportunities in nanocatalysis. *Anais da Academia Brasileira de Ciências, 90*(1 Suppl 1), 719–744. https://doi.org/10.1590/0001-3765201820170561

Gupta, R., Chauhan, H., Garg, V. K., & Kataria, N. (2022a). Chemical and physical properties of nanoparticles and hybrid materials. In *Sustainable nanotechnology for environmental remediation* (pp. 199–220). Elsevier. https://doi.org/10.1016/B978-0-12-824547-7.00024-2

Gupta, R. K., Guha, P., & Srivastav, P. P. (2022b). Natural polymers in bio-degradable/edible film: A review on environmental concerns, cold plasma technology and nanotechnology application on food packaging-A recent trends. *Food Chemistry Advances, 1*, 100135.

Gupta, R. K., Pipliya, S., Karunanithi, S., Eswaran, U., Kumar, S., Mandliya, S., & Kovács, B. (2024a). Migration of chemical compounds from packaging materials into packaged foods: Interaction, mechanism, assessment, and regulations. *Foods, 13(19)*, 3125.

Gupta, R. K., Abd El Gawad, F., Ali, E. A., Karunanithi, S., Yugiani, P., & Srivastav, P. P. (2024b). Nanotechnology: Current applications and future scope in food packaging systems. *Measurement: Food, 13*, 100131.

Gupta, R. K., Guha, P., & Srivastav, P. P. (2024c). Investigating the toxicological effects of nanomaterials in food packaging associated with human health and the environment. *Journal of Hazardous Materials Letters, 5*, 100125.

Hefny, M. E., El-Zamek, F. I., El-Fattah, A., & Mahgoub, S. (2019). Biosynthesis of zinc nanoparticles using culture filtrates of Aspergillus, Fusarium and Penicillium fungal species and their antibacterial properties against gram-positive and gram-negative bacteria. Zagazig. *Journal of Agricultural Research, 46*(6), 2009–2021. https://doi.org/10.21608/zjar.2019.51920

Hemmati, M., Rajabi, M., & Asghari, A. (2018). Magnetic nanoparticle based solid-phase extraction of heavy metal ions: A review on recent advances. *Microchimica Acta, 185*, 1–32. https://doi.org/10.1007/s00604-018-2670-4

Hossain, S. I., Kukushkina, E. A., Izzi, M., Sportelli, M. C., Picca, R. A., Ditaranto, N., & Cioffi, N. (2023). A review on montmorillonite-based nanoantimicrobials: State of the art. *Nanomaterials, 2023*(13), 848. https://doi.org/10.3390/nano13050848

Hussain, I., Singh, N., Singh, A., Singh, H., & Singh, S. (2016). Green synthesis of nanoparticles and its potential application. *Biotechnology Letters, 38*, 545–560. https://doi.org/10.1007/s10529-015-2026-7

Jin, X., Liu, Y., Tan, J., Owens, G., & Chen, Z. (2018). Removal of Cr (VI) from aqueous solutions via reduction and absorption by green synthesized iron nanoparticles. *Journal of Cleaner Production, 176*, 929–936. https://doi.org/10.1016/j.jclepro.2017.12.026

Karaduman, I., Güngör, A. A., Nadaroğlu, H., Altundaş, A., & Acar, S. (2017). Green synthesis of γ-Fe 2 O 3 nanoparticles for methane gas sensing. *Journal of Materials Science: Materials in Electronics, 28*, 16094–16105. https://doi.org/10.1007/s10854-017-7510-5

Karak, N. (2019). Fundamentals of nanomaterials and polymer nanocomposites. In *Nanomaterials and polymer nanocomposites* (pp. 1–45). Elsevier. https://doi.org/10.1016/B978-0-12-814615-6.00001-1

Kokabi, M., & Yousefzadi, M. (2022). Algal nanoparticles and their potential application in agriculture. In *Nano-enabled agrochemicals in agriculture* (pp. 189–198). Academic. https://doi.org/10.1016/B978-0-323-91009-5.00021-5

Korbekandi, H., Iravani, S., & Abbasi, S. (2009). Production of nanoparticles using organisms. *Critical Reviews in Biotechnology, 29*(4), 279–306. https://doi.org/10.3109/07388550903062462

Kumar, A., Choudhary, A., Kaur, H., Mehta, S., & Husen, A. (2021). Smart nanomaterial and nanocomposite with advanced agrochemical activities. *Nanoscale Research Letters, 16*, 1–26. https://doi.org/10.1186/s11671-021-03612-0

Kumar, S., Khan, S. I., Neidhe, M. M. R., Islam, M., & Hasan, M. M. (2024). Nanomaterials in industry: A review of emerging applications and development. *Control Systems and Optimization Letters, 2*(3), 342–348. https://doi.org/10.59247/csol.v2i3.135

Liu, Y., Jin, X., & Chen, Z. (2018). The formation of iron nanoparticles by Eucalyptus leaf extract and used to remove Cr (VI). *Science of the Total Environment, 627*, 470–479. https://doi.org/10.1016/j.scitotenv.2018.01.241

Madlala, N. C., Khanyile, N., & Masenya, A. (2024). Examining the correlation between the inorganic nano-fertilizer physical properties and their impact on crop performance and nutrient uptake efficiency. *Nanomaterials, 14*(15), 1263. https://doi.org/10.3390/nano14151263

Mahakham, W., Sarmah, A. K., Maensiri, S., & Theerakulpisut, P. (2017). Nanopriming technology for enhancing germination and starch metabolism of aged rice seeds using phytosynthesized silver nanoparticles. *Scientific Reports, 7*(1), 8263. https://doi.org/10.1038/s41598-017-08669-5

Majumdar, D., Singha, A., Mondal, P. K., & Kundu, S. (2013). DNA-mediated wirelike clusters of silver nanoparticles: An ultrasensitive SERS substrate. *ACS Applied Materials & Interfaces, 5*(16), 7798–7807. https://doi.org/10.1021/am402448j

Malik, M. I., Hussain, D., Shah, M. R., & Guo, D.-S. (2024). Handbook of nanomaterials. In *Volume 1: Electronics, information technology, energy, transportation, and consumer products* (Vol. 1). Elsevier.

Markov, A. N., Kapinos, A. A., Petukhov, A. N., Dokin, E. S., Emelyanov, A. V., Abarbanel, N. V., Zarubin, D. M., Golovacheva, A. A., Suvorov, S. S., & Barysheva, A. V. (2024). Synthesis of zinc nanoparticles by the gas condensation method in a non-contact crucible and their physical–chemical characterization. *Nanomaterials, 14*(2), 163. https://doi.org/10.3390/nano14020163

McNeil, S. E. (2005). Nanotechnology for the biologist. *Journal of Leukocyte Biology, 78*(3), 585–594. https://doi.org/10.1189/jlb.0205074

Mehta, J., Dhaka, R. K., Dilbaghi, N., Lim, D. K., Hassan, A. A., Kim, K. H., & Kumar, S. (2024). Recent advancements in adsorptive removal of organophosphate pesticides from aqueous phase using nanomaterials. *Journal of Nanostructure in Chemistry, 14*(1), 53–70. https://doi.org/10.1007/s40097-022-00516-y

Meyers, M. A., Mishra, A., & Benson, D. J. (2006). Mechanical properties of nanocrystalline materials. *Progress in Materials Science, 51*(4), 427–556. https://doi.org/10.1016/j.pmatsci.2005.08.003

Milani, P., França, D., Balieiro, A. G., & Faez, R. (2017). Polymers and its applications in agriculture. *Polímeros, 27*, 256–266. https://doi.org/10.1590/0104-1428.09316

Mohamed, S. A., El-Sakhawy, M., & El-Sakhawy, M. A. M. (2020). Polysaccharides, protein and lipid-based natural edible films in food packaging: A review. *Carbohydrate Polymers, 238*, 116178. https://doi.org/10.1016/j.carbpol.2020.116178

Mohammadi, H., Esmailpour, M., & Gheranpaye, A. (2016). Effects of TiO2 nanoparticles and water-deficit stress on morpho-physiological characteristics of dragonhead (Dracocephalummoldavica L.) plants. *Acta Agriculturae Slovenica, 107*(2), 385–396. https://doi.org/10.14720/aas.2016.107.2.11

Mujtaba, M., Wang, D., Carvalho, L. B., Oliveira, J. L., Espirito Santo Pereira, A. D., Sharif, R., & Fraceto, L. F. (2021). Nanocarrier-mediated delivery of miRNA, RNAi, and CRISPR-Cas for plant protection: Current trends and future directions. *ACS Agricultural Science & Technology, 1*(5), 417–435. https://doi.org/10.1021/acsagscitech.1c00146

Nabil, M., Perez-Quintana, I., Acosta, M., Mendez-Gamboa, J., & Castro-Rodriguez, R. (2021). Morphological, structural, and optical bandgap characterization of extracted zno nanoparticles from commercial paste. *Advances in Materials Science and Engineering, 2021*(1), 9926544. https://doi.org/10.1155/2021/9926544

Naeem, M., Hoque, M. Z., Ovais, M., Basheer, C., & Ahmad, I. (2021). Stimulus-responsive smart nanoparticles-based CRISPR-cas delivery for therapeutic genome editing. *International Journal of Molecular Sciences, 22*(20), 11300. https://doi.org/10.3390/ijms222011300

Naik, A., Lee, H. S., Herrington, J., Barandun, G., Flock, G., Güder, F., & Gonzalez-Macia, L. (2024). Enabling packaging integration of paper-based electrical gas sensors for the monitoring of spoilage in fresh spinach. *bioRxiv*, 2024–2007. https://doi.org/10.1101/2024.07.22.604629

Nasrollahzadeh, M., Sajadi, S. M., Sajjadi, M., & Issaabadi, Z. (2019). An introduction to nanotechnology. In *Interface science and technology* (Vol. 28, pp. 1–27). Elsevier. https://doi.org/10.1016/B978-0-12-813586-0.00001-8

Park, S., Sung, H. K., & Kim, Y. (2016). Green synthesis of metal nanoparticles using sprout plants: Pros and cons. *Journal of Nanoscience and Nanotechnology, 16*(5), 4444–4449. https://doi.org/10.1166/jnn.2016.10970

Rai, M., Gupta, I., Bonde, S., Ingle, P., Shende, S., Gaikwad, S., & Gade, A. (2022). Industrial applications of nanomaterials produced from Aspergillus species. In *The Genus Aspergillus-pathogenicity, mycotoxin production and industrial applications*. IntechOpen. https://doi.org/10.5772/intechopen.98780

Rajan, R., & Mathew, S. (2025). Future scope of nanotechnology in agriculture: Ethical concerns and international policies. In *Agricultural sustainability through nanotechnology* (pp. 256–260). CRC Press.

Raliya, R., Rathore, I., & Tarafdar, J. (2013). Development of microbial nanofactory for zinc, magnesium, and titanium nanoparticles production using soil fungi. *Journal of Bionanoscience, 7*(5), 590–596. https://doi.org/10.1166/jbns.2013.1182

Rawtani, D., Khatri, N., Tyagi, S., & Pandey, G. (2018). Nanotechnology-based recent approaches for sensing and remediation of pesticides. *Journal of Environmental Management, 206*, 749–762. https://doi.org/10.1016/j.jenvman.2017.11.037

Regmi, A., Bhandari, J., Bhattarai, S., & Gautam, S. K. (2019). Synthesis, characterizations and antimicrobial activity of cuprous oxide (Cu2O) nanoparticles. *Journal of Nepal Chemical Society, 40*, 5–10. https://doi.org/10.3126/jncs.v40i0.27271

Roy, S., & Rhim, J. W. (2021). Carrageenan/agar-based functional film integrated with zinc sulfide nanoparticles and Pickering emulsion of tea tree essential oil for active packaging applications. *International Journal of Biological Macromolecules, 193*, 2038–2046. https://doi.org/10.1016/j.ijbiomac.2021.11.035

Rubina, M. S., Vasil'kov, A. Y., Naumkin, A. V., Shtykova, E. V., Abramchuk, S. S., Alghuthaymi, M. A., & Abd-Elsalam, K. A. (2017). Synthesis and characterization of chitosan–copper nanocomposites and their fungicidal activity against two sclerotia-forming plant pathogenic fungi. *Journal of Nanostructure in Chemistry, 7*, 249–258. https://doi.org/10.1007/s40097-017-0235-4

Rukmanikrishnan, B., Ramalingam, S., Kim, S. S., & Lee, J. (2021). Rheological and anti-microbial study of silica and silver nanoparticles-reinforced k-carrageenan/hydroxyethyl cellulose composites for food packaging applications. *Cellulose, 28*(9), 5577–5590. https://doi.org/10.1007/s10570-021-03873-z

Safdar, M., Kim, W., Park, S., YonghyunGwon, Y.-O. K., & Kim, J. (2022). Engineering plants with carbon nanotubes: A sustainable agriculture approach. *Journal of Nanobiotechnology, 20*(1), 275. https://doi.org/10.1186/s12951-022-01483-w

Sani, M. N. H., Amin, M., Siddique, A. B., Nasif, S. O., Ghaley, B. B., Ge, L., & Yong, J. W. H. (2023). Waste-derived nanobiochar: A new avenue towards sustainable agriculture, environment, and circular bioeconomy. *Science of the Total Environment, 905*, 166881. https://doi.org/10.1016/j.scitotenv.2023.166881

Sankareswaran, M., Vanitha, M., Periakaruppan, R., & Anbukumaran, A. (2022). Phyllanthus emblica mediated silica nanomaterials: Biosynthesis, structural and stability analysis. *Silicon, 14*(15), 10123–10127. https://doi.org/10.1007/s12633-022-01724-5

Santhoshkumar, J., Rajeshkumar, S., & Kumar, S. V. (2017). Phyto-assisted synthesis, characterization and applications of gold nanoparticles – A review. *Biochemistry and Biophysics Reports, 11*, 46–57. https://doi.org/10.1016/j.bbrep.2017.06.004

Saraswat, P., Singh, S., Prasad, M., Misra, R., Rajput, V. D., & Ranjan, R. (2023). Applications of bio-based nanomaterials in environment and agriculture: A review on recent progresses. *Hybrid Advances, 4*, 100097. https://doi.org/10.1016/j.hybadv.2023.100097

Sarma, I. D., Bhowmick, D., Bhilkar, P., Sharma, R., Chaudhary, R. G., Sarma, I. D., & Chaudhary, R. G. (2023). Nanotechnology: Fundamental aspects and biomedical and technological applications. *Nanobiomaterials – Perspectives for Medical Applications in the Diagnosis and Treatment of Diseases, 145*, 1–18.

Sehsahb, M. D., & Salama, N. H. (2023). Biosynthesis of silver, zinc, titanium nanoparticles using pathogenic fungi: An overview on plant disease supress, toxicity and safety. *Environment, Biodiversity and Soil Security, 7*(2023), 61–84. https://doi.org/10.21608/jenvbs.2023.197433.1211

Shankar, S. S., Ahmad, A., & Sastry, M. (2003). Geranium leaf assisted biosynthesis of silver nanoparticles. *Biotechnology Progress, 19*(6), 1627–1631. https://doi.org/10.1021/bp034070w

Shankar, S. S., Rai, A., Ankamwar, B., Singh, A., Ahmad, A., & Sastry, M. (2004). Biological synthesis of triangular gold nanoprisms. *Nature Materials, 3*(7), 482–488. https://doi.org/10.1038/nmat1152

Sharma, M., Chauhan, P., Jangid, N. K., & Sharma, R. (2024). Optical properties of nanofillers. In *Handbook of nanofillers* (pp. 1–16). Springer.

Shukla, A. K., & Iravani, S. (2018). *Green synthesis, characterization and applications of nanoparticles*. Elsevier.

Shukla, Y. K., Pandey, P., Pandey, J. P., Shukla, A., & Kumar, J. (2025). Duckweed extract-mediated green synthesis of ZnO nanoparticles and its antibacterial, antioxidant, and photocatalytic properties. *Biomass Conversion and Biorefinery*, 1–15. https://doi.org/10.1007/s13399-025-06706-2

Sikder, A., Pearce, A. K., Parkinson, S. J., Napier, R., & O'Reilly, R. K. (2021). Recent trends in advanced polymer materials in agriculture related applications. *ACS Applied Polymer Materials, 3*(3), 1203–1217.

Stachowska, J. D., Gamża, M. B., Mellor, C., Gibbons, E. N., Krysmann, M. J., Kelarakis, A., Gumieniczek-Chłopek, E., Strączek, T., Kapusta, C., & Szwajca, A. (2022). Carbon dots/iron oxide nanoparticles with tuneable composition and properties. *Nanomaterials, 12*(4), 674. https://doi.org/10.3390/nano12040674

Sur, U. K., Ankamwar, B., Karmakar, S., Halder, A., & Das, P. (2018). Green synthesis of Silver nanoparticles using the plant extract of Shikakai and Reetha. *Materials Today: Proceedings, 5*(1), 2321–2329. https://doi.org/10.1016/j.matpr.2017.09.236

Tarafdar, J., Xiang, Y., Wang, W., Dong, Q., & Biswas, P. (2012). Nanoparticle synthesis characterization and application to solve some chronic agricultural problems. *Applied Biological Research, 14*, 138–144.

Tawfik, M. M., Mohamed, M. H., Sadak, M. S., & Thalooth, A. T. (2021). Iron oxide nanoparticles effect on growth, physiological traits and nutritional contents of Moringa oleifera grown in saline environment. *Bulletin of the National Research Centre, 45*, 1–9. https://doi.org/10.1186/s42269-021-00624-9

Thakkar, K. N., Mhatre, S. S., & Parikh, R. Y. (2010). Biological synthesis of metallic nanoparticles. *Nanomedicine: Nanotechnology, Biology and Medicine, 6*(2), 257–262. https://doi.org/10.1016/j.nano.2009.07.002

Tonelli, F. M. P., Delgado, V. M. S., Silva, C. S., da Costa Pinto, G., & Tonelli, F. C. P. (2024). Are biogenic nanomaterials biocompatible? In *Sustainable green nanotechnology* (pp. 35–53). CRC Press.

Yaseen, R., & Amin, B. H. (2021). The use of phosphorus nanoparticles synthesized by rhizospheric fungus Aspergillius fumigatus as a nanofertilizer for flax plant. *Malaysian Journal of Microbiology, 17*(3) https://doi.org/10.21161/mjm.200852

Yeasmin, M. N., Sultana, M., Siddika, A., Tabassum, S., Ullah, S. M., & Bashar, M. S. (2022). Structural, optical, and morphological characterization of silica nanoparticles prepared by sol-gel process. *Journal of the Turkish Chemical Society Section A: Chemistry, 9*(4), 1323–1334. https://doi.org/10.18596/jotcsa.1071086

Zhang, B. (2018). *Physical fundamentals of nanomaterials*. William Andrew.

Zhang, K., Li, Z., Meng, Z., Zhou, W., Gao, N., & Zhang, Z. (2022). UHF-HF integrated RFID sensor: A novel information interface for industrial integration. *IEEE Sensors Journal, 22*(15), 15477–15487. https://doi.org/10.1109/JSEN.2022.3185998

Zhao, Y., Liu, X., Yang, Y., Kang, L., Yang, Z., Liu, W., & Chen, L. (2015). Carbon dots: From intense absorption in visible range to excitation-independent and excitation-dependent photoluminescence. *Fullerenes, Nanotubes and Carbon Nanostructures, 23*(11), 922–929. https://doi.org/10.1080/1536383X.2015.1018413

Zinchenko, A., Miwa, Y., Lopatina, L. I., Sergeyev, V. G., & Murata, S. (2014). DNA hydrogel as a template for synthesis of ultrasmall gold nanoparticles for catalytic applications. *ACS Applied Materials & Interfaces, 6*(5), 3226–3232. https://doi.org/10.1021/am5008886

Chapter 2
Principles and Methods of Green Synthesis

M. L. Jiménez-Gonzalez, Alicia Rosales-Pérez, Adriana Hernández-Rangel, and Rafael Hernández-Rangel

2.1 Introduction

Green synthesis has emerged as a critical approach in sustainable chemistry, addressing the increasing need for environmentally friendly and efficient methods of chemical production. Based on the green chemistry principles, it aims to minimize environmental and health hazards while preserving or even enhancing the yields of chemical processes (Kharissova et al., 2019). As defined by Anastas and Warner, green chemistry is based on twelve guiding principles that advocate for designing safer chemicals, reducing waste, utilizing renewable resources, and maximizing energy efficiency (Zimmerman et al., 2020; Ganesh et al., 2021). By focusing on waste prevention rather than post-treatment, these principles provide practical guidance that aids researchers and industries toward eco-conscious practices (Huston et al., 2021).

In order to decrease the environmental impact of industrial and chemical processes, green chemistry was developed as a novel field. Creating an essential interface between sustainability and chemistry (Sánchez Morales et al., 2024). The fundamental ideal of green chemistry requires new concepts in which the consumer is not considered in the production scheme. Rather, the "ideal consumer"

M. L. Jiménez-Gonzalez · R. Hernández-Rangel (✉)
Chemistry Department, Universidad Autónoma de Aguascalientes, Aguascalientes, Mexico

Universidad Tecnologica Metropolitana de Aguascalientes, Aguascalientes, Mexico
e-mail: mjimenez@utma.edu.mx; rrangel@utma.edu.mx

A. Rosales-Pérez
Universidad Tecnologica Metropolitana de Aguascalientes, Aguascalientes, Mexico
e-mail: brosales@utma.edu.mx

A. Hernández-Rangel
Instituto Politécnico Nacional, ESIQIE, Mexico City, Mexico
e-mail: adhernandezra@ipn.mx

K. Esquivel Escalante et al. (eds.), *Green Nanomaterials*, Nanotechnology in the Life Sciences, https://doi.org/10.1007/978-3-032-04492-1_2

understands the need to preserve the environment and uses the least amount of goods (Soltys et al., 2021). Thus, indirectly reducing the waste generation, contributing to waste minimization. Likewise, avoiding the use of hazardous chemicals, making processes less toxic and more environmentally friendly (Soltys et al., 2021; Kar et al., 2022; Sánchez Morales et al., 2024).

To avoid expensive, hazardous reagents, biosynthesis proposes the use of biological entities to achieve chemical reactions (Chopra et al., 2022). A wide range of substances can be synthesized, including elemental metals and organometallic compounds varying from molecular to supramolecular systems (Mikhailov & Mikhailova, 2022; Alsaiari et al., 2023). The main reaction mechanism accepted for this approach can be categorized into reactions that change the oxidation state and reactions that do not change the oxidation state of chemical elements, thus, making it suitable for the synthesis of materials such as nanoparticles, pharmaceuticals, and bio-based polymers (Md Ishak et al., 2019; Kar et al., 2022). By exploiting natural processes, biosynthesis provides a framework for innovative solutions that follow green chemistry principles while improving scientific and industrial capabilities (Desai et al., 2024; Ali et al., 2024).

Incorporating microorganisms such as bacteria, fungi, and algae, enzymes and plants into chemical processes that operate efficiently at mild conditions, decreasing energy consumption along with waste reduction, allows green synthesis to address key challenges in modern industry (Cano-Flores et al., 2020; Chopra et al., 2022). These natural compounds not only replace synthetic chemicals but also offer renewable and biodegradable alternatives for chemical synthesis. The availability, renewability, and minimal environmental impact of plant-based synthesis make it a cornerstone of green synthesis methodologies (Ratti, 2020; Alsaiari et al., 2023).

To sum it up, green chemistry processes aided by biological sources highlight the potential of green synthesis. By combining practical methodologies with theoretical frameworks, green synthesis assesses environmental challenges, expanding the material science and chemistry boundaries (Zimmerman et al., 2020; Ganesh et al., 2021; Alsaiari et al., 2023). Accordingly, this chapter offers a comprehensive overview of green chemistry principles as a way to minimize energy consumption and waste generation; afterwards, a biosynthesis overview is assessed with special attention to the future trend of green chemistry; and, finally, the key biological sources will be discussed, underscoring their role in driving the upcoming generation of sustainable scientific advancements.

2.2 Green Chemistry Principles

Recent advancements in nanoparticles (NPs) synthesis and methodologies have enabled the way for more sustainable and environmentally responsible practices. This development aligns with green chemistry principles, which offer a clear

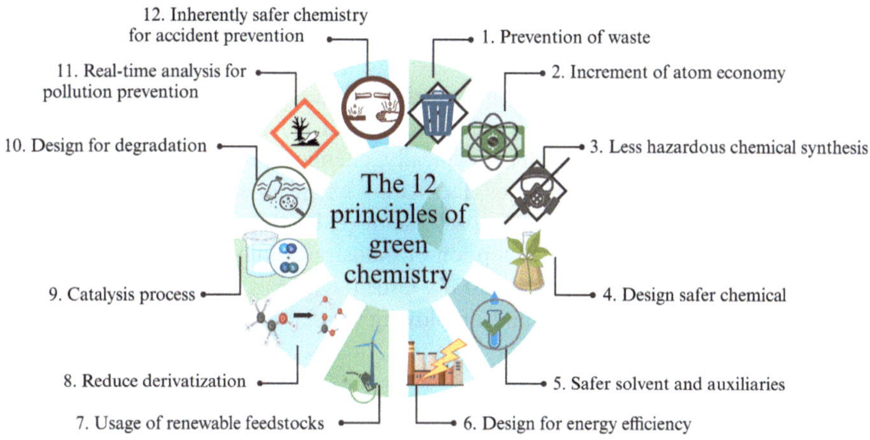

Fig. 2.1 Twelve principles of green chemistry

framework to ensure assimilation and implementation. (Duan et al., 2015; Ardila-Fierro & Hernández, 2021).

During the 1990s, green chemistry gained prominence by offering a safer alternative to traditional chemical synthesis for both workers and the environment. However, it was not until 1998 that Anastas and Warner formally defined green chemistry as "the utilization of a set of principles that reduces or eliminates the use or generation of hazardous substances in the design, manufacturing, and application of chemical products" (Anastas & Warner, 2000). In addition, they introduce the 12 Principles of Green Chemistry in collaboration with the US Environmental Protection Agency and the Organization for Economic Cooperation and Development (OECD). The OECD workshop on sustainable chemistry further supported these principles by compiling data on sustainable practices among member nations, encouraging worldwide initiatives (Zuin et al., 2021). Finally, in 2017, the International Union of Pure and Applied Chemistry (IUPAC) established the Interdivisional Committee on Green Chemistry for Sustainable Development (Kernaghan et al., 2024).

To summarize the 12 Principles of Green Chemistry, the PRODUCTIVELY mnemonic has been used, outlining key concepts such as preventing waste, promoting renewable materials, and designing safer chemical processes. Each letter represent a principle: P—prevent wastes; R—renewable resources; O—omit derivatization steps; D—degradable chemical products; U—use of safe synthetic methods; C—catalytic reagents; T—temperature, pressure ambient; I—in-process monitoring; V—very few auxiliary materials; E—E-factor, maximize feed in product; L—low toxicity of chemical products; Y—"Yes", it is safe (Tang et al., 2005; Tang et al., 2008).

Green Chemistry principles are playing a growing role in the synthesis of NPs for agricultural applications, with the aim of reducing environmental impact and promoting sustainability. As shown in Fig. 2.1, these principles focus on encouraging the minimization of harsh substances at the design, manufacture, and application steps of chemical processes.

Prevention: Avoiding waste is a core principle of green chemistry, encouraging the decrease of non-essential reactants such as solvents, catalysts, and auxiliaries that do not become part of the final product. Typically, chemical processes rely on solvents for controlling reaction temperatures, facilitating chemical equilibria, and influencing reactivity. The Environmental Impact Factor (E-Factor) measures the waste per kilogram of product, promoting adoption of industrial ecology strategies to reduce unavoidable by-products (Sheldon, 2007). Green synthesis addresses this issue by optimizing reaction conditions and employing biodegradable materials (Gour & Jain, 2019).

Atom Economy: Aims to ensure all resources used in reactions are integrated into the final product, thereby reducing raw material depletion and waste generation (Trost, 1995). In nanoparticle synthesis, this principle applies to multifunctional materials that integrate functional units while minimizing inactive components (Freund et al., 2018). This approach provides a safer and more sustainable alternative to conventional techniques (Jadoun et al., 2021). Some chemical reactions are inherently atom-economical, meaning that 100% of the atoms from the reactants are retained in the final product (Lim & Dolzhenko, 2020).

Less Hazardous Chemical Synthesis: Developing methodologies to use and generate less toxic substances is critical for green chemistry. Driving the development of new green reactions. This includes using safe solvents like water or supercritical fluids, and plant-derived reducing agents, which enhance safety, selectivity, and reduce environmental impact (Singh et al., 2018). For instance, natural compounds from plants can mediate the synthesis of metal oxide and metal nanoparticles, including gold (Santhosh et al., 2022), silver (Alharbi et al., 2022), zinc oxide (Bandeira et al., 2020), and titanium dioxide (Sunny et al., 2022). This approach is environmentally friendly, simple, and cost-effective, as it avoids hazardous chemicals and harmful byproducts (Soltys et al., 2021).

Designing Safer Chemicals: This principle emphasizes creating chemical products that are both effective and low in toxicity. This is particularly relevant in nanoparticle synthesis due to the potential hazards to human well-being and the environment (Duan et al., 2015). Knowledge in toxicology has evolved, integrating structure–property relationships to design safer chemicals (Anastas, 2016). For instance, enhancing the biocompatibility of nanoparticles by incorporating biological components such as phytochemicals and biomolecules in green synthesis, making them appropriate for biomedical applications (Radulescu et al., 2023). Integrating green toxicology early in the design process reduces hazards and ensures safer chemical products (Maertens et al., 2024).

Safer Solvents and Auxiliaries: This principle focuses on minimizing or replacing hazardous solvents with safer alternatives, including solventless systems (Shaoo & Banik, 2024), water (Mondal & Mukhopadhyay, 2024), supercritical fluids (Elmaaty

& Sorour, 2023), and ionic liquids (Sharma et al., 2022). The decision to avoid solvents is determined by the physicochemical properties of the reagents and the final product. While the ideal green solution is to eliminate solvents entirely, this often requires reengineering chemical processes to ensure the chemical reaction proceeds as effectively as it would with a traditional solvent (Avila-Ortiz & Juaristi, 2020). Among the safer alternatives, water is one of the most widely used. Its unique physicochemical properties improve reaction rates through the hydrophobic effect and simplify separation processes (Machado et al., 2021).

Design for Energy Efficiency: Minimizing intensive energy steps in the design of chemical reactions is a key principle of green chemistry. Excessive energy use is considered wasteful and counterproductive. Reducing energy consumption or optimizing energy requirements is, therefore, essential. Developing energy-efficient synthetic methods is a core focus of green chemistry. This involves decreasing the energy input of chemical reactions by utilizing appropriate reactants, allowing the process to occur at room temperature. In addition, the use of efficient green energy supplies, including microwave heating, ultrasound, and hydrothermal processing, has been recognized as a sustainable strategy to reduce energy consumption in chemical synthesis (Kharissova et al., 2019). For instance, the use of sonochemistry mediates and provides an ideal chemical environment, reduces reaction times, and facilitates processes at room temperature (Rosales Pérez & Esquivel Escalante, 2024).

Use of Renewable Feedstocks: The use of renewable resources is essential for reducing environmental impact, improving economic challenges, and decreasing the reliance on traditional fossil nonrenewable sources (Razmjoo et al., 2021). Green synthesis harnesses renewable materials, including plant extracts and other biological materials, as reducing and stabilizing agents. By doing so, it not only reduces the need for hazardous chemicals but also utilizes sustainable and eco-friendly materials (Madani et al., 2022).

Biomass is one of the most significant renewable feedstocks, serving as both a material and energy source. Derived from living organisms, biomass encompasses a wide range of resources, including wood, crops, agricultural residues, food, and more (Yang et al., 2021). The pursuit of green synthetic strategies to utilize renewable energy as starting materials is particularly critical for large-scale applications (Madani et al., 2022).

Reduce Derivatives: The strategy to reduce derivatization steps in chemical synthesis focuses on minimizing the use of additional reagents and preventing secondary reactions, thereby decreasing waste production. By simplifying these steps, the process becomes more efficient, less wasteful, and more environmentally friendly (Anastas & Eghbali, 2009; Duan et al., 2015).

In green nanoparticle synthesis, this principle promotes the use of direct synthesis methods that eliminate the need for intermediate derivatives. This approach reduces chemical and energy consumption while minimizing waste and potential environmental impact. Implementing this principle involves selecting non-toxic capping and reducing agents, using safe solvents, and developing energy-efficient methods, as outlined in recent studies on green chemistry and nanoparticle synthesis (Radulescu et al., 2023).

Catalysis: Catalysis improves reaction yields by reducing the required energy, eliminating the need for large reactant quantities, and increasing product selectivity. Thus, reducing energy consumption, feedstock usage, and waste generation (Delidovich & Palkovits, 2016; Devadasa & Laxminarayana, 2022). In particular, green nano catalysts are designed for high selectivity, minimizing unwanted by-product formation and maximizing the efficiency of chemical processes (Shen et al., 2021). In addition to its unique properties of high surface area-to-volume ratio, catalytic activity is further improved (Kidwai, 2010). Additionally, biocatalysis uses natural, biomimetic pathways for more sustainable and efficient catalytic processes (Woodley, 2022).

Design for Degradation: The design-for-degradation principle underscores the need to develop chemical products that can degrade into harmless by-products after use, thereby minimizing their environmental impact (Anastas & Eghbali, 2009). This principle is particularly relevant in green nanoparticle synthesis, where the focus is to develop materials that degrade both safely and efficiently through the use of biodegradable materials, plant-mediated synthesis, and the development of nanoparticles with enhanced degradation properties (Duan et al., 2015).

Real-Time Analysis for Pollution Prevention: One of the main challenges of green chemistry is minimizing the generation of byproducts and unsafe substances. This principle is focused on in-process monitoring and control to avoid the generation of unsafe substances. The concept of real-time refers to the ability to monitor chemical processes in real time and take immediate action to prevent undesirable outcomes. This is achieved through the application of advanced sensor technologies (Antonacci et al., 2023). In addition, this approach optimizes reagent use, identifies waste and effluent composition, and minimizes environmental impact (Brett, 2009; Anastas & Eghbali, 2009).

Inherently Safer Chemistry for Accident Prevention: This principle is of key importance for nanoparticle synthesis, where the aim is to limit risks related to both chemical reactions and materials involved. By centering on the design and implementation of chemical processes that inherently reduce the potential for chemical accidents, the synthesis of nanoparticles becomes more sustainable and less hazardous (Savitskaya et al., 2021). In practice, this involves the careful selection of non-toxic capping and reducing agents, the use of innocuous solvents, and the adoption of energy-efficient synthetic methods (Anastas & Eghbali, 2009). These strategies align with green chemistry's broader goal of creating safer, more sustainable processes (Kharissova et al., 2019).

Green chemistry has expanded beyond reaction and product design to include concepts like circular chemistry, a core element of the circular economy (Ardila-

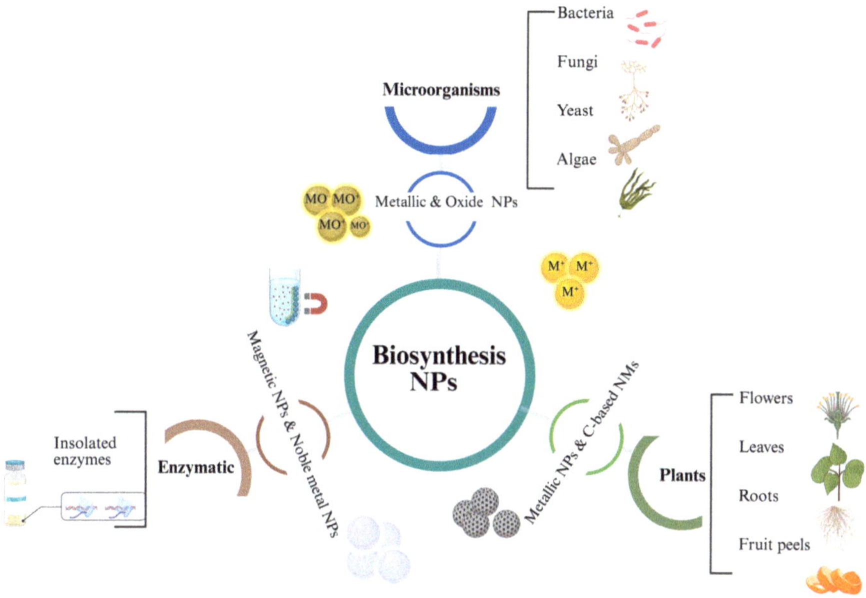

Fig. 2.2 Schematic of biosynthesis of nanoparticles

Fierro & Hernández, 2021). Sustainable chemistry drives innovation across industries, improving product performance while prioritizing environmental and human health. By using renewable resources and avoiding hazardous materials, green chemistry offers significant environmental and societal benefits (Zuin et al., 2021).

In the context of nanoparticle synthesis, adhering to green chemistry principles supports sustainable agricultural practices. These methods reduce environmental impact, promote renewable resources, and provide safer alternatives to traditional agricultural inputs. Despite challenges in optimizing NP synthesis for agricultural applications, green chemistry offers promising pathways for sustainable development (Bahadur et al., 2023). Furthermore, biosynthesis processes follow the green chemistry principles

Overview of Biosynthesis
Biosynthesis of nanomaterials is expanding quickly, merging nanotechnology with biological sciences to develop environmentally friendly and sustainable approaches for producing nanomaterials. Unlike conventional physical and chemical synthesis methods, which often require harsh conditions, toxic reagents, and significant energy input, biosynthesis leverages the natural capabilities of biological systems, such as microorganisms, plants, and enzymes, to produce nanomaterials under mild, eco-friendly conditions.

In this process, biological entities act as catalysts, utilizing their metabolic pathways and reducing agents to generate nanoparticles with unique physicochemical properties. For instance, metal ions can be reduced by bacteria and fungi;

likewise, plant extracts rich in phytochemicals can mediate a wide range of nanomaterials synthesis. Figure 2.2 shows a schematic of the current paths and microorganisms used in nanomaterials biosynthesis. These biologically derived nanomaterials often exhibit enhanced biocompatibility, reduced toxicity, and controlled morphology, making them highly desirable for various applications.

2.2.1 *Biosynthesis of Nanoparticles*

Biological entities offer a different alternative to conventional physical and chemical methods. Herein, two types of biogenic methods are aboard: (a) synthesis by microorganisms (bacteria and fungi) and (b) plant extracts. These approaches offer an environmentally friendly, low-toxicity, and efficient framework for the synthesis and fabrication of NPs.

Biosynthesis of nanoparticles can be classified based on both the type of nanomaterial and the biological system employed. In this context, microbial synthesis in which several biological entities react with the metal solutions to form the nanoparticles can be divided into two types:

1. Biosorption

   The metal cations' biosorption involves their binding to the cell walls of organisms. Stable nanoparticle formation occurs through cell or peptide interactions, with key mechanisms including physisorption, precipitation, ion exchange, and complexation (Saravanan et al., 2021). In the case of microorganisms, they secrete extracellular polysaccharides containing glycoproteins, lipopolysaccharides, and other compounds with anionic functional groups that attract cations from aqueous or pollutant solutions (Vijayaraghavan & Yun, 2008; Wang et al., 2018). For plants, it is widely recognized that certain plants accumulate metals, which are subsequently reduced to form nanoparticles within their cells (Vijayaraghavan & Ashokkumar, 2017).
2. Bioreduction

   The process of bioreduction refers to the conversion of metal ions into stable forms through the action of microorganisms and their enzymes. These biologically inert enzymes can be safely removed from polluted environments once nanoparticle formation is complete. Several agents in biological cells are able to initiate nanoparticle synthesis, including amides, amines, carbonyl groups, pigments, proteins, alkaloids, among others (Sampath et al., 2022). Additionally, plants and microorganisms (including bacteria, fungi, and algae) secrete compounds capable of oxidizing or reducing metal ions to create metallic or magnetic nanoparticles. These organisms are easy to manage and can produce nanoparticles intracellularly or extracellularly (Patil & Kim, 2018; Gahlawat & Choudhury, 2019).

Enzymes and Proteins: Protein-mediated bioreduction is a complex process due to the intricate structure of proteins. Free amino or carboxylate groups are known to have the main interaction with nanoparticles. Proteins and enzymes, due to their strong reducing activity, attract metal ions to specific molecular sites responsible for reduction, enabling NP formation. However, their chelating activity remains moderate. The size, morphology, and yield of nanoparticles can be influenced by amino acids within proteins; thus, playing a crucial role in determining its physicochemical properties (Álvarez-Chimal et al., 2023).

Saccharides: Saccharides, particularly monosaccharides like glucose, can also contribute to nanoparticle synthesis. This process is mediated through the oxidation of the aldehyde group of glucose by hydroxyl group addition into a carboxyl group. Further enhancing the reduction of metal ions and leading to nanoparticle formation (Álvarez-Chimal et al., 2023).

Phytochemicals: Phytochemicals are plant-derived secondary metabolites that serve as effective bioreducing agents. They can be divided into organic acids, phenolic acids, and flavonoids groups. Organic acids are characterized as secondary metabolites that can act as reducing agents in several nanoparticle fabrication processes (Dauthal & Mukhopadhyay, 2016). Phenolic acids, a part of the polyphenol family consisting of a phenolic ring and an organic carboxylic acid group, are known as one of the most widely used reducing agents in nanoparticle synthesis (Al-Snafi et al., 2022). Finally, the flavonoid group are large polyphenolic compound with multiple functional groups that actively chelate and reduce metal ions. During structural transformations, flavonoids generate protons that reduce metal ions, contributing to nanoparticle nucleation, formation, and subsequent aggregation (Thatyana et al., 2023).

2.2.2 Methodologies for the Biosynthesis of Nanoparticles Using Microorganisms

This approach is based on the natural metabolic pathways of microbes to convert metal ions into stable nanoparticles under mild conditions. Depending on where the reaction takes place, it can be divided into two classifications, namely, intracellular and extracellular (Tsekhmistrenko et al., 2020). In the intracellular mechanism, metal ions of the precursor enter microbial cells, where the reaction takes place, mediated by enzymes. In contrast, extracellular synthesis requires the binding of metal ions to the cell surface, followed by their enzymatic reduction (Dhand et al., 2015; Tsekhmistrenko et al., 2020). Microorganisms mediate the reduction of metal ions by providing different enzymes, including nitrate reductase. For instance, it has been proven that *Bacillus stearothermophilus* can produce nitrate reductase in the culture supernatant and significantly reduce silver ions, leading to the silver nanoparticles (AgNPs) formation (El-Batal et al., 2013). Among microorganisms employed for nanoparticle synthesis, bacteria and fungi have attracted attention in the research field.

Intracellular Synthesis
Intracellular synthesis occurs within living organisms, using their metabolic processes and growth conditions to form nanoparticles (Álvarez-Chimal et al., 2023). In this method, metal ions and other molecules are transported into microbial cells, where enzymes mediate their transformation into nanoparticles. This process relies on the organism's metabolic activity, including the role of electron donors, which reduce metal ions to their metallic state (zero charge). The controlled biosorption of metallic ions by yeast and bacterial cells during their metabolism illustrates the close connection between cellular processes and nanoparticle formation (Yang et al., 2022). Although intracellular synthesis has its benefits, it tends to be more complex than extracellular methods. Additional steps, such as cell disruption (via ultrasound or chemical detergents), are often required to release the synthesized nanoparticles from within the cells (Siddiqi et al., 2018). This adds an extra layer of complexity to downstream processing, making it less favorable for large-scale production. However, a key advantage of intracellular synthesis is its ability to produce nanoparticles with specific characteristics, since the internal cellular environment can influence the size, shape, and stability of the nanoparticles formed.

Extracellular Synthesis
The extracellular synthesis process relies on the use of bioresources as enzymes, proteins, and cell extract, produced from several organisms, to generate and stabilize nanoparticles in a single step (Faramarzi & Sadighi, 2013; Álvarez-Chimal et al., 2023). However, aggregation of nanoparticles is a common finding in this method, which simplifies recovery and purification (Singh et al., 2016).

Extracellular synthesis offers several advantages, including faster production rates, simpler downstream processing, and no extra steps to recover nanoparticles (Álvarez-Chimal et al., 2023). These features make it particularly suited for large-scale production, making it a preferred method in many applications (Siddiqi et al., 2018). Additionally, the scalability and ease of use highlight its potential for producing metal nanoparticles in a sustainable and efficient manner (Yang et al., 2022). Understanding and optimizing these biological synthesis processes can pave the way for sustainable and efficient production of nanoparticles for diverse applications.

2.2.3 Challenges and Future Directions

Currently, the future of research involving green chemistry principles is evolving at an incredible rate, providing both great opportunities and notable challenges. In this context, green synthesis combined with biosynthesis offers a promising, eco-friendly alternative to traditional chemical and physical synthesis methods. According to literature, the main obstacles requiring further investigation are: (i) the reproducibility and standardization of biological sources for ensuring uniform and predictable nanomaterial characteristics; (ii) ensuring biocompatibility, stability, and bioavailability of microbial nanoparticles, particularly for therapeutic applications, and (iii) the limited understanding of the biological processes involved in nanomaterial production (Grasso et al., 2019; Ahmed et al., 2022; Yang et al., 2022). Nevertheless, efforts toward overcoming these drawbacks are already being made.

The future of nanomaterials biosynthesis holds immense potential, supported by innovations across multiple fields that address current challenges and unlock new possibilities. Genetic engineering stands out as a transformative tool, enabling the enhancement of biosynthetic pathways for more efficient and scalable production of nanomaterials (Grasso et al., 2019). Meanwhile, the exploration of bio-inspired methods promises to integrate nanomaterials into micro- and nanodevices, with significant implications for drug delivery systems and biosensing technologies (Harun-Ur-Rashid et al., 2023). On an industrial scale, the use of recombinant strains like *Escherichia coli* for biosynthesis under mild conditions offers a sustainable and cost-effective pathway to produce nanomaterials (Choi et al., 2018). These advancements collectively underscore a future where biosynthesis not only addresses ecological concerns but also revolutionizes the fields of medicine, technology, and industry.

2.3 Essential Bio-Sources in Green Synthesis

Traditional physical and chemical methods for nanoparticle synthesis frequently require considerable time and energy expenditures, while also relying on harmful chemicals. Consequently, there has been a shift in research toward cleaner, more eco-friendly synthesis methods (Bao et al., 2021). Bio-based approaches to nanomaterial production are emerging as a preferable alternative to traditional techniques, offering substantial potential for NP production. Biological synthesis ensures the production of uniformly sized, well-defined nanoparticles and enhances their purity and safety. Moreover, they align with the green chemistry principles, making them an environmentally sustainable process (Osman et al., 2024).

2.3.1 Microorganisms Facilitating Nanoparticles Synthesis

Several microorganisms, including bacteria, fungi, yeast, viruses, and algae, act as catalysts for the sustainable, cost-effective synthesis of metal nanoparticles like gold, copper, silver, and palladium, and metal oxides such as zinc oxide, titanium oxide, among others. Furthermore, these nanoparticles could exist in different morphologies and sizes, such as nanotubes, nanoconjugates, nanorods, and nanowires (Gahlawat & Choudhury, 2019).

Table 2.1 Nanostructured systems synthesized by bacterial strains

Metal	Bacterial strain	Culture conditions		NPs characteristics		References
		Temperature (°C)	Culture time	Size (nm)	Morphology	
Silver	*Bacillus thuringiensis*	30	24 h	10–70	Spherical	Khaleghi et al. (2019)
	Serratia marcescens ssp *sakuensis*	30	48 h	10–20	Spherical	Akl et al. (2020)
	Shewanella sp. ARY1	35	6 h	38	Spherical	Mondal et al. (2020)
Gold	*Bacillus thuringiensis*			8–30	Spherical	Khaleghi et al. (2019)
	Vibrio alginolyticus	40	24 h	100–150	Irregular shape	Shunmugam et al. (2021)
	Paracoccus haeundaensis BC74171T	70	15 min	21	Spherical	Patil et al. (2019)
	Pseudomonas stutzeri	80	–	10–20	Spherical	Desai et al. (2021)

Bacteria

Bacteria, a diverse group of unicellular prokaryotic organisms, are found in a wide range of environments, including soil and water, and as symbionts in plants, animals, and other microorganisms (Vijayaraghavan & Ashokkumar, 2017). Their unique biological characteristics have drawn considerable interest in nanotechnology, particularly for nanoparticle synthesis. Through specific biochemical pathways, bacteria can reduce metal ions into nanoparticles, making them an attractive option for green, sustainable nanomaterial production (Patil & Kim, 2018; Kulkarni et al., 2023). This bacterial-mediated synthesis has gained popularity due to the ease of bacterial cultivation, cost-effectiveness, and scalability (Vijayaraghavan & Ashokkumar, 2017; Singh et al., 2018).

Bacteria can mediate nanoparticle synthesis via two main mechanisms: extracellular and intracellular (Narayanan & Sakthivel, 2010). Extracellular synthesis is characterized by the use of enzymes and metabolites located in the surrounding medium, which enables the metal ion reduction, resulting in nanoparticle formation outside the microorganisms (Kulkarni et al., 2023). On the other hand, the mechanism by which metal ions cross the cell membrane and are reduced by intracellular enzymes is called intracellular synthesis (Álvarez-Chimal et al., 2023).

Both mechanisms are influenced by factors such as bacterial species, pH, temperature, metal ion concentration, and incubation time, which can be optimized to achieve specific nanoparticle properties (Iravani, 2014; Singh et al., 2018). Moreover, bacteria can synthesize a wide variety of nanoparticles, including metallic nanoparticles (e.g., gold, silver, copper), metal oxide nanoparticles (e.g., titanium dioxide, zinc oxide), and magnetic nanoparticles (e.g., iron oxide).

In this context, Table 2.1 summarizes the conditions of some studies for silver and gold nanoparticles synthesis by different bacterial strains. For instance, Akl et al. (2020) optimized silver nanoparticle (AgNPs) production using *Serratia marcescens* ssp. *sakuensis*. According to their findings, optimal AgNPs required experimental conditions of pH 7, 2 mM of precursor ($AgNO_3$) concentration, 30 °C temperature, a mixing ratio of 20:50 ml (silver nitrate to culture supernatant), an incubation time of 48 h, and a continuous agitation speed of 140 rpm.

On the other hand, gold nanoparticles (AuNPs) were investigated by Patil et al. (2019). The key parameters influencing the synthesis were found to be temperature, incubation time, supernatant concentration, and chloroauric acid ($HAuCl_4$) concentration using the marine bacterium *Paracoccus haeundaensis* BC74171T. According to the results, the process depended mainly on 2 mM chloroauric acid and a temperature of 60 °C. Similarly, Desai et al. (2021) explored the effects of $HAuCl_4$ concentration, pH, incubation temperature, and incubation time on AuNP synthesis with *Pseudomonas stutzeri*. They identified the optimal conditions as a precursor concentration of 3 mM, an alkaline pH of 9, and an incubation temperature of 50 °C for 60 min.

Fungi

Fungi are characterized by its rigid cell walls made of glycoproteins and polysaccharides such as glucan and chitin (Yadav et al., 2015). The NPs green synthesis using fungi offers several significant advantages, including simplicity, scalability, economic feasibility, and efficient biomass processing (Hulkoti & Taranath, 2014). Besides, fungi-mediated biosynthesis of metal and metal oxide nanoparticles is particularly effective for yielding monodispersed NPs with well-defined morphologies. Compared to other organisms, fungi possess unique enzymatic and protein-based reducing agents on their cell surfaces, enabling superior nanoparticle synthesis and the production of larger quantities of NPs than bacteria (Singh et al., 2018). The distinctive composition of fungi enables both intracellular and extracellular nanoparticle synthesis; for intracellular synthesis, mycelial culture is incorporated with metal precursor and adsorbed in the biomass (Kulkarni et al., 2023). Although NPs synthesized intracellularly generally have a smaller size, the extracellular mechanism is preferred since it has distinct advantages, such as faster synthesis and avoiding contamination with unnecessary cellular components (Patil & Kim, 2018). Fungi are particularly regarded for extracellular NP production due to their significant secretory activity, which contributes to nanoparticle reduction, stabilization, and capping (Narayanan & Sakthivel, 2010).

Similarly to bacteria, fungi-intervened nanoparticle synthesis is also regulated by factors like temperature, pH, and precursor concentration (Table 2.2). Tyagi et al. (2019) explored the temperature and pH effect on the fabrication of silver nanoparticles (AgNPs) mediated by *Beauveria bassiana*. According to the investigation, the maximum amount of AgNPs was achieved at 30 °C with an optimal pH of 6. The results showed that the formation of nano crystals led to large nanoparticles mainly due to acidic environments, resulting in a slow rate of nucleation process. Wang et al. (2021) used *Aspergillus sydowii* to produce AgNP by varying

Table 2.2 Nanostructured systems synthesized by fungal strains

Metal	Fungal strain	Culture conditions: Temperature (°C)	Culture conditions: Culture time	NPs characteristics: Size (nm)	NPs characteristics: Morphology	References
Silver	*Alternaria* sp.	Room temperature	–	3–10	Spherical	Win et al. (2020)
	Penicillium oxalicum	37	72 h	60–80	Spherical	Feroze et al. (2020)
	Talaromyces purpureogenus	–	–	25	Spherical	Hu et al. (2019)
	Botryosphaeria rhodina	Room temperature	24 h	2–50	Spherical	Akther et al. (2019)
	Cladosporium halotolerans	–	–	20	Spherical	Ameen et al. (2023)
	Aspergillus sydowii	50	–	1–24	Spherical	Wang et al. (2021)
	Beauveria bassiana	25	30 min	10–50	Triangular, circular and hexagonal	Tyagi et al. (2019)
	Aspergillus fumigatus DSM819	30	–	80	Spherical	Othman et al. (2020)
Gold	*Fusarium solani*	28	48 h	40–45	Needle and flower-like	Clarance et al. (2020)
	Fusarium oxysporum	30	24 h	22–30	Spherical and hexagonal	Naimi-Shamel et al. (2019)
	Agaricus bisporus	Room temperature	–	10–50	Spherical	Krishnamoorthi et al. (2021)
	Agaricus bisporus	80	30 min	53	Oval, spherical, hexagonal, and triangular	Dheyab et al. (2020)
	Phoma sp.	28	48 h	10–100	Spherical	Nejad et al. (2022)

temperature (20–60 °C), and pH (5.0–8.0). They found an optimal temperature of 50 °C, with higher temperatures, the AgNPs synthesis was lower, and an alkaline medium (pH 8) favored its production. Similar results were found in another study (Othman et al., 2020), in which the optimum pH for AgNP production was 9.0 when using *A. fumigatus* DSM819 CFF. Moreover, it was noted that the threshold $AgNO_3$ concentration was at 1.5 mM, when higher quantities were probed, a sudden black precipitation from AgNPs occurred. In the case of AuNPs, optimization studies had been conducted with *Fusarium oxysporum* (Naimi-Shamel et al., 2019) and *Agaricus bisporus* (Krishnamoorthi et al., 2021) by changing the incubator temperature, pH, incubation time, and amount of the $HAuCl_4$ precursor.

According to literature, optimum parameters for synthesis of GNPs were determined as 3.5 mmol/L precursor concentration at pH 9, 48 h incubation, and 30 °C temperature. These conditions yield a higher quantity or quality of produced GNPs. Furthermore, a better approach for AuNPs synthesis was achieved by decreasing the precursor concentration to 1.0 mM at room temperature. The optimized ratio of mushroom extract to $HAuCl_4$ concentration was 1:9. According to the results, the reaction rate was higher, and particles were well dispersed in the solution.

Yeast

Yeasts, a type of fungi, are eukaryotic single-celled organisms; typically used for the biogenic synthesis of NPs due to their intrinsic capability to absorb and amass high amounts of metal ions from their environments and convert them into nontoxic complex polymer compounds (Shah et al., 2015). They show abundant secretion of extracellular enzymes responsible for reducing metal ions, allowing them to adapt under metal toxicity conditions using detoxification mechanisms like bio-precipitation, chelation, and intracellular sequestration (Boroumand Moghaddam et al., 2015). Moreover, yeast species have a high growth rate, making it feasible to preserve and collect them in a laboratory. Compared to bacteria, yeast has higher yields in nanoparticle production processes, which is one of the key benefits of using it as a source of biogenic nanoparticle synthesis (Kulkarni et al., 2023).

Algae

Among microorganisms, algae are an interesting option for nanoparticle production mainly due to its low toxicity, bioaccumulation ability, and chemical reduction potential. Algae mechanisms of producing nanoparticles can be either intracellular or extracellular, involving biological sources of polysaccharides, proteins, and pigments (Álvarez-Chimal et al., 2023). Furthermore, algae offer additional advantages in nanoparticle synthesis, like rapid growth, cost-effectiveness, and eco-friendliness. These properties underscore the potential of algae as an efficient and sustainable resource for nanotechnology applications (Sampath et al., 2022).

2.3.2 *Plants as Resource for Biosynthesis*

Using plants and derivatives, such as inactivated plant tissue, plant extracts, and living plants, for metal nanoparticle synthesis is a feasible option to traditional chemical and physical methods (Iravani, 2011). Plant-assisted synthesis of NPs is a straightforward, rapid, reliable, eco-friendly, and cost-effective method (Singh et al., 2016). Moreover, this green approach reduces the need for elaborate cell culture maintenance and harmful chemicals, making it a sustainable alternative (Ettadili et al., 2022). The fundamental mechanism for nanoparticle synthesis in plants is like that in microorganisms, wherein metal salts containing metal ions are reduced to neutral metal atoms by reducing agents. These atoms subsequently nucleate into small clusters, which grow into nanoparticles (Küünal et al., 2018).

Biosynthesis of nanoparticles using plants, like the biosynthesis using microorganisms, can proceed via intracellular (the whole plant) or extracellular (plant extract) pathways.

Plant-Mediated NP Synthesis
This method involves plant cultivation and growth in substrates under metal-rich natural media or metal-rich arrangements, where the metal ions are absorbed into plant cells (intracellular synthesis), and they are reduced by phytochemicals within the cell (Santhoshkumar et al., 2019). This method is rarely employed because the phytochemicals responsible for reduction and stabilization vary across different plant parts, leading to polydisperse nanoparticles with varying morphologies (shape and size). Besides, this approach necessitates additional steps for the separation and purification of the NPs. Furthermore, intracellularly synthesized nanoparticles are often associated with undesirable bio-fragments, reducing their viability and practical utility. Consequently, this approach is considered outdated and less favored (Khan et al., 2025).

Plant-Extract NPs Synthesis
Extracellular NP synthesis uses plant extracts, and it's preferred over intracellular methods due to its numerous upsides, including simple handling, non-toxicity, cost-effectiveness, and wide availability (Iravani, 2011; Bao et al., 2021). Unlike microorganisms, plants have been exhaustively utilized, mainly due to the greater stabilization and reduction that plant phytochemicals show (Khan et al., 2025). Leaves, stems, roots, seeds, and even fruits can be used to obtain plant extracts and can be prepared in an aqueous solution that facilitates the reaction, providing better manipulation over the crystal structure and size of nanoparticles (Thomas et al., 2024).

Biomolecules involved in reducing and promoting nanoparticle formation while acting as capping agents and preventing agglomeration are secondary metabolites, including flavonoids, terpenoids, polyphenols, alkaloids, acids, vitamins, polysaccharides, and organic acids. Additionally, synthesis mediated by extracts enables less toxic, improved water solubility, and longer lifespans of nanoparticles (Osman et al., 2024). Furthermore, surface functionalization of the nanoparticles by biomolecules involved in the synthesis process induces synergistic effects for antimicrobial or cancer treatment applications (Küünal et al., 2018).

The biogenic route offers several advantages over other techniques, including the capability to utilize all parts of the plant, with each part yielding nanoparticles with distinct properties due to differences in phytochemical composition and concentration (Küünal et al., 2018; Bao et al., 2021). Polyphenols, flavonoids, terpenoids, and proteins have been identified as key agents in reducing metal ions from their salt solutions to metallic nanoparticles while stabilizing them (Khan et al., 2025). However, the origin of the plant extract significantly influences the synthesized nanoparticles' type, size, shape, and morphology, since different plant extracts contain varying concentrations of biochemical reducing agents, affecting the reproducibility of NPs synthesized across different batches (Logaranjan et al., 2016; Pinto

Table 2.3 Nanostructured systems synthesized by plant extracts

Metal	Plant extract	NPs characteristics		References
		Size (nm)	Morphology	
Silver	*Camellia sinensis*	–	Spherical	Rolim et al. (2019)
	Syzigyum cymosum	17–35	Spherical	Mahmud et al. (2022)
	Olive leaf wastes	28	Spherical	Alowaiesh et al. (2023)
	Theobroma cacao	6–18	Spherical	Thatikayala et al. (2019)
	Cinnamomum tamala leaf	10–12	Spherical	Dash et al. (2020)
	Calligonum comosum roots	100	Spherical	Algebaly et al. (2020)
	Azadirachta indica leaf	34	Spherical	Algebaly et al. (2020)
	Gongronema latifolium	19	Spherical	Aisida et al. (2019)
	Myrtus communis	15	Spherical	Alyousef et al. (2019)
	Terminalia mantaly	11–83	Anisotropic shapes	Majoumouo et al. (2019)
	Phyllanthus emblica fruit	20–90	Spherical	Masum et al. (2019)
Gold	*Simarouba glauca*	–	Prism and spherical	Thangamani and Bhuvaneshwari (2019)
	Tinospora cordifolia stem	16	Spherical	Ali et al. (2020)
	Jasminum auriculatum leaf	8–37	Spherical	Balasubramanian et al. (2020)
	Camellia japonica L. leaf	20	Spherical	Sharma et al. (2019)
	Terminalia mantaly	22–43	Spherical	Majoumouo et al. (2020)
	Lactuca indica leaf	14.5	Multi shapes	Vo et al. (2020)

et al., 2017; Salayová et al., 2021). Furthermore, synthesis parameters like metal salt concentration, pH, sunlight radiation, temperature, and contact time are known to modify the NP production rate, quantity, and other physical characteristics (Mittal et al., 2013; Küünal et al., 2018).

In addition to the mentioned parameters, plant extract-mediated synthesis is influenced by the specific plant part and the solvent used for extraction. In this context, Table 2.3 summarizes some NPs synthesized by plant extracts. Thatikayala et al. (2019) investigated extracts from different parts of *Theobroma cacao*—husk, pulp, and seed—and observed variations in particle size, dispersion, and morphology of the resulting AgNPs. Uniform spherical AgNPs (6–9 nm) with no aggregation were obtained using pulp extracts. Husk extracts produced slightly aggregated AgNPs with sizes ranging from 10 to 14 nm, while seed extracts yielded non-uniform AgNPs (12–18 nm) with clear aggregation. Similarly, Majoumouo et al. (2019, 2020) examined *Terminalia mantaly* extracts from leaves, stem bark, and roots, prepared with aqueous or methanolic solvents for the synthesis of AgNPs and AuNPs. Methanolic root extracts and aqueous extracts from leaves and stem bark generated smaller, more uniform AgNPs compared to their counterparts. In the case of AuNPs the water leaf extract produced the smallest size when compared to the other aqueous and alcoholic extracts. In another study, Algebaly et al. (2020) demonstrated that aqueous and ethanolic extracts from *C. comosum* roots and *A. indica* leaves significantly affected AgNP size. For *C. comosum*, AgNP sizes were 440 nm for aqueous extracts and 470 nm for ethanolic extracts, whereas *A. indica* extracts produced AgNPs of 430 nm and 440 nm, respectively, for water and ethanol solvents.

2.3.3 Synthesis of Nanoparticles from Bio-Waste

Bio-waste represents a valuable, underutilized resource with significant potential for transformation into nanoparticles. This approach aligns with the principles of sustainability and waste valorization, enabling the reclamation and repurposing of biological waste materials. By harnessing the inherent biochemical properties of bio-waste, nanoparticles can be synthesized in an eco-friendly manner (Bhatt & Saraswat, 2024), contributing to the circular economy and reducing environmental impact.

2.3.4 Aspects to Consider for NP biosynthesis

Selection of the Biocatalysts
The organism and the route of synthesis selection play a crucial role in green nanoparticle production, either directly or indirectly. To choose the best candidate for NP synthesis, it is important to focus on the intrinsic properties of the organisms. For intracellular synthesis, properties like biochemical pathways and enzyme activities should be considered. In the case of extracellular synthesis, extracts obtained from different bio sources such as plants, fungi, algae, and distinct microorganisms are used as reducing and stabilizing agents. These extracts have different amounts and types of biologically active compounds that affect the synthesis (Velgosova et al., 2024).

Ideal Conditions for Cell Proliferation and Enzyme Activity
Proper development and proper conditions are essential to optimize growth conditions and enhance enzyme activity. These conditions include pH, light, temperature, reaction time, composition of the medium, and metallic salt concentration. Nevertheless, physical properties such as composition, morphology, and size may be affected by said environmental parameters (Hulkoti & Taranath, 2014; Gahlawat & Choudhury, 2019). The major impact conditions are temperature and pH. Due to

its nature, enzyme-mediated nanoparticle synthesis is prone to being temperature dependent; therefore, the maximum feasible temperature must be determined for an efficient synthesis. Finally, pH significantly affects nanoparticle synthesis, enabling particular nanoparticle formation at specific pH values (Singh et al., 2016).

Ideal Reaction Conditions

The green synthesis of nanoparticles is influenced by several key factors, which play a crucial role in determining the size, shape, and yield of the nanoparticles:

- Temperature: One of the key parameters influencing reaction rates and the formation of nucleation centers. As the temperature increases, the reaction accelerates, resulting in higher yields and larger nanoparticles (Jadoun et al., 2021). Additionally, varying temperatures can promote different interactions between reactants, leading to nanoparticles with diverse shapes (Álvarez-Chimal et al., 2023). However, excessive temperature increases may lead to aggregation or undesired growth.
- pH: The pH can affect the size of NPs produced; at higher values, typically, more nucleation centers are generated and therefore smaller-sized NPs are obtained (Handayani et al., 2020). Additionally, some nanoparticles can only be synthesized under specific pH conditions, such as acidic or alkaline media (Jadoun et al., 2020).
- Time: Reaction time plays a critical role in determining the size and yield of nanoparticles. Prolonged interaction between reactants during extended reaction times often leads to increased particle size and higher yields. However, excessively long reaction times may result in aggregation or instability (Jadoun et al., 2020, 2021).
- Concentration of biosource: An optimal amount of biosource improves the synthesis yield of nanoparticles while maintaining uniform size and shape distribution (Thatyana et al., 2023).

By optimizing these parameters, the green synthesis process can be tailored to produce nanoparticles with desired characteristics, enhancing their applicability across different fields (Osman et al., 2024).

2.4 Concluding Remarks

To summarize, green chemistry offers a promising alternative to traditional methods; combined with biosynthesis and microorganisms-mediated synthesis, it is a cornerstone of further research toward an environmental and economic future. Continued research and development in this field are essential to overcome current challenges and fully realize the potential of biosynthesized nanomaterials in various applications.

The future of "green" nanoparticles lies in overcoming current drawbacks, including the standardization and optimization of synthesis protocols, expanding knowledge in bioreduction process mechanisms, and ensuring industrial scalability of green processes. To assess and overcome such challenges, interdisciplinary collaborations among researchers, engineers, and industries will be essential to refine and integrate future technologies. Additionally, balancing economic viability with environmental sustainability is critical, with a focus on lifecycle management to minimize ecological risks and promote resource efficiency. By addressing these challenges, green synthesis has the potential to revolutionize nanoparticle production, enabling sustainable solutions across medicine, agriculture, and environmental remediation.

Acknowledgments The authors would like to acknowledge SECIHTI (Former CONAHCYT) (México) for the postdoctoral fellowship granted to Rafael Hernández and Martha Jiménez.

Competing Interests The authors have no conflicts of interest to declare that are relevant to the content of this chapter.

References

Ahmed, S. F., Mofijur, M., Rafa, N., et al. (2022). Green approaches in synthesising nanomaterials for environmental nanobioremediation: Technological advancements, applications, benefits and challenges. *Environmental Research, 204*, 111967. https://doi.org/10.1016/J.ENVRES.2021.111967

Aisida, S. O., Ugwu, K., Akpa, P. A., et al. (2019). Biogenic synthesis and antibacterial activity of controlled silver nanoparticles using an extract of Gongronema Latifolium. *Materials Chemistry and Physics*, 237, 121859. https://doi.org/10.1016/j.matchemphys.2019.121859

Akl, B., Nader, M., & El-Saadony, M. (2020). Biosynthesis of Silver nanoparticles by Serratia marcescens ssp sakuensis and its Antibacterial Application against some Pathogenic Bacteria. *Journal of Agricultural Chemistry and Biotechnology, 11*, 1–8. https://doi.org/10.21608/jacb.2020.76656

Akther, T., Mathipi, V., Kumar, N. S., et al. (2019). Fungal-mediated synthesis of pharmaceutically active silver nanoparticles and anticancer property against A549 cells through apoptosis. *Environmental Science and Pollution Research*, 26, 13649–13657. https://doi.org/10.1007/s11356-019-04718-w

Algebaly, A. S., Mohammed, A. E., Abutaha, N., & Elobeid, M. M. (2020). Biogenic synthesis of silver nanoparticles: antibacterial and cytotoxic potential. *Saudi Journal of Biological Sciences, 27*, 1340–1351. https://doi.org/10.1016/j.sjbs.2019.12.014

Alharbi, N. S., Alsubhi, N. S., & Felimban, A. I. (2022). Green synthesis of silver nanoparticles using medicinal plants: Characterization and application. *Journal of Radiation Research and Applied Science, 15*, 109–124. https://doi.org/10.1016/J.JRRAS.2022.06.012

Ali, S. S. L., Selvaraj, S., Batoo, K. M., et al. (2024). Green synthesis of cubic spinel ferrites and their potential biomedical applications. *Ceramics International, 50*, 52159–52189. https://doi.org/10.1016/j.ceramint.2024.10.084

Ali, S. G., Ansari, M. A., Alzohairy, M. A., et al. (2020). Biogenic Gold Nanoparticles as Potent Antibacterial and Antibiofilm Nano-Antibiotics against Pseudomonas aeruginosa. *Antibiotics*, 9, 100. https://doi.org/10.3390/antibiotics9030100

Alowaiesh, B. F., Alhaithloul, H. A. S., Saad, A. M., & Hassanin, A. A. (2023). Green Biogenic of Silver Nanoparticles Using Polyphenolic Extract of Olive Leaf Wastes with Focus on Their Anticancer and Antimicrobial Activities. *Plants*, 12. https://doi.org/10.3390/plants12061410

Alyousef, A. A., Arshad, M., AlAkeel, R., & Alqasim, A. (2019). Biogenic silver nanoparticles by Myrtus communis plant extract: biosynthesis, characterization and antibacterial activity. *Biotechnology and Biotechnological Equipment*, 33, 931–936. https://doi.org/10.1080/13102818.2019.1629840

Alsaiari, N. S., Alzahrani, F. M., Amari, A., et al. (2023). Plant and microbial approaches as green methods for the synthesis of nanomaterials: Synthesis, applications, and future perspectives. *Molecules, 28*, 463. https://doi.org/10.3390/molecules28010463

Al-Snafi, A. E., Al-Sa'idy, H. A. H., & Hamid, H. K. (2022). The utilization of plant extracts/biomaterials for the green synthesis of nanoparticles, their biological activity and mode of action. *Open Access Research Journal of Biology and Pharmacy, 6*, 017–046. https://doi.org/10.53022/oarjbp.2022.6.1.0063

Álvarez-Chimal, R., Arenas-Alatorre, J. Á., Álvarez-Chimal, R., & Arenas-Alatorre, J. Á. (2023). Green synthesis of nanoparticles: A biological approach. Green chemistry for environmental sustainability. *Prevention-Assurance-Sustainability (P-A-S) Approach*. https://doi.org/10.5772/INTECHOPEN.1002203

Ameen, F., Al-Homaidan, A. A., Al-Sabri, A., et al. (2023). Anti-oxidant, anti-fungal and cytotoxic effects of silver nanoparticles synthesized using marine fungus Cladosporium halotolerans. *Applied Nanoscience*, 13, 623–631. https://doi.org/10.1007/s13204-021-01874-9

Anastas, N. D. (2016). Connecting toxicology and chemistry to ensure safer chemical design. *Green Chemistry, 18*, 4325–4331. https://doi.org/10.1039/C6GC00758A

Anastas, P., & Eghbali, N. (2009). Green chemistry: Principles and practice. *Chemical Society Reviews, 39*, 301–312. https://doi.org/10.1039/B918763B

Anastas, P. T., & Warner, J. C. (2000). What is green chemistry? In P. T. Anastas & J. C. Warner (Eds.), *Green chemistry: Theory and practice*. Oxford University Press.

Antonacci, A., Frisulli, V., Carvalho, L. B., et al. (2023). An All-Green Photo-electrochemical biosensor using microalgae immobilized on eco-designed lignin-based screen-printed electrodes to detect sustainable nanoherbicides. *International Journal of Molecular Sciences, 24*, 10088. https://doi.org/10.3390/ijms241210088

Ardila-Fierro, K. J., & Hernández, J. G. (2021). Sustainability assessment of mechanochemistry by using the twelve principles of green chemistry. *ChemSusChem, 14*, 2145–2162. https://doi.org/10.1002/cssc.202100478

Avila-Ortiz, C. G., & Juaristi, E. (2020). Novel methodologies for chemical activation in organic synthesis under solvent-free reaction conditions. *Molecules, 25*, 3579. https://doi.org/10.3390/molecules25163579

Bahadur, R., Kumar, A., Trivedi, A., et al. (2023). An overview of green nanotechnology as a contemporary instrument for sustainable agriculture. *International Journal of Environment and Climate Change, 13*, 704–715. https://doi.org/10.9734/ijecc/2023/v13i123732

Balasubramanian, S., Kala, S. M. J., & Pushparaj, T. L. (2020). Biogenic synthesis of gold nanoparticles using Jasminum auriculatum leaf extract and their catalytic, antimicrobial and anticancer activities. *Journal of Drug Delivery Science and Technology*, 57, 101620. https://doi.org/10.1016/j.jddst.2020.101620

Bandeira, M., Giovanela, M., Roesch-Ely, M., et al. (2020). Green synthesis of zinc oxide nanoparticles: A review of the synthesis methodology and mechanism of formation. *Sustainable Chemistry and Pharmacy, 15*, 100223. https://doi.org/10.1016/j.scp.2020.100223

Bao, Y., He, J., Song, K., et al. (2021). Plant-extract-mediated synthesis of metal nanoparticles. *Journal of Chemistry, 2021*, 1–14. https://doi.org/10.1155/2021/6562687

Bhatt, S., & Saraswat, S. (2024). A review on phytochemical mediated synthesis of nanoparticles through fruits and vegetables extract and their potential applications. *Nanotechnology For Environmental Engineering, 9*, 359–374. https://doi.org/10.1007/s41204-024-00370-z

Boroumand Moghaddam, A., Namvar, F., Moniri, M., et al. (2015). Nanoparticles biosynthesized by fungi and yeast: A review of their preparation, properties, and medical applications. *Molecules, 20*, 16540–16565. https://doi.org/10.3390/molecules200916540

Brett, C. M. A. (2009). Novel sensor devices and monitoring strategies for green and sustainable chemistry processes. *Pure and Applied Chemistry, 79*, 1969–1980. https://doi.org/10.1351/pac200779111969

Cano-Flores, A., Gómez, J., Escalona-Torres, I. S., et al. (2020). Microorganisms as biocatalysts and enzyme sources. *Microorganisms*. https://doi.org/10.5772/INTECHOPEN.90338

Choi, Y., Park, T. J., Lee, D. C., & Lee, S. Y. (2018). Recombinant Escherichia coli as a biofactory for various single- and multi-element nanomaterials. *Proceedings of the National Academy of Sciences of the United States of America, 115*, 5944–5949. https://doi.org/10.1073/PNAS.1804543115/SUPPL_FILE/PNAS.1804543115.SD02.PDF

Chopra, H., Bibi, S., Singh, I., et al. (2022). Green metallic nanoparticles: biosynthesis to applications. *Frontiers in Bioengineering and Biotechnology, 10*, 874742. https://doi.org/10.3389/FBIOE.2022.874742/BIBTEX

Clarance, P., Luvankar, B., Sales, J., et al. (2020). Green synthesis and characterization of gold nanoparticles using endophytic fungi Fusarium solani and its in-vitro anticancer and biomedical applications. *Saudi Journal of Biological Sciences*, 27, 706–712. https://doi.org/10.1016/j.sjbs.2019.12.026

Dash, S. S., Samanta, S., Dey, S., et al. (2020). Rapid Green Synthesis of Biogenic Silver Nanoparticles Using Cinnamomum tamala Leaf Extract and its Potential Antimicrobial Application Against Clinically Isolated Multidrug-Resistant Bacterial Strains. *Biological Trace Element Research*, 198, 681–696. https://doi.org/10.1007/s12011-020-02107-w

Dauthal, P., & Mukhopadhyay, M. (2016). Noble metal nanoparticles: Plant-mediated synthesis, mechanistic aspects of synthesis, and applications. *Industrial & Engineering Chemistry Research, 55*, 9557–9577. https://doi.org/10.1021/acs.iecr.6b00861

Delidovich, I., & Palkovits, R. (2016). Catalytic versus stoichiometric reagents as a key concept for Green Chemistry. *Green Chemistry, 18*, 590–593. https://doi.org/10.1039/C5GC90070K

Desai, M. P., Patil, R. V., Harke, S. S., & Pawar, K. D. (2021). Bacterium mediated facile and green method for optimized biosynthesis of gold nanoparticles for simple and visual detection of two metal ions. *Journal of Cluster Science, 32*, 341–350. https://doi.org/10.1007/s10876-020-01793-9

Desai, A. V., Lizundia, E., Laybourn, A., et al. (2024). Green synthesis of reticular materials. *Advanced Functional Materials, 34*. https://doi.org/10.1002/adfm.202304660

Devadasa, K., & Laxminarayana, N. H. (2022). Green catalysis: Innovations and applications in sustainable chemistry. *ShodhKosh: Journal of Visual and Performing Arts, 3*, 651–658. https://doi.org/10.29121/shodhkosh.v3.i2.2022.2486

Dhand, C., Dwivedi, N., Loh, X. J., et al. (2015). Methods and strategies for the synthesis of diverse nanoparticles and their applications: A comprehensive overview. *RSC Advances, 5*, 105003–105037. https://doi.org/10.1039/C5RA19388E

Dheyab, M. A., Owaid, M. N., Rabeea, M. A., et al. (2020). Mycosynthesis of gold nanoparticles by the Portabello mushroom extract, Agaricaceae, and their efficacy for decolorization of Azo dye. *Environmental Nanotechnology, Monitoring and Management*, 14, 100312. https://doi.org/10.1016/j.enmm.2020.100312

Duan, H., Wang, D., & Li, Y. (2015). Green chemistry for nanoparticle synthesis. *Chemical Society Reviews, 44*, 5778–5792. https://doi.org/10.1039/C4CS00363B

El-Batal, A. I., Amin, M. A., Shehata, M. M. K., & Hallol, M. M. A. (2013). Synthesis of silver nanoparticles by Bacillus stearothermophilus using gamma radiation and their antimicrobial activity. *World Applied Sciences Journal, 22*, 1–16. https://doi.org/10.5829/idosi.wasj.2013.22.01.2956

Elmaaty, T. A., & Sorour, H. (2023). *Extraction of natural pigments using supercritical carbon dioxide and its application on different fabrics* (pp. 231–264). https://doi.org/10.1007/978-3-031-47471-2_11

Ettadili, F. E., Aghris, S., Laghrib, F., et al. (2022). Recent advances in the nanoparticles synthesis using plant extract: Applications and future recommendations. *Journal of Molecular Structure, 1248*, 131538. https://doi.org/10.1016/j.molstruc.2021.131538

Faramarzi, M. A., & Sadighi, A. (2013). Insights into biogenic and chemical production of inorganic nanomaterials and nanostructures. *Advances in Colloid and Interface Science, 189–190*, 1–20. https://doi.org/10.1016/j.cis.2012.12.001

Feroze, N., Arshad, B., Younas, M., et al. (2020). Fungal mediated synthesis of silver nanoparticles and evaluation of antibacterial activity. *Microscopy Research and Technique*, 83, 72–80. https://doi.org/10.1002/jemt.23390

Freund, R., Lächelt, U., Gruber, T., et al. (2018). Multifunctional efficiency: Extending the concept of atom economy to functional nanomaterials. *ACS Nano, 12*, 2094–2105. https://doi.org/10.1021/acsnano.8b00932

Gahlawat, G., & Choudhury, A. R. (2019). A review on the biosynthesis of metal and metal salt nanoparticles by microbes. *RSC Advances, 9*, 12944–12967. https://doi.org/10.1039/c8ra10483b

Ganesh, K. N., Zhang, D., Miller, S. J., et al. (2021). Green chemistry: A framework for a sustainable future. *Environmental Science & Technology Letters, 8*, 487–491. https://doi.org/10.1021/ACS.ESTLETT.1C00434

Gour, A., & Jain, N. K. (2019). *Advances in green synthesis of nanoparticles*. Artif Cells Nanomed Biotechnol.

Grasso, G., Zane, D., & Dragone, R. (2019). Microbial nanotechnology: Challenges and prospects for green biocatalytic synthesis of nanoscale materials for sensoristic and biomedical applications. *Nanomaterials, 10*, 11. https://doi.org/10.3390/NANO10010011

Handayani, W., Ningrum, A. S., & Imawan, C. (2020). The role of pH in synthesis silver nanoparticles using Pometia pinnata (Matoa) leaves extract as bioreductor. *Journal of Physics Conference Series, 1428*, 012021. https://doi.org/10.1088/1742-6596/1428/1/012021

Harun-Ur-Rashid M, Jahan I, Foyez T, Imran A Bin (2023) Bio-inspired nanomaterials for micro/nanodevices: A new era in biomedical applications. Micromachines 14, 1786. doi:https://doi.org/10.3390/MI14091786

Hulkoti, N. I., & Taranath, T. C. (2014). Biosynthesis of nanoparticles using microbes – A review. *Colloids and Surfaces. B, Biointerfaces, 121*, 474–483. https://doi.org/10.1016/j.colsurfb.2014.05.027

Huston, M., Debella, M., Dibella, M., & Gupta, A. (2021). Green synthesis of nanomaterials. *Nanomaterials, 11*, 2130. https://doi.org/10.3390/NANO11082130

Hu, X., Saravanakumar, K., Jin, T., & Wang, M. H. (2019). Mycosynthesis, characterization, anticancer and antibacterial activity of silver nanoparticles from endophytic fungus talaromyces purpureogenus. *International Journal of Nanomedicine*, 14, 3427–3438. https://doi.org/10.2147/IJN.S200817

Iravani, S. (2011). Green synthesis of metal nanoparticles using plants. *Green Chemistry, 13*, 2638. https://doi.org/10.1039/c1gc15386b

Iravani, S. (2014). Bacteria in nanoparticle synthesis: Current status and future prospects. *International Scholarly Research Notices, 2014*, 359316. https://doi.org/10.1155/2014/359316

Jadoun, S., Arif, R., Jangid, N. K., & Meena, R. K. (2020). Green synthesis of nanoparticles using plant extracts: A review. *Environmental Chemistry Letters, 19*, 355–374. https://doi.org/10.1007/s10311-020-01074-x

Jadoun, S., Arif, R., Jangid, N. K., & Meena, R. K. (2021). Green synthesis of nanoparticles using plant extracts: A review. *Environmental Chemistry Letters, 19*, 355–374. https://doi.org/10.1007/S10311-020-01074-X/METRICS

Kar, S., Sanderson, H., Roy, K., et al. (2022). Green chemistry in the synthesis of pharmaceuticals. *Chemical Reviews, 122*, 3637–3710. https://doi.org/10.1021/ACS.CHEMREV.1C00631/ASSET/IMAGES/MEDIUM/CR1C00631_0213.GIF

Kernaghan, S. M., Coady, T., Kinsella, M., & Lennon, C. M. (2024). A tutorial review for research laboratories to support the vital path toward inherently sustainable and green synthetic chemistry. *RSC Sustainability, 2*, 578–607. https://doi.org/10.1039/D3SU00324H

Khaleghi, M., Khorrami, S., & Ravan, H. (2019). Identification of Bacillus thuringiensis bacterial strain isolated from the mine soil as a robust agent in the biosynthesis of silver nanoparticles

with strong antibacterial and anti-biofilm activities. *Biocatalysis and Agricultural Biotechnology*, 18, 101047. https://doi.org/10.1016/j.bcab.2019.101047
Khan, T., Raza, S., Hashmi, K., & Khan, A. R. (2025). Recent advancements in the plant-based synthesis and mechanistic insights of noble metal nanoparticles and their therapeutic applications. *Inorganic Chemistry Communications, 171*, 113594. https://doi.org/10.1016/j.inoche.2024.113594
Kharissova, O. V., Kharisov, B. I., González, C. M. O., et al. (2019). Greener synthesis of chemical compounds and materials. *Royal Society Open Science, 6*. https://doi.org/10.1098/RSOS.191378
Kidwai, M. (2010). Nanoparticles in green catalysis. In *Handbook of green chemistry* (pp. 81–92). Wiley.
Krishnamoorthi, R., Bharathakumar, S., Malaikozhundan, B., & Mahalingam, P. U. (2021). Mycofabrication of gold nanoparticles: Optimization, characterization, stabilization and evaluation of its antimicrobial potential on selected human pathogens. *Biocatalysis and Agricultural Biotechnology, 35*, 102107. https://doi.org/10.1016/j.bcab.2021.102107
Kulkarni, D., Sherkar, R., Shirsathe, C., et al. (2023). Biofabrication of nanoparticles: Sources, synthesis, and biomedical applications. *Frontiers in Bioengineering and Biotechnology, 11*, 1159193. https://doi.org/10.3389/fbioe.2023.1159193
Küünal, S., Rauwel, P., & Rauwel, E. (2018). Plant extract mediated synthesis of nanoparticles. *Emerging Applications of Nanoparticles and Architectural Nanostructures: Current Prospects and Future Trends*, 411–446. https://doi.org/10.1016/B978-0-323-51254-1.00014-2
Lim, F. P. L., & Dolzhenko, A. V. (2020). Chapter 1 – Atom economy in green organic synthesis. In Inamuddin, R. Boddula, & A. M. Asiri (Eds.), *Green sustainable process for chemical and environmental engineering and science* (pp. 1–12). Elsevier.
Logaranjan, K., Raiza, A. J., Gopinath, S. C. B., et al. (2016). Shape- and size-controlled synthesis of silver nanoparticles using Aloe vera plant extract and their antimicrobial activity. *Nanoscale Research Letters, 11*, 1–9. https://doi.org/10.1186/S11671-016-1725-X/FIGURES/9
Machado, I. V., dos Santos, J. R. N., Januario, M. A. P., & Corrêa, A. G. (2021). Greener organic synthetic methods: Sonochemistry and heterogeneous catalysis promoted multicomponent reactions. *Ultrasonics Sonochemistry, 78*, 105704. https://doi.org/10.1016/j.ultsonch.2021.105704
Madani, M., Hosny, S., Alshangiti, D. M., et al. (2022). Green synthesis of nanoparticles for varied applications: Green renewable resources and energy-efficient synthetic routes. *Nanotechnology Reviews, 11*, 731–759. https://doi.org/10.1515/ntrev-2022-0034
Maertens, A., Luechtefeld, T., Knight, J., & Hartung, T. (2024). Alternative methods go green! Green toxicology as a sustainable approach for assessing chemical safety and designing safer chemicals. *ALTEX – Alternatives to Animal Experimentation, 41*, 3–19. https://doi.org/10.14573/altex.2312291
Majoumouo, M. S., Sibuyi, N. R. S., Tincho, M. B., et al. (2019). Enhanced anti-bacterial activity of biogenic silver nanoparticles synthesized from Terminalia mantaly extracts. *International Journal of Nanomedicine, 14*, 9031–9046. https://doi.org/10.2147/IJN.S223447
Majoumouo, M. S., Sharma, J. R., Sibuyi, N. R. S., et al. (2020). Synthesis of biogenic gold nanoparticles from terminalia mantaly extracts and the evaluation of their in vitro cytotoxic effects in cancer cells. *Molecules, 25*, 1–18. https://doi.org/10.3390/molecules25194469
Mahmud, K. M., Hossain, M. M., Polash, S. A., et al. (2022). Investigation of Antimicrobial Activity and Biocompatibility of Biogenic Silver Nanoparticles Synthesized using Syzigyum cymosum Extract. *ACS Omega*, 7, 27216–27229. https://doi.org/10.1021/acsomega.2c01922
Masum, M. I., Siddiqa, M. M., Ali, K. A., et al. (2019). Biogenic synthesis of silver nanoparticles using phyllanthus emblicafruit extract and its inhibitory action against the pathogen acidovorax oryzaestrain RS-2 of rice bacterial brown stripe. *Frontiers in Microbiology*, 10, 1–18. https://doi.org/10.3389/fmicb.2019.00820

Md Ishak, N. A. I., Kamarudin, S. K., & Timmiati, S. N. (2019). Green synthesis of metal and metal oxide nanoparticles via plant extracts: An overview. *Materials Research Express, 6*, 112004. https://doi.org/10.1088/2053-1591/ab4458

Mikhailov, O. V., & Mikhailova, E. O. (2022). Introduction to problematics of the special issue "Green biosynthesis of nanomaterials for biomedical applications". *Journal of Functional Biomaterials, 13*, 307. https://doi.org/10.3390/jfb13040307

Mittal, A. K., Chisti, Y., & Banerjee, U. C. (2013). Synthesis of metallic nanoparticles using plant extracts. *Biotechnology Advances, 31*, 346–356. https://doi.org/10.1016/j.biotechadv.2013.01.003

Mondal, A., & Mukhopadhyay, C. (2024). Solvent-free microwave reactions towards significant organic transformations: A green approach. *Tetrahedron Green Chem, 4*, 100054. https://doi.org/10.1016/J.TGCHEM.2024.100054

Mondal, A. H., Yadav, D., Mitra, S., & Mukhopadhyay, K. (2020). Biosynthesis of Silver Nanoparticles Using Culture Supernatant of Shewanella sp. ARY1 and Their Antibacterial Activity. *International Journal of Nanomedicine*, 15, 8295–8310. https://doi.org/10.2147/IJN.S274535

Naimi-Shamel, N., Pourali, P., & Dolatabadi, S. (2019). Green synthesis of gold nanoparticles using Fusarium oxysporum and antibacterial activity of its tetracycline conjugant. *Journal de Mycologie Médicale, 29*, 7–13. https://doi.org/10.1016/j.mycmed.2019.01.005

Narayanan, K. B., & Sakthivel, N. (2010). Biological synthesis of metal nanoparticles by microbes. *Advances in Colloid and Interface Science, 156*, 1–13. https://doi.org/10.1016/j.cis.2010.02.001

Nejad, M. S., Najafabadi, N. S., Aghighi, S., et al. (2022). Evaluation of Phoma sp. Biomass as an Endophytic Fungus for Synthesis of Extracellular Gold Nanoparticles with Antibacterial and Antifungal Properties. *Molecules*, 27. https://doi.org/10.3390/molecules27041181

Osman, A. I., Zhang, Y., Farghali, M., et al. (2024). Synthesis of green nanoparticles for energy, biomedical, environmental, agricultural, and food applications: A review. *Environmental Chemistry Letters, 22*, 841–887. https://doi.org/10.1007/s10311-023-01682-3

Othman, A. M., Elsayed, M. A., Al-Balakocy, N. G., et al. (2020). Correction to: Biosynthesis and characterization of silver nanoparticles induced by fungal proteins and its application in different biological activities. *Journal of Genetic Engineering and Biotechnology*, 18, 7. https://doi.org/10.1186/s43141-020-0022-3

Patil, M. P., & Kim, G.-D. (2018). Marine microorganisms for synthesis of metallic nanoparticles and their biomedical applications. *Colloids and Surfaces. B, Biointerfaces, 172*, 487–495. https://doi.org/10.1016/j.colsurfb.2018.09.007

Patil, M. P., Jae, K. M., Niyonizigiye, I., et al. (2019). Extracellular synthesis of gold nanoparticles using the marine bacterium Paracoccus haeundaensis BC74171T and evaluation of their antioxidant activity and antiproliferative effect on normal and cancer cell lines. *Colloids and Surfaces. B, Biointerfaces, 183*, 110455. https://doi.org/10.1016/j.colsurfb.2019.110455

Pinto, R. J. B., Lucas, J. M. F., Morais, M. P., et al. (2017). Demystifying the morphology and size control on the biosynthesis of gold nanoparticles using Eucalyptus globulus bark extract. *Industrial Crops and Products, 105*, 83–92. https://doi.org/10.1016/J.INDCROP.2017.05.003

Radulescu, D. M., Surdu, V. A., Ficai, A., et al. (2023). Green synthesis of metal and metal oxide nanoparticles: A review of the principles and biomedical applications. *International Journal of Molecular Sciences, 24*, 15397. https://doi.org/10.3390/IJMS242015397

Ratti, R. (2020). Industrial applications of green chemistry: Status, challenges and prospects. *SN Applied Sciences, 2*, 1–7. https://doi.org/10.1007/S42452-020-2019-6/SCHEMES/3

Razmjoo, A., Gakenia Kaigutha, L., Vaziri Rad, M. A., et al. (2021). A Technical analysis investigating energy sustainability utilizing reliable renewable energy sources to reduce CO2 emissions in a high potential area. *Renewable Energy, 164*, 46–57. https://doi.org/10.1016/j.renene.2020.09.042

Rolim, W. R., Pelegrino, M. T., de Araújo Lima, B., et al. (2019). Green tea extract mediated biogenic synthesis of silver nanoparticles: Characterization, cytotoxicity evaluation and antibacterial activity. *Applied Surface Science*, 463, 66–74. https://doi.org/10.1016/j.apsusc.2018.08.203

Rosales Pérez, A., & Esquivel Escalante, K. (2024). The evolution of sonochemistry: From the beginnings to novel applications. *ChemPlusChem, 89*, e202300660. https://doi.org/10.1002/cplu.202300660

Salayová, A., Bedlovičová, Z., Daneu, N., et al. (2021). Green synthesis of silver nanoparticles with antibacterial activity using various medicinal plant extracts: Morphology and antibacterial efficacy. *Nanomaterials, 11*, 1005. https://doi.org/10.3390/NANO11041005/S1

Sampath, S., Madhavan, Y., Muralidharan, M., et al. (2022). A review on algal mediated synthesis of metal and metal oxide nanoparticles and their emerging biomedical potential. *Journal of Biotechnology, 360*, 92–109. https://doi.org/10.1016/j.jbiotec.2022.10.009

Sánchez Morales, R., Sáenz-López, P., & de las Heras Perez, M. A. (2024). Green chemistry and its impact on the transition towards sustainable chemistry: A systematic review. *Sustainability, 16*, 6526. https://doi.org/10.3390/su16156526

Santhosh, P. B., Genova, J., & Chamati, H. (2022). Green synthesis of gold nanoparticles: An eco-friendly approach. *Chemistry (Easton), 4*, 345–369. https://doi.org/10.3390/chemistry4020026

Santhoshkumar, J., Agarwal, H., Menon, S., et al. (2019). A biological synthesis of copper nanoparticles and its potential applications. *Green Synthesis, Characterization and Applications of Nanoparticles*, 199–221. https://doi.org/10.1016/B978-0-08-102579-6.00009-5

Saravanan, A., Kumar, P. S., Karishma, S., et al. (2021). A review on biosynthesis of metal nanoparticles and its environmental applications. *Chemosphere, 264*, 128580. https://doi.org/10.1016/j.chemosphere.2020.128580

Savitskaya, T., Kimlenka, I., Lu, Y., et al. (2021). Applications of Green Chemistry. In T. Savitskaya, I. Kimlenka, Y. Lu, et al. (Eds.), *Green chemistry: Process technology and sustainable development* (pp. 31–92). Springer.

Shah, M., Fawcett, D., Sharma, S., et al. (2015). Green synthesis of metallic nanoparticles via biological entities. *Materials (Basel), 8*, 7278–7308. https://doi.org/10.3390/ma8115377

Shaoo, B. M., & Banik, B. K. (2024). Chapter 17 – Solvent-less reactions: Green and sustainable approaches in medicinal chemistry. In B. K. Banik (Ed.), *Green approaches in medicinal chemistry for sustainable drug design* (2nd ed., pp. 387–408). Elsevier.

Sharma, J., Kumar, P., Sillanpaa, M., et al. (2022). Immobilized ionic liquids on Fe3O4 nanoparticles: A potential catalyst for organic synthesis. *Inorganic Chemistry Communications, 145*, 110055. https://doi.org/10.1016/j.inoche.2022.110055

Sharma, T. S. K., Selvakumar, K., Hwa, K. Y., et al. (2019). Biogenic fabrication of gold nanoparticles using Camellia japonica L. leaf extract and its biological evaluation. *Journal of Materials Research and Technology*, 8, 1412–1418. https://doi.org/10.1016/j.jmrt.2018.10.006

Sheldon, R. A. (2007). The E Factor: Fifteen years on. *Green Chemistry, 9*, 1273–1283. https://doi.org/10.1039/B713736M

Shen, M., Yu, C., Guan, H., et al. (2021). Nanoparticle-catalyzed green chemistry synthesis of polybenzoxazole. *Journal of the American Chemical Society, 143*, 2115–2122. https://doi.org/10.1021/jacs.0c12488

Shunmugam, R., Renukadevi Balusamy, S., Kumar, V., et al. (2021). Biosynthesis of gold nanoparticles using marine microbe (Vibrio alginolyticus) and its anticancer and antioxidant analysis. *Journal of King Saud University - Science*, 33, 101260. https://doi.org/10.1016/j.jksus.2020.101260

Siddiqi, K. S., Husen, A., & Rao, R. A. K. (2018). A review on biosynthesis of silver nanoparticles and their biocidal properties. *Journal of Nanobiotechnology, 16*, 14. https://doi.org/10.1186/s12951-018-0334-5

Singh, P., Kim, Y.-J., Zhang, D., & Yang, D.-C. (2016). Biological synthesis of nanoparticles from plants and microorganisms. *Trends in Biotechnology, 34*, 588–599. https://doi.org/10.1016/j.tibtech.2016.02.006

Singh, J., Dutta, T., Kim, K. H., et al. (2018). 'Green' synthesis of metals and their oxide nanoparticles: Applications for environmental remediation. *Journal of Nanobiotechnology, 16*, 1–24. https://doi.org/10.1186/S12951-018-0408-4

Soltys, L., Olkhovyy, O., Tatarchuk, T., & Naushad, M. (2021). Green synthesis of metal and metal oxide nanoparticles: Principles of green chemistry and raw materials. *Magnetochemistry, 7*, 145. https://doi.org/10.3390/MAGNETOCHEMISTRY7110145

Sunny, N. E., Mathew, S. S., Chandel, N., et al. (2022). Green synthesis of titanium dioxide nanoparticles using plant biomass and their applications – A review. *Chemosphere, 300*, 134612. https://doi.org/10.1016/j.chemosphere.2022.134612

Tang, S. L. Y., Smith, R. L., & Poliakoff, M. (2005). Principles of green chemistry: Productively. *Green Chemistry, 7*, 761–762. https://doi.org/10.1039/B513020B

Tang, S. Y., Bourne, R. A., Smith, R. L., & Poliakoff, M. (2008). The 24 principles of green engineering and green chemistry: "Improvements Productively". *Green Chemistry, 10*, 268–269. https://doi.org/10.1039/B719469M

Thangamani, N., & Bhuvaneshwari, N. (2019). Green synthesis of gold nanoparticles using Simarouba glauca leaf extract and their biological activity of micro-organism. *Chemical Physics Letters*, 732, 136587. https://doi.org/10.1016/j.cplett.2019.07.015

Thatikayala, D., Jayarambabu, N., Banothu, V., et al. (2019). Biogenic synthesis of silver nanoparticles mediated by Theobroma cacao extract: Enhanced antibacterial and photocatalytic activities. *Journal of Materials Science: Materials in Electronics, 30*, 17303–17313. https://doi.org/10.1007/s10854-019-02077-3

Thatyana, M., Dube, N. P., Kemboi, D., et al. (2023). Advances in phytonanotechnology: A plant-mediated green synthesis of metal nanoparticles using phyllanthus plant extracts and their antimicrobial and anticancer applications. *Nanomaterials (Basel), 13*, 2616. https://doi.org/10.3390/nano13192616

Thomas, S., Gonsalves, R. A., Jose, J., et al. (2024). Plant-based synthesis, characterization approaches, applications and toxicity of silver nanoparticles: A comprehensive review. *Journal of Biotechnology, 394*, 135–149. https://doi.org/10.1016/j.jbiotec.2024.08.009

Trost, B. M. (1995). Atom economy – A challenge for organic synthesis: Homogeneous catalysis leads the way. *Angewandte Chemie International Edition in English, 34*, 259–281. https://doi.org/10.1002/anie.199502591

Tsekhmistrenko, S. I., Bityutskyy, V. S., Tsekhmistrenko, O. S., et al. (2020). Bacterial synthesis of nanoparticles: A green approach. *Biosystems Diversity, 28*, 9–17. https://doi.org/10.15421/012002

Tyagi, S., Tyagi, P. K., Gola, D., et al. (2019). Extracellular synthesis of silver nanoparticles using entomopathogenic fungus: Characterization and antibacterial potential. *SN Applied Sciences, 1*, 1–9. https://doi.org/10.1007/s42452-019-1593-y

Velgosova, O., Dolinská, S., Podolská, H., et al. (2024). Impact of plant extract phytochemicals on the synthesis of silver nanoparticles. *Materials (Basel), 17*, 2252. https://doi.org/10.3390/ma17102252

Vijayaraghavan, K., & Ashokkumar, T. (2017). Plant-mediated biosynthesis of metallic nanoparticles: A review of literature, factors affecting synthesis, characterization techniques and applications. *Journal of Environmental Chemical Engineering, 5*, 4866–4883. https://doi.org/10.1016/j.jece.2017.09.026

Vijayaraghavan, K., & Yun, Y.-S. (2008). Bacterial biosorbents and biosorption. *Biotechnology Advances, 26*, 266–291. https://doi.org/10.1016/j.biotechadv.2008.02.002

Vo, T. T., Dang, C. H., Doan, V. D., et al. (2020). Biogenic Synthesis of Silver and Gold Nanoparticles from Lactuca indica Leaf Extract and Their Application in Catalytic Degradation of Toxic Compounds. *Journal of Inorganic and Organometallic Polymers and Materials*, 30, 388–399. https://doi.org/10.1007/s10904-019-01197-x

Wang, L., Liu, X., Lee, D.-J., et al. (2018). Recent advances on biosorption by aerobic granular sludge. *Journal of Hazardous Materials, 357*, 253–270. https://doi.org/10.1016/j.jhazmat.2018.06.010

Wang, D., Xue, B., Wang, L., et al. (2021). Fungus-mediated green synthesis of nano-silver using Aspergillus sydowii and its antifungal/antiproliferative activities. *Scientific Reports, 11*, 1–9. https://doi.org/10.1038/s41598-021-89854-5

Win, T. T., Khan, S., & Fu, P. (2020). Fungus- (Alternaria sp.) Mediated Silver Nanoparticles Synthesis, Characterization, and Screening of Antifungal Activity against Some Phytopathogens. *Journal of Nanotechnology*, 2020, 1–9. https://doi.org/10.1155/2020/8828878

Woodley, J. M. (2022). Ensuring the sustainability of biocatalysis. *ChemSusChem, 15*, e202102683. https://doi.org/10.1002/cssc.202102683

Yadav, A., Kon, K., Kratosova, G., et al. (2015). Fungi as an efficient mycosystem for the synthesis of metal nanoparticles: Progress and key aspects of research. *Biotechnology Letters, 37*, 2099–2120. https://doi.org/10.1007/s10529-015-1901-6

Yang, L., Wang, X. C., Dai, M., et al. (2021). Shifting from fossil-based economy to bio-based economy: Status quo, challenges, and prospects. *Energy, 228*, 120533. https://doi.org/10.1016/J.ENERGY.2021.120533

Yang, Y., Waterhouse, G. I. N., Chen, Y., et al. (2022). Microbial-enabled green biosynthesis of nanomaterials: Current status and future prospects. *Biotechnology Advances, 55*, 107914. https://doi.org/10.1016/j.biotechadv.2022.107914

Zimmerman, J. B., Anastas, P. T., Erythropel, H. C., & Leitner, W. (2020). Designing for a green chemistry future. *Science (1979), 367*, 397–400. https://doi.org/10.1126/SCIENCE.AAY3060/ASSET/8997170C-4499-4F22-A12F-B46943929ED6/ASSETS/GRAPHIC/367_397_F3.JPEG

Zuin, V. G., Eilks, I., Elschami, M., & Kümmerer, K. (2021). Education in green chemistry and in sustainable chemistry: Perspectives towards sustainability. *Green Chemistry, 23*, 1594–1608. https://doi.org/10.1039/D0GC03313H

Chapter 3
Green Nanomaterials: An Agriculture and Circular Economy Alternative

Tharindu D. Nuwarapaksha, Shashi S. Udumann, Nuwandhya S. Dissanayaka, and Anjana J. Atapattu

3.1 Introduction

3.1.1 Brief Overview of Nanotechnology in Agriculture

Nanotechnology in agriculture introduced a revolutionary period for contemporary farming, which stands as the biggest technological breakthrough in agricultural science after the Green Revolution (Malik et al., 2023). The nanoscale technology operates between 1 and 100 nanometers to exploit material characteristics at this microscopic level for a revolutionary modification of traditional agriculture. Figure 3.1 shows types of nanomaterials used in agriculture, revealing how different nanomaterials and techniques can be used to enhance plant growth, enhance soil conditions, and protect crops. The unique properties of nanomaterials with their greater surface area-to-volume ratios and enhanced reactivity, as well as physical property modifications, create unprecedented solutions for enduring agricultural issues. Agricultural nanotechnology produces its effects across three essential fields, which include precise crop management and smart delivery technologies and advanced detection systems (Pramanik et al., 2020). Engineered nanomaterials provide flexible resources to modern agriculture because they act as nano-fertilizers, which enhance both nutrient distribution and optimization of nutrient usage. Nanoscale fertilizers penetrate soils better and regulate fertilizer release through delicate mechanisms, which conserve nutrients by maximizing their uptake from the plants. Nanopesticides offer an environmentally friendly pest management solution because they need reduced application amounts yet perform at least as well as traditional pesticides (Vurro et al., 2019).

Modern agriculture detects and diagnoses agricultural parameters through the implementation of nanosensors and nanobiosensors systems (Kumar et al., 2025).

T. D. Nuwarapaksha · S. S. Udumann · N. S. Dissanayaka · A. J. Atapattu (✉)
Agronomy Division, Coconut Research Institute, Lunuwila, Sri Lanka

K. Esquivel Escalante et al. (eds.), *Green Nanomaterials*, Nanotechnology in the Life Sciences, https://doi.org/10.1007/978-3-032-04492-1_3

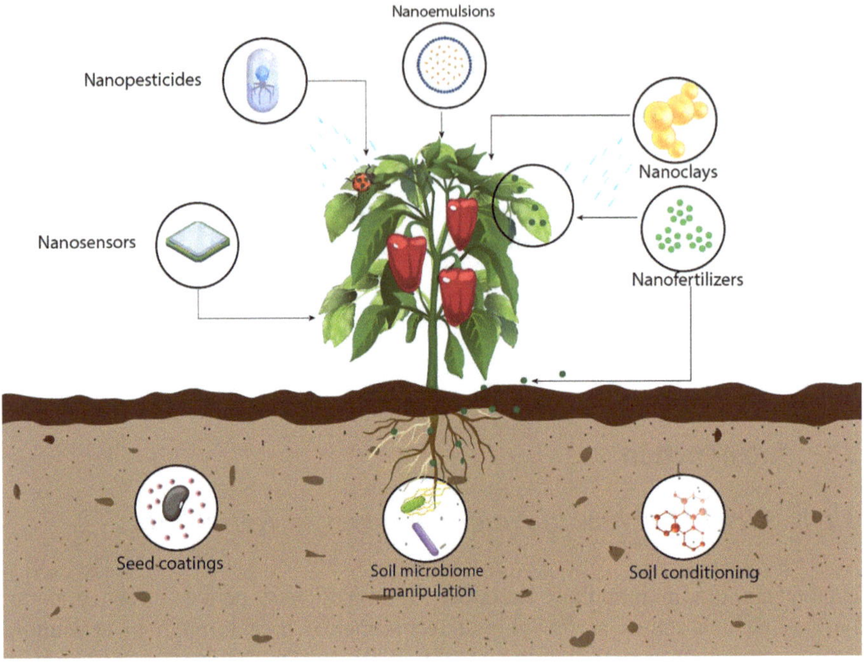

Fig. 3.1 Types of nanomaterials used in agriculture. (Constructed by authors)

These advanced devices provide immediate surveillance of multiple agricultural factors, including soil water content, together with nutritional concentrations and plant sicknesses and insect problems. The exact monitoring functions enable farmers to base their decisions on data, which leads to better resource management and faster intervention strategies. Crop farmers greatly benefit from the technology's exceptional early warning system because it prevents yield decreases and minimizes the need for reactive plant treatments (Yadav et al. 2023a, b). The field of genetic improvement in agriculture benefits from nanotechnology through nanoparticle-mediated systems, which deliver genes. These genetic enhancement systems use efficient methods to achieve precise gene modifications for creating crops that show improved stress adaptation along with enhanced nutrition and increased crop yields (Abdallah et al., 2015). Nanoscale delivery systems reduce unintentional impacts during transformations at the same time that they improve the rate of successful genetic modifications.

3.1.2 *Importance of Green Synthesis in Producing Eco-friendly Nanomaterials*

Nanomaterial synthesis through green methods marks a fundamental change in nanotechnology because it resolves issues stemming from established synthesis procedures (Jain et al., 2024). The environmentally friendly manufacturing method synergizes with international sustainability objectives and retains nanomaterial functionality throughout agricultural applications. Green synthesis techniques provide more advantages than sustainable practices because they deliver economic savings together with enhanced safety aspects for produced materials. Green synthesis methodology primarily uses biological resources as reducing and capping agents, which include plant extracts together with microorganisms and naturally occurring biomolecules (Alsaiari et al., 2023). Plant-mediated synthesis represents a beneficial synthetic method because it surpasses classical chemical techniques. Plants possess numerous bioactive components, such as polyphenols together with flavonoids and proteins that effectively lower metal ions and maintain nanoparticle stability (Soni et al., 2021). The biological approach creates nanomaterials that show both lower environmental toxicity while ensuring biocompatibility and eliminates the requirement for chemical reducing agents.

Green synthesis operates under gentle process conditions, which lead to lower energy requirements than conventional methods. The environmentally sustainable process conditions create economic advantages for large-scale production. The synthesis of nanomaterials under green conditions eliminates hazardous waste generation because toxic chemicals and harsh conditions are absent throughout the process (Shafey, 2020). During green synthesis, biological molecules perform two essential functions by reducing agents and modifying the surface functionality. Synthesizing nanomaterials through intrinsic functionalization leads to enhanced agricultural applications due to better plant system and soil environment interactions (Wahab et al., 2024). The synthesis of nanoparticles with plant extracts allows them to acquire beneficial properties that originate from the original source material, including antimicrobial resistance and growth-stimulating features. The techniques for green synthesis enable versatile production of different nanomaterials, such as metal nanoparticles, together with metal oxide nanoparticles, alongside hybrid nanostructures. Green synthesis stands out because it delivers multiple material options through adjustable particle sizes and natural template applications, which makes it ideal for agricultural requirements (Bahrulolum et al., 2021). Nanotechnology combined with green principles shows a sustainable development path for agriculture through its ability to create environmentally friendly, technologically advanced agricultural solutions (Atapattu et al., 2024b). The approach combines solutions for present-day agricultural difficulties with the sustainability principles of development and environmental protection.

3.1.3 Objectives and Scope of the Chapter

The chapter explores green synthesized nanomaterials in agriculture through a detailed examination of their various categories and applications and their current difficulties and potential future uses. The main focus includes studying the wide range of nanomaterials, including metal-based and hybrid systems, alongside their agricultural usage and sustainable environmental synthesis methods. This study performs a systematic examination of four fundamental nanomaterial categories, including metal-based, carbon-based, polymer-based and hybrid materials, through which scientists evaluate their functions in nutrient delivery and pest control and soil remediation and crop protection systems. This section examines sustainable synthesis approaches together with operational mechanisms of these materials in agricultural environments, while discussing field implementation hurdles, which include security matters and technological obstacles and regulatory complications. The discussion focuses on sustainable methods and upcoming breakthroughs in agricultural nanotechnology through an analysis of intelligent nutrient systems and enhanced crop defense technologies and advanced farming precision methods. The chapter provides comprehensive details about green synthesis and agricultural applications yet omits specific chemical synthesis procedures together with non-agricultural applications, conventional synthesis methods, economic assessments of commercial production, and precise product formulations. The detailed method focuses on delivering comprehensive knowledge about green-synthesized nanomaterials, which advance sustainable agricultural practices.

3.2 Types of Nanomaterials in Agriculture

3.2.1 Metal-Based Nanomaterials in Agriculture

The group of engineered particles known as metal-based nanomaterials contains at least one dimension between 1 and 100 nm that consists of pure metals or metal compounds (Zare et al., 2021). These materials instruct various physicochemical reactions that emerge because of their high surface area to volume ratio and distinctive quantum effects and reactivity properties. The agricultural sector values metal-based nanomaterials because they serve multiple purposes, including crop protection together with nutrient delivery and environmental cleanup. The main metals used in green synthesis applications are silver (Ag), along with gold (Au) and zinc (Zn) and iron (Fe) due to their unique features (Kumar et al., 2021). The antimicrobial features of silver nanoparticles parallel zinc oxide nanoparticles' functions as photocatalysts and nutritional elements. The biosensing field utilizes gold nanoparticles for superior performance yet iron oxide nanoparticles deliver magnetic properties that match their usage in targeted delivery systems. The green synthesis of nanoparticles uses biological systems combined with natural compounds as stabilizing reducing agents (Hussain et al., 2016). The biomolecules found in plant extracts work as reducing agents to convert metal ions into nanoparticles while maintaining their stability. Microorganisms use enzymatic reduction and cellular mechanisms to synthesize nanoparticles, as well as bacteria and fungi (Qamar & Ahmad, 2021). Natural reducing agents, which include ascorbic acid and citric acid, and amino acids, serve as environmentally friendly alternative methods to chemical nanoparticle synthesis.

3.2.1.1 Applications in Agriculture

3.2.1.1.1 Plant Growth Enhancement

Nanomaterials that contain metals function as effective delivery systems for essential nutrients, which boost their absorption and uptake properties (Elemike et al., 2019). The photosynthetic efficiency and biomass production benefit from zinc oxide nanoparticles that serve as micronutrient supplements. Iron oxide nanoparticles solve plant iron deficiency by using controlled-release delivery systems (Shirsat, 2024). The materials boost plant growth through dual mechanisms that control hormone activity and increase water absorption capability. The nano-formulations penetrate plant tissues better and deliver nutrients in controlled, sustained manner, thus requiring less frequent application of nutrients without diminishing their value. The nanosized dimension of these materials interacts with plant cells down to their molecular structures to improve nutrient absorption while decreasing nutrient waste (Faizan et al., 2024). Nano-sized fertilizers deliver nutrients slowly to plants over an extended period while decreasing the amount of nutrient that drains away from the soil. The precise delivery system enhances agricultural production and promotes sustainable norms by reducing fertilizer consumption.

3.2.1.1.2 Pest and Disease Control

Metal nanoparticles with antimicrobial properties including silver and copper serve as effective pathogens controllers for plants (Krishnani et al., 2022). These materials destroy pathogenic organisms through three distinct mechanisms, which involve breaking membranes and causing DNA damage and producing reactive oxygen species. Nanocarrier systems serve as targeted pesticide delivery vehicles, which protect the environment while preserving their active properties. The enlarged surface area improves pest protection along with better plant surface attachment. Metal nanoparticles work synergistically with conventional pesticides to possibly reduce the amount of pesticide needed for applications. The extensive use of silver nanoparticles occurs because they display antimicrobial effects against a wide range of bacterial, fungal, and viral pathogens (Khezerlou et al., 2018). Copper nanoparticles show a strong ability to combat fungal diseases and serve as an environmentally friendly option instead of conventional copper-based fungicides.

The penetrating capabilities of these nanoparticles let them destroy microbial communities within biofilms to provide extended crop protection. Metal nanoparticles help establish environmentally friendly pest management systems because they decrease the need for chemical pesticides (Ramezani et al., 2019).

3.2.1.1.3 Soil Remediation

The removal of pollutants from soil and enhancement of human health depend heavily on metal-based nanomaterials (Aragaw & Ayalew, 2023). Iron oxide nanoparticles eliminate heavy metals by using adsorption and reduction processes. The degradation of organic pollutants and immobilization of toxic metals occur exceptionally well through nanoscale zero-valent iron (nZVI). The materials improve both water retention and nutrient retention in the soil, which enhances its structural composition. Rapid contaminant removal occurs through the high reactivity and surface area properties without disrupting soil ecological systems. These nanomaterials demonstrate their ability to clean agricultural soils that contain pesticides and herbicides together with organic pollutants (Lu & Astruc, 2020). Nanoscale zerovalent iron (nZVI) demonstrates specific strength in transforming persistent organic pollutants (POPs) into harmless substances through its application. Metal nanoparticles added to soil enhance both its oxygen content and microbial ecosystem, which leads to better root growth and overall plant development (Zhou et al., 2020). Metal-based nanomaterials help maintain sustainable agriculture and preserve long-term soil fertility by resolving soil contamination problems.

3.2.2 Carbon-Based Nanomaterials in Agriculture

The diverse group of advanced materials known as carbon-based nanomaterials (CBNs) consists of carbon atoms arranged into different nanoscale structures. The carbon nanomaterials family consists of graphene and carbon nanotubes (CNTs) and carbon dots (CDs), along with their derivatives, which show distinct characteristics in their structure and mechanical properties as well as thermal and chemical properties (Devi et al., 2023). Graphene exists as a two-dimensional honeycomb lattice structure made of sp^2-hybridized carbon atoms in a single layer or few layers which provides outstanding electrical conductivity and mechanical strength together with extensive surface area. Carbon nanotubes exist as cylindrical shapes that develop from graphene sheet roll-ups and are divided into single-walled and multi-walled (Daisy Priscillal et al., 2024). The material demonstrates superior tensile resistance alongside exceptional electrical conduction and expansive surface area per volume measurement. Spherical nanoparticles known as carbon dots measure less than 10 nm and possess three distinct characteristics, including fluorescence properties and photostability and biocompatibility (Jorns & Pappas, 2021). The combination of these characteristics makes CBNs suitable for multiple applications in agricultural technology advancement. The production of CBNs through sustainable and eco-friendly approaches follows green chemistry principles to decrease environmental effects. The green synthesis method uses plant-derived biomass as its renewable and cost-effective carbon source, which contains cellulose together with hemicellulose and lignin (Dissanayaka et al., 2025; Mohammadinejad et al., 2016). Rice husks, together with coconut shells and corn stover and fruit peels, represent suitable raw materials for common agricultural biowastes (Dissanayaka et al., 2023; Ekanayaka et al., 2023). The materials experience hydrothermal carbonization and pyrolysis and microwave-assisted processes to generate high-quality CBNs (Dissanayake et al., 2023). Hydrothermal carbonization uses heat and water pressure to create carbon nanoparticles from biomass materials, whereas pyrolysis produces these nanoparticles through limited oxygen decomposition (Nizamuddin et al., 2018; Nuwarapaksha et al., 2024a). The surface reactivity and dispersibility of nanomaterials produced through natural compound or eco-friendly agent functionalization enable property tailoring for agricultural applications.

3.2.2.1 Applications in Agriculture

3.2.2.1.1 Improving Fertilizer Use Efficiency

Nanomaterials demonstrate promising capabilities to improve the use of fertilizers within agricultural systems, which resolves current production and environmental factors linked to traditional agricultural methods (Yadav et al. 2023a, b). The high surface area of graphene oxide combined with carbon nanotubes allows them to act as efficient nutrient carriers through their ability to absorb and release nutrients in controlled ways. The nanomaterials create stable nutrient complexes with nitrogen, phosphorus, and potassium that stop early environmental leakage (Babu et al., 2022). The nanoscale dimensions enable these materials to penetrate plant cell walls, which leads to the direct delivery of nutrients into the cytoplasm. Research investigations show that nano-carriers based on CBNs lower fertilizer requirements when used for plant fertilization without compromising or improving crop growth (Hassanisaadi et al., 2022).

3.2.2.1.2 Water Purification and Retention

The agricultural areas facing water scarcity benefit from CBNs, which provide modern solutions for water cleaning and storage (Mishra & Hussain, 2018). Because of its excellent adsorption capabilities, graphene-based materials retrieve heavy metals together with organic pollutants and agricultural residues from different water resources. The filtration efficiency of water receives improvement through carbon nanotubes because they trap all forms of particles together with microorganisms and toxic ions, to provide clean irrigation water (Joseph et al., 2023). The nanomaterials create 3D structures that capture water before distributing it slowly to plant roots through their networks. Drought-prone areas benefit from this property because it enhances water use efficiency along with crop resilience (Lankford et al., 2023).

3.2.2.1.3 Pest Management

Carbon nanostructures hold effective pest and pathogen control properties because of their physicochemical traits, which combine edge sharpness with surface reaction potential (Khan et al., 2025). The shape of CBNs allows them to break insect cuticles while damaging microbial membranes, which results in higher pest death rates. Carbon dots demonstrate effective photodynamic antimicrobial properties under sunlight conditions by producing reactive oxygen species that attack pathogens (Zhao et al., 2023). The addition of biocompatible agents or pesticides to CBNs improves their selectivity and efficiency during applications. CBNs decrease the need for traditional pesticides through controlled deliveries while using reduced application limits and reducing environmental hazards (Hegde et al., 2024). The combined method promotes environmentally friendly pest control systems for agricultural purposes.

3.2.3 Polymer-Based Nanomaterials

Materials science has witnessed a revolutionary shift through polymer-based nanomaterials, which unite the flexible nature of polymers with exclusive nanoscale phenomena. These nanomaterials exist between 1 and 100 nanometers in size and possess remarkable features such as a large surface-to-volume ratio and adjustable properties together with heightened chemical reactivity (Khan et al., 2022). The nanomaterials exist in four fundamental structural types: nanoparticles, nanofibers, nanogels, and polymer brushes that provide unique performance benefits for different applications. Green synthesis approaches rely heavily on chitosan because this natural polymer extracts chitin from crustacean shells and provides biodegradability and biocompatibility, as well as availability in nature. The polymer demonstrates exceptional features, including antimicrobial defense alongside metal absorption and film development abilities, which make it essential for agricultural purposes (Álvarez-Paino et al., 2017). The synthetic biodegradable polymer polylactic acid (PLA) has become popular for sustainable nanomaterial synthesis because it derives from renewable resources, including corn starch or sugarcane. PLA offers excellent potential to develop eco-friendly nanomaterials because of its controlled degradation properties along with its minimal environmental impact. The development of nanomaterials benefits from various important polymers, which include cellulose-based materials along with alginate and polyhydroxyalkanoates (PHAs) that contribute different properties to the process (Blanco et al., 2021). Nanomaterial synthesis benefits from these polymers' integration, which creates opportunities to resolve agricultural problems through environmentally sustainable methods. The choice of suitable polymers depends on several factors, such as application requirements and environmental conditions and degradation needs and cost-effectiveness. The development of efficient nanomaterial solutions for agriculture requires detailed knowledge about molecular structures, chemical modifications, and interaction mechanisms of these polymers.

3.2.3.1 Applications in Agriculture

3.2.3.1.1 Controlled Release of Agrochemicals

New agrochemical delivery systems have been transformed since the application of nanomaterials based on polymers in controlled release systems (An et al., 2022). Systems built with advanced polymer matrices function to control the delivery of active components, which include fertilizers together with pesticides, and growth regulators. The controlled release mechanism functions by utilizing principles which include diffusion and erosion and stimuli-responsive behavior (Klinger & Landfester, 2012). Customizable designs of polymer nanocarriers react to particular environmental signals, which include pH changes and temperature variations, and enzymes, to ensure precise substance delivery during specific conditions (Such et al., 2015). The targeted delivery method enables fewer applications, which in turn allows effective concentrations to persist at the desired locations. The improved performance results in decreased environmental pollution since it reduces both runoff and leaching occurrences. The three commonly utilized polymers in these systems are chitosan, along with alginate and poly lactic-co-glycolic acid (PLGA), which provide distinct benefits for degradation rates and release profiles (Kapoor et al., 2015). These systems become more effective when nanoscale features are incorporated because they improve adherence to plant surfaces and enhance penetration through biological barriers.

3.2.3.1.2 Encapsulation Techniques for Bioactive Agents

The sophisticated protection of essential bioactive agricultural compounds happens through encapsulation methods using polymer-based nanomaterials (Mondéjar-López et al., 2024). Multiple architectural methods are used within the encapsulation process, which range from basic core-shell structures to elaborate multilayered designs. The choice of polymers together with delivery techniques depends on the characteristics of bioactive agents, along with the targeted usages. The most frequently used techniques for encapsulation include emulsion-based methods and spray drying and electrospinning (Abdul-Al et al., 2023). Nanoscale packaging shields active ingredients from environmental harm such as ultraviolet (UV) rays and harsh temperatures and moisture and preserves their availability for use. Polymer systems at an advanced level can be engineered to have multiple chambers that contain different active agents which produce complementary results (Alrifaiy et al., 2012). The release behavior and efficiency of encapsulated materials can be controlled through modification of polymer characteristics, which includes molecular weight, surface properties, and crystalline structures.

3.2.3.1.3 Soil Amendment and Fertility Enhancement

Nanomaterials made from polymers function as vital agents to enhance both soil properties and fertility (Guleria et al., 2023). The integration of these materials in soils results in significant improvements in water retention abilities and better access to nutrients, along with modifications to soil structure. The nanoscale alliance between polymer materials and soil particles forms lasting aggregates that boost air circulation and water drainage in the soil (Corsi et al., 2023). The hydrogel nanomaterials serve as excellent water retainers, which support soil hydration throughout drought situations. Soil health gains support through beneficial microorganisms that these materials can transport for enhancing plant development while maintaining soil health. Soil remediation and nutrient uptake efficiency benefit significantly through the implementation of polymer-based nanomaterials, which demonstrate successful performance in contamination mitigation (Mukhopadhyay et al., 2022).

3.2.4 Hybrid Nanomaterials

Hybrid nanomaterials function as novel advanced materials because they unite two or more distinct nanoscale components, which generate properties that exceed the properties of their individual components (Park et al., 2020). Different material types unite strategically to make hybrid nanomaterials through the combination of metals with carbon-based structures and polymers that contribute particular features toward multifunctional systems. Different hybridization mechanisms, such as chemical bonding and physical entanglement and surface modification, create materials that demonstrate improved functionality as well as enhanced stability and performance (Yang et al., 2022). The distinctive features of hybrid nanomaterials emerge from the contact effects between their distinct components that researchers can customize for particular operational needs. Combining carbon nanotubes with metal nanoparticles results in systems that both possess better electrical conductivity and stronger catalytic functions, alongside polymer-metal hybrids that demonstrate advanced mechanical structures together with controlled release functions. The materials display exceptional properties because they possess expanded surface area alongside stronger mechanical properties, along with elevated thermal stability and better chemical reactivity (Jiang et al., 2010). Metal elements with polymer elements and carbon elements offer combined electric conductivity and catalytic functions and structural strength along with thermal properties and flexibility and biological compatibility. Multiple functionalities required in agricultural applications find great value in hybrid nanomaterials due to their ability to adapt to diverse needs (Singh et al., 2021). The fabrication process of these materials needs controlled manipulation of component composition and material structures alongside interface properties to achieve target performance and environmental stability.

3.2.4.1 Applications in Agriculture

3.2.4.1.1 Multi-functional Systems in Agriculture

Agricultural operations benefit from hybrid nanomaterials because they have introduced all-in-one systems able to handle several challenges at once (Wongkaew et al., 2019). Pest management solutions with comprehensive capabilities now exist through integrated systems built of multiple components. Metal nanoparticles embedded into polymer carriers function as effectively enriched pest management tools that both transmit pesticides and demonstrate antimicrobial activity (Kaur et al., 2020). The polymer substance enables gradual delivery, and the metal nanoparticles boost the effectiveness of incorporated compounds. Hybrid delivery systems, which integrate carbon-based materials with metal oxides, develop efficient carrier systems that perform simultaneous delivery and gradual essential nutrient release functions (Easwaran et al., 2024). The systems use pH-sensitive polymers that allow controlled delivery of substances in specific soil environments. Soil health advancement comes from hybrid materials that unite inorganic and organic components for enhancing water retention and soil structure and microbial growth promotion (Alori et al., 2017). Such hybrid systems produce combined actions between multiple components, which yield efficient, sustainable agricultural methods.

3.2.4.1.2 Sensing and Crop Monitoring

Advanced sensing and monitoring systems achieve their purpose in precision agriculture by utilizing hybrid nanomaterials (Ashique et al., 2024). Different components integrated into these materials enable the development of sensitive and selective sensors for agricultural parameter detection. Hybrid sensors made from metal–polymer combinations measure soil moisture accurately alongside nutrients, whereas carbon-based hybrids with metal nanoparticles serve well for plant pathology detection and stress signal identification (Hajipour et al., 2021). Smart farming systems become more effective through the addition of these materials because they allow real-time inspection of agricultural parameters, including the health status of crops and soil state and environmental elements (Atapattu et al. 2024a, b). The detection of trace amounts of analytes becomes possible through hybrid materials because they provide both high surface area and heightened sensitivity, thus making them suitable for crop protection early warning systems (Nuwarapaksha et al., 2024c). The integration of these systems with wireless networks enables remote monitoring as well as automated response systems, which ensure efficient resource management for agricultural purposes.

3.2.4.1.3 Enhancing Biocompatibility

Biocompatible hybrid nanomaterials developed through green synthesis methods work to resolve environmental problems caused by agricultural inputs (Ahmed et al., 2022). Hybrid nanomaterials incorporate eco-friendly designs that combine minimal toxicity measures with operational capacity through the selection of components and synthesis processes. Green synthesis achieves metal nanoparticle reduction and stabilization through plant extracts or biological systems and follows with natural polymers or carbon-based materials incorporation (Samuel et al., 2022). Such methods decrease environmental damage while improving the biological system compatibility of these materials. The protective coating from polymer components helps decrease metal nanoparticle toxicity while preserving their functional applications. The synthesized hybrid materials maintain better biological stability and exhibit degradation behavior that fulfills agricultural requirements (Castillo-Henríquez et al., 2020). Scientists working in this field maintain their focus on synthesizing novel methods that minimize environmental effects and improve both safety and performance characteristics of hybrid nanomaterials used in agricultural practices.

3.3 Challenges in Green Synthesized Nanomaterials in Agriculture

3.3.1 Safety and Environmental Concerns

The main obstacle regarding agricultural nanomaterials emerges from their possible environmental risks and safety concerns (Usman et al., 2020). Metal nanoparticles create environmental risks for soil bioaccumulation, which disrupts the natural food chain dynamics and disrupts biological equilibrium. Beneficial soil microorganisms face toxicity from carbon nanomaterials, which disrupts the natural processes occurring in soil (Mubeen et al., 2023). The environmental breakdown of marketed biodegradable polymer nanomaterials might not be total, so accumulation occurs. The environmental interactions between hybrid materials exhibit unpredictable and challenging behaviors to track. Scientific knowledge about how these materials influence soil vitality alongside water systems, as well as human health from the food sequence, is insufficient and requires intensive investigation and tracking systems.

3.3.2 Technical and Manufacturing Challenges

The manufacturing process for nanomaterials used in agriculture encounters major technological obstacles (Kumar et al., 2020). Nanomaterials present exceptional challenges when producers aim to maintain uniform size aspects and properties in industrial-scale manufacturing. Commercial production at industrial scales often experiences multiple efficiency problems paired with quality control issues during production scale-ups, which start in laboratory environments. The production expenses for hybrid materials remain high, which restricts their general application across industries (Yu et al., 2013). The stability of nanomaterials during storage, along with their shelf life, presents difficulties because nanomaterials tend to aggregate or transform their properties throughout the storage period. The distribution uniformity of nanomaterials in soil or foliar sprays for agricultural needs demands complex formulation approaches (Chowdhury et al., 2023). Technical issues represent major barriers to nanomaterial-based agricultural solutions because they affect both implementation feasibility and production economics.

3.3.3 Regulatory and Standardization Issues

Current legislation that governs nano-materials in agricultural systems shows both difficulties and inadequate coverage. Current agricultural nanotechnology practices lack standardized testing methods for this field (Mishra et al., 2017). Different countries implement distinct regulatory systems, which create trade obstacles for both international trade and implementation. Nanomaterial properties with multiple agricultural applications make risk assessment procedures challenging to implement (Lavicoli et al., 2017). Specific guidelines about different nanomaterials remain absent, which leads to unsafe compliance practices and unresolved safety verification. Technical obstacles in monitoring and enforcement make it hard for regulators to guarantee the safe application and practical effectiveness of agricultural nanomaterials.

3.3.4 Application and Performance Challenges

Various operational barriers exist when nanomaterials are used in field applications. Under different environmental settings, their behavior becomes unpredictable, which impacts both their performance and reliability. The control of active ingredient release from nano-carriers becomes challenging under different field conditions (Katouzian & Jafari, 2016). Implementing nanomaterials into current agricultural methods demands substantial changes, together with new operational procedures. Research has not yet clarified the mechanisms through which nanomaterials interact with various plant species which resulting in inconsistent outcomes between different crops. The effectiveness of nanomaterials remains challenging to predict because weather conditions affect their performance levels (Loannou et al., 2020). The unpredictable nature of nanomaterials in the field makes it difficult to develop stable application methods and forecast results.

3.3.5 *Economic and Social Barriers*

Social and economic barriers create major obstacles to the implementation of agricultural nanomaterials (Mukhopadhyay, 2014). Manufacturers, along with farmers, face major obstacles to begin operations because of their high initial capital requirements. The adoption of nanotechnology needs extensive training and education programs because farmers show limited understanding and acceptance of this technology (Handford et al., 2015). Specialized equipment, together with handling procedures, increases both implementation costs and manufacturer expenses. Market conditions along with perceptual risks determine both investment choices and use rates of agricultural nanomaterials (Ravichandran, 2010). The greater complexity in supply chains for storage, together with transportation needs and handling matters, drives up operational expenses (Nuwarapaksha et al., 2024b). The cost-benefit analysis of small-scale agricultural producers leads to technology access gaps that separate various farming communities. The way people view nanotechnology in food production determines how much it will be accepted by markets and how feasible its commercialization will become.

3.4 Future Potential of Green Synthesized Nanomaterials in Agriculture

3.4.1 *Smart Nutrient Management*

Different nanomaterials combined in agricultural systems continue to transform how farmers handle nutrients. Precise micronutrient delivery to plant tissues becomes possible through metal-based nanoparticles, especially zinc and iron oxide products, which improve absorption efficiency (Singh et al., 2024). Carbon nanotubes act as effective transportation mechanisms to shield nutrients from breakage while allowing regulators to activate nutrient delivery according to plant requirements. The biodegradable matrices of polymer nano-fertilizers support extended nutrient release together with hybrid nano-composite systems, which integrate the entire set of release characteristics (Kalia et al., 2020). By implementing this advanced nutrient management system, farmers gain two-fold benefits of increased crop production coupled with decreased nutrient contamination of the environment. Not

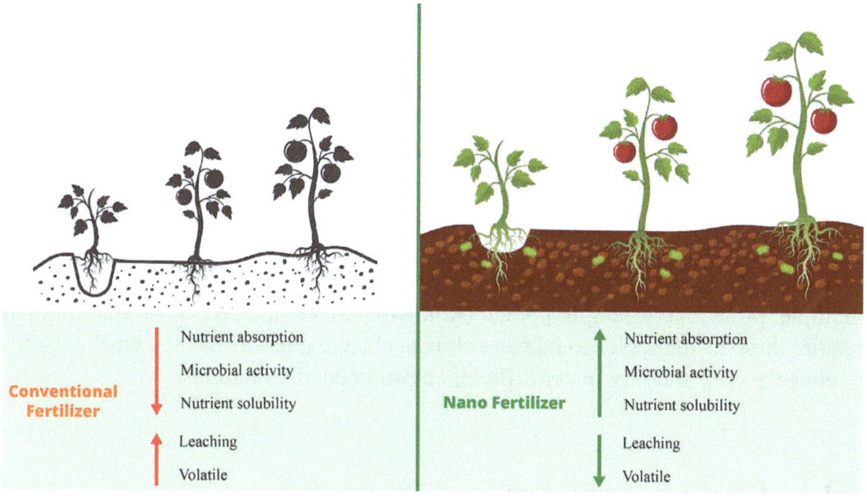

Fig. 3.2 Comparison of conventional and nano fertilizers: impact on plant growth and soil health (constructed by authors)

only does it add economic sense to the way fertilizer is utilized but it also reduces the possible negative effects that come with high nutrient leaching or emission of nitrogen gases. However, in the summary of the work, nano fertilizers have more enhancement when compared to conventional fertilizers (Fig. 3.2).

3.4.2 *Advanced Crop Protection*

The distinctive properties of nanomaterials power upcoming protection systems for agricultural crops that fight pests and diseases more effectively. The antimicrobial presence of silver nanoparticles brings excellent resistance against various plant pathogens (Sharma et al., 2018). Carbon-based materials strengthen plants' defensive capabilities, and polymer nanoparticles minimize pesticide impact on the environment by allowing better delivery control. The protective properties of metal–polymer hybrids combine two protection methods by using active antimicrobial elements and a mechanism that enables controlled compound delivery (Khalid & Nazir, 2021). By using this consolidated strategy, the need for conventional agrochemicals decreases substantially while the protective effect against crop threats grows or remains at equal levels.

3.4.3 Crop Stress Management

The use of nanomaterials brings about substantial changes in how agricultural crops fight against environmental challenges. Plants retain more water while better extracting water from their environment with the help of metal oxide nanoparticles (Nair et al., 2022). Improved cellular protection mechanisms along with better energy management through carbon nanomaterials enable plants to tolerate heat stress conditions. Soil water reservoirs form from polymer-based materials although hybrid nanomaterials defend plants against various environmental stresses due to multiple protective strength points (Kathalingam et al., 2023). Arable farmers require these technologies to manage climate change pressure as they work together to enhance crop stability under difficult environmental conditions.

3.4.4 Precision Agriculture and Monitoring

The adoption of nanomaterials in agricultural monitoring systems enables farmers to reach levels of productivity that were previously impossible (Prasad et al., 2017). Time-sensitive soil nutrient analyses become possible through metal nanoparticle sensors, whereas plant health parameters are monitored using carbon nanotube sensors. Smart delivery systems made from polymers use environmental triggers to function in an optimal fashion. The combined technology of nano-biosensors allows full monitoring of numerous parameters at once (Sheikh et al., 2024). Farmers rely on this data-driven strategy to optimize irrigation along with fertilization while controlling pests through data-based decisions, thus using resources more efficiently and obtaining superior harvests (Nuwarapaksha et al., 2024d).

3.4.5 Sustainable Agricultural Solutions

Nanomaterials employed strategically serve to upgrade agricultural sustainability. The implementation of biodegradable polymer nanomaterials leads to reduced environmental consequences as they deliver farming advantages (Pascoli et al., 2018). Soil improvement and plant growth enhancement occur through carbon-based materials, which remain friendly to the environment by causing no persistent contamination. Nanoparticles consisting of low-toxic metals deliver necessary agricultural functions without negatively impacting the environment (Otari et al., 2024). The integration of multiple benefits between different sustainable solutions enables multi-functional hybrid systems to create holistic sustainable systems. Merger-based technologies have enabled agriculture to move toward environmentally responsible practices while keeping essential high productivity levels for worldwide food security (Deutscher & Makris, 2024).

3.5 Conclusion

Green-synthesized nanomaterials have brought a substantial breakthrough to sustainable agriculture by providing modern solutions to enduring agricultural difficulties. Nanomaterials across four categories, including metals, carbon and polymers, and combinations thereof, exhibit exceptional capabilities for improving agricultural output and resource management while sustaining the environment. The implementation of bio-based green synthesis procedures combined with friendly environmental operation methods leads agriculture toward sustainable technological advancements. The successful deployment of these nanomaterials in agriculture faces significant obstacles that need resolution for their complete implementation across different applications. Sustainability efforts require ongoing research because safety issues about environmental effects and technical difficulties for large-scale production and regulatory confusion, and economic limitations exist. Agriculture must solve existing difficulties before integrating these materials at scale, since they need to maintain their environmental value and material efficiency. The critical development for agricultural nanotechnology will be creating advanced systems capable of addressing diverse agricultural problems at once. Adaptation of agriculture to climate change and food security maintenance requires the development of smart nutrient management systems combined with improved crop protection techniques and expanded stress management capabilities. The ongoing research, combined with technological progress, needs to keep environmental sustainability in equal focus with technological advancement. Achieving widespread adoption of green synthesized nanomaterials in agriculture will need research partnerships between scientists and government representatives and business leaders who are building helpful guidelines that guarantee farmer access to safe, effective nanomaterial methodologies worldwide.

Acknowledgments We would like to express our sincere appreciation to the dedicated team at the Agronomy Division of the Coconut Research Institute of Sri Lanka for their invaluable assistance.

Conflicts of Interest All authors declared that there are no conflicts of interest.

References

Abdallah, N. A., Prakash, C. S., & McHughen, A. G. (2015). Genome editing for crop improvement: Challenges and opportunities. *GM Crops & Food, 6*, 183–205. https://doi.org/10.1080/21645698.2015.1129937

Abdul-Al, M., Saeinasab, M., & Sefat, F. (2023). Encapsulation techniques overview. In *Principles of biomaterials encapsulation : Volume one* (pp. 13–36). Elsevier.

Ahmed, S. F., Mofijur, M., Rafa, N., et al. (2022). Green approaches in synthesising nanomaterials for environmental nanobioremediation: Technological advancements, applications, benefits and challenges. *Environmental Research, 204*, 111967. https://doi.org/10.1016/j.envres.2021.111967

Alori, E. T., Dare, M. O., & Babalola, O. O. (2017). Microbial inoculants for soil quality and plant health. In E. Lichtfouse (Ed.), *Sustainable agriculture reviews. Sustainable agriculture reviews* (Vol. 22, pp. 281–307). Springer. https://doi.org/10.1007/978-3-319-48006-0_9

Alrifaiy, A., Lindahl, O. A., & Ramser, K. (2012). Polymer-based microfluidic devices for pharmacy, biology and tissue engineering. *Polymers (Basel), 4*, 1349–1398. https://doi.org/10.3390/polym4031349

Alsaiari, N. S., Alzahrani, F. M., Amari, A., et al. (2023). Plant and microbial approaches as green methods for the synthesis of nanomaterials: Synthesis, applications, and future perspectives. *Molecules, 28*, 463. https://doi.org/10.3390/molecules28010463

Álvarez-Paino, M., Muñoz-Bonilla, A., & Fernández-García, M. (2017). Antimicrobial polymers in the nano-world. *Nanomaterials, 7*, 48. https://doi.org/10.3390/nano7020048

An, C., Sun, C., Li, N., et al. (2022). Nanomaterials and nanotechnology for the delivery of agrochemicals: Strategies towards sustainable agriculture. *Journal of Nanobiotechnology, 20*, 11. https://doi.org/10.1186/s12951-021-01214-7

Aragaw, T. A., & Ayalew, A. A. (2023). Application of metal-based nanoparticles for metal removal for treatments of wastewater — A review. In *Emerging techniques for treatment of toxic metals from wastewater* (pp. 183–231). Elsevier.

Ashique, S., Raikar, A., Jamil, S., et al. (2024). Artificial intelligence integration with nanotechnology: A new frontier for sustainable and precision agriculture. *Current Nanoscience, 21*, 242–273. https://doi.org/10.2174/0115734137275111231206072049

Atapattu, A. J., Perera, L. K., Nuwarapaksha, T. D., et al. (2024a). Challenges in achieving artificial intelligence in agriculture. In *Artificial Intelligence Techniques in Smart Agriculture* (pp. 7–34). Springer. https://doi.org/10.1007/978-981-97-5878-4_2

Atapattu, A. J., Ranasinghe, C. S., Nuwarapaksha, T. D., et al. (2024b). Sustainable agriculture and sustainable development goals (SDGs). In *Emerging technologies and marketing strategies for sustainable agriculture* (pp. 1–27). IGI Global. https://doi.org/10.4018/979-8-3693-4864-2.ch001

Babu, S., Singh, R., Yadav, D., et al. (2022). Nanofertilizers for agricultural and environmental sustainability. *Chemosphere, 292*, 133451. https://doi.org/10.1016/j.chemosphere.2021.133451

Bahrulolum, H., Nooraei, S., Javanshir, N., et al. (2021). Green synthesis of metal nanoparticles using microorganisms and their application in the agrifood sector. *Journal of Nanobiotechnology, 19*, 86. https://doi.org/10.1186/s12951-021-00834-3

Blanco, F. G., Hernández, N., Rivero-Buceta, V., et al. (2021). From residues to added-value bacterial biopolymers as nanomaterials for biomedical applications. *Nanomaterials, 11*, 1492. https://doi.org/10.3390/nano11061492

Castillo-Henríquez, L., Alfaro-Aguilar, K., Ugalde-Álvarez, J., et al. (2020). Green synthesis of gold and silver nanoparticles from plant extracts and their possible applications as antimicrobial agents in the agricultural area. *Nanomaterials, 10*, 1763. https://doi.org/10.3390/nano10091763

Chowdhury, M., Kushwah, A., Satpute, A. N., et al. (2023). A comprehensive review on potential application of nanomaterials in the field of agricultural engineering. *Journal of Biosystems Engineering, 48*, 457–477. https://doi.org/10.1007/s42853-023-00204-x

Corsi, I., Venditti, I., Trotta, F., & Punta, C. (2023). Environmental safety of nanotechnologies: The eco-design of manufactured nanomaterials for environmental remediation. *Science of the Total Environment, 864*, 161181. https://doi.org/10.1016/j.scitotenv.2022.161181

Daisy Priscillal, I. J., Wang, S.-F., & Kameoka, S. (2024). Influence of platinum content (X) in the LaNi5PtX catalyst and reaction conditions on the properties of ferromagnetic Ni-filled CNTs. *Ceramics International, 50*, 8907–8916. https://doi.org/10.1016/j.ceramint.2023.12.206

Deutscher, E., & Makris, S. (2024). Making sustainability visible: A new framework and operationalisation tests for merger control. In *Research handbook on sustainability and competition law* (pp. 375–411). Edward Elgar Publishing.

Devi, N., Kumar, R., Chen, Y.-S., & Singh, R. K. (2023). Carbon-based nanomaterials: Carbon nanotube, fullerene, and carbon dots. In *Nanomaterials* (pp. 27–57). Springer.

Dissanayaka, D. M. N. S., Udumann, S. S., Nuwarapaksha, T. D., & Atapattu, A. J. (2023). Effects of pyrolysis temperature on chemical composition of coconut-husk biochar for agricultural applications: A characterization study. *Technology in Agronomy, 3*, 13. https://doi.org/10.48130/TIA-2023-0013

Dissanayaka, N. S., Rajaratnam, K., Udumann, S. S., et al. (2025). Evaluation of the nutritional composition of king coconut husk waste biochar and ash soil conditioners: A comprehensive analysis. *Technology in Agronomy, 5*, e003. https://doi.org/10.48130/tia-0024-0034

Dissanayake, D. K. R. P. L., Dissanayaka, D. M. N. S., Udumann, S. S., et al. (2023). Is biochar a promising soil amendment to enhance perennial crop yield and soil quality in the tropics? *Technology in Agronomy, 3*, 4. https://doi.org/10.48130/TIA-2023-0004

Easwaran, C., Christopher, S. R., Moorthy, G., et al. (2024). Nano hybrid fertilizers: A review on the state of the art in sustainable agriculture. *Science of the Total Environment, 929*, 172533. https://doi.org/10.1016/j.scitotenv.2024.172533

Ekanayaka, E. M. G. N., Dissanayake, D. K. R. P. L., Udumann, S. S., et al. (2023). Sustainable utilization of king coconut husk as a feedstock in biochar production with the highest conversion efficiency and desirable properties. *IOP Conference Series: Earth and Environmental Science, 1235*, 012009. https://doi.org/10.1088/1755-1315/1235/1/012009

Elemike, E. E., Uzoh, I. M., Onwudiwe, D. C., & Babalola, O. O. (2019). The role of nanotechnology in the fortification of plant nutrients and improvement of crop production. *Applied Sciences, 9*, 499. https://doi.org/10.3390/app9030499

Faizan, M., Singh, A., Eren, A., et al. (2024). Small molecule, big impacts: Nano-nutrients for sustainable agriculture and food security. *Journal of Plant Physiology, 301*, 154305. https://doi.org/10.1016/j.jplph.2024.154305

Guleria, G., Thakur, S., Shandilya, M., et al. (2023). Nanotechnology for sustainable agro-food systems: The need and role of nanoparticles in protecting plants and improving crop productivity. *Plant Physiology and Biochemistry, 194*, 533–549. https://doi.org/10.1016/j.plaphy.2022.12.004

Hajipour, M. J., Saei, A. A., Walker, E. D., et al. (2021). Nanotechnology for targeted detection and removal of bacteria: Opportunities and challenges. *Advanced Science, 8*, e2100556. https://doi.org/10.1002/advs.202100556

Handford, C. E., Dean, M., Spence, M., et al. (2015). Awareness and attitudes towards the emerging use of nanotechnology in the agri-food sector. *Food Control, 57*, 24–34. https://doi.org/10.1016/j.foodcont.2015.03.033

Hassanisaadi, M., Barani, M., Rahdar, A., et al. (2022). Role of agrochemical-based nanomaterials in plants: Biotic and abiotic stress with germination improvement of seeds. *Plant Growth Regulation, 97*, 375–418. https://doi.org/10.1007/s10725-021-00782-w

Hegde, V., Bhat, M. P., Lee, J.-H., et al. (2024). Carbon-based nanomaterials: Multifaceted role in agrochemical recognition, remediation, and release. *Nano Today, 57*, 102388. https://doi.org/10.1016/j.nantod.2024.102388

Hussain, I., Singh, N. B., Singh, A., et al. (2016). Green synthesis of nanoparticles and its potential application. *Biotechnology Letters, 38*, 545–560. https://doi.org/10.1007/s10529-015-2026-7

Jain, K., Takuli, A., Gupta, T. K., & Gupta, D. (2024). Rethinking nanoparticle synthesis: A sustainable approach vs. Traditional methods. *Chemistry: An Asian Journal, 19*, 21. https://doi.org/10.1002/asia.202400701

Jiang, Z., Kuang, Q., Xie, Z., & Zheng, L. (2010). Syntheses and properties of micro/nanostructured crystallites with high-energy surfaces. *Advanced Functional Materials, 20*, 3634–3645. https://doi.org/10.1002/adfm.201001243

Jorns, M., & Pappas, D. (2021). A review of fluorescent carbon dots, their synthesis, physical and chemical characteristics, and applications. *Nanomaterials, 11*, 1448. https://doi.org/10.3390/nano11061448

Joseph, T. M., Al-Hazmi, H. E., Śniatała, B., et al. (2023). Nanoparticles and nanofiltration for wastewater treatment: From polluted to fresh water. *Environmental Research, 238*, 117114. https://doi.org/10.1016/j.envres.2023.117114

Kalia, A., Sharma, S. P., Kaur, H., & Kaur, H. (2020). Novel nanocomposite-based controlled-release fertilizer and pesticide formulations: Prospects and challenges. In *Multifunctional hybrid nanomaterials for sustainable agri-food and ecosystems* (pp. 99–134). Elsevier.
Kapoor, D. N., Bhatia, A., Kaur, R., et al. (2015). PLGA: A unique polymer for drug delivery. *Therapeutic Delivery, 6*, 41–58. https://doi.org/10.4155/tde.14.91
Kathalingam, A., Santhoshkumar, P., Ramesh, S., et al. (2023). Biogenic polymer nanoparticles to remove hydrophobic organic contaminants from water. *Water Environment Research, 95*, e10935. https://doi.org/10.1002/wer.10935
Katouzian, I., & Jafari, S. M. (2016). Nano-encapsulation as a promising approach for targeted delivery and controlled release of vitamins. *Trends in Food Science and Technology, 53*, 34–48. https://doi.org/10.1016/j.tifs.2016.05.002
Kaur, P., Choudhary, R., Pal, A., et al. (2020). Polymer-metal Nanocomplexes based delivery system: A boon for agriculture revolution. *Current Topics in Medicinal Chemistry, 20*, 1009–1028. https://doi.org/10.2174/1568026620666200330160810
Khalid, S., & Nazir, R. (2021). Hybrid metal-polymer nanocomposites: Synthesis, characterization, and applications. In *Handbook of polymer and ceramic nanotechnology* (pp. 565–600). Springer.
Khan, Y., Sadia, H., Ali Shah, S. Z., et al. (2022). Classification, synthetic, and characterization approaches to nanoparticles, and their applications in various fields of nanotechnology: A review. *Catalysts, 12*, 1386. https://doi.org/10.3390/catal12111386
Khan, T., Hashmi, K., Jabeen, S., et al. (2025). *Carbon-based nanomaterials in crop plant disease management* (pp. 271–301). https://doi.org/10.1007/978-981-97-5104-4_14
Khezerlou, A., Alizadeh-Sani, M., Azizi-Lalabadi, M., & Ehsani, A. (2018). Nanoparticles and their antimicrobial properties against pathogens including bacteria, fungi, parasites and viruses. *Microbial Pathogenesis, 123*, 505–526. https://doi.org/10.1016/j.micpath.2018.08.008
Klinger, D., & Landfester, K. (2012). Stimuli-responsive microgels for the loading and release of functional compounds: Fundamental concepts and applications. *Polymer (Guildf), 53*, 5209–5231. https://doi.org/10.1016/j.polymer.2012.08.053
Krishnani, K. K., Boddu, V. M., Chadha, N. K., et al. (2022). Metallic and non-metallic nanoparticles from plant, animal, and fisheries wastes: Potential and valorization for application in agriculture. *Environmental Science and Pollution Research, 29*, 81130–81165. https://doi.org/10.1007/s11356-022-23301-4
Kumar, P., Mahajan, P., Kaur, R., & Gautam, S. (2020). Nanotechnology and its challenges in the food sector: A review. *Materials Today Chemistry, 17*, 100332. https://doi.org/10.1016/j.mtchem.2020.100332
Kumar, J. A., Krithiga, T., Manigandan, S., et al. (2021). A focus to green synthesis of metal/metal based oxide nanoparticles: Various mechanisms and applications towards ecological approach. *Journal of Cleaner Production, 324*, 129198. https://doi.org/10.1016/j.jclepro.2021.129198
Kumar, P., Chugh, P., Ali, S. S., et al. (2025). Trends of Nanobiosensors in modern agriculture systems. *Applied Biochemistry and Biotechnology, 197*, 667–690. https://doi.org/10.1007/s12010-024-05039-6
Lankford, B., Pringle, C., McCosh, J., et al. (2023). Irrigation area, efficiency and water storage mediate the drought resilience of irrigated agriculture in a semi-arid catchment. *Science of the Total Environment, 859*, 160263. https://doi.org/10.1016/j.scitotenv.2022.160263
Lavicoli, I., Leso, V., Beezhold, D. H., & Shvedova, A. A. (2017). Nanotechnology in agriculture: Opportunities, toxicological implications, and occupational risks. *Toxicology and Applied Pharmacology, 329*, 96–111. https://doi.org/10.1016/j.taap.2017.05.025
Loannou, A., Gohari, G., Papaphilippou, P., et al. (2020). Advanced nanomaterials in agriculture under a changing climate: The way to the future? *Environmental and Experimental Botany, 176*, 104048. https://doi.org/10.1016/j.envexpbot.2020.104048
Lu, F., & Astruc, D. (2020). Nanocatalysts and other nanomaterials for water remediation from organic pollutants. *Coordination Chemistry Reviews, 408*, 213180. https://doi.org/10.1016/j.ccr.2020.213180

Malik, S., Muhammad, K., & Waheed, Y. (2023). Nanotechnology: A revolution in modern industry. *Molecules, 28*, 661. https://doi.org/10.3390/molecules28020661

Mishra, A. K., & Hussain, C. M. (Eds.). (2018). *Nanotechnology for sustainable water resources*. Wiley. https://doi.org/10.1002/9781119323655

Mishra, S., Keswani, C., Abhilash, P. C., et al. (2017). Integrated approach of Agri-nanotechnology: Challenges and future trends. *Frontiers in Plant Science, 8*, 471. https://doi.org/10.3389/fpls.2017.00471

Mohammadinejad, R., Karimi, S., Iravani, S., & Varma, R. S. (2016). Plant-derived nanostructures: Types and applications. *Green Chemistry, 18*, 20–52. https://doi.org/10.1039/C5GC01403D

Mondéjar-López, M., García-Simarro, M. P., Navarro-Simarro, P., et al. (2024). A review on the encapsulation of "eco-friendly" compounds in natural polymer-based nanoparticles as next generation nano-agrochemicals for sustainable agriculture and crop management. *International Journal of Biological Macromolecules, 280*, 136030. https://doi.org/10.1016/j.ijbiomac.2024.136030

Mubeen, B., Hasnain, A., Wang, J., et al. (2023). Current Progress and open challenges for combined toxic effects of manufactured Nano-sized objects (MNO'S) on soil biota and microbial community. *Coatings, 13*, 212. https://doi.org/10.3390/coatings13010212

Mukhopadhyay, S. S. (2014). Nanotechnology in agriculture: Prospects and constraints. *Nanotechnology, Science and Applications, 7*, 63. https://doi.org/10.2147/NSA.S39409

Mukhopadhyay, R., Sarkar, B., Khan, E., et al. (2022). Nanomaterials for sustainable remediation of chemical contaminants in water and soil. *Critical Reviews in Environmental Science and Technology, 52*, 2611–2660. https://doi.org/10.1080/10643389.2021.1886891

Nair, G. M., Sajini, T., & Mathew, B. (2022). Advanced green approaches for metal and metal oxide nanoparticles synthesis and their environmental applications. *Talanta Open, 5*, 100080. https://doi.org/10.1016/j.talo.2021.100080

Nizamuddin, S., Siddiqui, M. T. H., Mubarak, N. M., et al. (2018). Advanced nanomaterials synthesis from pyrolysis and hydrothermal carbonization: A review. *Current Organic Chemistry, 22*, 446–461. https://doi.org/10.2174/1385272821666171026153215

Nuwarapaksha, T. D., Udumann, S. S., Dissanayaka, N. S., et al. (2024c). Revolutionizing agriculture by advanced water and irrigation management technologies. In *Emerging trends and technologies in water management and conservation* (pp. 285–318). IGI Global Scientific Publishing. https://doi.org/10.4018/979-8-3693-6920-3.ch009

Nuwarapaksha, T. D., Dissanayaka, D. M. N. S., Udumann, S. S., et al. (2024a). Exploring the impact of pyrolysis temperature on nutrient composition of *Gliricidia sepium* biochar: A comprehensive study. *Technology in Agronomy, 4*, e016. https://doi.org/10.48130/tia-0024-0014

Nuwarapaksha, T. D., Udumann, S. S., Dissanayaka, N. S., et al. (2024d). AI-driven solutions for sustainable irrigation. In *Integrating agriculture, green marketing strategies, and artificial intelligence* (pp. 1–32). IGI Global Scientific Publishing. https://doi.org/10.4018/979-8-3693-6468-0.ch001

Nuwarapaksha, T. D., Udumann, S. S., Dissanayaka, N. S., & Atapattu, A. J. (2024b). Leveraging Blockchain for food supply transparency. In *Emerging trends in food and agribusiness marketing* (pp. 75–100). IGI Global. https://doi.org/10.4018/979-8-3693-6715-5.ch003

Otari, S., Bapat, V. A., Lakkakula, J., et al. (2024). Advancements in bionanotechnological applications for climate-smart agriculture and food production. *Biocatalysis and Agricultural Biotechnology, 57*, 103117. https://doi.org/10.1016/j.bcab.2024.103117

Park, W., Shin, H., Choi, B., et al. (2020). Advanced hybrid nanomaterials for biomedical applications. *Progress in Materials Science, 114*, 100686. https://doi.org/10.1016/j.pmatsci.2020.100686

Pascoli, M., Lopes-Oliveira, P. J., Fraceto, L. F., et al. (2018). State of the art of polymeric nanoparticles as carrier systems with agricultural applications: A minireview. *Energy, Ecology and Environment, 3*, 137–148. https://doi.org/10.1007/s40974-018-0090-2

Pramanik, P., Krishnan, P., Maity, A., et al. (2020). Application of nanotechnology in agriculture. In *Environmental nanotechnology volume 4* (pp. 317–348). https://doi.org/10.1007/978-3-030-26668-4_9

Prasad, R., Bhattacharyya, A., & Nguyen, Q. D. (2017). Nanotechnology in sustainable agriculture: Recent developments, challenges, and perspectives. *Frontiers in Microbiology, 8*, 1014. https://doi.org/10.3389/fmicb.2017.01014

Qamar, S. U. R., & Ahmad, J. N. (2021). Nanoparticles: Mechanism of biosynthesis using plant extracts, bacteria, fungi, and their applications. *Journal of Molecular Liquids, 334*, 116040. https://doi.org/10.1016/j.molliq.2021.116040

Ramezani, M., Ramezani, F., & Gerami, M. (2019). Nanoparticles in Pest incidences and plant disease control. In *Nanotechnology for agriculture: Crop production & protection* (pp. 233–272). Springer.

Ravichandran, R. (2010). Nanotechnology applications in food and food processing: Innovative green approaches, opportunities and uncertainties for global market. *International Journal of Green Nanotechnology: Physics and Chemistry, 1*, 72–P96. https://doi.org/10.1080/19430871003684440

Samuel, M. S., Ravikumar, M., John, J. A., et al. (2022). A review on green synthesis of nanoparticles and their diverse biomedical and environmental applications. *Catalysts, 12*, 459. https://doi.org/10.3390/catal12050459

Shafey, A. M. E. (2020). Green synthesis of metal and metal oxide nanoparticles from plant leaf extracts and their applications: A review. *Green Processing and Synthesis, 9*, 304–339. https://doi.org/10.1515/gps-2020-0031

Sharma, J., Singh, V. K., Kumar, A., et al. (2018). Role of silver nanoparticles in treatment of plant diseases. In *Microbial biotechnology* (pp. 435–454). Springer.

Sheikh, Z. N., Raina, S., Bhat, U. H., et al. (2024). Deciphering the potential role of nano-based biosensors in growth, development and stress management of crop plants. *South African Journal of Botany, 167*, 542–553. https://doi.org/10.1016/j.sajb.2024.02.035

Shirsat, S., & Suthindhiran, K. (2024). Iron oxide nanoparticles as iron micronutrient fertilizer—Opportunities and limitations. *Journal of Plant Nutrition and Soil Science, 187*, 565–588. https://doi.org/10.1002/jpln.202300203

Singh, R. P., Handa, R., & Manchanda, G. (2021). Nanoparticles in sustainable agriculture: An emerging opportunity. *Journal of Controlled Release, 329*, 1234–1248. https://doi.org/10.1016/j.jconrel.2020.10.051

Singh, A., Sharma, A., Singh, O., et al. (2024). In-depth exploration of nanoparticles for enhanced nutrient use efficiency and abiotic stresses management: Present insights and future horizons. *Plant Stress, 14*, 100576. https://doi.org/10.1016/j.stress.2024.100576

Soni, V., Raizada, P., Singh, P., et al. (2021). Sustainable and green trends in using plant extracts for the synthesis of biogenic metal nanoparticles toward environmental and pharmaceutical advances: A review. *Environmental Research, 202*, 111622. https://doi.org/10.1016/j.envres.2021.111622

Such, G. K., Yan, Y., Johnston, A. P. R., et al. (2015). Interfacing materials science and biology for drug carrier design. *Advanced Materials, 27*, 2278–2297. https://doi.org/10.1002/adma.201405084

Usman, M., Farooq, M., Wakeel, A., et al. (2020). Nanotechnology in agriculture: Current status, challenges and future opportunities. *Science of the Total Environment, 721*, 137778. https://doi.org/10.1016/j.scitotenv.2020.137778

Vurro, M., Miguel-Rojas, C., & Pérez-de-Luque, A. (2019). Safe nanotechnologies for increasing the effectiveness of environmentally friendly natural agrochemicals. *Pest Management Science, 75*, 2403–2412. https://doi.org/10.1002/ps.5348

Wahab, A., Muhammad, M., Ullah, S., et al. (2024). Agriculture and environmental management through nanotechnology: Eco-friendly nanomaterial synthesis for soil-plant systems, food safety, and sustainability. *Science of the Total Environment, 926*, 171862. https://doi.org/10.1016/j.scitotenv.2024.171862

Wongkaew, N., Simsek, M., Griesche, C., & Baeumner, A. J. (2019). Functional nanomaterials and nanostructures enhancing electrochemical biosensors and lab-on-a-Chip performances: Recent Progress, applications, and future perspective. *Chemical Reviews, 119*, 120–194. https://doi.org/10.1021/acs.chemrev.8b00172

Yadav, N., Garg, V. K., Chhillar, A. K., & Rana, J. S. (2023a). Recent advances in nanotechnology for the improvement of conventional agricultural systems: A review. *The Plant Nano Biology, 4*, 100032. https://doi.org/10.1016/j.plana.2023.100032

Yadav, A., Yadav, K., Ahmad, R., & Abd-Elsalam, K. (2023b). Emerging Frontiers in nanotechnology for precision agriculture: Advancements, hurdles and prospects. *Agrochemicals, 2*, 220–256. https://doi.org/10.3390/agrochemicals2020016

Yang, G., Kong, H., Chen, Y., et al. (2022). Recent advances in the hybridization of cellulose and carbon nanomaterials: Interactions, structural design, functional tailoring, and applications. *Carbohydrate Polymers, 279*, 118947. https://doi.org/10.1016/j.carbpol.2021.118947

Yu, G., Xie, X., Pan, L., et al. (2013). Hybrid nanostructured materials for high-performance electrochemical capacitors. *Nano Energy, 2*, 213–234. https://doi.org/10.1016/j.nanoen.2012.10.006

Zare, E. N., Zheng, X., Makvandi, P., et al. (2021). Nonspherical metal-based Nanoarchitectures: Synthesis and impact of size, shape, and composition on their biological activity. *Small, 17*, 17. https://doi.org/10.1002/smll.202007073

Zhao, W., Liu, K., Wang, Y., et al. (2023). Antibacterial carbon dots: Mechanisms, design, and applications. *Advanced Healthcare Materials, 12*, e2300324. https://doi.org/10.1002/adhm.202300324

Zhou, P., Adeel, M., Shakoor, N., et al. (2020). Application of nanoparticles alleviates heavy metals stress and promotes plant growth: An overview. *Nanomaterials, 11*, 26. https://doi.org/10.3390/nano11010026

Chapter 4
Green Nanomaterials: Mechanisms of Action in Plant Systems

Yolanda González-García and Antonio Juárez-Maldonado

4.1 Introduction

The quest for higher agricultural yields and the demand for ecological balance and sustainability are linked to the ever-changing world of agriculture. The degradation of the environment by traditional practices has led to the emergence of some innovative techniques such as the use of nanotechnology (Mahra et al., 2025). One of the technologies that is currently available is the one that is responsible for studying materials on a scale of 1–100 nm. This is known as nanotechnology. At this size, materials have different and special properties that can be extremely useful for different applications. In addition to other properties such as shape, size, chemical composition, concentration, surface structure, aggregation, and solubility (Bai et al., 2021). Nanomaterials have a high surface-to-volume ratio, which gives them a higher surface free energy and therefore greater reactivity (Juárez-Maldonado et al., 2019).

In particular, the application of NMs in plant species has demonstrated tremendous beneficial impacts, ranging from increased plant growth and development, as well as increased tolerance of different conditions of biotic and abiotic stresses (Awasthi et al., 2020; Kumar et al., 2019; Rizwan et al., 2017). One of the main effects of using NMs in plants is to modify the antioxidant defense system, resulting

Y. González-García
Instituto Nacional de Investigaciones Forestales, Agrícolas y Pecuarias, Centro de Investigación Regional Noroeste, La Paz, Mexico
e-mail: gonzalez.yolanda@inifap.gob.mx

A. Juárez-Maldonado (✉)
Departamento de Botánica, Universidad Autónoma Agraria Antonio Narro, Saltillo, Mexico

Laboratorio Nacional Conahcyt de Ecofisiología Vegetal y Seguridad Alimentaria (LANCEVSA), Universidad Autónoma Agraria Antonio Narro, Saltillo, Mexico
e-mail: antonio.juarez@uaaan.edu.mx

K. Esquivel Escalante et al. (eds.), *Green Nanomaterials*, Nanotechnology in the Life Sciences, https://doi.org/10.1007/978-3-032-04492-1_4

in stimulating the production of enzymatic and non-enzymatic antioxidants (Juárez-Maldonado et al., 2021; Lala, 2021). This is accomplished by being capable of modifying cell metabolism and gene expression (Khodakovskaya et al., 2012; Yan et al., 2013). Different NM based on elements like Fe, Zn, Cu, and so on, were tested and showed different responses to fertilizing these elements. These responses mainly included stimulation of plant growth, increased fruit productivity, increased production of fruit antioxidants and secondary metabolites such as phenols and flavonoids (Awasthi et al., 2020; Kumar et al., 2019; Rizwan et al., 2017). NMs have the ability to modify the antioxidant defense system and stimulate the production of ROS, inducing the production of enzymatic and nonenzymatic antioxidant compounds (Juárez-Maldonado et al., 2021; Lala, 2021). Furthermore, the activity of enzymes involved in antioxidant defense has been linked to secondary metabolism and secondary metabolite accumulation (Lala, 2021).

Concerns about the impact of new technologies on the environment have recently led to the emergence of so-called green nanotechnology. This is a relative and multidisciplinary field that has emerged as a rapidly developing area of research and serves as an important technique that focuses on making the process clean, safe, and especially environmentally friendly, unlike the chemical and physical methods currently used to synthesize nanomaterials (Saratale et al., 2018). In this field, the focus is on using natural ingredients to synthesize nanomaterials and designing environmentally friendly synthetic processes (Lu & Ozcan, 2015). Therefore, green nanomaterials are synthesized from compounds such as those derived from plants, e.g., phytochemicals such as alkaloids, terpenoids, saponins, and polyphenols, which are present in different parts of plants (leaves, stems, roots, seeds, fruits, and flowers) and act as surfactants or reducing agents during the synthesis of nanoparticles (Rani & Keshu, 2021).

In view of the above, it is important to analyze the mechanisms of action of different nanomaterials in plants in order to achieve the expected benefits from their application and, above all, to strive for sustainability in plant production systems.

4.2 Methods of Synthesis of Metal Nanoparticles

There are several methods of nanomaterial synthesis, both traditional and green, which are basically divided into two approaches, namely, bottom-up and top-down. Each of them has certain characteristics, which can ultimately impact the properties of the synthesized nanomaterials. The properties of the nanomaterials are fundamental as it depends on these, the impact they can have on plants through nanomaterial–plant interaction.

4.2.1 *Bottom-Up Approach*

The bottom-up method consists of obtaining nanoparticles through a construction or self-assembly technique, from the union of atoms or molecules, which can grow to the desired sizes. It is a very versatile method and the most used for the synthesis of nanoparticles because it allows controlling their shape, size, and chemical composition (Abid et al., 2022). These methods are very useful for very specific agricultural applications that require control of the properties of the materials. Following the self-assembly technique, various synthesis techniques have been developed, some of which are mentioned below.

4.2.1.1 Polyol Method

The polyol method allows the synthesis of well-defined nanoparticles, in terms of size, morphology, composition, and crystallinity. It is a very versatile liquid phase method that uses high temperatures and multivalent alcohols to produce a large number of inorganic compounds, ranging from metallic nanoparticles to alloys (Fiévet et al., 2018). The term polyol was first introduced in 1989 by Fievets for synthesizing metallic NPs. Several types of polyols have been included in this process: ethylene glycol (EG), propylene glycol (PG), butylene glycol (BG), diethylene glycol (DG), triethylene glycol (TrEG), tetraethylene glycol (TEG), and polyethylene glycol (PEG). Polyols can allow synthesis at high temperatures (200–320 °C), solubilizing simple and cheap metal precursors. In addition, it serves as a reducing agent, solvent, and ligand to avoid the agglomeration of NPs and control their nucleation and growth. The polyol selected for the NP synthesis will depend on the polyol's boiling point and reduction potential. Some nanoparticles that can be synthesized by this method are Y_2O_3, V_xO_y, Mn_3O_4, ZnO, $CoTiO_3$, SnO_2, PbO, and TiO_2. The most widely used polyol as a solvent to obtain metallic oxide nanoparticles is ethylene glycol because it is a strong reducing agent with a high dielectric constant and high boiling point. Also, this method has been used for the synthesis of bimetallic alloys and core-shell nanoparticles (Ammar & Fiévet, 2020; Dang-Bao et al., 2021; Fiévet et al., 2018).

4.2.1.2 Sol–gel

The sol–gel method is a widely used technique for the synthesis of high-quality metal-oxide-based nanomaterials. It is called the sol–gel method because, during the synthesis of nanoparticles, the initial precursor forms a sol, which later becomes a network-like structure, called a gel. A sol refers to colloidal particles dispersed in a liquid phase or polymers in a solvent. A gel consists of a three-dimensional continuous network, which is built from the agglomeration of colloidal particles. This method allows you to control the textual and surface properties of materials (Baig et al., 2021). The sol-gel method consists of hydrolysis, condensation, and drying. Initially, the metal precursor is rapidly hydrolyzed, producing a metal hydroxide solution, followed by immediate condensation, which increases the viscosity to form the gel. Finally, the gel is subjected to an aging time and drying process, on which the resulting product will depend (xerogel or aerogel) (Rao et al., 2017). Two variables of the sol–gel method can be used, depending on the reaction medium used during the synthesis: aqueous (water) or non-aqueous (organic solvents) sol–gel method. However, the final product obtained can change its characteristics depending on the type of precursor, hydrolysis speed, aging time, and pH, among others.

4.2.1.3 Thermal Decomposition

The thermal decomposition method is a technique used to obtain metallic nanoparticles. It consists of the chemical decomposition of organometallic precursors using heat and causing a break in the chemical bonds. This synthesis route is simple, cheap, and friendly to the environment. It also allows obtaining nanoparticles with controlled morphology and size distribution, using the appropriate surfactants during the nucleation and growth processes in the synthesis (Rao et al., 2017).

4.2.1.4 Microemulsions

According to Schulman in 1959, the term microemulsion applies to a mixture of three optically transparent and thermodynamically stable components, which include an oil phase, an aqueous phase, and surface-active species (known as surfactants). Depending on the relationship between the components, we can have drops of water dispersed in oil (w/o) or drops of oil dispersed in water (o/w). The microstructure of the mixtures changes from spherical to cylindrical, tubular, and continuous phases of water and oil separated by thin layers of surfactants. It is important to differentiate a microemulsion from an emulsion, in terms of structure and stability, where emulsions are unstable systems that present phase separation in the absence of agitation. In addition, droplet sizes are in the micrometer range, while in microemulsions the size of micelles is between 5 and 100 nm. However, this size can change depending on the type and concentration of the surfactant (Tartaro et al., 2020). A microemulsion is a versatile method that allows control of particle size, morphology, homogeneity, and surface area. This technique allows for obtaining particles of organic and inorganic compounds (Malik et al., 2012).

4.2.1.5 Coprecipitation

The synthesis by chemical coprecipitation is a simple and low-cost method, consisting of precipitating insoluble products in aqueous solutions, which are governed by two important processes that define their morphology and particle sizes: *nucleation and growth.* All precipitation reactions have the following characteristics; (1) the products formed during precipitation reactions are insoluble species formed under high saturation conditions, (2) nucleation is a key step of the precipitation process that allows the formation of small particles, (3) the ripening and aggregation processes affect the size, morphology, and properties of the products, (4) supersaturation conditions are necessary to induce the precipitation of chemical species. Any reaction conditions that influence the synthesis process, e.g., the rate of precursor addition and the rate of stirring, are important factors affecting the size, morphology, and particle size distribution of the products obtained (Cushing et al., 2004). Coprecipitation is one of the most widely used syntheses for obtaining magnetic nanoparticles (especially magnetite), where two components are usually required: iron salts as precursors and precipitating agents. The salts can be chlorides, sulfates, nitrates, and acetoacetate, while the precipitating agents can be NH_4OH, NaOH, or KOH (Al-Madhagi et al., 2023).

4.2.2 *Top-Down Approach*

Top-down methods consist of the transformation of bulk materials into nanometric particles. They are very easy-to-use techniques that allow large-scale production and do not require chemical purification. Its greatest limitation is not having control of the particle size or regularity in its shapes, considering that the properties of the materials depend strongly on their structure and size. Additionally, they may contain significant amounts of impurities. However, it could be useful for large producers of crops that do not require strict control of the properties of the materials. Some of the most commonly used synthesis methods are mentioned below.

4.2.2.1 Mechanochemical Synthesis

Mechanochemical synthesis consists of carrying out chemical reactions and structural changes using mechanical energy. It is possible to obtain various structures from milling materials through processes of deformation and fracture of particles accompanied by powder collision events. Tsuzuki and McCormick (2000) demonstrated that mechanochemical synthesis involves solid-state reactions that occur at low temperatures in a ball mill and without the need for external heating, obtaining materials of different types with uniform structures and nanometric sizes. Currently, this synthesis method is used at the industrial level for the large-scale production of various nanoparticles, among them the oxide-metal type. In addition, with technological progress, different types of ball mills have been built, for example, attrition mills and planetary mills, with energy high enough to induce many chemical reactions during the process (Tsuzuki, 2021).

4.2.2.2 Laser Ablation

The synthesis by laser ablation consists of the generation of nanoparticles from a powerful laser beam that irradiates the precursor materials at extremely high temperatures, vaporizing them into atoms and clusters, which are quickly cooled to room temperature, to carry out the formation of the particles. Nanoparticles (by supersaturated vapor nucleation). This synthesis method allows for obtaining metallic nanoparticles, carbon nanomaterials, oxide compounds, and ceramics. In addition, it also allows the modification of the particle size by adjusting the fluence of the pulsed laser and the wavelength (Baig et al., 2021).

4.2.3 Green Nanosynthesis

Green synthesis methods present a viable solution using bio-based materials for the synthesis of nanoparticles based on the following principles: the implementation of a reliable and environmentally friendly route, the use of non-toxic precursors, and the production of a smaller amount of waste, with the aim of ensuring a sustainable environment (Afzal et al., 2021). Green methods of nanoparticle synthesis have multiple advantages over traditional physical and chemical methods as they are environmentally friendly, non-toxic, and sustainable because they do not require hazardous chemicals, high pressure, energy, or heat (Altuwirqi et al., 2020). Furthermore, they offer the ability to produce nanoparticles of various shapes and sizes, which can be tailored to specific applications (Gupta et al., 2023). They are also ideal for large-scale production and are more economical compared to other synthesis methods (Altuwirqi et al., 2020).

The use of biological agents such as microorganisms (bacteria, algae, fungi, yeasts) and their enzymes and extracts, isolated sugars, agricultural residues, or plants such as shoots, leaves, roots, stems, seeds, bark, and flowers is highly compatible with the synthesis of nanoparticles since these agents produce functional biomolecules that actively reduce metal ions (Altuwirqi et al., 2020). Also, as natural coating agents, which contribute to the ecological nature of the synthesis process (Khalaj et al., 2020).

4.3 Impact of NMs Characteristics on Plant Uptake and Transport

Nanoparticles can be absorbed and transported by plants through different plant organs. Their uptake depends mainly on the size of the nanoparticle, as well as other factors such as shape, surface charge, concentration, aggregation, partial or complete coating, exposure time, and plant species (Afzal et al., 2024; Wang et al., 2023). While transport can be influenced by various elements such as the size and shape of the nanoparticles, their surface properties, the pH of the solution, the presence of other ions or compounds in the solution, the presence of other ions or compounds in the solution, and the presence of other ions or compounds in the solution, it can also influence the transport of nanoparticles (Prakash et al., 2021; Wang et al., 2023). In addition, there are anatomical and physiological characteristics of the plants, as well as environmental factors that interfere with plant physiology. For example, water movement and transpiration flow are affected by temperature, relative humidity, and light (Afzal et al., 2024; Bhaskar et al., 2024). All these factors that influence the uptake and transport of nanoparticles in plants present diverse relationships; however, all of them are interconnected (Fig. 4.1).

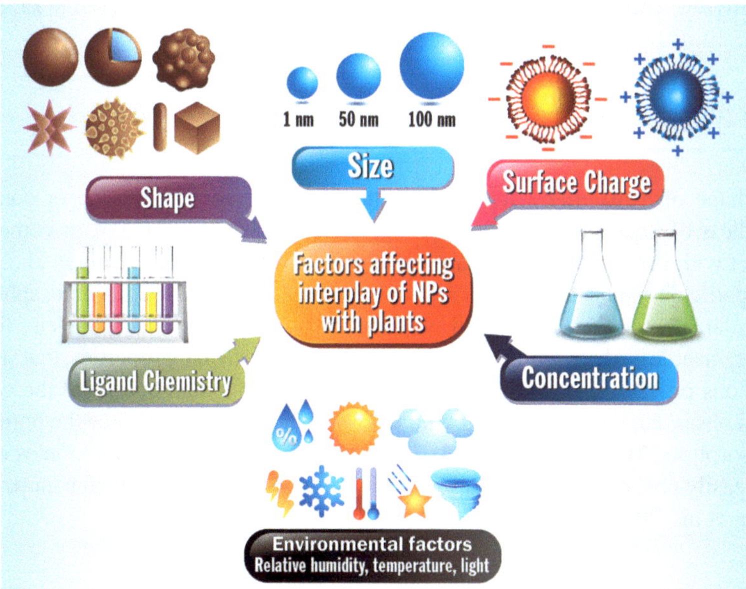

Fig. 4.1 Factors affecting the interplay of NPs with plants. (Constructed by Y. González-García)

4.3.1 Size of NMs

The size of the NP has a direct influence on its ability to penetrate and be absorbed by a plant system (Przemieniecki et al., 2025). Due to their small size, NPs can be distributed in a more controlled and effective way to a specific site, which increases the efficiency of plant uptake and decreases nutrient loss (Bhaskar et al., 2024). NP size and surface-to-volume ratio or surface area are highly correlated. Surface area refers to the complete outer covering of a material, while volume represents the amount of space occupied by the material (Joudeh & Linke, 2022). The smaller the NP size, the higher the surface area-to-volume ratio, and as the particle size increases, the surface area-to-volume ratio decreases (Madlala et al., 2024).

Nanoparticles smaller than 1 nm can enter cell membranes along with water through permeation (Lin et al., 2020). NPs in between 1 and 5 nm can enter the cell through endocytosis, which includes phagocytosis, pinocytosis, or receptor-mediated endocytosis, where NPs are taken up by cells, forming small vesicles and fusing with the cell membrane (Bhaskar et al., 2024). It has also been reported that in this size range, NPs can enter through the cuticular route (Lv et al., 2019), or by the formation of pores, with the help of carrier proteins and through ion channels (Afzal et al., 2024).

NPs between 1 and 50 nm can enter the plant through the stomata, depending on the plant species, size, and stomatal density (Afzal et al., 2021; Munir et al., 2023). The length and width of stomata are usually in micrometer scale, with variations among plant species. For example, the length and width of stomata in rice are approximately 30 μm and 15 μm and in barley 27 μm wide and 4 μm and length, respectively. This size is large enough to allow the passage of NPs (Wu & Li, 2022).

4.3.2 Shape of NMs

The shape of a material refers to its external characteristics, outline or contours, regardless of its actual size, since as the size of a particle decreases, or comes into contact with the plant cell, its shape undergoes a transformation (Baig et al., 2021). Nanoparticles can be designed to adopt various shapes (cylindrical, cubic, spherical, triangular, pentagonal, hexagonal, star, helical, rod, amorphous, platelets, flower, cluster, needle) which confer a wide range of interfacial properties that induce variations or modifications in the surface area of the nanoparticles and the contact angles observed when interacting with the plant surface that influence the regulation of adsorption (Madlala et al., 2024). For example, rod-shaped nanoparticles can be more easily absorbed and internalized by leaves compared to spherical nanoparticles (Wang et al., 2023).

4.3.3 Surface Charge of NMs

Surface charge also plays an important role in the uptake of nanoparticles in plants. The negative charge of the cell walls of plant roots determines the interaction with the surface charge of nanoparticles (Wang et al., 2021a). Unlike leaves, positively charged nanoparticles tend to accumulate on the root surface due to electrostatic attraction with the negatively charged cell wall, without entering the root tissue (Francis et al., 2024). The effect of surface charge on nanoparticle uptake in plant roots is slightly different from that in leaves, as positively charged nanoparticles tend to accumulate on the root surface due to electrostatic attraction between the negative charge of the cell wall and the positively charged nanoparticle, but cannot enter the root tissue (Parkinson et al., 2022).

It was demonstrated that uncharged and/or negatively charged nanoparticles can be taken up by *Arabidopsis thaliana* root cells and subsequently migrate into the root xylem. In contrast, positively charged nanoparticles simply accumulate in the root epidermis, unable to penetrate the root tissue (Wang et al., 2020). In leaf tissues, the surface charge of nanoparticles affects their accumulation and transport since the accumulation of positively charged nanoparticles is significantly higher than that of negatively charged nanoparticles (Javed et al., 2024). This is because the pectin in the cell wall has carboxyl groups that make the cell wall negatively charged; therefore, the positively charged nanoparticles will be attracted to the negative charge in the cell wall, which makes the positively charged nanoparticles easier to transport through the extracellular pathway (Azeem et al., 2021).

4.3.4 Ligand Chemistry of NMs

The chemistry of the ligand is relevant because if it is hydrophilic in nature it will allow the nanoparticle to pass through the membrane rather than embedding itself inside or adhering to the surface (Lin et al., 2020). Increasing the hydrophobicity of the ligand will increase the enthalpic interaction between the non-polar interior of the membrane and the ligands, which helps to trap the nanoparticles within the membrane. In addition, nanoparticles that have more hydrophobic ligands tend to have a higher free energy gain and larger surface area compared to nanoparticles with less hydrophobic ligands (da Silva et al., 2025).

4.3.5 Concentration of NMs

There is a lot of research on the beneficial effect of NP application at low concentrations, while they become harmful at higher concentrations, which is known as the hormesis effect (Liu et al., 2020). Hormesis is defined as a biphasic dose-response relationship showing low-dose stimulation and high-dose inhibition (Trela-Makowej et al., 2024). Hormone dose–response in NP penetration or entry into the plant may occur as a result of direct stimulation or as a modest overcompensation response following a disruption in homeostasis or a moderate induction of toxicity (Agathokleous et al., 2019). Regardless of the process by which the hormone dose–response is induced, the efficiency of NP penetration into the plant is similar, so dose–response characteristics may describe biological plasticity and the ability of organisms to adapt to stress, achieving greater resilience (Agathokleous et al., 2019).

The way of application of the NPs is a characteristic that influences their concentration, since the concentration of nanoparticles decreases as the distance from the foliar application site increases. Karny et al. (2018) applied 100 nm nanoparticles on cherry-tomato plants (*Solanum lycopersicum* L.) and showed that at short distances of 1 and 2 meters, the nanoparticles in the plant retained their size and only decreased by 5% of the concentration applied. While at longer distances (8 m), it was not possible to detect the nanoparticles in the plant.

4.3.6 Impact of NMs Application Pathways

The routes of exposure of plants to NPs are mainly soil application or in hydroponic media, where they interact directly with the root system, foliar spraying where they are absorbed through leaves and stems, and foliar spraying, where they are absorbed through leaves and stems (Francis et al., 2024). Seed preparation and encapsulation, where they are absorbed by the seed coat, composed mainly of epidermal and sclerenchyma cells, is another route (Mahra et al., 2025). In addition, NMs have been applied in plant tissue culture transformation techniques such as callus induction, organogenesis, somatic embryogenesis, protoplast culture, somaclonal variation, and secondary metabolites synthesis (Gunasena et al., 2024).

When NPs are applied to the soil or in hydroponic media, the first contact with the plant is the rough outer layer of the root, which is made up of exudates from root hairs, mucilage, and organic acids that have positive and negative surface charges (Afzal et al., 2024). Such charges aid in the attachment of NMs to the outside of the roots. Generally, the positively charged layer on the root surface attracts oppositely charged or neutral NMs (Afzal et al., 2024; Wang et al., 2021b).

NPs can be transported from the root, through the extracellular spaces of the cells via apoplastic to the endodermis or vascular system, the xylem being the most important vehicle in the distribution and translocation of NPs (Castro-González et al., 2019). The rate of internalization of NMs through the root system occurs mainly through two processes: first, where NMs are sequestered by the mucilage, as seen in the case of the absorption of gold NMs which are easily absorbed by the plants and the second is the case where NMs are not trapped by the root cap but a direct translocation occurs through the root tissues, especially the NMs that have a negative charge by the root cap (Ali et al., 2021; Wang et al., 2021b).

In addition, it has been seen that, like other solute particles, NMs, for their radial movement, take apoplastic pathways and continue to use the same route until they are hindered by the Casparian fringe present longitudinally in the endodermis of roots, and a switch to the symplastic pathway is required (Wang et al., 2021b). Movement of NPs in the root can occur through the cell wall (apoplastic pathway) to move larger particles (200 nm) and through the plasmodesmata (symplastic pathway) to move smaller particles around 50 nm. NPs move through the epidermis and cortex and enter endodermal cells via the Caspari band. After crossing the endodermis, the NPs enter the xylem and travel through the transpiration stream to various parts of the plant, including the stems, leaves, and fruits (Fallah et al., 2024).

Application of nanoparticles by foliar spray can enter plants through the epidermis or leaf stomata and then travel through apoplastic or symplastic pathways. When absorbed by plants, the NPs are transported through the phloem and xylem and stored in vacuoles and cell walls (Ahmad et al., 2024). The cuticle offers a potentially more efficient route for NP entry due to its larger surface area, in contrast to stomata. After passage through the cuticle, they are transported to epidermal and mesophyll tissues and subsequently transported to other parts of the plant via the xylem and phloem systems (Fallah et al., 2024). Basically, once inside the plant, NMs can be moved by three types of transport: (1) macroscopic transport, which facilitates movement through plant organelles such as leaf cuticles and vascular bundles, (2) microscopic transport, which enables passage through cell walls, and (3) molecular transport, which involves crossing lipid bilayers of cells and organelles (Hubbard et al., 2020).

In addition, the concentration of nanoparticles decreases as the distance from the foliar application site increases. Karny et al. (2018) applied 100 nm nanoparticles on plants and showed that at short distances of 1 and 2 m, the nanoparticles in the plant retained their size and only decreased by 5% of the concentration applied. While at longer distances (8 m), it was not possible to detect the nanoparticles in the plant.

4.4 Interaction of NMs and Plants and Their Impacts at Different Levels

Cellular uptake and absorption of NMs are intricate processes involving several pathways and mechanisms, determined by the physicochemical properties of the NMs (such as size, shape, surface charge, and coating (partial or complete), as well as by the cell type (leaf, root, stem, flower, fruit) and environmental factors involved (Afzal et al., 2024). In other words, the characteristics and properties of the nanomaterials themselves will ultimately have an impact on the plants at the morphological, biochemical, physiological, genetic, etc. levels, which will result in plant responses, which may be positive, negative, or null at different levels (Juárez-Maldonado et al., 2021).

4.4.1 Mechanisms of Action of NMs

In addition to stimulating plant growth, development, and productivity, NM has been applied to enhance the synthesis of antioxidant compounds and secondary compounds (phenolic and flavonoid) (Awasthi et al., 2020; Kumar et al., 2019; Rizwan et al., 2017). Due to their aforementioned physical and chemical properties, NMs have the ability to induce biostimulation in plants. Upon coming into contact with different biotic or abiotic materials, NM can absorb organic components or biomolecules and form a corolla (Juárez-Maldonado et al., 2019). Therefore, the surface properties of NMs and the corona primarily influence plant responses, since the first interaction between NMs and plants occurs through the surface of the corona (Juárez-Maldonado et al., 2021). The entire process of stimulation by NMs can be divided into two phases, one of physicochemical nature and the other of chemical nature. During NMs' surface interaction with cell walls and membranes, receptor, transporter, and protein activities can be modified, which in turn can alter metabolite and ion transport in cells. Intracellular NMs also trigger other responses at the cellular level due to their interactions with metabolites in the cytoplasm or with organelle membranes, inducing oxidative stress (Juárez-Maldonado et al., 2019). However, these responses depend on the concentration of NMs applied, their type and size, and even the plant species to which they are applied (Juárez-Maldonado et al., 2019, 2021). This interaction between NM and different cellular organelles can improve physiological processes, as well as affect nutritive quality and postharvest life of crops (Mali et al., 2020). NMs can induce the production of secondary metabolites and bioactive compounds in plants (Lala, 2021; de Rivero-Montejo et al., 2021), and through the induction of oxidative stress at the cellular level, they can have an impact on the physiological processes and on the nutritional quality and postharvest life of plants (Mali et al., 2020).

4.4.2 Processes Modified by NMs

NMs can alter photosynthetic capacity, antioxidant defense systems, and gene expression. They can also alter other characteristics, such as root uptake of water and nutrients, that promote plant growth (Jordan et al., 2020). Photosynthesis is the energy source for the entire functioning of the plant and is a fundamental process for plants. When applied to plants, NMs have been shown to move into the chloroplast and incorporate into the thylakoid membranes, modifying the activity of this process. This results in increased electron transport and therefore photosynthesis, so that there is more energy to be used for different plant processes such as growth or secondary metabolism (Giraldo et al., 2014). In addition, NMs can positively influence photosynthesis from other pathways. NMs can affect the chloroplast enzymes involved in photosynthetic dye synthesis, increasing chlorophyll content and thereby enhancing photochemistry efficiency (Rahmani et al., 2020). NMs can also increase the activity of photosystem I and RUBISCO (ribulose bisphosphate carboxylase oxygenase), resulting in higher photosynthesis (Wang et al., 2018). On the other hand, carbon nanomaterials (CNMs), in addition to inducing the production of photosynthetic pigments (chlorophyll, carotenoid), are also a source of carbon, favoring fixation, as well as increasing the rate of electron transport (Lahiani et al., 2018).

The antioxidant defense system involves producing enzymatic and nonenzymatic compounds that scavenge ROS, thereby preventing damage (Choudhary et al., 2020). ROS are essential because their absence can have negative effects on various structures at the cellular level, resulting in oxidative stress, cytotoxic and genotoxic effects, lipid peroxidation, apoptosis, intracellular proteolysis, and dysregulation of miRNAs (Rahmani et al., 2020). In addition, ROS are important because they can act as stress signaling molecules and stimulate the production of other compounds such as phytohormones (abscisic acid, salicylic acid, jasmonic acid), which can induce tolerance to various stress conditions (Mertens et al., 2021). Modification of the antioxidant defense system is one of the key traits that plants modify to defend themselves against stress (Hasanuzzaman et al., 2020; Mahto et al., 2021). NMs can modify the antioxidant defense system by directly interacting with genetic material and indirectly affecting intermediary molecules, triggering ROS and/or reactive nitrogen species (RNS) production (Mahto et al., 2021; Rahmani et al., 2020). However, NM can stimulate the production of various metabolites as well, since it alters the metabolism of carbohydrates, energy, amino acids, lipids, and secondary metabolites (X. Li et al., 2019).

Some important metabolic pathways related to the synthesis of non-enzymatic antioxidants, such as ascorbic acid, phenols, and flavonoids, were positively modified by the application of Fe NPs in Zea mays. It was observed that the ascorbate and aldarate metabolism and the tricarboxylic acid cycle were activated among the main changes due to the application of Fe NPs (Li et al., 2020). The metabolism of reserve sugars and tocopherol, phenylalanine and tryptophan pathways, aspartate family pathway together with serine, alanine, and valine metabolism, and glycerolipid biosynthesis was upregulated in wheat plants (Triticum aestivum) by application of TiO_2 NPs (Silva et al., 2020).

Another important pathway that may be modified is the plastid isoprenoid biosynthetic pathway, which leads to the synthesis of carotenoids, which are very important compounds that serve as chloroplast accessory pigments, energy scavengers, and antioxidants (Joyard et al., 2009). There is evidence that the application of various NMs can induce the accumulation of carotenoids in different plant species, such as Al_2O_3 NPs in *Hibiscus sabdariffa* (Abdel Latef et al., 2020), CeO_2 NPs in *Dracocephalum moldavica* (Mohammadi et al., 2021), chitosan nanoparticles (CS NPs) in *Vigna radiata* (Sen et al., 2020), Fe_2O_3 NPs in Lactuca sativa (Jurkow et al., 2020), even multilayer carbon nanotubes (MWCNT) in *Catharanthus roseus* (Ghasempour et al., 2019).

The effect of NM application on the production of non-enzymatic antioxidant compounds by altering gene expression has been demonstrated in several studies. In addition, it has been shown that the induction of these compounds is independent of the route of application (Wang et al., 2021c).

MWCNT application in Satureja khuzestanica resulted in an increase in phenolic, flavonoid, rosmarinic (RA), and caffeinic (CA) contents, as well as in polyphenoloxidase (PPO), L-phenylalanine ammonia-lyase (PAL), and peroxidase (POD) activities. The authors proposed that through the induction of ROS that oxidize polyunsaturated fatty acids (PUFA) to polyunsaturated fatty acid hydroperoxide (PUFA-OOH), the application of MWCNTs induces the biosynthesis of secondary metabolites. These are further converted to oxylipins, leading to the expression of genes involved in biosynthesizing and accumulating secondary metabolites in plant cells (Ghorbanpour & Hadian, 2015).

NMs are able to interact with DNA and modify the expression of a variety of genes that are involved in different metabolic pathways (Khodakovskaya et al., 2012; Yan et al., 2013) (Rahmani et al., 2020). However, the production and accumulation of proteins may differ because changes at the transcriptional level do not always coincide with changes at the protein level. But it can also induce positive responses in the plant (Guo et al., 2021; Khodakovskaya et al., 2013; Smirnova et al., 2012).

Nickel oxide nanoparticles (NiO NPs) were found to increase malondialdehyde, hydrogen peroxide, ROS levels, as well as peroxidase (POD) enzyme activity in seeds of Chinese cabbage (*Brassica rapa* ssp. Pekinensis var. Seoul). Genetically, oxidative stress-related genes (*CAT, POD*, and *GST*), MYB transcription factors (*BrMYB28, BrMYB29, BrMYB34*, and *BrMYB51*), and phenolic compounds (*ANS, PAP1*, and *PAL*) were overexpressed, and higher levels of secondary metabolites such as anthocyanins, glucosinolates, and phenolic compounds were observed. (Chung et al., 2019).

The foliar application of silicon NPs in maize (*Zea mays*) induced the expression of genes (*PAL, C4H, 4CL, C3H,* and *HCT*) related to the biosynthesis of chlorogenic acid. This resulted in an increased production of these compounds and of all phenolics in the maize husk (Wang et al., 2021a, b, c). Foliar application of chitosan NPs in *Catharanthus roseus* increased gene expression of mitogen-activated protein kinase (*MAPK3*), geissoschizine synthase (*GS*), and octadecanoid-responsive AP2 (*ORCA3*) domain, resulting in greater accumulation of alkaloids in both shoot and root, and greater tolerance to salinity (Hassan et al., 2021).

Fe_3O_4 nanoparticles applied to *Nicotiana benthamiana* leaves increased *PR1* and *PR2* gene expression, which led to increased salicylic acid synthesis and increased TMV tolerance (Cai et al., 2020). In fenugreek (*Trigonella foenum-graecum*), Co NPs applied in MS medium modified the expression of genes involved in diosgenin biosynthesis (squalane monooxygenase (*SQS*), cycloartenol synthase (*CYC*), 26-beta-glucosidase (*BGL*)) and improved the production of diosgenin in both roots and shoots (Motevalli et al., 2021). Application of Zn-Ag NP complex through MS medium in *Withania somnifera* modified terpenoid skeleton biosynthesis, steroid biosynthesis, starch and sucrose metabolism, fructose and mannose metabolism by overexpressing genes related to photosynthesis, carbohydrate metabolism, and withanolide biosynthesis; showed more abundant transcripts of *SQS, SQM, STE1, ERG6, DWF1, DWF5, SMT1,* and *SMT2* genes. Consequently, a higher content of stigmasterol, sitosterol, withaferin A, withastamonoside, withanolide A, and withanone was observed. In addition, the rate of photosynthesis, the rate of transpiration, and some carbohydrates were also found to be increased (Singh et al., 2019).

Overexpression of phenylalanine ammonia-lyase (*PAL*) and rosmarinic acid synthase (*RAS*) genes related to the rosmarinic acid (RA) biosynthetic pathway and increased flavonoid content (xanthomicrol, cirsimaritin, and isokaempferide) were induced in *Dracocephalum kotschyi* Boiss. by application of silicon dioxide NPs (SiO_2 NPs) in the MS culture medium (Nourozi et al., 2019).

LOX, PAL, GLU, and *SOD* genes, which are involved in plant defense against pathogens, were overexpressed in tomato (*Solanum lycopersicon*) after application of selenium NPs (Se NPs) by seed priming. The result was a higher activity of the enzymes lipoxygenase (LOX), phenylalanine ammonia lyase (PAL), beta-1,3-glucanase (GLU), and superoxide dismutase (SOD) and a higher tolerance of tomato plants to *Phytophthora infestans* (Joshi et al., 2021).

The application of nickel oxide NPs (NiO NPs) in seeds of Chinese cabbage (*B. rapa* ssp. Pekinensis var. Seoul) overexpressed oxidative stress-related genes (*CAT, POD,* and *GST*), MYB transcription factors (*BrMYB28, BrMYB29, BrMYB34,* and *BrMYB51*), and phenolic compounds (ANS, PAP1, and PAL), resulting in higher levels of anthocyanins, glucosinolates, and phenolic compounds, as well as malondialdehyde, hydrogen peroxide and ROS, and peroxidase (POD) enzyme activity (Chung et al., 2019). The application of silicon dioxide NPs (SiO_2 NPs) directly to the root of *Hyoscyamus reticulatus* and *H. pusillus* increased the expression of *pmt* (putrescine N-methyltransferase) and *h6h* (hyoscyamine 6 <beta> -hydroxylase) genes. This resulted in increased production of phenols and flavonoids and accumulation of tropane alkaloids (hyoscyamine and scopolamine) (Hedayati et al., 2020).

4.5 Conclusions

One of the main concerns of the application of nanotechnology in plants is the environmental impact caused by the different synthesis methods. Fortunately, currently available methods allow practically eliminating the environmental impact when synthesizing nanoparticles and/or nanomaterials, thanks to green nanosynthesis. Through this method, green nanomaterials are obtained, so-called because botanical extracts or microorganisms are commonly used as reducing agents.

Considering this, what follows is to determine the impact of these nanomaterials on plants. Basically, it is the physicochemical characteristics of the nanomaterials that will ultimately determine the effects on plants. While the specific mechanisms will essentially result from the particular characteristics of each nanomaterial used. These characteristics are the core of the nanomaterial or nanoparticle (i.e., the element or elements that form it), shape, size, surface charge, ligand chemistry, concentration, and environmental factors.

Conflict of Interest The authors have no conflicts of interest to declare that are relevant to the content of this chapter.

References

Abdel Latef, A. A. H., Zaid, A., Abu Alhmad, M. F., & Abdelfattah, K. E. (2020). The impact of priming with al2o3 nanoparticles on growth, pigments, osmolytes, and antioxidant enzymes of Egyptian Roselle (Hibiscus sabdariffa L.) cultivar. *Agronomy, 10*, 681. https://doi.org/10.3390/agronomy10050681

Abid, N., Khan, A. M., Shujait, S., Chaudhary, K., Ikram, M., Imran, M., Haider, J., Khan, M., Khan, Q., & Maqbool, M. (2022). Synthesis of nanomaterials using various top-down and bottom-up approaches, influencing factors, advantages, and disadvantages: A review. *Advances in Colloid and Interface Science, 300*, 102597. https://doi.org/10.1016/j.cis.2021.102597

Afzal, S., Sharma, D., & Singh, N. K. (2021). Eco-friendly synthesis of phytochemical-capped iron oxide nanoparticles as nano-priming agent for boosting seed germination in rice (Oryza sativa L.). *Environmental Science and Pollution Research, 28*, 40275–40287. https://doi.org/10.1007/s11356-020-12056-5

Afzal, S., Singh, N. K., Lal, A. F., Sohrab, S., Singh, N., Gupta, P. S., Mishra, S. K., Adeel, M., & Faizan, M. (2024). Nanostructure and plant uptake: Assessing the ecological footprint and root-to-leaf dynamics. *Plant Nano Biology, 10*, 100122. https://doi.org/10.1016/j.plana.2024.100122

Agathokleous, E., Feng, Z., Iavicoli, I., & Calabrese, E. J. (2019). The two faces of nanomaterials: A quantification of hormesis in algae and plants. *Environment International, 131*, 105044. https://doi.org/10.1016/j.envint.2019.105044

Ahmad, A., Javad, S., Iqbal, S., Shahid, T., Naz, S., Shah, A. A., Shaffique, S., & Gatasheh, M. K. (2024). Efficacy of soil drench and foliar application of iron nanoparticles on the growth and physiology of Solanum lycopersicum L. exposed to cadmium stress. *Scientific Reports, 14*, 27920. https://doi.org/10.1038/s41598-024-79270-w

Ali, S., Mehmood, A., & Khan, N. (2021). Uptake, translocation, and consequences of nanomaterials on plant growth and stress adaptation. *Journal of Nanomaterials, 2021*, 1–17. https://doi.org/10.1155/2021/6677616

Al-Madhagi, H., Yazbik, V., Abdelwahed, W., & Alchab, L. (2023). Magnetite nanoparticle co-precipitation synthesis, characterization, and applications: Mini review. *BioNanoScience, 13*, 853–859. https://doi.org/10.1007/s12668-023-01113-1

Altuwirqi, R. M., Albakri, A. S., Al-Jawhari, H., & Ganash, E. A. (2020). Green synthesis of copper oxide nanoparticles by pulsed laser ablation in spinach leaves extract. Optik, 219(April), 165280. https://doi.org/10.1016/j.ijleo.2020.165280

Ammar, S., & Fiévet, F. (2020). Polyol synthesis: A versatile wet-chemistry route for the design and production of functional inorganic nanoparticles. *Nanomaterials, 10*, 1–8. https://doi.org/10.3390/nano10061217

Awasthi, G., Singh, T., Tiwari, Y., Awasthi, A., Tripathi, R. D., Shrivastava, S., Vajpayee, P., Kumar, A., & Awasthi, K. K. (2020). A review on nanotechnological interventions for plant growth and production. *Materials Today Proceedings, 31*, 685–693. https://doi.org/10.1016/j.matpr.2020.07.255

Azeem, I., Adeel, M., Ahmad, M. A., Shakoor, N., Jiangcuo, G. D., Azeem, K., Ishfaq, M., Shakoor, A., Ayaz, M., Xu, M., & Rui, Y. (2021). Uptake and accumulation of nano/microplastics in plants: A critical review. *Nanomaterials, 11*, 2935. https://doi.org/10.3390/nano11112935

Bai, T., Zhang, P., Guo, Z., Chetwynd, A. J., Zhang, M., Adeel, M., Li, M., Guo, K., Gao, R., Li, J., Hao, Y., & Rui, Y. (2021). Different physiological responses of C3 and C4 plants to nanomaterials. *Environmental Science and Pollution Research, 28*(20), 25542–25551. https://doi.org/10.1007/s11356-021-12507-7

Baig, N., Kammakakam, I., Falath, W., & Kammakakam, I. (2021). Nanomaterials: A review of synthesis methods, properties, recent progress, and challenges. *Materials Advances, 2*, 1821–1871. https://doi.org/10.1039/d0ma00807a

Bhaskar, R., Pandey, S. P., Kumar, U., Kim, H., Jayakodi, S. K., Gupta, M. K., & Han, S. S. (2024). Nanobionics for sustainable crop production: Recent development to regulate plant growth and protection strategies from pests. *OpenNano, 15*, 100198. https://doi.org/10.1016/j.onano.2023.100198

Cai, L., Cai, L., Jia, H., Liu, C., Wang, D., & Sun, X. (2020). Foliar exposure of Fe3O4 nanoparticles on Nicotiana benthamiana: Evidence for nanoparticles uptake, plant growth promoter and defense response elicitor against plant virus. *Journal of Hazardous Materials, 393*, 122415. https://doi.org/10.1016/j.jhazmat.2020.122415

Castro-González, C. G., Sánchez-Segura, L., Gómez-Merino, F. C., & Bello-Bello, J. J. (2019). Exposure of stevia (Stevia rebaudiana B.) to silver nanoparticles in vitro: transport and accumulation. *Scientific Reports, 9*, 10372. https://doi.org/10.1038/s41598-019-46828-y

Choudhary, A., Kumar, A., & Kaur, N. (2020). ROS and oxidative burst: Roots in plant development. *Plant Diversity, 42*, 33–43. https://doi.org/10.1016/j.pld.2019.10.002

Chung, I.-M., Venkidasamy, B., & Thiruvengadam, M. (2019). Nickel oxide nanoparticles cause substantial physiological, phytochemical, and molecular-level changes in Chinese cabbage seedlings. *Plant Physiology and Biochemistry, 139*, 92–101. https://doi.org/10.1016/j.plaphy.2019.03.010

Cushing, B. L., Kolesnichenko, V. L., & O'Connor, C. J. (2004). Recent advances in the liquid-phase syntheses of inorganic nanoparticles. *Chemical Reviews, 104*, 3893–3946. https://doi.org/10.1021/cr030027b

da Silva, E. F., dos Santos, F. A. L., Pires, H. M., Bastos, L. M., & de Ribeiro, L. N. M. (2025). Lipid nanoparticles carrying essential oils for multiple applications as antimicrobials. *Pharmaceutics, 17*, 178. https://doi.org/10.3390/pharmaceutics17020178

Dang-Bao, T., Favier, I., & Gómez, M. (2021). Metal nanoparticles in polyols: Bottom-up and top-down syntheses and catalytic applications. In *Nanoparticles in catalysis* (pp. 99–122). Wiley. https://doi.org/10.1002/9783527821761.ch5

de Rivero-Montejo, S. J., Vargas-Hernandez, M., & Torres-Pacheco, I. (2021). Nanoparticles as novel elicitors to improve bioactive compounds in plants. *Agriculture (Switzerland), 11*, 1–16. https://doi.org/10.3390/agriculture11020134

Fallah, S., Yusefi-Tanha, E., & Peralta-Videa, J. R. (2024). Interaction of nanoparticles and reactive oxygen species and their impact on macromolecules and plant production. *Plant Nano Biology, 10*, 100105. https://doi.org/10.1016/j.plana.2024.100105

Fiévet, F., Ammar-Merah, S., Brayner, R., Chau, F., Giraud, M., Mammeri, F., Peron, J., Piquemal, J.-Y., Sicard, L., & Viau, G. (2018). The polyol process: a unique method for easy access to metal nanoparticles with tailored sizes, shapes and compositions. *Chemical Society Reviews, 47*, 5187–5233. https://doi.org/10.1039/C7CS00777A

Francis, D. V., Abdalla, A. K., Mahakham, W., Sarmah, A. K., & Ahmed, Z. F. R. (2024). Interaction of plants and metal nanoparticles: Exploring its molecular mechanisms for sustainable agriculture and crop improvement. *Environment International, 190*, 108859. https://doi.org/10.1016/j.envint.2024.108859

Ghasempour, M., Iranbakhsh, A., Ebadi, M., & Oraghi Ardebili, Z. (2019). Multi-walled carbon nanotubes improved growth, anatomy, physiology, secondary metabolism, and callus performance in Catharanthus roseus: An in vitro study. *3 Biotech, 9*, 404. https://doi.org/10.1007/s13205-019-1934-y

Ghorbanpour, M., & Hadian, J. (2015). Multi-walled carbon nanotubes stimulate callus induction, secondary metabolites biosynthesis and antioxidant capacity in medicinal plant Satureja khuzestanica grown in vitro. *Carbon N Y, 94*, 749–759. https://doi.org/10.1016/J.CARBON.2015.07.056

Giraldo, J. P., Landry, M. P., Faltermeier, S. M., McNicholas, T. P., Iverson, N. M., Boghossian, A. A., Reuel, N. F., Hilmer, A. J., Sen, F., Brew, J. A., & Strano, M. S. (2014). Plant nanobionics approach to augment photosynthesis and biochemical sensing. *Nature Materials, 13*, 400–408. https://doi.org/10.1038/nmat3890

Gunasena, M. D. K. M., Alahakoon, A. M. P. D., Polwaththa, K. P. G. D. M., Galpaya, G. D. C. P., Priyanjani, H. A. S. A., Koswattage, K. R., & Senarath, W. T. P. S. K. (2024). Transforming plant tissue culture with nanoparticles: A review of current applications. *Plant Nano Biology, 10*, 100102. https://doi.org/10.1016/j.plana.2024.100102

Guo, X., Zhao, J., Wang, R., Zhang, H., Xing, B., Naeem, M., Yao, T., Li, R., Xu, R., Zhang, Z., & Wu, J. (2021). Effects of graphene oxide on tomato growth in different stages. *Plant Physiology and Biochemistry, 162*, 447–455. https://doi.org/10.1016/j.plaphy.2021.03.013

Gupta D., Boora A., Thakur A., Gupta T.K. 2023. Green and sustainable synthesis of nanomaterials: Recent advancements and limitations. Environmental Research, 231(3), 116316. https://doi.org/10.1016/j.envres.2023.116316

Hasanuzzaman, M., Bhuyan, M. H. M., Zulfiqar, F., Raza, A., Mohsin, S., Mahmud, J., Fujita, M., & Fotopoulos, V. (2020). Reactive oxygen species and antioxidant defense in plants under abiotic stress: Revisiting the crucial role of a universal defense regulator. *Antioxidants, 9*, 1–52. https://doi.org/10.3390/antiox9080681

Hassan, F. A. S., Ali, E., Gaber, A., Fetouh, M. I., & Mazrou, R. (2021). Chitosan nanoparticles effectively combat salinity stress by enhancing antioxidant activity and alkaloid biosynthesis in Catharanthus roseus (L.) G. Don. *Plant Physiology and Biochemistry, 162*, 291–300. https://doi.org/10.1016/j.plaphy.2021.03.004

Hedayati, A., Hosseini, B., Palazon, J., & Maleki, R. (2020). Improved tropane alkaloid production and changes in gene expression in hairy root cultures of two Hyoscyamus species elicited by silicon dioxide nanoparticles. *Plant Physiology and Biochemistry, 155*, 416–428. https://doi.org/10.1016/j.plaphy.2020.07.029

Hubbard, J. D., Lui, A., & Landry, M. P. (2020). Multiscale and multidisciplinary approach to understanding nanoparticle transport in plants. *Current Opinion in Chemical Engineering, 30*, 135–143. https://doi.org/10.1016/j.coche.2020.100659

Javed, R., Khan, B., Sharafat, U., Bilal, M., Galagedara, L., Abbey, L., & Cheema, M. (2024). Dynamic interplay of metal and metal oxide nanoparticles with plants: Influencing factors, action mechanisms, and assessment of stimulatory and inhibitory effects. *Ecotoxicology and Environmental Safety, 271*, 115992. https://doi.org/10.1016/j.ecoenv.2024.115992

Jordan, J. T., Oates, R. P., Subbiah, S., Payton, P. R., Singh, K. P., Shah, S. A., Green, M. J., Klein, D. M., & Cañas-Carrell, J. E. (2020). Carbon nanotubes affect early growth, flowering time and phytohormones in tomato. *Chemosphere, 256*, 127042. https://doi.org/10.1016/j.chemosphere.2020.127042

Joshi, S. M., De Britto, S., & Jogaiah, S. (2021). Myco-engineered selenium nanoparticles elicit resistance against tomato late blight disease by regulating differential expression of cellular, biochemical and defense responsive genes. *Journal of Biotechnology, 325*, 196–206. https://doi.org/10.1016/j.jbiotec.2020.10.023

Joudeh, N., & Linke, D. (2022). Nanoparticle classification, physicochemical properties, characterization, and applications: a comprehensive review for biologists. *Journal of Nanobiotechnology, 20*, 262. https://doi.org/10.1186/s12951-022-01477-8

Joyard, J., Ferro, M., Masselon, C., Seigneurin-Berny, D., Salvi, D., Garin, J., & Rolland, N. (2009). Chloroplast proteomics and the compartmentation of plastidial isoprenoid biosynthetic pathways. *Molecular Plant, 2*, 1154–1180. https://doi.org/10.1093/mp/ssp088

Juárez-Maldonado, A., Ortega-Ortiz, H., González-Morales, S., Morelos-Moreno, Á., Cabrera-de la Fuente, M., Sandoval-Rangel, A., Cadenas-Pliego, G., & Benavides-Mendoza, A. (2019). Nanoparticles and nanomaterials as plant biostimulants. *International Journal of Molecular Sciences, 20*, 1–19. https://doi.org/10.3390/ijms20010162

Juárez-Maldonado, A., Tortella, G., Rubilar, O., Fincheira, P., & Benavides-Mendoza, A. (2021). Biostimulation and toxicity: The magnitude of the impact of nanomaterials in microorganisms and plants. *Journal of Advanced Research, 31*, 113–126. https://doi.org/10.1016/j.jare.2020.12.011

Jurkow, R., Agnieszka, S., Pokluda, R., Smole, S., & Kalisz, A. (2020). Biochemical response of Oakleaf Lettuce seedlings to different concentrations of some metal (oid).

Karny, A., Zinger, A., Kajal, A., Shainsky-Roitman, J., & Schroeder, A. (2018). Therapeutic nanoparticles penetrate leaves and deliver nutrients to agricultural crops. *Scientific Reports, 8*, 7589. https://doi.org/10.1038/s41598-018-25197-y

Khalaj, M., Kamali, M., Costa, M. E. V., & Capela, I. (2020). Green synthesis of nanomaterials - A scientometric assessment. Journal of Cleaner Production, 267, 122036. https://doi.org/10.1016/j.jclepro.2020.122036

Khodakovskaya, M. V., de Silva, K., Biris, A. S., Dervishi, E., & Villagarcia, H. (2012). Carbon nanotubes induce growth enhancement of tobacco cells. *ACS Nano, 6*, 2128–2135. https://doi.org/10.1021/nn204643g

Khodakovskaya, M. V., Kim, B.-S., Kim, J. N., Alimohammadi, M., Dervishi, E., Mustafa, T., & Cernigla, C. E. (2013). Carbon nanotubes as plant growth regulators: Effects on tomato growth, reproductive system, and soil microbial community. *Small, 9*, 115–123. https://doi.org/10.1002/smll.201201225

Kumar, A., Gupta, K., Dixit, S., Mishra, K., & Srivastava, S. (2019). A review on positive and negative impacts of nanotechnology in agriculture. *International journal of Environmental Science and Technology, 16*, 2175–2184. https://doi.org/10.1007/s13762-018-2119-7

Lahiani, M. H., Nima, Z. A., Villagarcia, H., Biris, A. S., & Khodakovskaya, M. V. (2018). Assessment of effects of the long-term exposure of agricultural crops to carbon nanotubes. *Journal of Agricultural and Food Chemistry, 66*, 6654–6662. https://doi.org/10.1021/acs.jafc.7b01863

Lala, S. (2021). Nanoparticles as elicitors and harvesters of economically important secondary metabolites in higher plants: A review. *IET Nanobiotechnology, 15*, 28–57. https://doi.org/10.1049/nbt2.12005

Li, X., Peng, T., Mu, L., & Hu, X. (2019). Phytotoxicity induced by engineered nanomaterials as explored by metabolomics: Perspectives and challenges. *Ecotoxicology and Environmental Safety, 184*, 109602. https://doi.org/10.1016/j.ecoenv.2019.109602

Li, P., Wang, A., Du, W., Mao, L., Wei, Z., Wang, S., Yuan, H., Ji, R., & Zhao, L. (2020). Insight into the interaction between Fe-based nanomaterials and maize (Zea mays) plants at metabolic level. *Science of the Total Environment, 738*, 139795. https://doi.org/10.1016/j.scitotenv.2020.139795

Lin, J., Miao, L., Zhong, G., Lin, C.-H., Dargazangy, R., & Alexander-Katz, A. (2020). Understanding the synergistic effect of physicochemical properties of nanoparticles and their cellular entry pathways. *Communications Biology, 3*, 205. https://doi.org/10.1038/s42003-020-0917-1

Liu, Y., Pan, B., Li, H., Lang, D., Zhao, Q., Zhang, D., Wu, M., Christian, E., Steinberg, W., & Xing, B. (2020). Ecotoxicology and Environmental Safety Can the properties of engineered nanoparticles be indicative of their functions and effects in plants ? *Ecotoxicology and Environmental Safety, 205*, 111128. https://doi.org/10.1016/j.ecoenv.2020.111128

Lu, Y., & Ozcan, S. (2015). Green nanomaterials: On track for a sustainable future. *Nano Today, 10*, 417–420. https://doi.org/10.1016/j.nantod.2015.04.010

Lv, J., Christie, P., & Zhang, S. (2019). Uptake, translocation, and transformation of metal-based nanoparticles in plants: recent advances and methodological challenges. *Environmental Science. Nano, 6*, 41–59. https://doi.org/10.1039/C8EN00645H

Madlala, N. C., Khanyile, N., & Masenya, A. (2024). Examining the correlation between the inorganic nano-fertilizer physical properties and their impact on crop performance and nutrient uptake efficiency. *Nanomaterials, 14*, 1263. https://doi.org/10.3390/nano14151263

Mahra, S., Tripathi, S., Tiwari, K., Sharma, S., Mathew, S., Kumar, V., & Sharma, S. (2025). Harnessing nanotechnology for sustainable agriculture: From seed priming to encapsulation. *Plant Nano Biology, 11*, 100124. https://doi.org/10.1016/j.plana.2024.100124

Mahto, R., Rani, P., Bhardwaj, R., Singh, R. K., Prasad, S. K., & Rakshit, A. (2021). Nanotechnology: A potential approach for abiotic stress management. In *Food science, technology and nutrition* (pp. 249–264). Woodhead Publishing. https://doi.org/10.1016/B978-0-12-820092-6.00010-0

Mali, S. C., Raj, S., & Trivedi, R. (2020). Nanotechnology a novel approach to enhance crop productivity. *Biochemistry and Biophysics Reports, 24*, 100821. https://doi.org/10.1016/j.bbrep.2020.100821

Malik, M. A., Wani, M. Y., & Hashim, M. A. (2012). Microemulsion method: A novel route to synthesize organic and inorganic nanomaterials. 1st Nano update. *Arabian Journal of Chemistry, 5*, 397–417. https://doi.org/10.1016/j.arabjc.2010.09.027

Mertens, D., Boege, K., Kessler, A., Koricheva, J., Thaler, J. S., Whiteman, N. K., & Poelman, E. H. (2021). Predictability of biotic stress structures plant defence evolution. *Trends in Ecology & Evolution, 36*, 444–456. https://doi.org/10.1016/j.tree.2020.12.009

Mohammadi, M. H. Z., Panahirad, S., Navai, A., Bahrami, M. K., Kulak, M., & Gohari, G. (2021). Cerium oxide nanoparticles (CeO2-NPs) improve growth parameters and antioxidant defense system in Moldavian Balm (Dracocephalum moldavica L.) under salinity stress. Plant. *Stress, 1*, 100006. https://doi.org/10.1016/j.stress.2021.100006

Motevalli, S., Hassani, S. B., Ghalamboran, M. R., & Rezadoost Chahardeh, H. (2021). Increase of diosgenin in fenugreek seedlings by cobalt nanoparticles. *Rhizosphere, 18*, 100335. https://doi.org/10.1016/j.rhisph.2021.100335

Munir, N., Gulzar, W., Abideen, Z., Hasanuzzaman, M., El-Keblawy, A., & Zhao, F. (2023). Plant–Nanoparticle interactions: Transcriptomic and proteomic insights. *Agronomy, 13*, 2112. https://doi.org/10.3390/agronomy13082112

Nourozi, E., Hosseini, B., Maleki, R., & Mandoulakani, B. A. (2019). Pharmaceutical important phenolic compounds overproduction and gene expression analysis in Dracocephalum kotschyi hairy roots elicited by SiO2 nanoparticles. *Industrial Crops and Products, 133*, 435–446. https://doi.org/10.1016/j.indcrop.2019.03.053

Parkinson, S. J., Tungsirisurp, S., Joshi, C., Richmond, B. L., Gifford, M. L., Sikder, A., Lynch, I., O'Reilly, R. K., & Napier, R. M. (2022). Polymer nanoparticles pass the plant interface. *Nature Communications, 13*, 7385. https://doi.org/10.1038/s41467-022-35066-y

Prakash, V., Peralta-Videa, J., Tripathi, D. K., Ma, X., & Sharma, S. (2021). Recent insights into the impact, fate and transport of cerium oxide nanoparticles in the plant-soil continuum. *Ecotoxicology and Environmental Safety, 221*, 112403. https://doi.org/10.1016/j.ecoenv.2021.112403

Przemieniecki, S. W., Damszel, M., & Kosewska, O. (2025). The impact of nanoparticles and molecular forms of TiO2 on the rhizosphere of plants in the example of common wheat (Triticum aestivum L.)—Shifts in microbiome structure and predicted microbial metabolic functions. *International Journal of Molecular Sciences, 26*, 685. https://doi.org/10.3390/ijms26020685

Rahmani, N., Radjabian, T., & Soltani, B. M. (2020). Impacts of foliar exposure to multi-walled carbon nanotubes on physiological and molecular traits of Salvia verticillata L., as a medicinal plant. *Plant Physiology and Biochemistry, 150*, 27–38. https://doi.org/10.1016/j.plaphy.2020.02.022

Rani, M., Keshu, & Shanker, U. (2021). Green nanomaterials: An overview. In *Green functionalized nanomaterials for environmental applications* (pp. 43–80). Elsevier. https://doi.org/10.1016/B978-0-12-823137-1.00026-9

Rao, B. G., Mukherjee, D., & Reddy, B. M. (2017). Nanostructures for Novel Therapy NOVEL APPROACHES FOR PREPARATION OF NANOPARTICLES. In *Nanostructures for novel therapy*.

Rizwan, M., Ali, S., Qayyum, M. F., Ok, Y. S., Adrees, M., Ibrahim, M., Zia-ur-Rehman, M., Farid, M., & Abbas, F. (2017). Effect of metal and metal oxide nanoparticles on growth and physiology of globally important food crops: A critical review. *Journal of Hazardous Materials, 322*, 2–16. https://doi.org/10.1016/j.jhazmat.2016.05.061

Saratale, R. G., Karuppusamy, I., Saratale, G. D., Pugazhendhi, A., Kumar, G., Park, Y., Ghodake, G. S., Bharagava, R. N., Banu, J. R., & Shin, H. S. (2018). A comprehensive review on green nanomaterials using biological systems: Recent perception and their future applications. *Colloids and Surfaces. B, Biointerfaces, 170*, 20–35. https://doi.org/10.1016/j.colsurfb.2018.05.045

Sen, S. K., Chouhan, D., Das, D., Ghosh, R., & Mandal, P. (2020). Improvisation of salinity stress response in mung bean through solid matrix priming with normal and nano-sized chitosan. *International Journal of Biological Macromolecules, 145*, 108–123. https://doi.org/10.1016/j.ijbiomac.2019.12.170

Silva, S., Ribeiro, T. P., Santos, C., Pinto, D. C. G. A., & Silva, A. M. S. (2020). TiO2 nanoparticles induced sugar impairments and metabolic pathway shift towards amino acid metabolism in wheat. *Journal of Hazardous Materials, 399*, 122982. https://doi.org/10.1016/j.jhazmat.2020.122982

Singh, R., Singh, D. P., Gupta, P., Jain, P., Sanchita, M. T., Kumar, A., Dhawan, S. S., & Shirke, P. A. (2019). Nanoparticles alter the withanolide biosynthesis and carbohydrate metabolism in Withania somnifera (Dunal). *Industrial Crops and Products, 127*, 94–109. https://doi.org/10.1016/j.indcrop.2018.10.049

Smirnova, E., Gusev, A., Zaytseva, O., Sheina, O., Tkachev, A., Kuznetsova, E., Lazareva, E., Onishchenko, G., Feofanov, A., & Kirpichnikov, M. (2012). Uptake and accumulation of multiwalled carbon nanotubes change the morphometric and biochemical characteristics of Onobrychis arenaria seedlings. *Frontiers of Chemical Science and Engineering, 6*, 132–138. https://doi.org/10.1007/s11705-012-1290-5

Tartaro, G., Mateos, H., Schirone, D., Angelico, R., & Palazzo, G. (2020). Microemulsion microstructure(s): A tutorial review. *Nanomaterials, 10*, 1–40. https://doi.org/10.3390/nano10091657

Trela-Makowej, A., Orzechowska, A., & Szymańska, R. (2024). Less is more: The hormetic effect of titanium dioxide nanoparticles on plants. *Science of the Total Environment, 910*, 168669. https://doi.org/10.1016/j.scitotenv.2023.168669

Tsuzuki, T. (2021). Mechanochemical synthesis of metal oxide nanoparticles. *Communications Chemistry, 4*, 143. https://doi.org/10.1038/s42004-021-00582-3

Tsuzuki, T., & McCormick, P. G. (2000). Synthesis of metal-oxide nanoparticles by mechanochemical processing. *Journal of Metastable and Nanocrystalline Materials, 8*, 383–388. https://doi.org/10.4028/www.scientific.net/JMNM.8.383

Wang, H., Zhang, M., Song, Y., Li, H., Huang, H., Shao, M., Liu, Y., & Kang, Z. (2018). Carbon dots promote the growth and photosynthesis of mung bean sprouts. *Carbon N Y, 136*, 94–102. https://doi.org/10.1016/j.carbon.2018.04.051

Wang, W., Ren, Y., He, J., Zhang, L., Wang, X., & Cui, Z. (2020). Impact of copper oxide nanoparticles on the germination, seedling growth, and physiological responses in Brassica pekinensis L. *Environmental Science and Pollution Research, 27*, 31505–31515. https://doi.org/10.1007/s11356-020-09338-3

Wang, C., Liu, X., Li, J., Yue, L., Yang, H., Zou, H., Wang, Z., & Xing, B. (2021a). Copper nanoclusters promote tomato (Solanum lycopersicum L.) yield and quality through improving photosynthesis and roots growth. *Environmental Pollution, 289*, 117912. https://doi.org/10.1016/j.envpol.2021.117912

Wang, J. W., Cunningham, F. J., Goh, N. S., Boozarpour, N. N., Pham, M., & Landry, M. P. (2021b). Nanoparticles for protein delivery in planta. *Current Opinion in Plant Biology, 60*, 102052. https://doi.org/10.1016/j.pbi.2021.102052

Wang, Z., Zhu, W., Chen, F., Yue, L., Ding, Y., Xu, H., Rasmann, S., & Xiao, Z. (2021c). Nanosilicon enhances maize resistance against oriental armyworm (Mythimna separata) by activating the biosynthesis of chemical defenses. *Science of the Total Environment, 778*, 146378. https://doi.org/10.1016/j.scitotenv.2021.146378

Wang, X., Xie, H., Wang, P., & Yin, H. (2023). Nanoparticles in plants: Uptake, transport and physiological activity in leaf and root. *Materials, 16*, 3097. https://doi.org/10.3390/ma16083097

Wu, H., & Li, Z. (2022). Nano-enabled agriculture: How do nanoparticles cross barriers in plants? *Plant Communications, 3*, 100346. https://doi.org/10.1016/j.xplc.2022.100346

Yan, S., Zhao, L., Li, H., Zhang, Q., Tan, J., Huang, M., He, S., & Li, L. (2013). Single-walled carbon nanotubes selectively influence maize root tissue development accompanied by the change in the related gene expression. *Journal of Hazardous Materials, 246–247*, 110–118. https://doi.org/10.1016/j.jhazmat.2012.12.013

Chapter 5
Plant Stress

Daniel Mendoza Jiménez, Edgardo Ulises Esquivel Naranjo, and José Antonio Cervantes Chávez

5.1 Introduction

Plants face many challenges during their life cycles owing to various environmental stressors in nature (biotic and abiotic). These stressors significantly impact plant's growth, development (flowering), and productivity (fruiting). Abiotic stresses add to factors such as extreme temperatures, drought, salinity, nutrient deficiencies, and heavy metal concentrations in soil. Biotic stresses appear from interactions with other organisms, including pathogens, pests, and competing plants.

When confronted with these stressors, plants undergo complex physiological and biochemical changes as they attempt to adapt and survive. These responses often involve alterations in gene expression, metabolism, and resource allocation. For instance, drought stress typically triggers mechanisms to conserve water, while pathogen attacks elicit defensive responses to limit infection.

The severity and time exposure of stress play crucial roles in resolving plant's outcomes. Mild, short-term stress may prompt adaptive responses that enhance resilience, while prolonged or intense stress can precedent to significant damage or plant death. Furthermore, plants often deal with multiple stressors simultaneously in natural environment, causing difficulties with their responses and potential management approaches.

Recognizing the intricate ways plants recognize, signal, and respond to various stressors is fundamental for developing strategies to enhance crop resilience and productivity in the face of changing environmental conditions. This knowledge can update breeding programs, cultural practices, and the improvement of novel

D. M. Jiménez · E. U. E. Naranjo · J. A. C. Chávez (✉)
Unit for Basic and Applied Microbiology, School of Natural Sciences, Autonomous, University of Queretaro, Santiago de Queretaro, Queretaro, Mexico
e-mail: dmendoza21@alumnos.uaq.mx; ulises.esquivel@uaq.mx; jose.antonio.cervantes@uaq.mx

K. Esquivel Escalante et al. (eds.), *Green Nanomaterials*, Nanotechnology in the Life Sciences, https://doi.org/10.1007/978-3-032-04492-1_5

technologies to moderate the negative influences of biotic and abiotic stresses on plant health and agricultural yields.

The latest advancements in molecular biology, genetics, and biotechnology have prolonged our perception of plant stress responses. This awareness is opening new ways to enhance plant stress tolerance through targeted genetic modifications, the application of beneficial microorganisms, and the development of stress-protective compounds. As global climate change continues to modify growing conditions worldwide, research into plant stress biology and mitigation strategies remains a critical area of convergence in plant science and agriculture.

5.2 Biotic and Abiotic Factors Influencing Stress in Plants

Plant stress is recognized as an abrupt alteration concerning environmental fitness that negatively affects their growth via the interference of many physio-morphological and metabolic processes. They are frequently subjected to a wide array of difficult relations concerning several environmental factors, ranging from disruptions in their metabolism and suffering physiological losses when faced with different stress factors. At the time to be exposed to biotic and abiotic stress, plants begin to experience disruptions in their metabolic processes, the principal is growth reduction, as well as more diseases and finally low productivity (Li et al., 2023; Nawaz et al., 2023).

Plants are frequently challenged against a great diversity of abiotic factors, e.g. climate, light, temperature, ultraviolet light (UV), drought, salinity, nutrient availability, and soil physicochemical properties (pH, soil moisture, aeration, cationic exchange capacity (CEC), soil organic matter, etc.). The growth reduction can go above 50% for most plant species, in the other hand, biotic stress carriages an additional challenge, as it exerts strong pressure on the plants and exacerbates the damage caused by attacks from pathogens; bacteria, viruses, fungi and nematodes or herbivores (pests) (Jha et al., 2024; Yang et al., 2018).

Through stressful environmental conditions, the plant endures countless metabolic and physiological variations such as adjustment in shoot/root biomass, reduction of photosynthesis and nutrient uptake, decrease of flowering and seed development, which eventually regulate the plant growth and productivity. The intensity of this damage is determined by the severity, frequency, and time under the stress condition (Nawaz et al., 2023).

One of the most detrimental environmental factors is the heat stress affecting plant growth and productivity. This phenomenon negatively impacts seed germination rates, reduces seedling survival, and hinders the overall germination process. During their vegetative growth phase, plants exposed to high temperatures suffer damage to their photosynthetic pigments, which are crucial for light perception. Additionally, carbon metabolism is disrupted, and the transport of organic compounds is impaired. Excessive heat also interferes with the pollen stage and seed filling, decreasing the reproductive period and grain production. Initially, plants attempt to adapt to heat by regulating certain physiological and morphological functions for a brief period. However, when heat stress continues, they activate more complex defense mechanisms, including osmotic regulation, antioxidant modulation, phytohormone stability, and signal transduction. Nevertheless, these self-defense systems may prove insufficient when temperatures reach extremely high levels. On plants under HS, exogenous salicylic acid (SA) demonstrates to maintain the cell osmotic potential, increase the cell water holding capacity, recover the cell membrane stability, moderate the osmotic and oxidative damage, and maintain healthy root vitality through increasing the content of osmotic protectants such as soluble sugars, proteins, and proline, referred as compatible solutes (Aguirre-Becerra et al., 2022; Arif et al., 2020; Feng et al., 2023).

Excess salinity in soil creates a significant risk to crop plants, causing injury at multiple levels from molecular to physiological and biochemical. This stress severely impairs their growth, reproduction, and survival fitness. The storage of sodium (Na^+) and chloride (Cl^-) ions within plant cells is a primary factor in stunting plant development. These ions, when present in high concentrations, can cause extensive cellular damage, leading to salt stress throughout the plant. As drought conditions set in, leaves respond by closing their stomata. This protective mechanism, while reducing water loss, also has the unintended consequence of inhibiting photosynthesis. The general impact includes a reduction in total leaf surface area, a decrease in water potential, and a slowdown in plant growth. These effects are accompanied by an increase in osmolites and reactive oxygen species (ROS). Two critical aspects of drought stress are its intensity and duration. The combined influence of these factors often correlates directly with decreased crop yields. The plant's ability to cope with these stressors can significantly impact its overall productivity and survival. On the other hand, heavy metals are known to disrupt redox homeostasis and promote the formation of free radicals and ROS, leading to the impairment of protein function (Arif et al., 2020; Li et al., 2023; Nawaz et al., 2023). Heavy metals have extreme and unfavorable effects on plant growth and development, impacting several stages of a plant's life cycle. When exposed to elevated concentration of heavy metals such as Pb, Cd, Hg, or As, plants often exhibit restricted growth, with reduced root and shoot length, and overall diminished biomass accumulation. At the same time, seed germination rates may be significantly lowered, and those seeds that do germinate often produce weaker seedlings. Furthermore, heavy metals exposure can alter hormonal balances, affecting various aspects of plant development, including flowering, fruit set, and seed production. The cumulative result of these impacts is often a significant reduction in plant productivity, vigor, and reproductive success, highlighting the serious threat that heavy metal's contamination poses to plant health and agricultural sustainability (El-Saadony et al., 2022; Mahto et al., 2020).

Understanding these impacts is crucial for assessing environmental risks, developing remediation strategies, and breeding plants with enhanced tolerance to heavy metal stress. It is also important for maintaining agricultural production in areas affected by heavy metals contamination.

5.3 Tolerance Mechanisms to Biotic and Abiotic Stress in Plants

Plants have elaborated rapid stress–response mechanisms to offset harmful effects and enhance their resilience. Because of their stationary nature, plants cannot simply relocate to escape unfavorable environmental conditions. Instead, they depend on a variety of specialized signaling pathways to cope with different types of stress. These pathways allow plants to detect and respond to several environmental challenges, triggering appropriate cellular responses that help them tolerate adverse conditions. The specific signaling routes activated depend on the stressor encountered. This adaptability enables plants to tailor their defensive strategies to the unique threats they face. By employing these diverse signaling networks, plants can initiate precise cellular reactions, allowing them to withstand stress stimuli and maintain their survival in challenging environments. In other words, plants respond to these stressful conditions by undergoing modifications in transcriptomic and metabolomic levels via signaling mechanisms that can perceive these environmental and biotic stimuli (Jha et al., 2024; Villavicencio-Pulido et al., 2024).

At the moment that a plant cell finds stress, second messengers, such as calcium, ROS, phospholipids and nitric oxide (NO), as well as several protein kinases, relay and amplify the signal. Consequently, they have acquired advanced signaling and protective systems to cope with stress conditions. ROS molecules primarily function as signal transduction that influences the nucleus through redox reactions and controls several pathways during plant adaptation to stress (Shiade et al., 2024).

When ROS levels rise, maintaining cellular redox balance becomes a significant challenge. This imbalance can destabilize cell membranes, damage nucleic acids, and disrupt protein structures. Consequently, photosynthetic efficiency declines, potentially leading to plant death if left unrestricted. Plants have evolved a range of defensive strategies. One key approach involves the production of various ROS-neutralizing compounds. Among these, the synthesis of anthocyanins plays a crucial role. These pigments serve as a shielding function, helping plants cope with stress conditions by mitigating the harmful effects of excessive ROS. This adaptive mechanism showcases the sophisticated ways in which plants have developed to safeguard themselves against oxidative stress. By deploying these molecular defenses, plants can improve environmental challenges and preserve their vital functions under adverse conditions. These molecules accumulated in different plant tissues and provide significant antioxidant activity to induce morpho-physiological and metabolic adaptations in plants (Jha et al., 2024; Kaur et al., 2023).

Primary and secondary metabolites (PMs, SMs) influence drastically as molecular signals in plant physiological responses to environmental stimuli. PMs are necessary molecules implicated in essential cellular systems, specifically energy production, growth, and development. So, these compounds can act as signaling molecules in plant responses to environmental stimuli; PMs can have a place as osmoprotectants, preserving cell turgor under conditions of water shortage or high levels of salinity. Additionally, SM plays a crucial role in the defense mechanisms of plants, protecting against stress, SM are specialized compounds that contribute to various plant defense mechanisms and adaptive responses, e.g.,: alkaloids, terpenoids, phenols and glucosinolates have antifungal, antimicrobial or insecticidal properties, these compounds are often synthesized and accumulated in response to pathogen attack or herbivore feeding (Nawaz et al., 2023; Shiade et al., 2024). Intensified production of phenols and flavonoids is a mutual action described in many plants under various abiotic stresses. The presence of phytochelatin in significant quantities was observed frequently in response to the presence of several heavy metals. At the same time, abscisic acid and ethylene also serve as initial signals for plant defense responses, such as stomata closing during drought stress. For example, in *Iris hexagona*, herbivory and salinity stress induce jasmonic acid (JA) and suppress salicylic acid (SA), even though methyl jasmonate (MeJA) is used to simulate plant defenses against herbivory and salinity, it indicates that *I. hexagona* responds to abiotic and biotic stress induced by hormonal pathways (Li et al., 2023; Nawaz et al., 2023).

5.4 Nanomaterials as Abiotic Stress

Nanomaterials are increasingly being explored as a novel approach to resist abiotic stress in crops, offering potential solutions to improve agricultural productivity in challenging environmental conditions. When crops are exposed to abiotic stressors such as drought, salinity, extreme temperatures, or heavy metals contamination, nanomaterials can play a crucial role in mitigating their negative impacts. NPs can also act as gene expression modulators in response to abiotic stress conditions (drought, salinity, low temperatures, UV rays, and HMs), providing plant defense by several mechanisms such as activation of the antioxidant enzyme system that alleviates oxidative stress, intervening in biosynthesis and cellular organization, as well as in electron and energy transport (Aguirre-Becerra et al., 2022). The production of ROS alters plant cellular metabolism at variable levels, including photosynthesis, carbon assimilation, biochemical cascades, and permeability of membranes.

Nevertheless, it is described that nanoparticles can have as positive and negative effects upon the plant system. At higher doses, numerous nanoparticles can cause oxidative damage to biomolecules, leading to plant cellular damage and even cell death. However, at optimal levels (lower concentrations), nanoparticles perform as crucial regulators of plant development and growth (Tariq et al., 2021).

ZnO and CuO NPs showed protection against drought stress in different plants. This protection may be persuaded to increase production of lateral root hairs, which causes appropriate water absorption. Cell wall lignification under lower concentrations of CuO may modify water flow, thus limiting cell wall elongation. The reaction of plants to drought stress is an increase in lignification. The interruption of water flow occurs due to the binding of copper ions to the pectin of the cell wall (Khalid et al., 2022).

5.5 Nanomaterials Behavior on Biotic and Abiotic Stressed Plant

In the last years, the application of nanotechnology in agronomy has become a fast-developing field and has acquired a considerable position, because of its ability to offer potential and rapid solutions for sustainable agriculture. Moreover, nanoparticles are also used in agriculture for the efficient and controlled delivery of fertilizers, herbicides, pesticides and for recovering soil physicochemical properties such as water holding capacity (Feng et al., 2023; Yadav et al., 2024).

Nanotechnology has begun as a hopeful approach to enhance plants' tolerance to abiotic stresses. There is increasing global interest in using nanomaterials to protect plant growth from various environmental challenges, including drought, salinity, heavy metal contamination, extreme temperatures, and flooding. Nanomaterials (NMs) are seen as valuable tools for boosting crop yield by improving plants' resistance to abiotic stressors, addressing both current and future agricultural production limitations in sustainable farming systems. Nanoparticles measuring around 3–5 nanometers can be immediately absorbed by plants. These microscopic particles can penetrate the epidermal cell wall of roots through small pores and subsequently enter the plant's vascular system. Interestingly, some NMs, even when slightly larger than typical pore sizes, have demonstrated the ability to create their own passages through the cell wall, facilitating their uptake. This modern application of nanotechnology in agriculture showcases its potential to revolutionize crop management strategies. Furthermore, the nanoparticles take both apoplastic and symplastic pathways to reach the target tissue (Alabdallah & Alzahrani, 2020; Singh et al., 2021; Yadav et al., 2024).

It is known that zinc plays a role as a necessary micronutrient for organisms including plants. It has many mandatory functions in the cell. Additionally, zinc is also integral in the operation of intra- and inter-cellular signaling, also regulates DNA transcription, and is thought to have innumerable pivotal roles in plants, specifically when they are exposed to stress conditions such as salinity. The use of nanotechnology in recent years demonstrated that ZnNPs are contemplated as an important metal oxide, developing multiple beneficial effects on plants. It was demonstrated that treatments based on ZnNPs improved plant morphological and physiological characteristics under normal and stress conditions, including salinity stress, in another research of Alabdallah tested bulk zinc oxide (ZnO) and ZnO-NPs application in *Abelmoschus esculentus* L. plants and observed that ZnO-NPs increased photosynthetic pigments and CAT and SOD (Superoxide Dismutase) activities, but lowered proline and total soluble sugars (Alabdallah & Alzahrani, 2020; Liu et al., 2024).

Worldwide flooding stress is contemplated as one of the most critical factors limiting plant production. Plants developed some adaptive schemes to cope with this stress like, restriction of respiration, ethylene and abscisic acid synthesis, cell wall loosening, and formation of parenchymatous cells in the root cortex. So it is known that AgNPs have the ability to inhibit the expression of ACC-CoA (Acetyl-CoA carboxylase) synthase and ACC (aminocyclopropanecarboxylate) oxidase, lowering in this way the ethylene biosynthesis rate (Aguirre-Becerra et al., 2022; Mahto et al., 2020).

Under heavy metal stress conditions, soil or foliar treatments of NPs can eliminate the adverse effects of stress, improve photosynthesis and plant development, also reduce oxidative stress. The implementation of SiNPs on maize plants under arsenic (As^{+3}) stress reduced the total chlorophyll, carotenoid, and protein content and additionally, mitigated the adverse effects of As stress on maximum quantum efficiency, photochemical quenching, and non-photochemical quenching of FS II. At the same time, TiO_2 NPs can efficiently limit Cd^{2+} toxicity by enhancing the physiological parameters and photosynthetic activity in soybean plants (El-Saadony et al., 2022; Singh et al., 2021)

Copper nanoparticles (CuNPs) are one of the most explored NMs in plants. CuNPs are recognized to promote antioxidant faculties in plants by collaborating with intracellular structures, which accelerate the formation of ROS and/or reactive nitrogen species (RNS). The outcomes indicate that foliar treatment of CuNPs develops a positive effect on the plant's salt-stress tolerance and then is a suggested approach to decrease salt-stress damage in tomato crops. When foliar application of CuNPs was established, an enhancement in components like phenols, vitamin C, flavonoids, lycopene, and glutathione in leaves and fruits was scored (Nitnavare et al., 2022; Singh et al., 2021).

Recent searches have significantly advanced our understanding of nanomaterials' effects on plants under biotic and abiotic stress. In the field of abiotic stress, in 2021, Alharby supported that zinc oxide nanoparticles enhanced drought tolerance in wheat by improving antioxidant enzyme activities and photosynthetic efficiency (Alharby et al., 2021). At the same time, Aouz (2023) informed the role of SiNPs in mitigating the adverse effects of salinity stress (SS) and heat stress (HS) on wheat crop. Their results expose that HS and SS caused a significant decrease in the growth and yield of wheat, however, when the crop is fertilized with SiNPs it offsets the negative effects of both salinity and HS and improves the growth and yield of wheat by increasing chlorophyll and carotenoid contents, and nutrient uptake ($Ca^{2+,}$ K^+, and Mg^{2+}) (Aouz et al., 2023)

Heping Shang in 2020 evaluated their copper sulfide nanoparticles (CuS NPs) because of their antifungal efficacy against the pathogen *Gibberella fujikuroi* (bakanae disease) in rice (*Oryza sativa* L.), at 50 mg L^{-1}, decreased the growth of *G. fujikuroi* to 35.7%. In the crop, the application of a foliar fertilizer decreased disease by 30% and modulated the shoot salicylic acid (SA) and jasmonic acid (JA) production to enhance the plant defense (Heping Shang, 2020). These recent studies emphasize the growing potential of nanomaterials in plant stress management, while also pointing up the need for careful optimization and consideration of potential ecological impacts.

5.6 Green Synthesized Nanomaterials Behavior on Biotic and Abiotic Stressed Plant

Nowadays, the green synthesized nanomaterials have obtained substantial attention in recent years because of their potential to mitigate mutually biotic and abiotic stresses in plants. It involves using biological resources like plant extracts, microorganisms, or biomolecules to produce nanomaterials. This approach is considered more environmentally friendly and green compared to traditional chemical synthesis methods. Recently, attention has been focused in synthesizing nanomaterials using environmentally friendly approaches, these involve biological molecules as metal reducing agents. Among the biological reducing agents, the green route by using plant extracts is useful over microorganisms because of the availability of an extensive range of secondary metabolites, such as sugars, terpenoids, polyphenols, alkaloids, enzymes, proteins, and other biomolecules (Nitnavare et al., 2022).

Nowadays, there has been a rising focus on green-synthesized nanomaterials due to their capacity to address both biotic and abiotic stresses in plants. This eco-friendly method applies biological resources such as plant extracts, microorganisms, or biomolecules to produce nanomaterials, presenting a more sustainable alternative to conventional (chemical synthesis methods). These methods use biological molecules as reduction agents for metal nanoparticles. Among the various biological reductants available, plant extracts have emerged as a particularly helpful option compared to microorganism-based approaches. This preference stems from the wide array of secondary metabolites found in plants, including sugars, terpenoids, polyphenols, alkaloids, enzymes, proteins, and other bioactive compounds. The advantage of plant extracts in nanomaterial synthesis corresponds to a significant advancement in green nanotechnology. This method not only reduces the environmental impact of nanoparticle production but also harnesses the natural reducing power of plant-derived compounds. Furthermore, the plant-mediated nanomaterial-synthesis approach is economically feasible, and the source of contaminants is highly reduced (El-Saadony et al., 2021, 2022; Noman et al., 2020)

A great diversity of plants and plant parts is used to synthesize AgNPs, demonstrating their broad antibacterial activity. To reduce silver ions to a nanometric scale, several secondary metabolites (biomolecules) are linked; these biomolecules act as reducing and capping agents, which later offer stability to the AgNPs. It has been observed that the fundamental reason for AgNPs being exposed to antibacterial agents is their capacity to produce free radicals such as hydrogen peroxide (H_2O_2), superoxide anion (O^{2-}), hydroxyl radical (OH^-), and singlet oxygen. In the same way, *Xanthomonas oryzae* pv. oryzae is a rice pathogen; metallic nanoparticles can control its growth. Novel AgNPs prepared using *Bacillus cereus* presented bactericidal activity against *X. oryzae* pv. oryzae treated with 5 to 20 μg mL^{-1} (García-Ovando et al., 2022; Nitnavare et al., 2022; Singh et al., 2021).

El-Saadony in 2021 showed high potential to produce Si-NPs using fungal biomass of *Aspergillus tubingensis.* They consider microorganisms convenient to be used as tools for metal and metal oxide NPs biosynthesis, outstanding due to their negative electrokinetic potential, which makes them able to attract cations; moreover, it is resistant to salt concentration up to 250 mg L^{-1}. Their treatments of Bio-Si-NPs and potassium silicate significantly upgraded plant growth and production, chlorophylls, carotenoids, transpiration rate, net photosynthetic rate, stomatal conductance, and membrane stability index (El-Saadony et al., 2021).

Bio-AuNPs were synthesized using the extract of the rhizome of *Alpinia galanga* L. This extract contains flavonoids, carbohydrates, terpenoids, and phenolic compounds. It is known that the mechanism of action of AuNPs is not clearly understood; it has been recounted that the increase of gold NPs causes puncture in the bacterial cells and leads to toxicity by working together with DNA replication machinery. At the same time, another report indicates that gold NPs in high concentrations are implicated in protein synthesis machinery and inhibit bacterial growth. Controlled size and morphology during the synthesis procedure play an important role in achieving antibacterial property (Aguirre-Becerra et al., 2022; Zulfiqar & Ashraf, 2021).

Titanium dioxide (TiO_2) has been described as an inactive, non-toxic and organic material, because of its high refractive index and high capability to absorb UV light, making it a promising and environmentally friendly catalyst. Literature review resulted that the TiO_2 nanoparticles were synthesized by green sources such as *Nyctanthes arbortristis*, *Catharanthus roseus*, *Eclipta* prostrate leaf extract and *Annona squamosa* L. peel (Armaković et al., 2023; Kumari et al., 2023)

Muhammad in 2020 made the biosynthesis and characterization of copper nanoparticles (CuNPs) using *Klebsiella pneumoniae*, results exposed that the CuNPs, at 25 and 50 mg kg^{-1} of soil, improved plant growth, biomass, and cellular antioxidants contents, whereas decreased the reactive oxygen species and Cr (chromium) translocation from soil to roots and shoots as compared to control plants, nevertheless, a soil adjustment of 100 mg kg^{-1} CuNPs was found to have phytotoxic effect on wheat plants and these effects were exposed by the compact growth and raised levels of ROS (Noman et al., 2020).

Green-synthesized NMs have demonstrated remarkable faculty to alleviate various abiotic stresses, such as drought, salinity, and extreme temperatures. They have also shown efficacy against biotic stressors such as fungal and bacterial pathogens. This dual-action capability makes them particularly valuable in agricultural applications. These nanomaterials operate through multiple mechanisms, including nutrient uptake, improving antioxidant and osmolyte production. Their ability to induce systemic resistance against pathogens adds to their versatility in stress management. The green synthesis approach significantly reduces the environmental impact associated with traditional nanomaterial production methods.

5.7 Conclusion

The harmful impacts of abiotic stresses like heat, salinity, and heavy metal contamination are clearly delineated, emphasizing their disruptive influence on critical physiological processes such as photosynthesis, nutrient uptake, and reproduction. Furthermore, the synergistic effects of combined stresses are hinted at, underscoring the complexity of the challenge faced by plants in diverse and often unpredictable environments.

The intricate signaling pathways involved in stress response, including the role of ROS, secondary metabolites, and hormonal regulation, showcase the sophistication of plant defenses. The efficacy of these natural defense systems is, however, often challenged by the intensity and duration of stress events. This underscores the vulnerability of agricultural systems to climate change and environmental degradation.

The exploration of nanomaterials as a potential tool for mitigating stress provides a promising chance for enhancing plant resilience. While the positive effects of certain nanomaterials on various aspects of plant growth and stress tolerance are encouraging, it is crucial to acknowledge the potential negative impacts associated with high concentrations or inappropriate applications. The shift towards green synthesis methods, using plant extracts as reducing agents, represents a significant advance in sustainability, minimizing the environmental footprint of nanomaterial production and mitigating potential toxic effects.

Acknowledgements The authors thank the Autonomous University of Queretaro-"Secretaría de Investigación, Innovación y Posgrado"- for the financial support granted through the FONFIVE-UAQ-2025 to J.A.C.CH. D.M.J. is indebt to CONAHCyT for fellowship: 823516

Competing Interests The authors declare no competing interests

References

Aguirre-Becerra, H., Feregrino-Pérez, A. A., Esquivel, K., Perez-Garcia, C. E., Vazquez-Hernandez, M. C., & Mariana-Alvarado, A. (2022). Nanomaterials as an alternative to increase plant resistance to abiotic stresses. In *Frontiers in plant science* (Vol. 13). Frontiers Media S.A. https://doi.org/10.3389/fpls.2022.1023636

Alabdallah, N. M., & Alzahrani, H. S. (2020). The potential mitigation effect of ZnO nanoparticles on [Abelmoschus esculentus L. Moench] metabolism under salt stress conditions. *Saudi Journal of Biological Sciences, 27*(11), 3132–3137. https://doi.org/10.1016/j.sjbs.2020.08.005

Alharby, H. F., Rizwan, M., Iftikhar, A., Hussaini, K. M., Zia ur Rehman, M., Bamagoos, A. A., Alharbi, B. M., Asrar, M., Yasmeen, T., & Ali, S. (2021). Effect of gibberellic acid and titanium dioxide nanoparticles on growth, antioxidant defense system and mineral nutrient uptake in wheat. *Ecotoxicology and Environmental Safety, 221*. https://doi.org/10.1016/j.ecoenv.2021.112436

Aouz, A., Khan, I., Chattha, M. B., Ahmad, S., Ali, M., Ali, I., Ali, A., Alqahtani, F. M., Hashem, M., Albishi, T. S., Qari, S. H., Chatta, M. U., & Hassan, M. U. (2023). Silicon induces heat and salinity tolerance in wheat by increasing antioxidant activities, photosynthetic activity, nutrient homeostasis, and osmo-protectant synthesis. *Plants, 12*(14). https://doi.org/10.3390/plants12142606

Arif, Y., Singh, P., Siddiqui, H., Bajguz, A., & Hayat, S. (2020). Salinity induced physiological and biochemical changes in plants: An omic approach towards salt stress tolerance. In *Plant physiology and biochemistry* (Vol. 156, pp. 64–77). Elsevier Masson SAS. https://doi.org/10.1016/j.plaphy.2020.08.042

Armaković, S. J., Savanović, M. M., & Armaković, S. (2023). Titanium dioxide as the most used photocatalyst for water purification: An overview. In *Catalysts* (Vol. 13, Issue 1). MDPI. https://doi.org/10.3390/catal13010026

El-Saadony, M. T., Desoky, E. S. M., Saad, A. M., Eid, R. S. M., Selem, E., & Elrys, A. S. (2021). Biological silicon nanoparticles improve Phaseolus vulgaris L. yield and minimize its contaminant contents on a heavy metals-contaminated saline soil. *Journal of Environmental Sciences (China), 106*, 1–14. https://doi.org/10.1016/j.jes.2021.01.012

El-Saadony, M. T., Saad, A. M., Soliman, S. M., Salem, H. M., Desoky, E. S. M., Babalghith, A. O., El-Tahan, A. M., Ibrahim, O. M., Ebrahim, A. A. M., Abd El-Mageed, T. A., Elrys, A. S., Elbadawi, A. A., El-Tarabily, K. A., & AbuQamar, S. F. (2022). Role of nanoparticles in enhancing crop tolerance to abiotic stress: A comprehensive review. In *Frontiers in Plant Science* (Vol. 13). Frontiers Media S.A. https://doi.org/10.3389/fpls.2022.946717

Feng, D., Jia, X., Yan, Z., Li, J., Gao, J., Xiao, W., Shen, X., & Sun, X. (2023). Underlying mechanisms of exogenous substances involved in alleviating plant heat stress. In *Plant Stress* (Vol. 10). Elsevier B.V. https://doi.org/10.1016/j.stress.2023.100288

García-Ovando, A. E., Ramírez Piña, J. E., Esquivel Naranjo, E. U., Cervantes Chávez, J. A., & Esquivel, K. (2022). Biosynthesized nanoparticles and implications by their use in crops: Effects over physiology, action mechanisms, plant stress responses and toxicity. *Plant Stress, 6*. https://doi.org/10.1016/j.stress.2022.100109

Jha, Y., Macwan, A. A., Ghanaim, A. M., & Mohamed, H. I. (2024). Management of abiotic and biotic stresses by microbiome-based engineering of the rhizosphere. *Biocatalysis and Agricultural Biotechnology, 61*, 103365. https://doi.org/10.1016/j.bcab.2024.103365

Kaur, S., Tiwari, V., Kumari, A., Chaudhary, E., Sharma, A., Ali, U., & Garg, M. (2023). Protective and defensive role of anthocyanins under plant abiotic and biotic stresses: An emerging application in sustainable agriculture. *Journal of Biotechnology, 361*, 12–29). Elsevier B.V. https://doi.org/10.1016/j.jbiotec.2022.11.009

Khalid, M. F., Iqbal Khan, R., Jawaid, M. Z., Shafqat, W., Hussain, S., Ahmed, T., Rizwan, M., Ercisli, S., Pop, O. L., & Alina Marc, R. (2022). Nanoparticles: The plant saviour under abiotic stresses. In *Nanomaterials* (Vol. 12, Issue 21). MDPI). https://doi.org/10.3390/nano12213915

Kumari, S., Raturi, S., Kulshrestha, S., Chauhan, K., Dhingra, S., András, K., Thu, K., Khargotra, R., & Singh, T. (2023). A comprehensive review on various techniques used for synthesizing nanoparticles. *Journal of Materials Research and Technology, 27*, 1739–1763). Elsevier Editora Ltda. https://doi.org/10.1016/j.jmrt.2023.09.291

Li, S., Yan, L., Venuste, M., Xu, F., Shi, L., White, P. J., Wang, X., & Ding, G. (2023). A critical review of plant adaptation to environmental boron stress: Uptake, utilization, and interplay with other abiotic and biotic factors. In *Chemosphere* (Vol. 338). Elsevier Ltd.. https://doi.org/10.1016/j.chemosphere.2023.139474

Liu, N., Li, C., Wu, F., Yang, Y., Yu, A., Wang, Z., Zhao, L., Zhang, X., Qu, F., Gao, L., Xia, T., & Wang, P. (2024). Genome-wide identification and expression pattern analysis of WRKY transcription factors in response to biotic and abiotic stresses in tea plants (Camellia sinensis). *Plant Physiology and Biochemistry, 211*. https://doi.org/10.1016/j.plaphy.2024.108670

Mahto, R., Rani, P., Bhardwaj, R., Singh, R. K., Prasad, S. K., & Rakshit, A. (2020). Nanotechnology: A potential approach for abiotic stress management. In *Advances in nano-fertilizers and nano-pesticides in agriculture: A smart delivery system for crop improvement* (pp. 249–264). Elsevier. https://doi.org/10.1016/B978-0-12-820092-6.00010-0

Nawaz, M., Sun, J., Shabbir, S., Khattak, W. A., Ren, G., Nie, X., Bo, Y., Javed, Q., Du, D., & Sonne, C. (2023). A review of plants strategies to resist biotic and abiotic environmental stressors. In *Science of the total environment* (Vol. 900). Elsevier B.V. https://doi.org/10.1016/j.scitotenv.2023.165832

Nitnavare, R., Bhattacharya, J., & Ghosh, S. (2022). Nanoparticles for effective management of salinity stress in plants. In *Agricultural nanobiotechnology: Biogenic nanoparticles, nanofertilizers and nanoscale biocontrol agents* (pp. 189–216). Elsevier. https://doi.org/10.1016/B978-0-323-91908-1.00013-4

Noman, M., Shahid, M., Ahmed, T., Tahir, M., Naqqash, T., Muhammad, S., Song, F., Abid, H. M. A., & Aslam, Z. (2020). Green copper nanoparticles from a native Klebsiella pneumoniae strain alleviated oxidative stress impairment of wheat plants by reducing the chromium bioavailability and increasing the growth. *Ecotoxicology and Environmental Safety, 192*. https://doi.org/10.1016/j.ecoenv.2020.110303

Shiade, S. R. G., Zand-Silakhoor, A., Fathi, A., Rahimi, R., Minkina, T., Rajput, V. D., Zulfiqar, U., & Chaudhary, T. (2024). Plant metabolites and signaling pathways in response to biotic and abiotic stresses: Exploring bio stimulant applications. In *Plant Stress* (Vol. 12). Elsevier B.V. https://doi.org/10.1016/j.stress.2024.100454

Singh, A., Tiwari, S., Pandey, J., Lata, C., & Singh, I. K. (2021). Role of nanoparticles in crop improvement and abiotic stress management. *Journal of Biotechnology, 337*, 57–70. https://doi.org/10.1016/j.jbiotec.2021.06.022

Tariq, M., Choudhary, S., Singh, H., Asif Siddiqui, M., Kumar, H., Amir, A., & Kapoor, N. (2021). Role of nanoparticles in abiotic stress. In *Technology in agriculture*. IntechOpen. https://doi.org/10.5772/intechopen.99928

Villavicencio-Pulido, J. G., Robles-Sámano, R., & Santos, B.-T. (2024). Sustainable scenarios in a plants-rhizobacteria-plant consumers system are in risk when biotic or abiotic factors change. *Mathematical Biosciences, 109261*. https://doi.org/10.1016/j.mbs.2024.109261

Yadav, N., Bora, S., Devi, B., Upadhyay, C., & Singh, P. (2024). Nanoparticle-mediated defense priming: A review of strategies for enhancing plant resilience against biotic and abiotic stresses. In *Plant physiology and biochemistry* (Vol. 213). Elsevier Masson s.r.l. https://doi.org/10.1016/j.plaphy.2024.108796

Yang, Y., Dou, Y., An, S., & Zhu, Z. (2018). Abiotic and biotic factors modulate plant biomass and root/shoot (R/S) ratios in grassland on the Loess Plateau, China. *Science of the Total Environment, 636*, 621–631. https://doi.org/10.1016/j.scitotenv.2018.04.260

Zulfiqar, F., & Ashraf, M. (2021). Nanoparticles potentially mediate salt stress tolerance in plants. *Plant Physiology and Biochemistry, 160*, 257–268. https://doi.org/10.1016/j.plaphy.2021.01.028

Chapter 6
Soil Interaction and Health

Alejandra Jimenez-Hernandez, Valeria Aldaz-Medina, Jesús Antonio Angole-Tierrablanca, Graciela Avila-Quezada, and Ramon Gerardo Guevara-Gonzalez

6.1 Nanomaterial Interaction with Soil Components

Nanotechnology will be the best option in the agricultural sector through its applications in several fields in a sustainable manner (Shahrajabian et al., 2021). For example, a lot of yield seeds and fruits will be natural throughout the year. Likewise helps in the protection of plants through the early detection of diseases and stresses. Just as nanostructured catalysts can be accessible to raise the effectiveness of pesticides and herbicides. It is necessary to comprehend the activity and impacts of nanoparticles (NPs) when being applied to crops (Rivero-Montejo et al., 2021). The surface structure, shape, chemical composition, size, concentration, solubility, and aggregation of engineered nanomaterials (NM) are fundamental qualities that ought to be considered in understanding the impact on plants. Additionally, plant species have various responses (enhancing or inhibiting) to different NPs (Fig. 6.1). The impact of engineered NM on plants can change with the phases of plant development and period of exposure (Abd El-Moneim et al., 2021). NM promotes various plant essential parameters upon absorption via roots and leaves, which are the vital supplement entry of plants (Shang et al., 2019). All progressions initiated by the NPs in plant physiological parameters, morphological traits, and gene expressions are vital to further enhance field yields to satisfy future food needs. Understanding both the positive and negative impacts of NPs is important to realize higher

A. Jimenez-Hernandez · J. A. Angole-Tierrablanca · R. G. Guevara-Gonzalez (✉)
Centro de Investigaciones Aplicadas en Biosistemas (CIAB). Facultad de Ingeniería-Campus Amazcala, Universidad Autónoma de Querétaro, Queretaro, Mexico
e-mail: alejandra.jimenez@uaq.mx; ramon.guevara@uaq.mx

V. Aldaz-Medina · G. Avila-Quezada
Universidad Autónoma de Chihuahua, Facultad de Ciencias Agrotecnológicas, Chihuahua, Chihuahua, Mexico
e-mail: gdavila@uach.mx

K. Esquivel Escalante et al. (eds.), *Green Nanomaterials*, Nanotechnology in the Life Sciences, https://doi.org/10.1007/978-3-032-04492-1_6

crop productivity. Despite what might be positively expected, hazards must be identified with long exposure of farmers to NM, various environmental interactions, and conceivable intensified bioaccumulation of NPs; all of these aspects ought to be taken into account before nanotechnology application (Abd El-Moneim et al., 2021). This chapter describes various properties of NM and their effects on crop production and highlights the importance of nanotechnology in agriculture. Furthermore, it highlights features of NPs as well as the differences between nanofertilizers (NFs) and conventional fertilizers, including the factors that affect their properties, and the impact of NPs on crop production quality and quantity.

6.1.1 Interaction with Physical Soil Components

It is well known that nanomaterials (NMs) are materials with a particle size less than 100 nm (Salah-Sheteiwy et al., 2021) as well as can be classified according to their origin as industrial and natural processes that are released to the environment (Martínez et al., 2021; Pérez-Hernández et al., 2020). Actually, the use of NMs has increased due to their physical–chemical properties as well as their wide range of applications (Suazo-Hernández et al., 2023), however, the study of the impact, effects, and consequences that NMs to the environment is essential (Martínez et al., 2021). In the case of agriculture, NMs have been used to make the application of agrochemicals more efficient in order to improve the growth rate of crops (Pérez-Hernández et al., 2020: Pratap-Singh et al., 2020). The release of these NMs, especially in the soil, has attracted attention due to its characteristics as well as its matrix that is considered abundant in natural NMs, which exist as primary particles and as aggregates (Rajput et al., 2019).

The soil is a potential system to accumulate pollution and drive it into the food through the production chains. Therefore, the quality and health of soil are very important, which are determined by its physical and chemical properties (Rajput et al., 2019; Suazo-Hernández et al., 2023). These properties can be influenced by the presence and accumulation of NMs, which can be absorbed onto soil particles through electrostatic interactions, affecting their mobility in the soil matrix (Zaib et al., 2023). These interactions make agglomerates that can fill the empty spaces of the soil matrix and therefore impact its physical properties, such as bulk density, texture, permeability, plasticity, total porosity, and particle density that could be altered definitely under the long-term conditions (Salah-Sheteiwy et al., 2021). For instance, NMs can promote aggregate stability and enhance soil structure, reducing erosion as well as modifying water-holding, reducing water runoff (Zaib et al., 2023). Some studies have shown the effect of some NMs on soil physical properties as the hydraulic conductivity and soil porosity (Table 6.1).

These effects of NMs on soil properties make a more rigid matrix, favoring agricultural productivity and making the produce processes safer to the environment and healthier for life (Suazo-Hernández et al., 2023).

Table 6.1 Effects of NMs on physical properties of soils. (Constructed by A.J-H, J.A.A-T, and R. G.G-G)

NPs	Effect on physical properties	References
Nano-ZnO and nano-SiO_2	Reduce bulk density and soil compaction. Increased soil hydraulic conductivity and organic matter content	Saad-Kheir et al. (2019)
Nano-Pt	Decreased the evaporation enthalpy of water in soil organic matter and supported soil desiccation. Additionally, improved the soil structural rigidity	Komendová et al. (2019)
Nano-MgO and nano-Fe_3O_4	Nano-MgO increased the total porosity, enhancing the stability of the aggregates Nano-Fe_3O_4 increased the soil bulk density	Bayat et al. (2019)

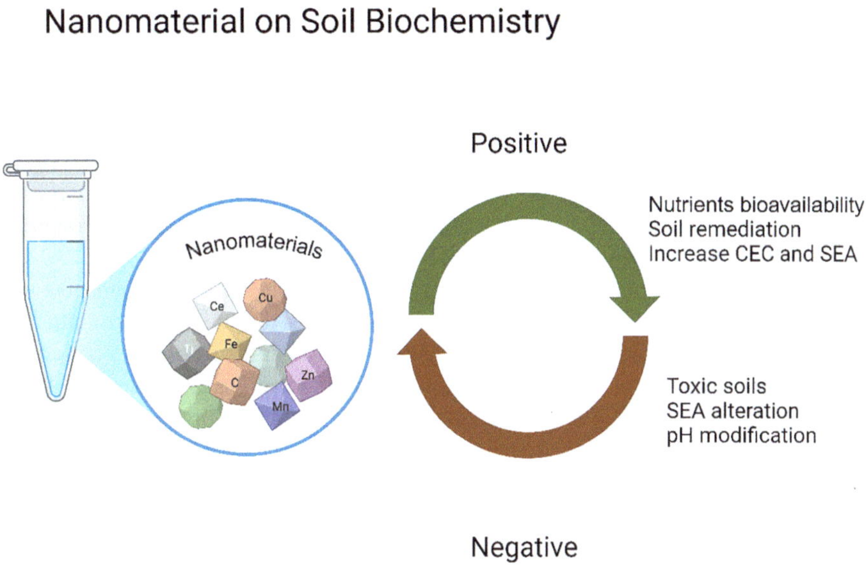

Fig. 6.1 Positive and negative biochemical interactions of NMs in soils. (Constructed by J.A.A-T)

6.1.2 Interaction with Biochemical Soil Components

The application of NMs in agriculture has generated significant interest due to their potential benefits and associated risks. Concerns primarily focus on their accumulation in soils, which can lead to toxicity for flora and fauna, alterations in soil pH, impacts on microbial communities and soil enzymatic activity (SEA), and changes in cation exchange capacity (CEC). However, extensive research has revealed strategies to harness NMs for positive outcomes in agricultural systems. These approaches improve nutrient bioavailability, facilitate the remediation of soils contaminated with heavy metals, enhance CEC, and increase SEA (Fig. 6.1).

6.1.2.1 Positive Effects

6.1.2.1.1 Nutrients Bioavailability

Nanotechnology is modernizing food processing production chains due to its significant impact on the agricultural sector. Several studies have reported that using NMs improves the solubility and bioavailability of nutrients, directly enhancing resource efficiency (Arshad et al., 2021). Nanomaterials are introduced into soils through natural processes or human interventions to improve soil properties transformed primarily via chemical reactions such as redox processes, sulfidation, phosphorylation, or chelation (Zhang et al., 2020).

Considerable efforts have been made to develop diverse applications for NMs, including nanopesticides, nano-carriers for agrochemicals, and nanofertilizers, all designed for ease of application (Li et al., 2021). For example, metallic oxide nanoparticles (MONPs) have been shown to enhance nutrient uptake, such as zinc (Zn) and copper (Cu), in potato plants. These nanoparticles also activate the antioxidant defense system without inducing toxicity (Peng et al., 2020; Ma et al., 2018). Another instance involves the use of nano-hydroxyapatite as a phosphorus carrier in soil amendments, resulting in up to a 30% increase in soybean yield and promoting early maize development (Liu & Lal, 2014; Yoon et al., 2020).

Organic NMs, such as graphene oxide, have also been studied for their potential in generating new materials. When graphene undergoes oxidation to form aromatic hydrocarbons, it can be reduced back to graphene through the respiratory activity of soil bacteria (Salas et al., 2010). The biotransformation and assimilation of NMs in soil appear inevitable due to the presence of a wide range of organisms, including plants, fungi, protozoa, and bacteria, that facilitate chemical reactions, thereby enabling the integration of these particles into the soil ecosystem (Zhang et al., 2020).

6.1.2.1.2 Soil Remediation

One of the main causes of soil contamination is the presence of heavy metals, which have been shown to negatively affect microbial diversity and biomass. In soils contaminated with heavy metals, microorganisms require more energy to survive, reducing their metabolic efficiency under such stress conditions (Giller et al., 2009). NMs have been reported to enhance microbial growth even in the presence of heavy metal stress, making them an excellent option for soil bioremediation (Rajput et al., 2023).

For instance, the bioremediation potential of certain microorganisms, such as *Halomonas* sp., *Staphylococcus aureus*, *Bacillus cereus*, and *Lysinibacillus macrolides* coated with metal oxide nanoparticles (MONPs) like FeO or ZnO has been demonstrated. These microorganisms can effectively clean contaminated areas by removing up to 100% of heavy metals like Pb, Cd, Cu, and Ni within 48 h (Mahmoud et al., 2016; Alabresm et al., 2018; Baragaño et al., 2020a, b).

Organic carbon also plays a crucial role in maintaining soil health and structure (Jatav et al., 2020). This is largely influenced by various biochemical processes mediated by microbial activities in association with plants and animals, often serving as indicators of soil health (Rajput et al., 2023). This becomes particularly relevant when considering studies involving root-associated microorganisms that use NMs, such as graphene oxide, to stimulate plant growth and activate nutrient cycles, thereby promoting soil health (Alamri et al., 2022).

6.1.2.1.3 Cation Exchange Capacity

Cation exchange capacity depends on soil charge, surface area, and its specific characteristics. It plays a crucial role in nutrient adsorption and their uptake by plants, serving as an indicator of soil fertility in agricultural systems (Suazo-Hernández et al., 2023). Studies have shown that metal oxide nanoparticles (MONPs) introduced into soils can elicit varying responses. For instance, Fe_3O_4 nanoparticles have been reported to increase CEC by up to 17% in clay soils (Baragaño et al., 2020a, b). Similarly, Ag-based MONPs significantly enhanced both CEC and soil porosity (Das et al., 2018).

6.1.2.1.4 Soil Enzymatic Activity Increase

The presence of NMs in soil directly affects soil organisms, plants, and microorganisms involved in soil improvement processes. NMs can also influence biogeochemical processes, such as soil enzymatic activity (SEA) and microbial biomass, which are critical for mediating nutrient cycles (Asadishad et al., 2018). In certain cases, metal and metal oxide nanoparticles (MNOPs) like TiO_2 have been shown to induce eustress responses in soil microorganisms at low concentrations, thereby enhancing enzymatic SEA. However, as the concentration of these MNOPs increases, enzymatic activity decreases (Alharby et al., 2021).

6.1.2.2 Negative Effects

6.1.2.2.1 Toxic Soils

The use of NMs can induce cellular toxicity due to the excessive production of reactive oxygen species (ROS) when present in high concentrations. This phenomenon, known as NM-induced stress, primarily occurs with nanomaterials dissolved in water or present in soil environments (Nel et al., 2006). However, it is noteworthy that not all NMs immediately generate ROS. Nonetheless, they can enter cells and cause damage, affecting mitochondria in animals or thylakoids in plants (Sharifi et al., 2012). ROS are indicators of cellular stress, as they can induce lipid peroxidation, protein alteration, DNA damage, and interference with vital signaling pathways (Oberdörster et al., 2005).

With the increasing application of nanotechnology in sectors such as agriculture, concerns are rising regarding the potential health implications and future impacts of NMs on agricultural soils. These materials often accumulate in the soil, which can adversely affect crops (Wang et al., 2024). Although NMs show promise when combined with microorganisms for heavy metal fixation in plants and contaminated soil remediation, their presence in high concentrations can negatively affect soil health and consequently inhibit the growth of plants and microorganisms, potentially disrupting bioremediation processes (Chen et al., 2017). For instance, nano-hydroxyapatite, at high concentrations, can increase soil pH (Xing et al., 2016), while nano-ZnO and FeO have been shown to inhibit biomass production in *Arabidopsis thaliana* (Wang et al., 2015; Huang et al., 2018). Similarly, improperly applied nanomaterials like CeO_2 can impair productivity processes, such as wheat cultivation, by directly damaging leaves and affecting grain quality (Rico et al., 2015). These insights contribute to understanding non-toxic doses that enable the efficient use of NMs in agricultural practices for food production and soil remediation in heavy metal-contaminated environments.

6.1.2.2.2 Soil pH Modification

Given that soils are the primary recipients of NMs, understanding their dynamics and interactions resulting from accumulation is essential (Conway & Keller, 2016). The adhesion of NMs can occur through various mechanisms, including soil porosity or microbial interactions. Certain NMs, such as ZnO or CuO, can release protons (H^+) and hydroxyl ions (OH^-) due to their metal oxide surfaces. These materials typically exhibit acidic characteristics, which can alter the pH of both water and soil environments (Boehm, 1971).

At high concentrations, these NMs can jeopardize the development of soil organisms by inducing toxicity and modifying resource availability. For example, they can covalently adsorb phosphate ions, reducing the availability of this essential nutrient for soil microbiota and plant development (Daou et al., 2007). Consequently, strategies are being developed to apply NMs without compromising soil health. Organic coatings on non-volatile NMs, such as TiO_2, CuO, and CeO_2, have shown promise in providing bio-stimulatory properties to plants without significantly affecting soil pH (Conway & Keller, 2016).

6.2 Effects of the NM on Soil Microbial Communities

6.2.1 *Soil Microbial Composition and Potential Changes in Response to NMs*

Soil microbial diversity is rich, as evidenced by metagenomic studies conducted on the rhizosphere of several crops, including pecan trees, apple trees, chili peppers, and grapevines. These studies have reported the presence of a wide array of fungal and bacterial phyla (Muñoz-Ramírez et al., 2024; González-Escobedo et al., 2022; Muñoz-Castellanos et al., 2024; Montes-Montes et al., 2024). Recent research indicates that the application of NMs to soil-whether for pathogen control or fertilization-could impact the diversity and activity of microbial communities. To understand how NMs may influence specific microbial taxa, it is important to investigate their effects at the level of cell wall or membrane components. The extent of these effects will depend on the type of nanomaterial used as well as the specific soil conditions.

Nanomaterials are a promising solution for controlling phytopathogens as well as environmental and human pathogens. Their in-situ application allows for large-scale, rapid remediation of contaminated sites, effectively reducing the need for extensive use of synthetic chemicals. As a result, contaminant concentrations can be decreased to near-zero levels in a short period.

6.2.2 *Abundance*

Metal nanoparticles have been shown to reduce the abundance of nitrogen-fixing bacteria. In a study by Lu et al. (2023), metal oxide nanoparticles, including TiO_2NPs, Al_2O_3NPs, and ZnONPs, inhibited the growth and colonization of *Pseudomonas stutzeri* in the rhizosphere. This effect was attributed to the induction of ROS in the bacteria, which increased membrane permeability. Similarly, in the study by Chai et al. (2015), ZnO and CeO_2 nanoparticles suppressed thermogenic metabolism and reduced the abundance of soil bacteria such as *Azotobacter,* as well as P- and K-solubilizing bacteria, while also inhibiting enzymatic activities. The authors also found that TiO_2 nanoparticles led to a decline in the population of functional bacteria and reduced enzymatic activity. In the research by Chavan and Nadanathangam (2019), AgNPs significantly altered the relative abundances of dominant agricultural soil phyla, such as Proteobacteria, Actinobacteria and Firmicutes, by up to 45%. However, the same study indicated that ZnONPs caused only minimal changes to these three phyla.

Liu et al. (2021) reported that low doses of CuONPs and ZnONPs released ions that stimulated the abundance and diversity of bacterial and fungal communities in soil. However, they also found that higher doses of CuNPs and ZnNPs, along with environmental factors, diminished the positive effects on bacteria. Therefore, it appears that low doses of these NP enhance both fungal and bacterial abundance and diversity.

However, it is important to note that not all NPs affect microorganisms in the same way. For instance, multi-walled carbon nanotubes had minimal effect on soil bacterial and fungal diversity. In the study by Chen et al. (2022), fungi were less sensitive to NPs than bacteria, suggesting that these nanomaterials may pose less risk to soil ecosystems, making them a potential option for use in agriculture.

Among the NPs with agricultural applications, such as those used to improve water retention and act as nutrient carriers, the following are considered to pose the least risk to non-target soil microorganisms:

Carbon nanoparticles including nanotubes and graphene, particularly those functionalized with biocompatible groups, exhibit antimicrobial properties. The effectiveness of carbon nanotubes (CNTs) depends on several factors, such as their size, aggregation degree, concentration, surface functionalization, exposure time (Maksimova, 2019), and bioaccumulation potential (Patil & Lekhak, 2020). Interestingly, there is also evidence suggesting that carbon nanomaterials may be degraded by microorganisms (Maksimova, 2019).

Gold nanoparticles, which are used as sensors and in controlled-release nutrient systems, are generally considered less toxic due to gold's inertness and biocompatibility. However, similar to other metal NPs, the smaller they are, the stronger their antibacterial effect (Abdussalam-Mohammed, 2020).

-*Calcium nanoparticles* (CaO, $CaCO_3$) are naturally occurring materials that have been shown to enhance microbial activity (Azeez et al., 2021). Jha et al. (2018) demonstrated that calcium oxide nanoparticles, when used at low concentration (less than 1500 μg/ml), can be applied as nanobiofertilizers mixed with rhizobacteria. At these concentrations, the NPs do not inhibit bacterial growth.

Biopolymer nanoparticles, such as polylactic acid (PLA) or polyhydroxyalkanoate (PHA) derivatives, are biodegradable and commonly used as carriers for the controlled release of fertilizers or pesticides (Wypij et al., 2023). Additionally, biopolymers like chitosan, gelatin, and alginate are used for the biological synthesis of nanoparticles, offering an eco-friendly alternative to toxic products (Rai et al., 2021).

6.2.3 Activity

Bacteria from the phyla Proteobacteria, Actinobacteria, and Firmicutes, as studied by Chavan and Nadanathangam (2019), play important roles in soil functions such as nitrogen fixation, phosphate solubilization, and biofilm formation. When the population of these bacteria is reduced by the effect of AgNPs, the soil's decomposition cycles are disrupted. Lu et al. (2023) found that TiO_2NPs, Al_2O_3NPs, and ZnONPs inhibited *Pseudomonas stutzeri* in soil, resulting in a reduction in nitrogen fixation, nitrogenase activity, and biofilm formation in the rhizosphere.

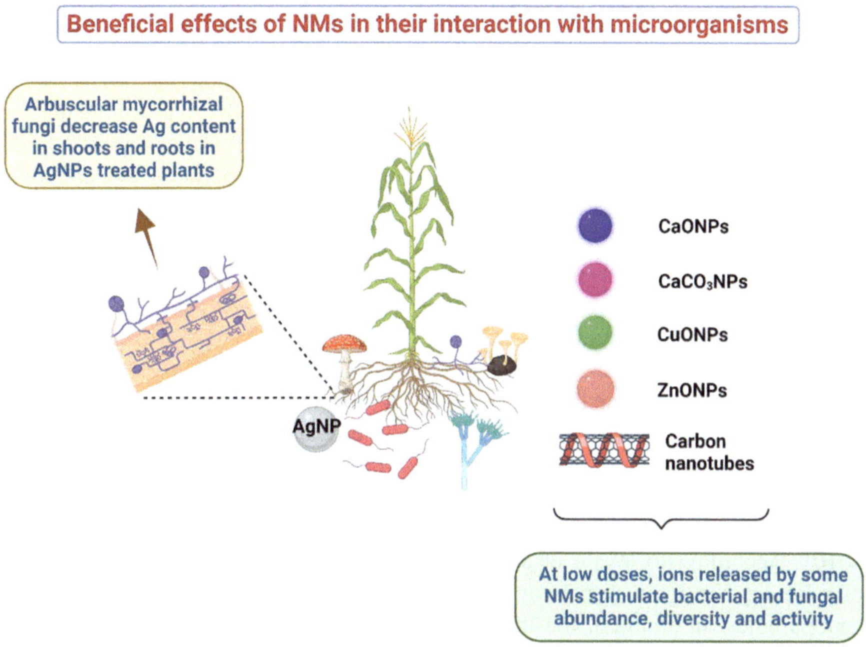

Fig. 6.2 Some beneficial effects of nanomaterials in their interaction with microorganisms. (Constructed by V.A-M and G.A-Q)

On the other hand, some studies highlight the positive effects of certain NMs on microbial communities and their activity. For instance, Chen et al. (2022) reported that multi-walled carbon nanotubes significantly enhanced rhizosphere microbial activity, while Chai et al. (2015) observed a slight increase in soil microbial activity with the application of SiO_2 nanoparticles.

The benefits of mycorrhizal fungi in soil and plants are well documented, particularly in enhancing phosphorus absorption, providing root protection against phytopathogens, and mitigating water stress (Madrid-Delgado et al., 2021; Lara-Capistrán et al., 2022; Avila-Quezada et al., 2022; Olivas-Tarango et al., 2024; Tarango-Rivero et al., 2022; Wu et al., 2023; Sáenz-Hidalgo et al., 2023).

In an experiment by Cao et al. (2020), arbuscular mycorrhizal (AM) fungi reduced the accumulation of Ag in the shoots and roots of maize plants exposed to AgNPs in soil. Additionally, the expression of genes associated with Ag transport and metal homeostasis was down-regulated. The study highlighted that AM fungal inoculation helped mitigate the toxicity of AgNPs by decreasing their impact on soil microbial activity and bacterial abundance. These findings suggest that the symbiosis relationship between AM fungi and plants plays an important role in reducing silver concentrations in plants treated with AgNPs (Saenz-Hidalgo et al., 2025) (Fig. 6.2).

6.2.4 *How Do NMs Affect Cells?*

NMs have a direct impact on cells due to their properties, including surface area, high reactivity, and small size. As a result, they can affect both phytopathogenic and beneficial microorganisms by damaging DNA, proteins, and microbial lipids (Adeyemi et al., 2020). This cell damage occurs when nanoparticles induce oxidative stress through the generation of reactive oxygen species (ROS) (Avila-Quezada et al., 2023). Specifically, metal NPs such as silver, copper, and oxide zinc release ions that contribute to this oxidative stress (Godoy-Gallardo et al., 2021).

The antimicrobial activity of AgNPs and other metal NPs is mainly influenced by factors such as their concentration, structure, surface charge, exposure time, and size (Maksimova, 2019). Nanoparticles in the range of 1 to 10 nm present higher bactericidal activity compared to those of 100 nm, due to their large surface area, which enhances their ability to interact with microbial cell walls, membranes, and DNA (Bolívar-Rodríguez et al., 2021). The most likely mechanism of NPs toxicity involves binding to single-stranded DNA during replication, leading to their incorporation into the DNA double helix and disrupting cellular function (An et al., 2010). According to Wang et al. (2016), the oxidation state, ligands, solubility, morphology, and the surrounding environment influence the toxicity of metal nanoparticles. In addition, the impact of NMs applied to soil can be affected by pH, ionic strength, moisture, and organic matter content (Gao et al., 2019). Sengul and Asmatulu (2020) suggest that nanoparticles, whether metallic or not, may accumulate in human organs such as the heart, liver, spleen, kidneys, and brain following inhalation, ingestion, or skin contact. This potential risk is supported by studies from Ahamed et al. (2010), Fu et al. (2014), Manke et al. (2013), Gatoo et al. (2014), Shvedova et al. (2012), and Gliga et al. (2014), who conclude that both the duration of nanoparticle exposure and their size are crucial in determining the damage to these organs.

6.3 Conclusion and Final Remarks

The ability of soils to sustain crop growth and yield is declining due to intensive cropping and pollution. From the perspective based on the previous sections in this chapter, it is clear that NMs might influence either positively or negatively the soil fertility and health based on the dose of the NMs, as well as the physico-chemical nature. Recent reports suggest that the management of NMs in agricultural applications as for instance, in soil as fertilizer, needs to include the phenomenon of hormesis in the study (Godínez-Mendoza et al., 2023). Both plants and microorganisms have been influenced by NMs in soils (Godinez-mendoza et al., 2023; Ferrusquía-Jiménez et al., 2022). Finally, all the potential benefits for soils when using NMs as soil conditioners, nutrients to promote plant growth and development, as soil remediation treatments, etc., urgently need serious studies regarding their impact on the environment when present in soils added for anthropocentric purposes.

Acknowledgements A.J-H, V.A-M, and J.A.A-T acknowledge to CONAHCyT (Mexico), for postgraduate studies grant during the preparation of this chapter.

Conflict of Interest The authors have no conflict of interest to declare that are relevant to the content of this chapter.

References

Abdussalam-Mohammed, W. (2020). Comparison of chemical and biological properties of metal nanoparticles (Au, Ag), with metal oxide nanoparticles (ZnO-NPs) and their applications. *Advanced Journal of Chemistry. Section A, 3*(2), 111–236.

Adeyemi, O. S., Shittu, E. O., Akpor, O. B., Rotimi, D., & Batiha, G. E. S. (2020). Silver nanoparticles restrict microbial growth by promoting oxidative stress and DNA damage. *EXCLI Journal, 19*, 492.

Ahamed, M., Siddiqui, M., Akhtar, M., Ahmad, I., Pant, A., & Alhadlaq, H. (2010). Genotoxic potential of copper oxide nanoparticles in human lung epithelial cells. *Biochemical and Biophysical Research Communications, 396*(2), 578–583. https://doi.org/10.1016/j.bbrc.2010.04.156

Alabresm, A., Chen, Y. P., Decho, A. W., & Lead, J. R. (2018). A novel method for the synergistic remediation of oil-water mixtures using nanoparticles and oil-degrading bacteria. *Science of the Total Environment, 630*, 1292–1297. https://doi.org/10.1016/j.scitotenv.2018.02.277

Alamri, S., Nafady, N. A., El-Sagheer, A. M., El-Aal, M. A., Mostafa, Y. S., Hashem, M., & Hassan, E. A. (2022). Current utility of arbuscular mycorrhizal fungi and hydroxyapatite nanoparticles in suppression of tomato root-knot nematode. *Agronomy, 12*, 671. https://doi.org/10.3390/agronomy12030671

Alharby, H. F., Rizwan, M., Iftikhar, A., Hussaini, K. M., ur Rehman, M. Z., Bamagoos, A. A., et al. (2021). Effect of gibberellic acid and titanium dioxide nanoparticles on growth, antioxidant defense system and mineral nutrient uptake in wheat. *Ecotoxicology and Environmental Safety, 221*, 112436. https://doi.org/10.1016/j.ecoenv.2021.112436

An, H., Liu, Q., Ji, Q., & Jin, B. (2010). DNA binding and aggregation by carbon nanoparticles. *Biochemical and Biophysical Research Communications, 393*, 571–576.

Arshad, R., Gulshad, L., Haq, I. U., Farooq, M. A., Al-Farga, A., Siddique, R., & Karrar, E. (2021). Nanotechnology: A novel tool to enhance the bioavailability of micronutrients. *Food Science & Nutrition, 9*(6), 3354–3361. https://doi.org/10.1002/fsn3.2311

Asadishad, B., Chahal, S., Akbari, A., Cianciarelli, V., Azodi, M., Ghoshal, S., & Tufenkji, N. (2018). Amendment of agricultural soil with metal nanoparticles: Effects on soil enzyme activity and microbial community composition. *Environmental Science & Technology, 52*(4), 1908–1918. https://doi.org/10.1021/acs.est.7b05389

Avila-Quezada, G. D., Ingle, A. P., Golińska, P., & Rai, M. (2022). Strategic applications of nano-fertilizers for sustainable agriculture: Benefits and bottlenecks. *Nanotechnology Reviews, 11*(1), 2123–2140.

Avila-Quezada, G., Rai, M., Orduño-Cruz, N., Rivas-Valencia, P., Golińska, P., Muñoz-Castellanos, L., et al. (2023). Nanophytopathology a new and emerging science. In *Nanophytopathology* (pp. 3–22). CRC Press.

Azeez, L., Lateef, A., Adetoro, R. O., & Adeleke, A. E. (2021). Responses of *Moringa oleifera* to alteration in soil properties induced by calcium nanoparticles (CaNPs) on mineral absorption,

physiological indices and photosynthetic indicators. *Beni-Suef University Journal of Basic and Applied Sciences, 10*, 1–15.
Baragaño, D., Forján, R., Welte, L., & Gallego, J. L. R. (2020a). Nanoremediation of As and metals polluted soils by means of graphene oxide nanoparticles. *Scientific Reports, 10*, 1896. https://doi.org/10.1038/s41598-020-58852-4
Baragaño, D., Gallego, J. L. R., Baleriola, G., & Forján, R. (2020b). Effects of different in situ remediation strategies for an As-polluted soil on human health risk, soil properties, and vegetation. *Agronomy, 6*, 759–776. https://doi.org/10.3390/agronomy10060759
Bayat, H., Kolahchi, Z., Valaey, S., Rastgou, M., & Mahdavi, S. (2019). Iron and magnesium nano-oxide on some physical and mechanical properties of a loamy hypocalcic cambisol. *Geoderma, 335*, 57–68. https://doi.org/10.1016/j.geoderma.2018.08.007
Boehm, H. P. (1971). Acidic and basic properties of hydroxylated metal oxide surfaces. *Discussions of the Faraday Society, 52*, 264–275.
Bolívar-Rodríguez, K. L., Agudelo-Amaya, F., & Vásquez-Ochoa, Y. (2021). Evaluación del efecto de las nanopartículas de plata sobre microorganismos patógenos. *Ingeciencia, 6*, 91–111.
Cao, J., Feng, Y., Lin, X., & Wang, J. (2020). A beneficial role of arbuscular mycorrhizal fungi in influencing the effects of silver nanoparticles on plant-microbe systems in a soil matrix. *Environmental Science and Pollution Research, 27*, 11782–11796.
Chai, H., Yao, J., Sun, J., et al. (2015). The effect of metal oxide nanoparticles on functional bacteria and metabolic profiles in agricultural soil. *Bulletin of Environmental Contamination and Toxicology, 94*, 490–495. https://doi.org/10.1007/s00128-015-1485-9
Chavan, S., & Nadanathangam, V. (2019). Effects of nanoparticles on plant growth-promoting bacteria in Indian agricultural soil. *Agronomy, 9*, 140. https://doi.org/10.3390/agronomy9030140
Chen, M., Qin, X., & Zeng, G. (2017). Biodegradation of carbon nanotubes, graphene, and their derivatives. *Trends in Biotechnology, 35*, 836–846. https://doi.org/10.1016/j.tibtech.2016.12.001
Chen, X., Wang, J., You, Y., Wang, R., Chu, S., Chi, Y., et al. (2022). When nanoparticle and microbes meet: The effect of multi-walled carbon nanotubes on microbial community and nutrient cycling in hyperaccumulator system. *Journal of Hazardous Materials, 423*, 126947.
Conway, J. R., & Keller, A. A. (2016). Gravity-driven transport of three engineered nanomaterials in unsaturated soils and their effects on soil pH and nutrient release. *Water Research, 98*, 250–260. https://doi.org/10.1016/j.watres.2016.04.021
Daou, T. J., Begin-Colin, S., Greneche, J. M., Thomas, F., Derory, A., Bernhardt, P., & Pourroy, G. (2007). Phosphate adsorption properties of magnetite-based nanoparticles. *Chemistry of Materials, 19*(18), 4494–4505. https://doi.org/10.1021/cm071046v
Das, P., Barua, S., Sarkar, S., Karak, N., Bhattacharyya, P., Raza, N., et al. (2018). Plant extract-mediated green silver nanoparticles: Efficacy as soil conditioner and plant growth promoter. *Journal of Hazardous Materials, 346*, 62–72. https://doi.org/10.1016/j.jhazmat.2017.12.020
D.A, El-Moneim., Dawood, M.F.A., Moursi, Y.S. et al. Positive and negative effects of nanoparticles on agricultural crops. Nanotechnol. Environ. Eng. 6, 21 (2021). https://doi.org/10.1007/s41204-021-00117-0
Ferrusquía-Jiménez, N. I., González-Arias, B., Rosales, A., Esquivel, K., Escamilla-Silva, E. M., Ortega-Torres, A. E., & Guevara-González, R. G. (2022). Elicitation of Bacillus cereus-Amazcala (B.c-A) with SiO2 nanoparticles improves its role as a plant growth promoting bacteria (PGPB) in chili pepper plants. *Plants, 11*(24), 3445. https://doi.org/10.3390/plants11243445
Fu, P., Xia, Q., Hwang, H., Ray, P., & Yu, H. (2014). Mechanisms of nanotoxicity: Generation of reactive oxygen species. *Journal of Food and Drug Analysis, 22*(1), 64–75. https://doi.org/10.1016/j.jfda.2014.01.005
Gao, X., Rodrigues, S. M., Spielman-Sun, E., Lopes, S., Rodrigues, S., Zhang, Y., et al. (2019). Effect of soil organic matter, soil pH, and moisture content on solubility and dissolution rate of CuONPs in soil. *Environmental Science and Technology, 53*(9), 4959–4967.

Gatoo, M. A., Naseem, S., Arfat, M. Y., Mahmood Dar, A., Qasim, K., & Zubair, S. (2014). Physicochemical properties of nanomaterials: implication in associated toxic manifestations. *BioMed Research International, 2014*, 498420. https://doi.org/10.1155/2014/498420

Giller, K. E., Witter, E., & McGrath, S. P. (2009). Heavy metals and soil microbes. *Soil Biology and Biochemistry, 41*(10), 2031–2037. https://doi.org/10.1016/j.soilbio.2009.04.026

Gliga, A., Skoglund, S., Wallinder, I., Fadeel, B., & Karlsson, H. (2014). Size-dependent cytotoxicity of silver nanoparticles in human lung cells: The role of cellular uptake, agglomeration and Ag release. *Particle and Fibre Toxicology, 11*(1), 11. https://doi.org/10.1186/1743-8977-11-11

Godínez-Mendoza, P. L., Rico-Chávez, A. K., Ferrusquía-Jimenez, N. I., Carbajal-Valenzuela, I. A., Villagómez-Aranda, A. L., Torres-Pacheco, I., & Guevara-González, R. G. (2023). Science of the Total environment plant hormesis: Revising of the concepts of biostimulation, elicitation and their application in a sustainable agricultural production. *Science of the Total Environment, 894*, 164883.

Godoy-Gallardo, M., Eckhard, U., Delgado, L. M., de Roo Puente, Y. J., Hoyos-Nogués, M., Gil, F. J., & Perez, R. A. (2021). Antibacterial approaches in tissue engineering using metal ions and nanoparticles: From mechanisms to applications. *Bioactive Materials, 6*(12), 4470–4490.

González-Escobedo, R., Muñoz-Castellanos, L. N., Muñoz-Ramirez, Z. Y., Guigón-López, C., & Avila-Quezada, G. D. (2022). Rhizosphere bacterial and fungal communities of healthy and wilted pepper (*Capsicum annuum* L.) in an organic farming system. *Ciência Rural, 53*(7), 20220072.

Huang, D., Qin, X., Peng, Z., Liu, Y., Gong, X., Zeng, G., et al. (2018). Nanoscale zero-valent iron-assisted phytoremediation of Pb in sediment: Impacts on metal accumulation and antioxidative system of *Lolium perenne*. *Ecotoxicology and Environmental Safety, 153*, 229–237. https://doi.org/10.1016/j.ecoenv.2018.01.060

Jatav, H. S., Sharma, L. D., Sadhukhan, R., Singh, S. K., Singh, S., Rajput, V. D., et al. (2020). An overview of micronutrients: Prospects and implication in crop production. In T. Aftab & K. R. Hakeem (Eds.), *Plant micronutrients: Deficiency and toxicity management* (pp. 1–30). Springer. https://doi.org/10.1007/978-3-030-49856-6_1

Jha, S., Singh, R., Pandey, A., Bhardwaj, M., Tripathi, S. K., Mishra, R. K., & Dikshit, A. (2018). Bacterial toxicological assay of calcium oxide nanoparticles against some plant growth-promoting rhizobacteria. *Int J Res Appl Sci Eng Technol, 6*(11), 460–466.

Komendová, R., Zidek, J., Berka, M., Jemelková, M., Rezácova, V., Conte, P., et al. (2019). Small-sized platinum nanoparticles in soil organic matter: Influence on water holding capacity, evaporation and structural rigidity. *Science of the Total Environment., 694*, 133822. https://doi.org/10.1016/j.scitotenv.2019.133822

Lara-Capistrán, L., Ledea-Rodríguez, J. L., Zulueta-Rodríguez, R., Ávila-Quezada, G. D., Gómez-Meriño, F. C., & Hernandez-Montiel, L. G. (2022). Effect of plastic padding and vermicompost on the efficiency of arbuscular mycorrhizal fungi in *Cucurbita pepo* L. *Tropical and Subtropical Agroecosystems, 25*, 022.

Li, Q., Ma, C., White, J. C., & Xing, B. (2021). Effects of phosphorus-ensembled nanomaterials on nutrient uptake and distribution in *Glycine max* L. under simulated precipitation. *Agronomy, 11*(6), 1086. https://doi.org/10.3390/agronomy11061086

Liu, R., & Lal, R. (2014). Synthetic apatite nanoparticles as a phosphorus fertilizer for soybean (*Glycine max*). *Scientific Reports, 4*, 5686. https://doi.org/10.1038/srep05686

Liu, Y., Li, Y., Pan, B., Zhang, X., Zhang, H., Steinberg, C. E., et al. (2021). Application of low dosage of copper oxide and zinc oxide nanoparticles boosts bacterial and fungal communities in soil. *Science of the Total Environment, 757*, 143807.

Lu, C., Hei, R., Song, X., Fan, Z., Guo, D., Luo, J., & Ma, Y. (2023). Metal oxide nanoparticles inhibit nitrogen fixation and rhizosphere colonization by inducing ROS in associative nitrogen-fixing bacteria *Pseudomonas stutzeri* A1501. *Chemosphere, 336*, 139223.

Ma, Y. H., Yao, Y., Yang, J., He, X., Ding, Y. Y., Zhang, P., et al. (2018). Trophic transfer and transformation of CeO_2 nanoparticles along a terrestrial food chain: Influence of exposure routes. *Environmental Science & Technology, 52*, 7921–7927. https://doi.org/10.1021/acs.est.8b00596

Madrid-Delgado, G., Orozco-Miranda, M., Cruz-Osorio, M., Hernández-Rodríguez, O. A., Rodríguez-Heredia, R., Roa-Huerta, M., & Avila-Quezada, G. D. (2021). Pathways of phosphorus absorption and early signaling between the mycorrhizal fungi and plants. *Phyton, 90*(5), 1321.

Mahmoud, M. E., Abdou, A. E. H., Mohamed, S. M. S., & Osman, M. M. (2016). Engineered *Staphylococcus aureus* via immobilization on magnetic Fe_3O_4-phthalate nanoparticles for biosorption of divalent ions from aqueous solutions. *Journal of Environmental Chemical Engineering, 4*, 3810–3824. https://doi.org/10.1016/j.jece.2016.08.022

Maksimova, Y. G. (2019). Microorganisms and carbon nanotubes: Interaction and applications. *Applied Biochemistry and Microbiology, 55*, 1–12.

Manke, A., Wang, L., & Rojanasakul, Y. (2013). Mechanisms of nanoparticle-induced oxidative stress and toxicity. *BioMed Research International, 2013*, 942916. https://doi.org/10.1155/2013/942916

Martínez, G., Merinero, M., Pérez-Aranda, M., Pérez-Soriano, E. M., Ortiz, T., Villamor, E., et al. (2021). Environmental impact of nanoparticles application as an emerging technology: A reveiw. *Materials, 14*, 166. https://doi.org/10.3390/ma14010166

Montes-Montes, G., González-Escobedo, R., Muñoz-Castellanos, L. N., Avila-Quezada, G. D., Ramírez-SánchezO, B.-L. A., et al. (2024). Whole genome analysis of *Streptomyces* spp. strains isolated from the rhizosphere of *Vitis vinifera* L. reveals their role in nitrogen and phosphorus metabolism. *Nitrogen, 5*(2), 301–314.

Muñoz-Castellanos, L. N., Avila-Quezada, G. D., Sáenz-De La Riva, G., Salas, E., Muñoz-Ramírez, Z. Y., & González-Escobedo, R. (2024). Revealing microbial patterns in the rhizosphere of pecan trees asymptomatic and symptomatic for Texas root rot using a high-throughput sequencing approach. *Rhizosphere, 29*, 100833.

Muñoz-Ramírez, Z. Y., González-Escobedo, R., Avila-Quezada, G. D., Ramírez-Sánchez, O., Higareda-Alvear, V. M., Zapata-Chávez, E., et al. (2024). Exploring microbial rhizosphere communities in asymptomatic and symptomatic apple trees using amplicon sequencing and shotgun metagenomics. *Agronomy, 14*(2), 357.

Nel, A., Xia, T., Madler, L., & Li, N. (2006). Toxic potential of materials at the nanolevel. *Science, 311*(5761), 622–627. https://doi.org/10.1126/science.1114397

Oberdörster, G., Oberdörster, E., & Oberdörster, J. (2005). Nanotoxicology: An emerging discipline evolving from studies of ultrafine particles. *Environmental Health Perspectives, 113*(7), 823–839. https://doi.org/10.1289/ehp.7339

Olivas-Tarango, A. L., Sáenz-Hidalgo, H. K., Porras-Chavira, A. A., de Jesus, R.-A. T., Leyva-Mir, G., Tarango-Rivero, S. H., & Avila-Quezada, G. D. (2024). El suelo vivo: un poco de lo que ocurre en este entorno. Un énfasis en los fitopatógenos. *Cultivos Tropicales, 45*(1), cu-id. https://ediciones.inca.edu.cu/index.php/ediciones/article/view/1762

Patil, S. S., & Lekhak, U. M. (2020). Toxic effects of engineered carbon nanoparticles on environment. In *Carbon nanomaterials for agri-food and environmental applications* (pp. 237–260). Elsevier.

Peng, C., Tong, H., Shen, C., Sun, L., Yuan, P., He, M., & Shi, J. (2020). Bioavailability and translocation of metal oxide nanoparticles in the soil-rice plant system. *Science of the Total Environment, 713*, 136662. https://doi.org/10.1016/j.scitotenv.2020.136662

Pérez-Hernández, H., Fernández-Luqueño, F., Huerta-Lwanga, E., Mendoza-Vega, J., & Álvarez-Solís, J. D. (2020). Effect of engineered nanoparticles on soil biota: Do they improve the soil quality and crop production or jeopardiza them? *Land Degradation & Development., 31*(16), 2213–2230. https://doi.org/10.1002/ldr.3595

Pratap-Singh, R., Handa, R., & Manchanda, G. (2020). Nanoparticles in sustainable agricultura: An emerging opportunity. *Journal of Controlled Release, 329*, 1234–1248. https://doi.org/10.1016/j.jconrel.2020.10.051

Rai, M., Yadav, A., Ingle, A., & dos Santos, C. A. (2021). An introduction to biopolymer-based nanofilms, their applications, and limitations. In *Biopolymer-based nano films* (pp. 3–17). Elsevier.

Rajput, V., Minkina, T., Ahmed, B., Sushkova, S., Singh, R., Soldatov, M., et al. (2019). Interaction of copper-based nanoparticles to soil, terrestrial, and aquatic systems: Critical review of the state of the science and future perspectives. *Reviews of Environmental Contamination and Toxicology., 252*, 51–96. https://doi.org/10.1007/398_2019_34

Rajput, Vishnu D., Arpna Kumari, Sudhir K. Upadhyay, Tatiana Minkina, Saglara Mandzhieva, Anuj Ranjan, Svetlana Sushkova, Marina Burachevskaya, Priyadarshani Rajput, Elizaveta Konstantinova, and et al. (2023). Can Nanomaterials Improve the Soil Microbiome.

Rico, C. M., Barrios, A. C., Tan, W., Rubenecia, R., Lee, S. C., Varela-Ramirez, A., et al. (2015). Physiological and biochemical response of soil-grown barley (*Hordeum vulgare* L.) to cerium oxide nanoparticles. *Environmental Science and Pollution Research, 22*, 10551–10558. https://doi.org/10.1007/s11356-015-4243-y

Rivero-Montejo, Samantha de Jesus, Marcela Vargas-Hernandez, and Irineo Torres-Pacheco. 2021. Nanoparticles as Novel Elicitors to Improve Bioactive Compounds in Plants Agriculture 11, no. 2: 134. https://doi.org/10.3390/agriculture11020134

Saad-Kheir, A. M., Mahmoud-Abouelsoud, H., Maher-Hafez, E., & Mohamed-Ali, O. A. (2019). Integrated effect of nano-Zn, nano-Si, and drainage using crop Straw-filled ditches on saline sodic soil properties and rice productivity. *Arabian Journal of Geosciences., 12*, 471. https://doi.org/10.1007/s12517-019-4653-0

Sáenz-Hidalgo, H. K., Jacobo-Cuellar, J. L., Zúñiga-Rodríguez, E., Avila-Quezada, G. D., et al. (2023). Soil structure and ectomycorrhizal root colonization of pecan orchards in northern Mexico. *Journal of Fungi, 9*(4), 440.

Saenz-Hidalgo, H. K., Sánchez-Chávez, E., Orduño-Cruz, N., Rai, M., Olalde-Portugal, V., & Avila-Quezada, G. D. (2025). Mitigating chili plant wilt: Synergistic effects of arbuscular endomycorrhizal fungi and silver nanoparticles. *Revista Mex Fitopatol, 42*. in press.

Salah-Sheteiwy, M., Shaghaleh, H., Alhaj-Hamoud, Y., Holford, P., Shao, H., Qi, W., et al. (2021). Zinc oxide nanoparticles: Potential effects on soil properties, crop production, food processing, and food quality. *Environmental Science and Pollution Research., 28*, 36942–36966. https://doi.org/10.1007/s11356-021-14542-w

Salas, E. C., Sun, Z., Lüttge, A., & Tour, J. M. (2010). Reduction of graphene oxide via bacterial respiration. *ACS Nano, 4*(8), 4852–4856. https://doi.org/10.1021/nn101081t

Sengul, A. B., & Asmatulu, E. (2020). Toxicity of metal and metal oxide nanoparticles: A review. *Environmental Chemistry Letters, 18*, 1659–1683. https://doi.org/10.1007/s10311-020-01033-6

Shang, Yifen, Md. Kamrul Hasan, Golam Jalal Ahammed, Mengqi Li, Hanqin Yin, and Jie Zhou. 2019. Applications of Nanotechnology in Plant Growth and Crop Protection: A Review Molecules 24, no. 14: 2558. https://doi.org/10.3390/molecules24142558

Sharifi, S., Behzadi, S., Laurent, S., Forrest, M. L., Stroeve, P., & Mahmoudi, M. (2012). Toxicity of nanomaterials. *Chemical Society Reviews, 41*(6), 2323–2343. https://doi.org/10.1039/C1CS15188F

Shahrajabian, Mohamad Hesam, Christina Chaski, Nikolaos Polyzos, and Spyridon A. Petropoulos. 2021. Biostimulants Application: A Low Input Cropping Management Tool for Sustainable Farming of Vegetables Biomolecules 11, no. 5: 698. https://doi.org/10.3390/biom11050698

Shvedova, A., Pietroiusti, A., Fadeel, B., & Kagan, V. (2012). Mechanisms of carbon nanotube-induced toxicity: Focus on oxidative stress. *Toxicology and Applied Pharmacology, 261*(2), 121–133. https://doi.org/10.1016/j.taap.2012.03.023

Suazo-Hernández, J., Aracibia-Miranda, N., Mlih, R., Cáceres-Jensen, L., Bolan, N., & Mora, M. (2023). Impact on some soil physical and chemical properties caused by metal and metallic oxide engineered nanoparticles: A review. *Nanomaterials, 13*, 572.

Tarango-Rivero, S. H., Ávila-Quezada, G. D., Jacobo-Cuellar, J. L., Ramírez-Valdespino, C. A., Orrantia-Borunda, E., Rodríguez-Heredia, R., & Olivas-Tarango, A. L. (2022). Chelated zinc and beneficial microorganisms: A sustainable fertilization option for pecan production. *Revista Chapingo Serie Horticultura, 28*(3), 145–159.

Wang, X., Yang, X., Chen, S., Li, Q., Wang, W., Hou, C., et al. (2015). Zinc oxide nanoparticles affect biomass accumulation and photosynthesis in *Arabidopsis*. *Frontiers in Plant Science, 6*, 1243. https://doi.org/10.3389/fpls.2015.01243

Wang, Z., Zhang, L., Zhao, J., & Xing, B. (2016). Environmental processes and toxicity of metallic nanoparticles in aquatic systems as affected by natural organic matter. *Environmental Science. Nano, 3*(2), 240–255.

Wang, C., Xu, J., Zhang, L., Yang, X., Zhang, X., Zhang, C., & Zhu, L. (2024). Environmental behaviors and toxic mechanisms of engineered nanomaterials in soil. *Environmental Research, 242*, 117820. https://doi.org/10.1016/j.envres.2023.117820

Wu, W. J., Zou, Y. N., Hashem, A., Avila-Quezada, G. D., AbduAllah, E. F., & Wu, Q. S. (2023). *Rhizoglomus intraradices* is more prominent in improving soil aggregate distribution and stability than in improving plant physiological activities. *Agronomy, 13*(5), 1427.

Wypij, M., Trzcińska-Wencel, J., Golińska, P., Avila-Quezada, G. D., Ingle, A. P., & Rai, M. (2023). The strategic applications of natural polymer nanocomposites in food packaging and agriculture: Chances, challenges, and consumers' perception. *Frontiers in Chemistry, 10*, 1106230.

Xing, J., Hu, T., Cang, L., & Zhou, D. (2016). Remediation of copper-contaminated soil using different particle sizes of apatite: A field experiment. *Springer Plus, 5*, 1182. https://doi.org/10.1186/s40064-016-2492-y

Yoon, H. Y., Lee, J. G., Esposti, L. D., Iafisco, M., Kim, P. J., Shin, S. G., et al. (2020). Synergistic release of crop nutrients and stimulants from hydroxyapatite nanoparticles functionalized with humic substances: Toward a multifunctional nanofertilizer. *ACS Omega, 5*, 6598–6610. https://doi.org/10.1021/acsomega.9b04354

Zaib, M., Raza, I., Zubair, M., Arif, Z., Mohsin-Mumtaz, M., Qamer-Abbas, M., et al. (2023). Nano-enabled soil amendments for improved soil structure and water holding capacity: An in-depth review. *International Research Journal of Education and Technology., 05*(08), 2581–7795.

Zhang, P., Guo, Z., Zhang, Z., Fu, H., White, J. C., & Lynch, I. (2020). Nanomaterial transformation in the soil–plant system: Implications for food safety and application in agriculture. *Small, 16*(21), 2000705. https://doi.org/10.1002/smll.202000705

Chapter 7
Green Synthesized Nanofertilizers and Plant Growth Promoters

Songita Sonowal and Ram Prasad

7.1 Introduction

Rapid population growth and diminishing agricultural land increase the overall food demand. Therefore, to fulfill the demand for food security, people practice inappropriate agricultural methods by using chemical fertilizers. However, using these harsh chemical fertilizers negatively impacts nature by degrading soil quality, declining crop yields in agriculture, and causing various health issues. Although chemical fertilizers increase the yielding property of crops, they cannot provide healthy and nutritious fruit to the people (Maity et al., 2024). All of the minerals and nutrients that plants need to grow and develop are taken up by them from the soil. However, toxic chemical fertilizer application reduces the soil nutrients by degrading their physico-chemical bonding and destroying microbial biodiversity. Moreover, the application of chemical fertilizers not only degrades the soil quality but also affects the aquatic system (Asadu et al., 2024). For example, due to the excess use of urea and manure in agriculture, the quantity of phosphorus and nitrogen increases in the aquatic system and later causes eutrophication. It not only pollutes the water but also causes harm to aquatic animals. Therefore, by looking upon these problems, next-generation agrochemical nanofertilizers are developed to reduce pollution and provide global food security (Gade et al., 2023).

Nanofertilizers are newly designed nanomaterials that encapsulate the fertilizers inside their cavity (Nongbet et al., 2022). Nowadays green synthesis nanofertilizers have become the most popular fertilizers instead of chemical fertilizers. Because it has many advantages like being rich in nutrients, controlled release property, soil fertility, adsorption, and increased photosynthesis and production of crops. Green synthesized nanofertilizers are produced from naturally available materials like plants, microorganisms, and other organic substances. Therefore, the use of green

S. Sonowal · R. Prasad (✉)
Department of Botany, Mahatma Gandhi Central University, Motihari, Bihar, India

K. Esquivel Escalante et al. (eds.), *Green Nanomaterials*, Nanotechnology in the Life Sciences, https://doi.org/10.1007/978-3-032-04492-1_7

synthesized nanofertilizers has no side effects on the environment (Ndaba et al., 2022). According to the study, applying green-synthesized zinc, manganese, and iron nano oxide nanofertilizers to squash plants dramatically raises their levels of organic material, biomolecules, lipids, protein, and energy (Shebl et al., 2019). Application of nanofertilizers increases the uptake of nutrients and decreases nitrogen leaching and runoff loss by influencing the rhizospheric microbes of soil. Some nanofertilizers like titanium, zinc oxide, nanourea, etc. restore the soil microorganisms, which help in the growth of plants. However, while nanofertilizers like amine-modified polystyrene nanospheres and sulfate-modified polystyrene nanospheres enhance rhizospheric microorganisms but have no beneficial effect on plants, their effect on plants varies depending on their size and concentration. Recent literature exhibits that the arbuscular mycorrhizal fungi (AMF) are crucial to plants' supply of water, phosphate, and nitrogen, but their colony size has decreased as a result of the use of silver nanoparticles (Zhao et al., 2024). Numerous varieties of nanofertilizers are available, including biological nanofertilizers, nanoscale nutrients, and nano-encapsulated fertilizers. These nanofertilizers are used in various agricultural practices, hydroponics, and aeroponics for biofortification in crops. Innovation of nanofertilizers brings a light of hope in modern agriculture technology by improving nutrient availability and reducing negative impacts on the environment (Yadav et al., 2023). Initially, scientists investigated the use of several nanoparticles in agriculture and discovered that they might have an impact on enhancing plant development. We often see that traditional fertilizers have various side effects in the environment, like soil degradation, loss of nutrients, and many other environmental side effects. Due to the use of harsh chemical fertilizers, the aquatic system becomes polluted and forms a toxic algal bloom, which negatively affects the aquatic animals. Moreover, chemical fertilizers fuel the increase of greenhouse gas emissions, which impact climate change, loss of biodiversity, and many more (Chausali et al., 2023). But by using the nanofertilizers we can overcome these problems. When plants get nanofertilizers, their availability of vital micro and macronutrients is increased (Jiang et al., 2022).

This chapter explores the different types of nano fertilizers, their synthesis methods, and their application in different agricultural crops for better yields. Furthermore, the chapter discusses the mechanism and field studies of green-synthesized nanofertilizers and plant growth promoters.

7.2 Development and Types of Nanofertilizers

The negative drawbacks of traditional fertilizers are addressed by the development of nanofertilizers. There are different types of nanofertilizers available in the market. Some examples of commercially available nanofertilizers are nano-gro™, nanogreen, nano-Ag Answer®, Biozar nano-fertilizers, etc. These commercially available nanofertilizers are produced from the extract of plants and microbes (Prasad et al., 2017). Depending on the base materials, the nanofertilizers can be divided into different types, e.g., metal based, carbon-based, polymer-based, and hybrid modes, etc. They are also classified into different broad categories depending on their nutrient delivery, actions performed, and consistency. Action-based nanofertilizers include targeted distribution, controlled release, promoting plant development, and preventing water and nutrient loss. Among the nutrient-based nanofertilizers are nutrient-loaded, inorganic, organic, and hybrid varieties. Surface-coated, synthetic polymer-coated, biological product-coated, and nanocarrier-based are the several types of consistency-based nanofertilizers. Three types of nanofertilizers are distinguished: nanoscale nutritional particles, nanocomposites, and nanocoated fertilizers. There are three different ways to synthesize them: top-down, bottom-up, and electrochemical. Chemical vapor deposition, laser ablation, and milling are examples of top-down techniques. Chemical and biological synthesis are included in the bottom-up approach (Toksha et al., 2021). Green synthesis or biogenic synthesis nanofertilizers have many advantages including economically feasible, eco-friendly, better stable, slow-release, and less time-consuming (Athanassiou et al., 2023).

7.3 Metal-Based Nanofertilizers

Everyone is aware that macro and micronutrients are crucial components needed by crops in order to grow and develop. Some deficiency signs, including chlorosis, necrosis, suppression of cell division, and delayed flowering, may result from a lack of these important components (Muneer et al., 2024). Therefore, farmers used harsh chemical fertilizers to supply those elements to the plants. But they cause environmental pollution due to their harsh chemical nature. However, green-synthesized metal-based nanofertilizers are environmentally benign and provide all the essential elements to the crops in a limited amount. Metal-based nanofertilizers' dynamic qualities allow them to release vital nutrients gradually over time, giving plants a more effective and consistent supply (Ďurišová et al., 2023). The controlled release feature of metal-based nanofertilizers may decrease nutrient loss through volatilization and leaching. They also improve plant development and minimize nutritional deficiencies. Because metal-based nanofertilizers can improve nutrient availability, boost resistance to stress, and reduce nutrient loss. Because of these dynamic qualities, metal-based nanofertilizers are widely used (Heinisch et al., 2019). Some examples of metal-based nanofertilizers are gold, silver, copper, zinc, iron, manganese, magnesium, nickel, calcium, titanium, cerium, and silicon oxide nanoparticles (Chen, 2018). The most commonly used metal based nanofertilizers for the controlled release and plant growth promoters are nano zinc, TiO_2, Si, and Fe (Prasad et al., 2014, 2017; Singhal et al., 2017). These metal-based nanofertilizers are synthesized physically, chemically, and biologically. Among all these methods, the biologically synthesized method is preferred most because they are less toxic, cost-effective, and produced from waste biomass by using non-toxic solvents (Prasad et al., 2016). The biologically synthesized method is also known as green synthesis. As in green synthesis, the metal-based nanofertilizer synthesis forms the plant extract, microbes, and waste organic matter (Thangavelu et al., 2024). Due to the presence of antimicrobial properties in silver application of silver-based nanofertilizers in plants helps to strengthen their defense against insects, bacterial, and fungal infections. They also help vegetative growth and seed germination. Zinc oxide is a metal-based nanofertilizer that regulates plant growth and antioxidant enzyme activity (Singhal et al., 2017; Kaningini et al., 2022). Nanofertilizers based on gold have proved to a very beneficial for agricultural crops. From the studies, it is found that the application of gold nanoparticles in crops increases the germination rate, photosynthesis, and enzymatic activity of plants (Bandi et al., 2023). In plants belonging to the families, i.e., Poaceae, Poneteriaceae, Ceratophyllaceae, Typhaceae, and Haloragaceae, copper is one of the most important micronutrients. From the studies, it is found that due to a deficiency of copper, plants suffer a decrease in chlorophyll, degraded thylakoid membranes, and depressed CO_2 fixation, which disturbs the anabolic process of plants. Therefore, using copper oxide nanoparticles in crops can overcome these deficiency symptoms (Xu et al., 2024). Moreover, copper oxide nanoparticles have an antifungal capacity which helps plants fight against the fungal pathogen (Kaningini et al., 2022). Like this silver is also well known for its antimicrobial properties (Raza et al., 2024). According to earlier research, silver nanoparticles in plants not only aid in the defense against dangerous pathogens but also boost agricultural productivity by promoting crop photosynthesis and seed germination (Pallavi et al., 2016). Although, the metal-based nanofertilizers have many positive impacts on plants but they also have some negative impacts on the crops. For example, if the concentration of silver nanoparticles becomes high then it breaks down the cell wall of *Oryza sativa* and causes harmful impacts on the plants (Rastogi et al., 2017). All metal-based nanofertilizers are suitable for all agricultural crops and if they are favorable on one side and the same time they may show negative impacts on the other side (Min et al., 2023). Likewise, although it helps in the defence system but it may cause oxidative stress and phytotoxicity in plants. In maize, soybean, and *Vicia narbonensis* plants due to higher concentration of titanium oxide decreases the plant development, germination rate, AMF colony, and phosphorus level in leaf (Ďurišová et al., 2023). Therefore, the synthesis method, size, and concentration are very important for the better yielding of crops (Kaningini et al., 2022). Some examples of metal-based nanofertilizers and their impacts on agricultural crops are described in Table 7.1.

7.3.1 Carbon-Based Nanofertilizers

Carbon atoms are the primary building blocks of carbon-based nanofertilizers. Carbon quantum dots, graphene oxide, carbon nanotubes, and many more are examples of carbon-based nanofertilizers (Yadav et al., 2023). Because of its fascinating physicochemical characteristics, including its tiny size, high surface area to volume ratio, and outstanding mechanical and thermal strength, it finds

Table 7.1 Some examples of metal based nanofertilizers with their function in plants

Metal-based nanofertilizers	Applied to crops	Function	References
Titanium dioxide	*Triticum aestivum*	Increase growth and yielding capacity of crops	Jaberzadeh et al. (2013)
	Cucumis sativus	Increase morphological parameters in plants	Servin et al. (2012)
	Dracocephalum moldavica	Increase antioxidant enzyme activities and promote plant growth in salt stress conditions	Gohari et al. (2020)
	Triticum aestivum	Increase production of wheat, increase antioxidant enzyme activity	Manzoor et al. (2023)
	Canola	Improve development of radicles and plumules, increase germination rate	Mahmoodzadeh et al. (2013)
Calcium carbonate	*Citrus*	Repel fruit flies and pests when sprayed on leaves	Hua et al. (2015)
	Glycine max	Increase yielding capacity	Abd El-Hady and Hussein (2021)
	Archis hypogaea	Increase seedling growth and dry biomass	Liu and Lal (2015)
Iron oxide nanoparticle	*Archis hypogaea*	Stimulate plant growth by controlling the levels of phytohormones and antioxidant activities	Lau et al. (2020)
Zinc oxide nanoparticle	*Cicer arietinum*	Increase the plant biomass	Pavani et al. (2014)
	Trigonella foenum-graecum	Enhanced plant growth, root development, and seed germination	Shaik et al. (2020)
	Brassica juncea	Enhanced seed germination	Mazumder et al. (2020)
	Zea mays	Boost photosynthetic rates and increase seed viability	Sabir et al. (2020)
	Prunus domestica	Suppress soil-borne disease	Malandrakis et al. (2019)
Aluminum oxide nanoparticle	*Solanum lycopersicum*	Control root rot disease	Shenashen et al. (2017)
Magnesium oxide nanoparticle	*Solanum lycopersicum*	Suppress pathogens like *Ralstonia solanacearum*	Imada et al. (2016)
	Vigna unguiculata	Boost the tissue culture plants' growth efficiency	Koçak et al. (2023)

extensive usage in crop production and sustainable agriculture. These characteristics of carbon-based nanofertilizers improve the plants' capability to retain water and help them thrive under stressful conditions. Green synthetic carbon quantum dots improve plant germination, growth, water solubility, stress tolerance, and photosynthetic pigments when they are used in agricultural crops (Dong et al., 2024). Several scientists are interested in carbon-based nanofertilizers (fullerenes, single-walled and multi-walled carbon nanotubes, fullerol), because they enhance plant growth, boost soil fertility by improving nutrient availability, and provide long-lasting fertilization. Another carbon-based nanofertilizer graphene helps to hold the soil water and enhance the water conservation capacity in plants. From the studies, it is found that the application of graphene nanofertilizers in *Solanum lycopersicum* and *Spinacia oleracea* speeds up the process of seed germination. Besides this, graphene nanofertilizers have the properties of site-specific cargo delivery, slow release of nutrients, and carrier of pesticides. The plants treated with the single and multi-walled carbon nanotubes have dense stomata and long roots and shoots. The apoplastic and symplastic pathways allow the nanofertilizers to readily enter plant cells due to their reduced size, altering the functional and metabolic progressions of plants (Chowmasundaram et al., 2023). The carbon-based nanofertilizers also increase the number of photosynthetic pigments, which boosts the maturation period of the plants. But if the concentration increases, then it may cause toxicity instead of a beneficial effect. Therefore, at the time of application, we should notice the concentration of carbon-based nanofertilizers (Mukherjee et al., 2016). For instance, the biomass of maize and soybean was decreased when the concentration of fullerenes exceeded 500 $mgkg^{-1}$. Additionally, increased concentrations of single and multi-walled carbon nanotubes result in cell death, cell reduction, and plasma membrane separation from the cell wall, as well as delayed blooming in tomatoes and rice (Lin et al., 2009; Tan et al., 2009). Nowadays, due to over-application of nanofertilizers in agriculture, it faces the leaching of soil nutrients and disturbs microbial diversity. Therefore, green synthesized nanofertilizers include control and slow-release drug delivery properties, which help to release the cargoes in a limited and targeted species (Haydar et al., 2024). Some functions of carbon-based nanofertilizers in different plants are described in Table 7.2.

7.3.2 Polymer-Based and Hybrid NMs

Polymer-based and hybrid nanomaterials are the combination of polymers with other organic and inorganic nanoparticles such as carbon nanotubes, nanowires, graphene, metal oxide nanoparticles, dendrimers, etc. (Ferreira Soares et al., 2020). It has a size from 1 to 1000 nm. Polymeric hybrid nanoparticles come in a variety of forms, including glue cell membrane camouflaged polymeric nanoparticles, polymer-caged nanobins, core–shell-type hollow lipid–polymer, and polymer core–lipid–shell type. In this case, the medicines are enclosed in polymers, and the nanomaterials serve as the core (Dave et al., 2019). Therefore, these polymer-based hybrid nanomaterials have all the physical, chemical, and mechanical properties of both components. Nowadays it has various applications in different fields. Especially in agriculture, polymer-based nanomaterials are used in drug delivery to a target-specific site, also they can easily entrap the lipophilic and hydrophilic drugs. The core shell of polymer-based nanomaterials helps to encapsulate the hydrophilic drugs in their

Table 7.2 Some examples of carbon-based nanofertilizers with roles in selected plants

Carbon-based nanofertilizers	Plants	Function	References
Graphene	*Vicia faba*, lettuce	Increased photosynthesis	Gao et al. (2020a)
	Paeonia ostii	Relieve drought stress	Zhao et al. (2020)
	Lactuca sativa	Improve biomass, reduce oxidative damage of Cd stress	Gao et al. (2020a, b)
	Oryza sativa	Increase stem and leaf fresh weight	Liu et al. (2015)
	Coriandrum sativum	Promote growth of roots, stem, leaves, flower, fruit	Chakravarty et al. (2015)
	Allium sativum	Enhanced growth rate	Chakravarty et al. (2015)
	Medicago sativa	Increase height of plant and biomass	Chen et al. (2023)
Multiwalled carbon nanotube	*Eysenhardtia polystachya*	Increase the rate of germination and length of the plants	Lara-Romero et al. (2017)
	Lupinus elegans	Increase rate of germination and biomass	Lara-Romero et al. (2017)
	Salvia verticillata	Improve the photosynthetic pigment and activated vital enzymes	Rahmani et al. (2020)
	Triticum aestivum	Accelerate plant growth, elongate roots and shoots, and improve nutrient intake by plants	Wang et al. (2012)
	Glycine max	Increase germination rate	Samadi et al. (2020)
	Hordeum vulgare		Samadi et al. (2020)
Single-walled carbon nanotube	*Malus hupehensis*	Enhance the metabolism of carbohydrates and photosynthetic activity	Wu et al. (2023)
Carbon quantum dot	Lettuce, lentil, pea	Increase seed germination, biomass, water-holding capacity by plants	Chauhan et al. (2022)
	Vigna radiata	Improvement of plant biomass, chlorophyll content, and rate of germination	Liang et al. (2023)
	Solanum lycopersicum	Increase biomass, mineral contents, and improve the quality of fruit	Wang et al. (2023)

cavity (Zhang et al., 2008). Moreover, the carbon-based nanomaterials give tensile strength and stiffness to the polymers, which increase the mechanical strength, ductility, and hardness of the polymer-based hybrid nanomaterials (Rashid et al., 2024). Also, the use of silica and aluminum oxide nanomaterials in polymer-based hybrid nanomaterials increases the stability and heat resistance of the compounds. Moreover, it has good electrical conductivity and optical properties which can be used in different photonic or optoelectronic devices (Li et al., 2022). It has various dynamic properties, because of which the polymer-based hybrid nanomaterials are used in drug delivery in agriculture. Application of cellulose nanocrystals hybrid nanofertilizers and polymeric nanocapsules loaded with iron nanoparticles increases the nutrient availability of plants, reduces the use of synthetic agrochemicals, and increases soil fertility naturally (Ashfaq et al., 2019). In agriculture most commonly used polymer-based hybrid nanofertilizers are NPK hybrid nanofertilizers, chitosan-based hybrid nanofertilizers, polymeric micelles encapsulating nutrients, polymer encapsulated nano nutrients, polymer stabilized nano silica fertilizers, nanocellulose-based fertilizers, polymer-encapsulated iron nanoparticles, polymeric hydrogels, biopolymer nanocomposites, polymer stabilized nano phosphorus fertilizers, etc. They play a variety of roles in the development and growth of crops. For instance, the use of polymer-based hybrid nanocomposites in agriculture, such as starch nanocomposite fertilizers, increases the rate at which plants germinate, release agrochemicals, and have access to nutrients (Dhiman et al., 2022). Sometimes due to direct use of metal-based nanofertilizers causes toxicity to the water and soil, but if it is applied by producing composites with polymers then it reduces toxicity in the environment (Chauhan et al., 2020). The controlled release behavior of polymer-based hybrid nanomaterials makes it easier to release the drugs and nutrients to the target-specific site of the plants. In the market, both natural and synthetic polymer based nanofertilizers are available, but the naturally used polymer-based nanofertilizers are most popular due to their low cost and eco-friendly nature. For example, chitosan-based silver nanoparticle composite fertilizers have antimicrobial properties; therefore, when it is used in agriculture, they decrease the prone of infection by pathogens and promote the growth and development of the plants (Ashfaq et al., 2019).

7.4 Plant Growth Promoters

The substance that promotes the enlargement of cells, elongation, cell division, and plant development is known as a plant growth promoter. Green synthesis of plant growth promoters is an eco-friendly method where the promoters are derived from plant extract, microbes, and other biological sources. In plants, the promoters are naturally present as a hormone (phytohormone) which helps in the physiological activities in plants (Fig. 7.1). But in the present day they are commercially available in the market at a large amount. The major plant growth regulator producers in the world are the USA, Germany, and Australia. In the last 2015–2020, the global market of plant growth regulators was $1.91 billion; now it has increased to $4.06 billion. In a report on the plant growth regulator market, it is predicted that the size of the global plant growth regulator market will increase up to USD 9.53 billion by 2031 (Plant Growth Regulators Market Size & Share – Industry Growth | 2031, n. d.). In the green synthesis method, the phytohormones (auxin, cytokinin, gibberellin, and abscisic acid) are synthesized by the fermentation of organic substrate using microorganisms (Pirog, 2018). *Pseudomonas solanacearum, Mycobacterium*

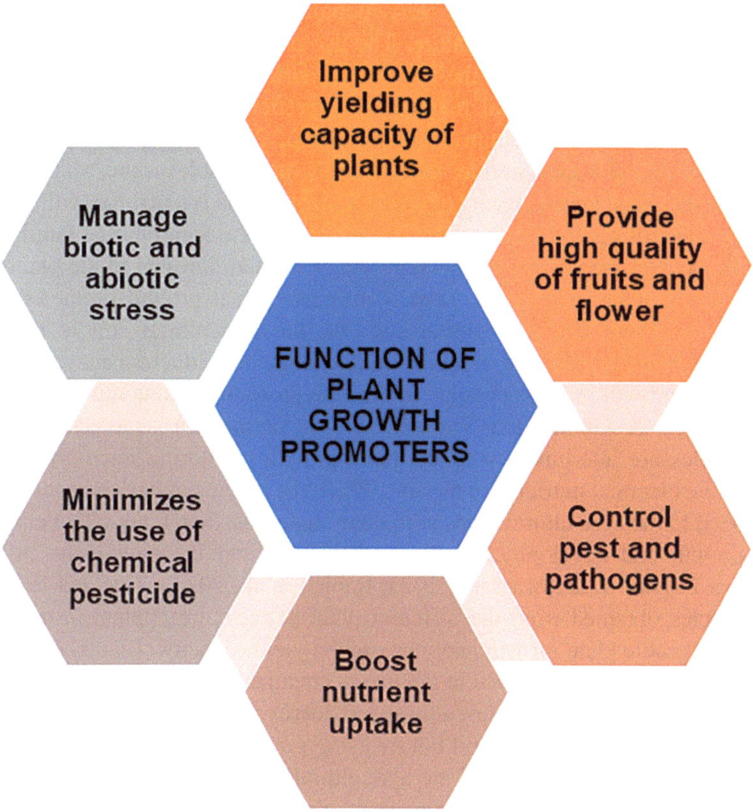

Fig. 7.1 Physiological functions of plant growth promoters

hiemalis are well known for the synthesis of ethylene hormone. This hormone is essential for fruit ripening in plants. Also, another major hormone is abscisic acid (responsible for the leaf fall), biologically synthesized from lichens, algae, and plant-pathogenic fungi (Patel et al., 2015). Different kinds of microorganisms that aid in plant growth can be found in the soil. They also enhance the immune system of plants against pathogens. For example, microbes like *Trichoderma, Bacillus,* and *Pseudomonas* show antifungal and antibacterial activity against some pathogenic microbes. Besides phytohormones, other plant growth promoters are microbes, and secondary metabolites like alkaloids, phenolics, flavonoids, terpenoids, and humic acids, etc. (Plant Growth Regulators, 2009). When these plant growth promoters are applied, the roots grow larger and the plants are better able to absorb vital nutrients. They also help to survive in stressful conditions. For example, in drought conditions, the foliar spray of salicylic acid in controlled amounts helps the plants to survive stress and treat the physiological damages of plants caused by abiotic stress (Khan et al., 2020). Biologically synthesized nanoparticles also act as plant growth promoters because they can directly deliver the essential nutrients and hormones to promote plant growth. They are very popular among audiences due to their eco-friendly and cost-effective nature. The use of green synthesized plant growth promoters not only improves the quality of agriculture but also minimizes the adverse influence of chemical manures and pesticides on the environment and human health (Rasmiya Begum & Jayawardana, 2023). They also increase the productivity of hydrogen sulfide, siderophores, and indole acetic acid in plants, which improves the yield capacity of crops. Rhizobacteria is a plant growth promoter that forms the nitrogen-fixing nodules in the roots of the leguminosae family. They boost the development of plants by supplying essential nitrogen to the plants. One more effective characteristic of green-synthesized plant growth promoters is the remediation of polluted agricultural soil. As the agricultural soil is frequently polluted by heavy metals and organic contaminants, but reductase enzyme secreted by the plant growth promoters helps to degrade those toxic contaminants from soil (Saberi Riseh & Gholizadeh Vazvani, 2024). In modern nanotechnology, the nanoparticles are interlinked with the plant growth-promoting microorganisms to increase the essential nutrients in the soil. When the green synthesized nanoparticles are applied to the agricultural crops with other materials, then it not only helps in the plant growth and development but also helps to fight the pathogenic microbes (Rasmiya Begum & Jayawardana, 2023; Ranjan et al., 2024). Zinc and iron oxide nanoparticles obtained from the different plant extracts are applied in cotton and barley to promote plant growth and increase the yield capacity (Ahmed et al., 2017; Rostamizadeh et al., 2020). Due to the use of chemically synthesized fertilizers and plant growth promoters the agricultural soil losses the physico-chemical bonding and water-holding capacity of soil but by using green synthesized nanoparticles we can revive the original texture of the agricultural soil (Prasad et al., 2014; Nayana et al., 2020). The activity of enzymes is also a very important part of plant physiology, which can be enhanced by the association of biogenic nanoparticles and PGPR (plant growth-promoting rhizobacteria) (Verma et al., 2024).

7.5 Mechanistic Approaches and Benefits

Nanofertilizers and plant growth promoters have a significant impact on how plants develop and alter their defence mechanisms against both abiotic and biotic stresses. Various nanoparticles can regulate protective responses through signaling cross-talk, affecting phytohormone responses, and influencing stress signals in plants. Figure 7.2 shows that nanofertilizers encapsulate the essential nutrients in their cavity. For example, nanoparticles like liposomes, polymers, and silica encapsulate the hydrophobic and hydrophilic drugs in their cavity. Due to this property of nanofertilizers wastage of nutrients is reduced. The application of these newly synthesized nanoparticles enhances the stability and bioavailability of nutrients in plants. Also, modification of surface structure in nanofertilizers improves the solubility and adhesion to plant surfaces. Due to the small size of nanofertilizers, they are uniformly sprayed in the soil when foliar spray and easily penetrate the plant through

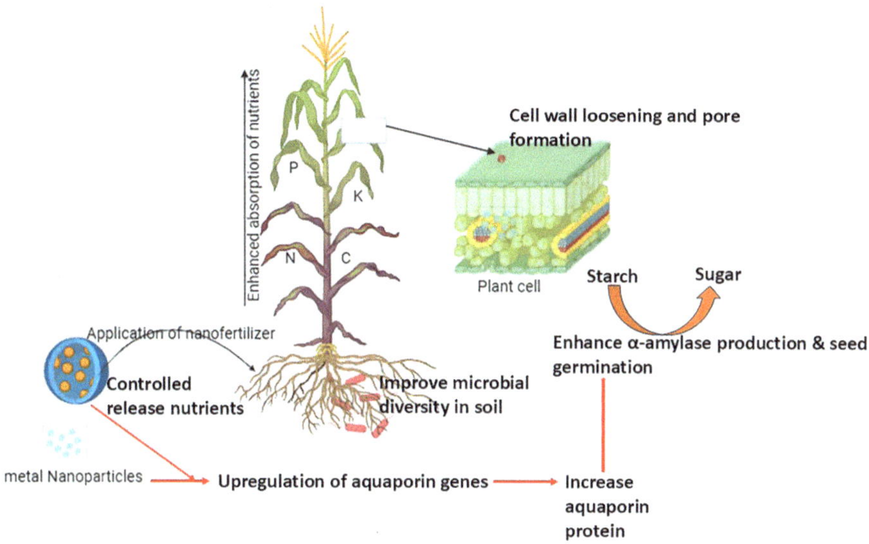

Fig. 7.2 Mechanism of nanofertilizers in plant growth and development

the cuticle and stomata. Due to the controlled release property, the nanofertilizers release nutrients depending on pH, and moisture in the soil and as per the requirements of plants. This mechanism helps the plants to survive in times of critical conditions. There are some genes and enzymes present in plants that support growth and development, and these are stimulated by the application of nanofertilizers (Ghosh & Bera, 2021).

7.6 Field Studies and Results

Field studies on green-synthesized nanofertilizers and plant growth promoters are being directed to investigate their efficacy in sustainable agriculture. Wheat, rice, maize, legumes, and fruits have all been the subjects of field studies (Abd El-Azeim et al., 2020). Typically, studies compare the performance of green-synthesized nanofertilizers to conventional fertilizers, with a focus on crop yield, plant growth parameters, soil health, and stress tolerance. According to previous studies, crop productivity gains can range from 15% to 30% (Fatima et al., 2021). Plants that receive nanofertilizers have higher levels of root, shoot, leaf area index, chlorophyll content, and total biomass. Additionally, the regulated release mechanism of green-synthesized nanofertilizers and plant growth promoters guarantees a steady supply of nutrients while reducing losses from leaching or volatilization. As a result, the efficiency with which nutrients are used increases by 20–40% (Haydar et al., 2024). It also has an impact on the environment by reducing greenhouse gas emissions and leaching losses, which helps to achieve sustainability goals. Despite these positive findings, issues like as scalability, cost-effectiveness, and standardization of green synthesis procedures remain. However, recent improvements in this subject highlight its transformational potential for finding a balance between agricultural output and environmental conservation. Farmers benefit from cheaper input costs due to reduced fertilizer usage and increased output per hectare.

7.7 Conclusion and Future Prospects

Green synthesis of nanofertilizers and plant growth promoters brings a new revolution to modern agricultural technology by providing sustainable development of crop productivity and high quality. By using these environmentally friendly methods, we can reduce the application of hazardous chemical fertilizers. As these hazardous chemicals not only deteriorate the environment but are also life-threatening to human beings. The green synthesized nanofertilizers and plant growth promoters have the property of nanoscale precision in nutrient delivery, optimize resource efficiency, reduce wastage, and minimize pollution. Therefore, this technology can mitigate the global goals for food security and environmental conservation. Introducing green synthesized nanofertilizers into smart agriculture technology can improve nutrient delivery and resource efficiency by lowering nutrient loss. Furthermore, the synthesis of multifunctional nanofertilizers that include pest control, stress management, and controlled release can address several difficulties in a single application while increasing agricultural productivity. Additionally, adopting standardized protocols for green manufacture and use of nanofertilizers will ensure their safety, efficacy, and scalability. Regulatory frameworks should encourage the global adoption of these sustainable solutions. The use of plant biomass, such as agricultural waste, algae, and fungi, to synthesize nanofertilizers is a low-cost method. Therefore, we need to investigate more to discover new biomaterials for nanofertilizer production. Moreover, the effect of the long-term existence of nanofertilizers in the environment and how they affect human health and animals should also be a study topic.

Green synthetic nanofertilizers and plant growth promoters have enormous potential to improve present agricultural technology by making it more sustainable. Overcoming the gap between research and practical applications, such as the commercialization of green nanofertilizers and farmer training programs, can ensure their widespread acceptance, particularly in resource-constrained countries. Also, innovation, collaboration, and investment in this subject will be critical in solving the twin challenges of feeding a growing population and conserving plant resources.

Acknowledgments The author wants to thank the Department of Botany, Mahatma Gandhi Central University, Motihari, for providing the infrastructure facility. Also, Songita Sonowal is always grateful to the Ministry of tribal affairs for providing the fellowship of NFST.

Conflict of Interest The authors have no conflicts of interest to declare that are relevant to the content of this chapter.

References

Abd El-Azeim, M. M., Sherif, M. A., Hussien, M. S., Tantawy, I. A. A., & Bashandy, S. O. (2020). Impacts of nano- and non-nanofertilizers on potato quality and productivity. *Acta Ecologica Sinica, 40*(5), 388–397. https://doi.org/10.1016/J.CHNAES.2019.12.007

Abd El-Hady, M. A., & Hussein, H. (2021). Effect of foliar nano fertilizers and irrigation intervals on soybean productivity and quality. *Journal of Plant Production, 12*(9), 1007–1014. https://doi.org/10.21608/JPP.2021.202836

Ahmed, S., Annu, C., & S. A., & Ikram, S. (2017). A review on biogenic synthesis of ZnO nanoparticles using plant extracts and microbes: A prospect towards green chemistry. *Journal of Photochemistry and Photobiology B: Biology, 166*, 272–284. https://doi.org/10.1016/J.JPHOTOBIOL.2016.12.011

Asadu, C. O., Ezema, C. A., Ekwueme, B. N., Onu, C. E., Onoh, I. M., Adejoh, T., Ezeorba, T. P. C., Ogbonna, C. C., Otuh, P. I., Okoye, J. O., & Emmanuel, U. O. (2024). Enhanced efficiency fertilizers: Overview of production methods, materials used, nutrients release mechanisms, benefits and considerations. *Environmental Pollution and Management, 1*, 32–48. https://doi.org/10.1016/J.EPM.2024.07.002

Ashfaq, M., Talreja, N., Chuahan, D., Srituravanich, W., Ashfaq, M., Talreja, N., Chuahan, D., & Srituravanich, W. (2019). Polymeric nanocomposite-based agriculture delivery system: Emerging technology for agriculture. In *Genetic engineering – A glimpse of techniques and applications*. IntechOpen. https://doi.org/10.5772/INTECHOPEN.89702

Athanassiou, G., Bairwa, P., Kumar, N., Devra, V., & Abd-Elsalam, K. A. (2023). Nanobiofertilizers synthesis and applications in agroecosystems. *Agrochemicals, 2*(1), 118–134. https://doi.org/10.3390/AGROCHEMICALS2010009

Bandi, R., Dadigala, R., & Alle, M. (2023). Emerging role of gold nanoparticles for healthier crop plants growth and enhanced yield. In *Engineered nanomaterials for sustainable agricultural production, soil improvement and stress management* (pp. 125–143). https://doi.org/10.1016/B978-0-323-91933-3.00014-3

Chakravarty, D., Erande, M. B., & Late, D. J. (2015). Graphene quantum dots as enhanced plant growth regulators: Effects on coriander and garlic plants. *Journal of the Science of Food and Agriculture, 95*(13), 2772–2778. https://doi.org/10.1002/JSFA.7106

Chauhan, D., Afreen, S., Talreja, N., & Ashfaq, M. (2020). Multifunctional copper polymer-based nanocomposite for environmental and agricultural applications. In *Multifunctional hybrid nanomaterials for sustainable agri-food and ecosystems* (pp. 189–211). Elsevier. https://doi.org/10.1016/B978-0-12-821354-4.00008-X

Chauhan, P., Kumari, M., Chaudhary, S., Chaudhary, G. R., Umar, A., & Baskoutas, S. (2022). Seed germination studies on chickpeas, barley, mung beans and wheat with natural biomass and plastic waste derived C-dots. *Science of the Total Environment, 837*. https://doi.org/10.1016/J.SCITOTENV.2022.155593

Chausali, N., Saxena, J., & Prasad, R. (2023). Nanotechnology as a sustainable approach for combating the environmental effects of climate change. *Journal of Agriculture and Food Research*. https://doi.org/10.1016/j.jafr.2023.100541

Chen, H. (2018). Metal based nanoparticles in agricultural system: Behavior, transport, and interaction with plants. *Chemical Speciation & Bioavailability, 30*(1), 123–134. https://doi.org/10.1080/09542299.2018.1520050

Chen, Z., Guo, Z., Xu, N., Cao, X., & Niu, J. (2023). Graphene nanoparticles improve alfalfa (Medicago sativa L.) growth through multiple metabolic pathways under salinity-stressed environment. *Journal of Plant Physiology, 289*, 154092. https://doi.org/10.1016/J.JPLPH.2023.154092

Chowmasundaram, Y. A. P., Tan, T. L., Nulit, R., Jusoh, M., & Rashid, S. A. (2023). Recent developments, applications and challenges for carbon quantum dots as a photosynthesis enhancer in agriculture. *RSC Advances, 13*(36), 25093–25117). Royal Society of Chemistry. https://doi.org/10.1039/d3ra01217d

Dave, V., Tak, K., Sohgaura, A., Gupta, A., Sadhu, V., & Reddy, K. R. (2019). Lipid-polymer hybrid nanoparticles: Synthesis strategies and biomedical applications. *Journal of Microbiological Methods, 160*, 130–142. https://doi.org/10.1016/J.MIMET.2019.03.017

Dhiman, A., Sharma, A. K., & Agrawal, G. (2022). Polymer based engineered materials for sustainable agriculture. *ACS Agricultural Science and Technology, 2*(4), 693–711. https://doi.org/10.1021/ACSAGSCITECH.1C00278/ASSET/IMAGES/MEDIUM/AS1C00278_0007.GIF

Dong, Z., Qi, J., Yue, L., Zhou, H., Chen, L., Gu, J., He, Y., & Wu, H. (2024). Biomass-based carbon quantum dots and their agricultural applications. In *Plant stress* (Vol. 11). https://doi.org/10.1016/j.stress.2024.100411

Ďurišová, Ľ., Ďúranová, H., Kšiňan, S., Ernst, D., Šebesta, M., Žitniak Čurná, V., Eliáš, P., Qian, Y., Straka, V., Feng, H., Tomovičová, L., Kotlárová, N., Kratošová, G., & Kolenčík, M. (2023). Exploring the impact of metal-based nanofertilizers: A case study on sunflower pollen morphology and yield in field conditions. *Agronomy, 13*(12), 2922. https://doi.org/10.3390/AGRONOMY13122922

Fatima, F., Hashim, A., & Anees, S. (2021). Efficacy of nanoparticles as nanofertilizer production: A review. *Environmental Science and Pollution Research, 28*(2), 1292–1303. https://doi.org/10.1007/S11356-020-11218-9/METRICS

Ferreira Soares, D. C., Domingues, S. C., Viana, D. B., & Tebaldi, M. L. (2020). Polymer-hybrid nanoparticles: Current advances in biomedical applications. *Biomedicine & Pharmacotherapy, 131*, 110695. https://doi.org/10.1016/J.BIOPHA.2020.110695

Gade, A., Ingle, P., Nimbalkar, U., Rai, M., Raut, R., Vedpathak, M., Jagtap, P., & Abd-Elsalam, K. A. (2023). Nanofertilizers: The next generation of agrochemicals for long-term impact on sustainability in farming systems. *Agrochemicals, 2*(2), 257–278. https://doi.org/10.3390/AGROCHEMICALS2020017

Gao, M., Chang, X., Yang, Y., & Song, Z. (2020a). Foliar graphene oxide treatment increases photosynthetic capacity and reduces oxidative stress in cadmium-stressed lettuce. *Plant Physiology and Biochemistry, 154*, 287–294. https://doi.org/10.1016/J.PLAPHY.2020.06.021

Gao, M., Xu, Y., Chang, X., Dong, Y., & Song, Z. (2020b). Effects of foliar application of graphene oxide on cadmium uptake by lettuce. *Journal of Hazardous Materials, 398*, 122859. https://doi.org/10.1016/J.JHAZMAT.2020.122859

Ghosh, S. K., & Bera, T. (2021). Molecular mechanism of nano-fertilizer in plant growth and development: A recent account. In *Advances in nano-fertilizers and nano-pesticides in agriculture: A smart delivery system for crop improvement* (pp. 535–560). https://doi.org/10.1016/B978-0-12-820092-6.00022-7

Gohari, G., Mohammadi, A., Akbari, A., Panahirad, S., Dadpour, M. R., Fotopoulos, V., & Kimura, S. (2020). Titanium dioxide nanoparticles (TiO_2 NPs) promote growth and ameliorate salinity stress effects on essential oil profile and biochemical attributes of *Dracocephalum moldavica*. *Scientific Reports, 10*(1), 1–14. https://doi.org/10.1038/s41598-020-57794-1

Haydar, M. S., Ghosh, D., & Roy, S. (2024). Slow and controlled release nanofertilizers as an efficient tool for sustainable agriculture: Recent understanding and concerns. *Plant Nano Biology, 7*, 100058. https://doi.org/10.1016/J.PLANA.2024.100058

Heinisch, M., Jácome, J., & Miricescu, D. (2019). Current experience with application of metal-based nanofertilizers. EDP Sciences. https://doi.org/10.1051/matecconf/20192

Hua, K. H., Wang, H. C., Chung, R. S., & Hsu, J. C. (2015). Calcium carbonate nanoparticles can enhance plant nutrition and insect pest tolerance. *Journal of Pesticide Science, 40*(4), 208–213. https://doi.org/10.1584/JPESTICS.D15-025

Imada, K., Sakai, S., Kajihara, H., Tanaka, S., Ito, S. (2016). Magnesium oxide nanoparticles induce systemic resistance in tomato against bacterial wilt disease. Plant Pathology, 65 (4): 551–560. https://doi.org/10.1111/ppa.12443

Jaberzadeh, A., Moaveni, P., Tohidi Moghadam, H. R., & Zahedi, H. (2013). Influence of bulk and nanoparticles titanium foliar application on some agronomic traits, seed gluten and starch contents of wheat subjected to water deficit stress. *Notulae Botanicae Horti Agrobotanici Cluj-Napoca, 41*(1), 201–207. https://doi.org/10.15835/NBHA4119093

Jiang, Y., Zhou, P., Zhang, P., Adeel, M., Shakoor, N., Li, Y., Li, M., Guo, M., Zhao, W., Lou, B., Wang, L., Lynch, I., & Rui, Y. (2022). Green synthesis of metal-based nanoparticles for sustainable agriculture. *Environmental Pollution, 309*, 119755. https://doi.org/10.1016/J.ENVPOL.2022.119755

Kaningini, A. G., Nelwamondo, A. M., Azizi, S., Maaza, M., & Mohale, K. C. (2022). Metal nanoparticles in agriculture: A review of possible use. *Coatings, 12*(10), 1586. MDPI. https://doi.org/10.3390/coatings12101586

Khan, N., Bano, A. M. D., & Babar, A. (2020). Impacts of plant growth promoters and plant growth regulators on rainfed agriculture. *PLoS One, 15*(4), e0231426. https://doi.org/10.1371/JOURNAL.PONE.0231426

Koçak, R., Okcu, M., Haliloğlu, K., Türkoğlu, A., Pour-Aboughadareh, A., Jamshidi, B., Janda, T., Alaylı, A., & Nadaroğlu, H. (2023). Magnesium oxide nanoparticles: An influential element in cowpea (*Vigna unguiculata* L. Walp.) tissue culture. *Agronomy, 13*(6), 1646. https://doi.org/10.3390/AGRONOMY13061646

Lara-Romero, J., Campos-García, J., Dasgupta-Schubert, N., Borjas-García, S., Tiwari, D. K., Paraguay-Delgado, F., Jiménez-Sandoval, S., Alonso-Nuñez, G., Gómez-Romero, M., Lindig-Cisneros, R., De la Cruz, H. R., & Villegas, J. A. (2017). Biological effects of carbon nanotubes generated in forest wildfire ecosystems rich in resinous trees on native plants. *Peer J, 2017*(8), e3658. https://doi.org/10.7717/PEERJ.3658/SUPP-2

Lau, E. C. H. T., Carvalho, L. B., Pereira, A. E. S., Montanha, G. S., Corrêa, C. G., Carvalho, H. W. P., Ganin, A. Y., Fraceto, L. F., & Yiu, H. H. P. (2020). Localization of coated iron oxide (Fe_3O_4) nanoparticles on tomato seeds and their effects on growth. *ACS Applied Bio Materials, 3*(7), 4109–4117. https://doi.org/10.1021/ACSABM.0C00216/ASSET/IMAGES/LARGE/MT0C00216_0007.JPEG

Li, H., Cheng, B., Gao, W., Feng, C., Huang, C., Liu, Y., Lu, P., & Zhao, H. (2022). Recent research progress and advanced applications of silica/polymer nanocomposites. *Nanotechnology Reviews, 11*(1), 2928–2964). De Gruyter Open Ltd. https://doi.org/10.1515/ntrev-2022-0484

Liang, L., Wong, S. C., & Lisak, G. (2023). Effects of plastic-derived carbon dots on germination and growth of pea (*Pisum sativum*) via seed nano-priming. *Chemosphere, 316*, 137868. https://doi.org/10.1016/J.CHEMOSPHERE.2023.137868

Lin, S., Reppert, J., Hu, Q., Hudson, J. S., Reid, M. L., Ratnikova, T. A., Rao, A. M., Luo, H., Chun Ke, P., Luo, H., Hu, Q., Ke, P. C., Lin, S., Reppert, J., Reid, M. L., Ratnikova, T. A., Rao, A. M., & Hudson, J. S. (2009). Cellular uptake uptake, translocation, and transmission of carbon nanomaterials in rice plants. *Small, 5*(10), 1128–1132. https://doi.org/10.1002/smll.200801556

Liu, R., & Lal, R. (2015). Potentials of engineered nanoparticles as fertilizers for increasing agronomic productions. *Science of the Total Environment, 514*, 131–139. https://doi.org/10.1016/J.SCITOTENV.2015.01.104

Liu, S., Wei, H., Li, Z., Li, S., Yan, H., He, Y., & Tian, Z. (2015). Effects of graphene on germination and seedling morphology in rice. *Journal of Nanoscience and Nanotechnology, 15*(4), 2695–2701. https://doi.org/10.1166/JNN.2015.9254

Mahmoodzadeh, H., Nabavi, M., & Kashefi, H. (2013). Effect of nanoscale titanium dioxide particles on the germination and growth of canola (*Brassica napus*). *Journal of Ornamental and Horticultural Plants, 3*(1) www.SID.ir

Maity, D., Gupta, U., Ajith, S., Arora, S., & Saha, S. (2024). Nanofertilizers for sustainable fruit growing: Improving productivity and quality. In *Nanofertilizer delivery, effects and application methods* (pp. 281–298). Elsevier. https://doi.org/10.1016/B978-0-443-13332-9.00003-4

Malandrakis, A. A., Kavroulakis, N., & Chrysikopoulos, C. V. (2019). Use of copper, silver and zinc nanoparticles against foliar and soil-borne plant pathogens. *Science of the Total Environment, 670*, 292–299. https://doi.org/10.1016/J.SCITOTENV.2019.03.210

Manzoor, N., Ali, L., AL-Huqail, A. A., Alghanem, S. M. S., Al-Haithloul, H. A. S., Abbas, T., Chen, G., Huan, L., Liu, Y., Wang, G. (2023). Comparative efficacy of silicon and iron oxide nanoparticles towards improving the plant growth and mitigating arsenic toxicity in wheat (Triticum aestivum L.). Ecotoxicology and Environmental Safety, 264, 115382. https://doi.org/10.1016/J.ECOENV.2023.115382

Mazumder, J. A., Khan, E., Perwez, M., Gupta, M., Kumar, S., Raza, K., & Sardar, M. (2020). Exposure of biosynthesized nanoscale ZnO to *Brassica juncea* crop plant: Morphological, biochemical and molecular aspects. *Scientific Reports, 10*(1), 1–13. https://doi.org/10.1038/s41598-020-65271-y

Min, Y., Suminda, G. G. D., Heo, Y., Kim, M., Ghosh, M., & Son, Y. O. (2023). Metal-based nanoparticles and their relevant consequences on cytotoxicity cascade and induced oxidative stress. *Antioxidants, 12*(3), 703. https://doi.org/10.3390/ANTIOX12030703

Mukherjee, A., Majumdar, S., Servin, A. D., Pagano, L., Dhankher, O. P., & White, J. C. (2016). Carbon nanomaterials in agriculture: A critical review. *Frontiers in Plant Science, 7*(FEB2016), 168246. https://doi.org/10.3389/FPLS.2016.00172/BIBTEX

Muneer, M. A., Afridi, M. S., Saddique, M. A. B., Chen, X., Zaib-Un-Nisa, Yan, X., Farooq, I., Munir, M. Z., Yang, W., Ji, B., Zheng, C., & Wu, L. (2024). Nutrient stress signals: Elucidating morphological, physiological, and molecular responses of fruit trees to macronutrients deficiency and their management strategies. *Scientia Horticulturae, 329*, 112985. https://doi.org/10.1016/J.SCIENTA.2024.112985

Nayana, A. R., Joseph, B. J., Jose, A., & Radhakrishnan, E. K. (2020). *Nanotechnological advances with PGPR applications* (pp. 163–180). Springer International Publishing. https://doi.org/10.1007/978-3-030-33996-8_9

Ndaba, B., Roopnarain, A., Rama, H., & Maaza, M. (2022). Biosynthesized metallic nanoparticles as fertilizers: An emerging precision agriculture strategy. *Journal of Integrative Agriculture, 21*(5), 1225–1242. https://doi.org/10.1016/S2095-3119(21)63751-6

Nongbet, A., Mishra, A. K., Mohanta, Y. K., Mahanta, S., Ray, M. K., Khan, M., Baek, K. H., & Chakrabartty, I. (2022). Nanofertilizers: A smart and sustainable attribute to modern agriculture. *Plants, 11*(19), 2587. https://doi.org/10.3390/PLANTS11192587

Pallavi, Mehta, C. M., Srivastava, R., Arora, S., & Sharma, A. K. (2016). Impact assessment of silver nanoparticles on plant growth and soil bacterial diversity. *3. Biotech, 6*(2), 1–10. https://doi.org/10.1007/S13205-016-0567-7/FIGURES/4

Patel, K., Goswami, D., Dhandhukia, P., & Thakker, J. (2015). *Techniques to study microbial Phytohormones*. In *Bacterial metabolites in sustainable agroecosystem* (pp. 1–27). Springer. https://doi.org/10.1007/978-3-319-24654-3_1

Pavani, K. V., Divya, V., Veena, I., Aditya, M., & Devakinandan, G. V. S. (2014). Influence of bioengineered zinc nanoparticles and zinc metal on *Cicer arietinum* seedlings growth. *Asian Journal of Agriculture and Biology, 2*(4), 216–223.

Pirog, T. P. (2018). Microbial synthesis of phytohormones. *Biotechnologia Acta, 11*(1), 5–24. https://doi.org/10.15407/BIOTECH11.01.005

Plant growth regulators. (2009). *Trease and Evans' pharmacognosy* (16th ed., pp. 93–97). Saunders/Elsevier. https://doi.org/10.1016/B978-0-7020-2933-2.00012-5

Plant Growth Regulators Market Size & Share – Industry Growth | 2031. (n.d.). Retrieved December 12, 2024, from https://www.skyquestt.com/report/plant-growth-regulators-market

Prasad, R., Kumar, V., & Prasad, K. S. (2014). Nanotechnology in sustainable agriculture: Present concerns and future aspects. *African Journal of Biotechnology, 13*(6), 705–713.

Prasad, R., Pandey, R., & Barman, I. (2016). Engineering tailored nanoparticles with microbes: Quo vadis. *WIREs Nanomedicine and Nanobiotechnology, 8*, 316–330. https://doi.org/10.1002/wnan.1363

Prasad, R., Bhattacharyya, A., & Nguyen, Q. D. (2017). Nanotechnology in sustainable agriculture: Recent developments, challenges, and perspectives. *Frontiers in Microbiology, 8*(Jun) Frontiers Media S.A. https://doi.org/10.3389/fmicb.2017.01014

Rahmani, N., Radjabian, T., & Soltani, B. M. (2020). Impacts of foliar exposure to multi-walled carbon nanotubes on physiological and molecular traits of *Salvia verticillata* L., as a medicinal plant. *Plant Physiology and Biochemistry, 150*, 27–38. https://doi.org/10.1016/J.PLAPHY.2020.02.022

Ranjan, A., Rajput, V. D., Prazdnova, E. V., Gurnani, M., Sharma, S., Bhardwaj, P., Shende, S. S., Mandzhieva, S. S., Sushkova, S., Minkina, T., Chauhan, A., Jindal, T., Prasad, R., & Wong, M. H. (2024). Augmenting abiotic stress tolerance and root architecture: The function of phytohormone-producing PGPR and their interaction with nanoparticles. *South African Journal of Botany, 167*, 612–629. https://doi.org/10.1016/j.sajb.2024.02.041

Rashid, A. B., Haque, M., Islam, S. M. M., & Uddin Labib, K. M. R. (2024). Nanotechnology-enhanced fiber-reinforced polymer composites: Recent advancements on processing techniques and applications. *Heliyon, 10*(2), e24692. https://doi.org/10.1016/J.HELIYON.2024.E24692

Rasmiya Begum, S. L., & Jayawardana, N. U. (2023). Green synthesized metal nanoparticles as an ecofriendly measure for plant growth stimulation and disease resistance. *Plant Nano Biology, 3*, 100028. https://doi.org/10.1016/J.PLANA.2023.100028

Rastogi, A., Zivcak, M., Sytar, O., Kalaji, H. M., He, X., Mbarki, S., & Brestic, M. (2017). Impact of metal and metal oxide nanoparticles on plant: A critical review. *Frontiers in Chemistry, 5*, 292528. https://doi.org/10.3389/FCHEM.2017.00078/BIBTEX

Raza, A., Khandelwal, K., Soumya Pandit, S., Singh, M., Kumar, S., Rustagi, S., Ranjan, N., Verma, R., Priya, K., & Prasad, R. (2024). Exploring the potential of metallic and metal oxide nanoparticles for reinforced disease management in agricultural systems: A comprehensive review. *Environmental Nanotechnology, Monitoring & Management*. https://doi.org/10.1016/j.enmm.2024.100998

Rostamizadeh, E., Iranbakhsh, A., Majd, A., Arbabian, S., & Mehregan, I. (2020). Green synthesis of Fe_2O_3 nanoparticles using fruit extract of *Cornus mas* L. and its growth-promoting roles in Barley. *Journal of Nanostructure in Chemistry, 10*(2), 125–130. https://doi.org/10.1007/S40097-020-00335-Z/METRICS

Saberi Riseh, R., & Gholizadeh Vazvani, M. (2024). Green synthesis of metal nanoparticles using plant growth promoting rhizobacteria and application in agriculture. *Plant Nano Biology, 10*, 100111. https://doi.org/10.1016/J.PLANA.2024.100111

Sabir, S., Zahoor, M. A., Waseem, M., Siddique, M. H., Shafique, M., Imran, M., Hayat, S., Malik, I. R., & Muzammil, S. (2020). Biosynthesis of ZnO nanoparticles using bacillus subtilis: Characterization and nutritive significance for promoting plant growth in *Zea mays* L. *Dose-Response, 18*(3). https://doi.org/10.1177/1559325820958911/ASSET/IMAGES/LARGE/10.1177_1559325820958911-FIG7.JPEG

Samadi, S., Saharkhiz, M. J., Azizi, M., Samiei, L., & Ghorbanpour, M. (2020). Multi-walled carbon nanotubes stimulate growth, redox reactions and biosynthesis of antioxidant metabolites in thymus daenensis celak. in vitro. *Chemosphere, 249*, 126069. https://doi.org/10.1016/J.CHEMOSPHERE.2020.126069

Servin, A. D., Castillo-Michel, H., Hernandez-Viezcas, J. A., Diaz, B. C., Peralta-Videa, J. R., & Gardea-Torresdey, J. L. (2012). Synchrotron micro-XRF and micro-XANES confirmation of the uptake and translocation of TiO_2 nanoparticles in cucumber (*Cucumis sativus*) plants. *Environmental Science and Technology, 46*(14), 7637–7643. https://doi.org/10.1021/ES300955B/ASSET/IMAGES/MEDIUM/ES-2012-00955B_0002.GIF

Shaik, A. M., David Raju, M., & Rama Sekhara Reddy, D. (2020). Green synthesis of zinc oxide nanoparticles using aqueous root extract of *Sphagneticola trilobata* Lin and investigate its role in toxic metal removal, sowing germination and fostering of plant growth. *Inorganic and Nano-Metal Chemistry, 50*(7), 569–579. https://doi.org/10.1080/24701556.2020.1722694

Shebl, A., Hassan, A. A., Salama, D. M., Abd El-Aziz, M. E., & Abd Elwahed, M. S. A. (2019). Green synthesis of nanofertilizers and their application as a foliar for *Cucurbita pepo* L. *Journal of Nanomaterials, 2019*(1), 3476347. https://doi.org/10.1155/2019/3476347

Shenashen, M., Derbalah, A., Hamza, A., Mohamed, A., & El Safty, S. (2017). Antifungal activity of fabricated mesoporous alumina nanoparticles against root rot disease of tomato caused by *Fusarium oxysporium. Pest Management Science, 73*(6), 1121–1126. https://doi.org/10.1002/PS.4420

Singhal, U., Khanuja, M., Prasad, R., & Varma, A. (2017). Impact of synergistic association of ZnO-nanorods and symbiotic fungus *Piriformospora indica* DSM 11827 on *Brassica oleracea* var. botrytis (Broccoli). *Frontiers in Microbiology, 8*, 1909. https://doi.org/10.3389/fmicb.2017.01909

Tan, X. M., Lin, C., & Fugetsu, B. (2009). Studies on toxicity of multi-walled carbon nanotubes on suspension rice cells. *Carbon, 47*(15), 3479–3487. https://doi.org/10.1016/J.CARBON.2009.08.018

Thangavelu, R. M., da Silva, W. L., Zuverza-Mena, N., Dimkpa, C. O., & White, J. C. (2024). Nano-sized metal oxide fertilizers for sustainable agriculture: Balancing benefits, risks, and risk management strategies. *Nanoscale, 16*, 19998–20026. https://doi.org/10.1039/D4NR01354A

Toksha, B., Sonawale, V. A. M., Vanarase, A., Bornare, D., Tonde, S., Hazra, C., Kundu, D., Satdive, A., Tayde, S., & Chatterjee, A. (2021). Nanofertilizers: A review on synthesis and impact of their use on crop yield and environment. *Environmental Technology & Innovation, 24*, 101986. https://doi.org/10.1016/J.ETI.2021.101986

Verma, K. K., Joshi, A., Song, X. P., Singh, S., Kumari, A., Arora, J., Singh, S. K., Solanki, M. K., Seth, C. S., & Li, Y. R. (2024). Synergistic interactions of nanoparticles and plant growth promoting rhizobacteria enhancing soil-plant systems: A multigenerational perspective. *Frontiers in Plant Science, 15*, 1376214. https://doi.org/10.3389/FPLS.2024.1376214/BIBTEX

Wang, X., Han, H., Liu, X., Gu, X., Chen, K., & Lu, D. (2012). Multi-walled carbon nanotubes can enhance root elongation of wheat (*Triticum aestivum*) plants. *Journal of Nanoparticle Research, 14*(6). https://doi.org/10.1007/S11051-012-0841-5

Wang, Z., Zhang, Y., Zhang, S., Ge, M., Zhang, H., Wang, S., Chen, Z., Li, S., & Yang, C. (2023). Natural xylose-derived carbon dots towards efficient semi-artificial photosynthesis. *Journal of Colloid and Interface Science, 629*, 12–21. https://doi.org/10.1016/J.JCIS.2022.09.044

Wu, M., Su, H., Li, C., Fu, Z., Wu, F., Yang, J., & Wang, L. (2023). Effects of foliar application of single-walled carbon nanotubes on carbohydrate metabolism in crabapple plants. *Plant Physiology and Biochemistry, 194*, 214–222. https://doi.org/10.1016/J.PLAPHY.2022.11.023

Xu, E., Liu, Y., Gu, D., Zhan, X., Li, J., Zhou, K., Zhang, P., & Zou, Y. (2024). Molecular mechanisms of plant responses to copper: From deficiency to excess. *International Journal of Molecular Sciences, 25*(13), 6993. https://doi.org/10.3390/IJMS25136993

Yadav, A., Yadav, K., & Abd-Elsalam, K. A. (2023). Nanofertilizers: Types, delivery and advantages in agricultural sustainability. *Agrochemicals, 2*(2), 296–336. https://doi.org/10.3390/AGROCHEMICALS2020019

Zhang, L., Chan, J. M., Gu, F. X., Rhee, J. W., Wang, A. Z., Radovic-Moreno, A. F., Alexis, F., Langer, R., & Farokhzad, O. C. (2008). Self-assembled lipid-polymer hybrid nanoparticles: A robust drug delivery platform. *ACS Nano, 2*(8), 1696–1702. https://doi.org/10.1021/NN800275R/ASSET/IMAGES/MEDIUM/NN-2008-00275R_0005.GIF

Zhao, D., Fang, Z., Tang, Y., & Tao, J. (2020). Graphene oxide as an effective soil water retention agent can confer drought stress tolerance to *Paeonia ostii* without toxicity. *Environmental Science and Technology, 54*(13), 8269–8279. https://doi.org/10.1021/ACS.EST.0C02040/SUPPL_FILE/ES0C02040_SI_001.PDF

Zhao, H., Liu, Z., Han, Y., & Cao, J. (2024). Impact of silver nanoparticles on arbuscular mycorrhizal fungi and glomalin-related soil proteins in the rhizosphere of maize seedlings. *Diversity, 16*(5), 273. https://doi.org/10.3390/D16050273/S1

Chapter 8
Green Synthesized Nanotechnology in Pest and Disease Management

M. Y. Hernández-Ramírez, E. Aguirre-García, and L. A. Martínez-Chávez

8.1 Introduction

Sustainable agriculture is essential for harmonizing food production with environmental conservation and economic sustainability (El-Saadony et al., 2022). It encompasses practices that preserve soil integrity, minimize ecological impacts, and guarantee long-term productivity without compromising natural resources (Usman et al., 2020).

The concept of sustainable agriculture is multifaceted and often lacks a precise definition, posing challenges for its implementation (Velten et al., 2015). It encompasses multiple dimensions—social, economic, and environmental—that must be harmoniously integrated to achieve effective sustainability (German et al., 2017). Sustainable agriculture seeks to minimize adverse environmental impacts, such as greenhouse gas emissions and soil degradation, while ensuring economic viability (Dönmez et al., 2024). This approach involves utilizing renewable energy, reducing chemical inputs, and adopting advanced technologies to improve resource efficiency and productivity (Bathaei & Štreimikienė, 2023; Santiteerakul et al., 2020).

The implementation of sustainable agriculture faces challenges such as scalability, economic limitations, and the necessity for policy reforms (Dönmez et al., 2024). New practitioners in sustainable farming often encounter obstacles like limited access to land, capital, and markets, highlighting the need for supportive frameworks and policies to address these barriers effectively (Carlisle et al., 2019).

Green nanotechnology is rapidly emerging as a transformative strategy in sustainable agriculture, providing innovative solutions to boost productivity while

M. Y. Hernández-Ramírez · E. Aguirre-García · L. A. Martínez-Chávez (✉)
División de Investigación y Posgrado, Facultad de Ingeniería, Universidad Autónoma de Querétaro, Cerro de las Campanas, Santiago de Querétaro, Querétaro, Mexico
e-mail: mhernandez392@alumnos.uaq.mx; eaguirre03@alumnos.uaq.mx; luis.alejandro.martinez@uaq.edu.mx; luis.alejandro.martinez@uaq.mx

K. Esquivel Escalante et al. (eds.), *Green Nanomaterials*, Nanotechnology in the Life Sciences, https://doi.org/10.1007/978-3-032-04492-1_8

reducing environmental impact (Khan et al., 2022). This field harnesses the distinctive properties of nanomaterials to optimize agricultural practices, tackling critical challenges such as resource depletion, climate change, and food security (Rai et al., 2018; Shafey, 2020; Singh et al., 2021). Nanotechnology has the potential to greatly enhance the efficiency of agricultural inputs such as fertilizers and pesticides (Singh et al., 2021). Innovations like nano-plant growth promoters, nanopesticides, and nanofertilizers are designed to boost crop yields, suppress plant pathogens, and manage weeds and pests while minimizing costs and waste generation (Periakaruppan et al., 2023). These nano-enabled products facilitate precise delivery and controlled release of nutrients and agrochemicals, promoting plant growth while significantly reducing environmental contamination (Prasad et al., 2017; Usman et al., 2020).

The integration of nanotechnology and the circular economy in sustainable agriculture presents promising solutions to boost productivity while minimizing environmental impact (El-Ramady et al., 2023). This synergy focuses on optimizing resource utilization, reducing waste, and advancing agricultural sustainability (Prasad et al., 2017). Nanotechnology facilitates the creation of nanofertilizers and nanopesticides, which improve nutrient uptake and pest control efficiency. These advancements reduce reliance on chemical inputs and significantly minimize environmental contamination (Fraceto et al., 2016; Shukla et al., 2024).

Despite their potential benefits, there are concerns about the environmental and health implications of nanoparticles and nanomaterials (Singh et al., 2021). Comprehensive research is required to examine the synthesis, application, and behavior of nanomaterials in agricultural systems, particularly their interactions with soil components and potential toxicity (Singh et al., 2021; Usman et al., 2020). A systems approach is crucial for the design and implementation of nanotechnologies in agriculture. This requires a thorough understanding of the mechanisms governing nanoparticle uptake and translocation within plants, as well as their impacts on plant physiology and soil ecosystems (Bahadur et al., 2023; Lowry et al., 2019).

The aim of this chapter is to explore the applications of nanomaterials in pest and disease management within agriculture, emphasizing their advantages over conventional pesticides. While this chapter does not delve deeply into green synthesis methods, it highlights that the nanomaterials discussed herein can be obtained through the sustainable processes described in earlier chapters. This reinforces the broader context of sustainability and links to the technologies detailed in other sections of the book, offering a comprehensive framework on their applicability and effectiveness in promoting sustainable agricultural practices.

8.2 Nanopesticides: Types and Mechanisms

8.2.1 Introduction to Nanopesticides

One of the principal challenges facing the agricultural sector today is the significant loss of crop production caused by pathogens, weeds, and pests. Combined with the decline in fertile land and the limited availability of nutrients, food security has been severely affected, making the adoption of new technologies a priority (Paradva & Kalla, 2023). Conventional agricultural methods exist to address these challenges, with pesticides playing a significant role in enhancing crop productivity and supporting sustainable agricultural production (Duhan et al., 2017; Singh et al., 2021). However, these practices have notable limitations, primarily due to the excessive use of agrochemicals, which adversely affect biodiversity, disrupt nitrogen fixation, foster disease resistance, and result in the bioaccumulation of toxic substances in aquatic ecosystems, thereby compromising ecosystems and posing serious risks to human health (Ding et al., 2023; Kannan et al., 2023).

Conventional pesticides are synthetically formulated chemicals developed to repel or eliminate pest organisms (Kannan et al., 2023). They are indispensable in modern agricultural practices for managing pathogens and pests that contribute to significant crop losses (Chhipa, 2017). Common types of pesticides include organochlorines, organophosphates, carbamates, and synthetic pyrethroids. Based on their target organisms, pesticides are categorized into several groups, the most prominent being insecticides (targeting insects), herbicides (targeting weeds), rodenticides (targeting rodents), fungicides (targeting fungi), and bactericides (targeting pathogens) (Nisha Raj et al., 2021).

It is estimated that only 0.1% of applied pesticides effectively reach their target organisms, as they are often lost through processes such as photolysis, volatilization, and degradation (Ding et al., 2023; Kannan et al., 2023). Furthermore, conventional pesticides exhibit limited dispersibility and biological activity, alter the genetic composition of pests, leading to increased resistance, and frequently disperse into the environment before reaching their intended targets (Bhaskar et al., 2023; Kumar et al., 2019). This leads to soil and water contamination and serious human health problems through the food chain (Chhipa, 2017; Ding et al., 2023; Kannan et al., 2023). Given their low efficiency, it is imperative to limit and control their use. The growing demand for food underscores the need for sustainable agricultural practices that maintain productivity (Chaud et al., 2021; Yin et al., 2023).

Nanotechnology emerges as a tool offering solutions to promote sustainable agriculture by mitigating the toxic effects of conventional pesticides using nanopesticides (Paradva & Kalla, 2023). These are formulations that allow the controlled release of active ingredients (AI) encapsulated at the nanoscale (Chhipa, 2017; Kannan et al., 2023). Nanopesticides improve water solubility, enhance bioavailability, and protect agrochemicals from environmental degradation (Fraceto et al., 2020; Yin et al., 2023). Nevertheless, the properties of nanomaterials must be carefully assessed, particularly concerning potential cytotoxicity and genotoxicity (Chaud et al., 2021; Dong et al., 2021a).

Nanopesticides offer improved delivery and efficacy, enhancing the stability of their dispersion. Slow or controlled-release formulations provide greater precision and, unlike conventional chemical pesticides, exhibit higher durability and potency while significantly reducing environmental impact (Chhipa, 2017; Kannan et al., 2023). These formulations are cost-effective alternatives that address critical challenges, such as pest resistance, low volatility, the elimination of toxic organic

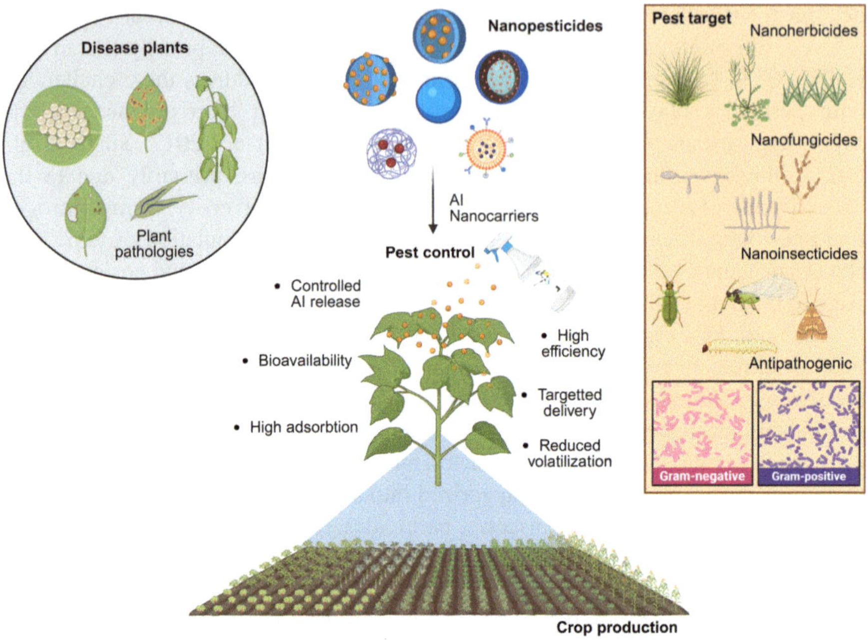

Fig. 8.1 Nanopesticides as protectors and delivery systems for agrochemical ingredients in crop production. (Constructed by M.Y. Hernández-Ramírez)

solvents, improved mobility, and increased insecticidal activity. Furthermore, they require lower doses, reduce runoff, and minimize eutrophication (Gahukar & Das, 2020; Nisha Raj et al., 2021). Nanopesticides represent an innovative technology for agriculture, employing nanomaterials to deliver agrochemical ingredients more efficiently, promoting agricultural productivity while minimizing environmental and health risks associated with conventional pesticides (Bocca et al., 2020; Nisha Raj et al., 2021) (Fig. 8.1).

8.2.2 Types of Nanopesticides

Nanopesticides significantly improve the efficacy and longevity of pesticides while minimizing the quantity of active ingredients required. In agricultural applications, they can be integrated in various ways: fixed onto hybrid substrates, encapsulated within protective matrices, or incorporated into functionalized nanocarriers designed to respond to external stimuli or enzyme-mediated triggers (Anjaneyulu et al., 2024; Bhaskar et al., 2023). Leveraging their distinctive shapes and properties, nanoparticles enhance pesticide performance through advanced nanocarrier systems composed of materials such as silica, lipids, polymers, copolymers, ceramics, metals, and carbon. Nanostructured matrix systems—such as nanocapsules, nanospheres, and nanomicelles—have been meticulously designed and

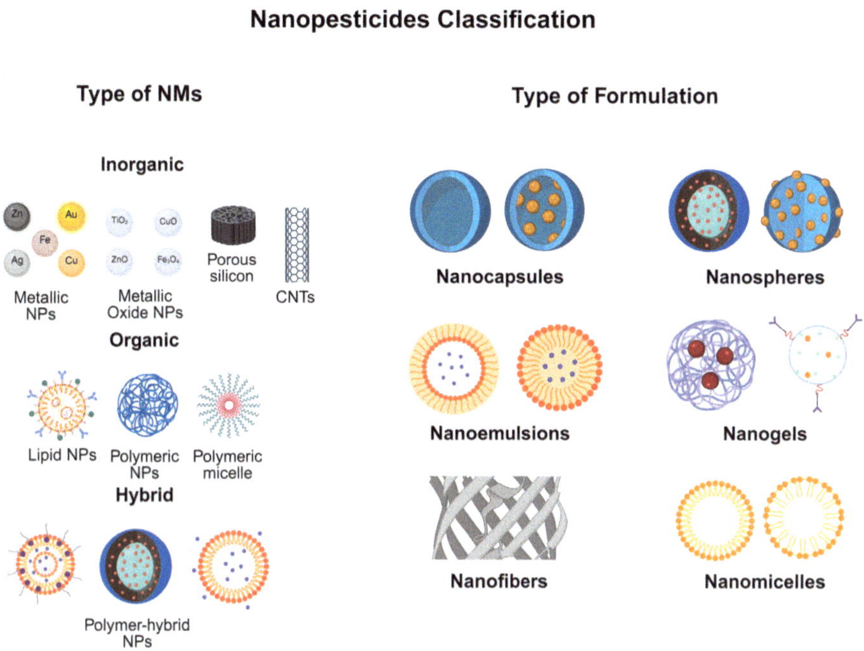

Fig. 8.2 Classification of nanopesticides according to the type of nanomaterials used and the specific formulation developed. (Constructed by M.Y. Hernández-Ramírez)

functionalized with responsive polymeric materials to enable precise and controlled pesticide delivery (Kumar et al., 2019). Therefore, their classification follows two main criteria: the type of nanomaterials used (organic, inorganic, or hybrid) and the specific formulation developed (e.g., nanogels, nanoemulsions, nanocapsules, nanomicelles) (Fig. 8.2).

8.2.3 *Classification by Nanomaterial*

Nanopesticides can be categorized as inorganic, organic, or hybrid (Kumar et al., 2019). Inorganic nanopesticides, often composed of metals and metal oxides, exhibit strong antimicrobial activity through mechanisms such as adhesion, dissolution, oxidative stress, cytotoxicity, and cell death induced by genotoxicity (Anjaneyulu et al., 2024; Dong et al., 2024). Organic nanopesticides rely on nanocarriers, which encapsulate active ingredients to enable controlled release, improve solubility, and protect the pesticide from environmental degradation. Hybrid nanopesticides combine the properties of both inorganic and organic materials to optimize efficiency and versatility. Most nanocarriers, including polymer- and clay-based systems, are biocompatible, cost-effective, and stimuli-responsive, offering tailored delivery solutions aligned with sustainable agriculture goals (Ding et al., 2023; Kannan et al., 2023). Table 8.1 presents the advances in the development of nanopesticides based on organic, inorganic, and hybrid nanoparticles for pest control.

Table 8.1 Advances in the development of nanopesticides based on inorganic, organic, and hybrid NMs for pest control

NMs type	NM/active ingredient	Concentration	Target pest/dual capacity	Efficiency/impact	Reference
Inorganic	Ag	170 mg mL^{-1}	**Insect pests:** *Cabbage loopers*	32.8 ± 1.2 significant effect on larval mortality.	Ashitha and Mathew (2020); Bapat et al. (2022); Jia et al. (2023); Jagadheeswari et al. (2020); Sánchez Reyna (2016); Santos et al. (2021); Sayed et al. (2017); Vidyasagar et al. (2024)
		25, 50, and 75 g/kg soil	*Cyst nematodes*	Showed good nematicidal effects.	Ashitha and Mathew (2020); Bocca et al. (2020); Fabiyi et al. (2018)
		215.76 ppm	*Phytophagous mites*	93.37% with minimal side effects, beneficial organisms.	Ashitha and Mathew (2020); Mhamoud et al. (2019)
		100 ppm	**Pathogens:** *Gram (+) and Gram (−) bacteria*	89.40% eco-friendly antibacterial agents.	Amin (2020); Bapat et al. (2022); Jia et al. (2023); Jagadheeswari et al. (2020); Sánchez Reyna (2016); Santos et al. (2021); Shahryari et al. (2020); Vidyasagar et al. (2024)
	Au	18.8–41.5 ppm, 150 μg/mL, respectively	**Insect pests:** *Mosquito vecctors and larvae*	50% kills the exposed population.	Bahattab et al. (2021); Benelli (2018); Li et al. (2023a)
	MWCNTs	25, 50, and 100 ppm	**Insect pests:** *Larvae of Trogoderma granarium (Khapra Beetle)*	Caused mortality in beetles and larvae 50–70%	Al-Esawy et al. (2021); Wang et al. (2022)
		0.054 mg/L	**Fungus pests:** *Pyricularia oryzae and P. oryzae*	97% fungicidal activity	
	CuO	10, 100, 300, and 500 ppm 5–25 μg/mL	**Insect pests:** *Larvae (S. frugiperda) and mosquito vectors*	70–100% larvicidal activity 8–98% aantifeedant	Kora (2022); Li et al. (2023a); Rahman et al. (2022); Vivekanandhan et al. (2021)

		15, 20, and 25 μL	**Pathogens:** *Gram (+) and Gram (−) bacteria*	13.6 ± 7 inhibition of bacterial action	Jia et al. (2023); Liu et al. (2023); Priya et al. (2023)
		15, 20, and 25 μL	**Fungus pests:** *Aspergillus flavus*, *Aspergillus niger*, and *Penicillium frequentans.*	13.1 ± 1.1, 14.7 ± 0.7, 16.2 ± 1.4, respectively.	Liu et al. (2023); Priya et al. (2023)
	TiO_2	100 ppm	**Insect pests:** *Larvae (Helicoverpa armigera)*	100% mortality, antifeedant activity	Chinnaperumal et al. (2018); Narayanan et al. (2021); Senbill et al. (2023)
		5.82 mg L^{-1}	*Spider mite (Tetranychus urticae Koch)*	92.4% mortality	
		100–1000 μg/mL	*Mosquito (Aedes aegypti)*	Larvicidal and pupicidal activity	
		0.1 mg mL^{-1}	**Pathogens:** *Ralstonia solanacearum (Gram-negative)*	85.99% inhibition of bacteria	Li et al. (2023a); Pan et al. (2023)
	ZnO	100–500 ppm	**Insect pests:** *Larvae (S. frugiperda)*	Reduce their population	El-Ansary et al. (2022); Li et al. (2023a); Malaikozhundan and Vinodhini (2018); Pittarate et al. (2021); Shukla et al. (2020)
		0.25–0.5 mg/mL	*Nematode (Meloidogyne incógnita)*	72.86% mortality	
		10–20 ppm	*White grubs (Holotrichia* sp.)	Larvicidal properties	
		25 ppm	*Pulse beetle (Callosobruchus maculatus)*	100% mortality	
		0.5, 1, 2, 4, 6, 8, and 10 mg/L	*Mosquito (Aedes aegypti)*		
		100 and 250 ppm	**Pathogens:** *Bactericera cockerelli (Hemiptera: Triozidae)*	Increased mortality of 99–100%	Gutiérrez-Ramírez et al. (2021)
	Fe	130.31 and 89.32 mg/L	**Insect pests:** *Tobacco cutworm (Spodoptera*	High mortality 92.83% and 91.41%,	Li et al. (2023a); Muthusamy et al. (2023); Ramkumar et al. (2021); Xu et al. (2020)

(continued)

Table 8.1 (continued)

NMs type	NM/active ingredient	Concentration	Target pest/dual capacity	Efficiency/impact	Reference
			litura) and cotton bollworm (Helicoverpa armigera)	respectively, antifeedant activity	
		100 μg mL^{-1}	*Tuta absoluta (Lepidoptera)*	Maximum 72% mortality	
		200 and 500 ppm	*Larvae (Spodoptera litura)*	80–98% mortality	
	Fe_3O_4	15, 30, 45, 60, 75, and 90 μg/mL	**Insect pests:** *Musca domestica*	High mortality by contact	Li et al. (2023a); Toto et al. (2022)
Organic	Chitosan	100 μg/mL	**Insect pests:** *Flie and larvae (Drosophila melanogaster)*	Increased toxicity and reduced climbing ability	Pontes et al. (2021); Sharma et al. (2019); Wu et al. (2022)
		50 ppm	*Plutella xylostella (Linnaeus)*	High mortality	
	Chitosan-lanthanum/ avermectin	1000 μg/mL, 0.5% Silwet L-77	**Fungus pests:** *Magnaporthe grisea*	Fungicidal activity	Liang et al. (2018)
	Chitosan-alginate/ imazapyr and imazapyc	400 g/ha	**Weed pests:** *Bidens pilosa*	Herbicidal activity	Maruyama et al. (2016)
	Polymer-valved MoS2-embedded mesoporous silica	2.0 mg L^{-1}	**Fungus pests:** *Rhizoctonia solani and Fusarium graminearum*	69–70% efficiency	Dong et al. (2021b)
	Polysuccinimide nanoparticles/ emamectin benzoate	2 μg mL^{-1}	**Insect pests:** *Plutella xylostella*	80% mortality	Ye et al. (2022)
	Rosin modified carboxymethyl cellulose	2.62, 2.40 and 2.39 mg/L	**Insect pests:** *Larvae (Plutella xylostella)*	80.00 ± 10% of pests control rate	Zhao et al. (2021)
	Sodium alginate	10 μg/mL	**Insect pests:** *Leafhoppers*	98% mortality	Kumar et al. (2014)
	Diacyl hydrazine-based nanoformulation	50–200 mg L^{-1}	**Fungus pests:** *Colletotrichum gloeosporioides,*	50% growth inhibition	Pandey et al. (2020)

			Rhizoctonia solani, Fusarium solani, and Alternaria solani		
		0.015 and 0.010 mg/L	**Insect pests:** *Spodoptera litura*		
Hybrid	Lambda-cyhalothrin/silver nanoparticles	0.01–0.32 ppm	**Insect pests:** *Larvae (Spodoptera littoralis)*	10–95% mortality	Ahmed et al. (2019)
	Silica/hydroxypropylcellulose hybrids doped with copper ions	*5 wt.%*	**Pathogens (bacteria and fungus):** *Bacillus subtilis, Escherichia coli and yeast*	94% mortality	Arjunan et al. (2017); Chauhan et al. (2020); Foteva and Georgieva (2022); Pho et al. (2020); Safaei et al. (2019)
	Chitosan-coated CuNPs	100, 500, and 1000 μg mL^{-1} (chitosan 500 μg mL^{-1})	*Staphylococcus aureus, Streptococcus pneumonia, Pseudomonas aeruginosa, Candida albican*	Inhibited the growth of both bacteria and fungus	Arjunan et al. (2017)
	CuO alginate-based composite	4 mg/mL (CuO NPs) and 1 mg/mL alginate	*Aspergillus niger*	Inhibition of fungal growth as 75.08%	Chauhan et al. (2020)
	Chitosan-porous carbon NPs/paraquat	150 μg mL^{-1}	**Fungus pests:** *Candida albicans*	Fungal inhibition	Dong et al. (2021c); Morsi et al. (2017); Sharifiaghdam et al. (2022); Song et al. (2019)
		1–3.0 mg mL^{-1}	**Weed pests:** *Cynodon dactylon*	99.7% damaged efficacy	
	Carboxymethyl chitosan modified carbon nanoparticle	4 mg/L	**Insect pests:** *Mythimna separata*	80% mortality	Pho et al. (2020)
	Chitosan coated metallic nanoparticles (Ag, Au, Se)	0.5–100 μg/mL	**Pathogens:** *E. coli, S. aureus, and S. epidermidis*	Antibacterial effects induce ROS	Dong et al. (2021c)
	Cu/Ag-CNT chitosan-based composites	1% and 2% initial concentration	*S. aureus and fungus A. flavus*	Antibacterial and antifungal activity	Song et al. (2019)

Inorganic nanomaterials such as noble metal ions (e.g., Ag, Au, Cu), exhibit unique insecticidal properties through surface plasmon resonance mechanisms, offering alternatives to traditional pesticides because they are effective in protecting active ingredients against degradation and volatilization while enhancing pesticide stability and efficiency (Chaud et al., 2021; Chhipa, 2017; Yin et al., 2023). Also, organic nanomaterials like polymers, both natural (e.g., chitosan, cellulose, polylactide) and synthetic (e.g., polylactic acid, polyethylene glycol), are widely used for nanocarrier formulations (Anjaneyulu et al., 2024; Bocca et al., 2020; Dong et al., 2024). These materials are biodegradable, environmentally friendly, and capable of encapsulating hydrophilic and hydrophobic active ingredients, prolonging their release and improving bioavailability (Mondéjar-López et al., 2024). Combining inorganic and organic components, hybrid systems aim to leverage the strengths of both material types, achieving higher efficiency in pesticide delivery while maintaining environmental safety (Jia et al., 2023; Rivero-Montejo et al., 2021; Shukla et al., 2016).

8.2.4 Classification by Formulation

Nanopesticides are also produced either through direct synthesis of nanoscale pesticides or by loading active ingredients into nanocarriers (Granetto et al., 2022; Mondéjar-López et al., 2024). In the latter, nanocarriers encapsulate, absorb, or bind active ingredients, protecting them during transport and releasing them at target sites under optimal conditions (Chhipa, 2017; Ding et al., 2023; Kannan et al., 2023; Yin et al., 2023) (Fig. 8.3). This approach ensures the active ingredients maintain stability against environmental degradation, such as high temperatures or ultraviolet radiation, reducing unnecessary losses and requiring smaller doses to achieve the desired efficacy. Controlled-release formulations prolong the active concentration at target sites, further reducing the quantity of active ingredients needed (Anjaneyulu et al., 2024; Bhaskar et al., 2023). Stimuli-responsive nanocarriers enhance delivery precision, minimizing environmental contamination and improving pest control effectiveness (Ding et al., 2023; Yin et al., 2023). Various nanocarriers, including

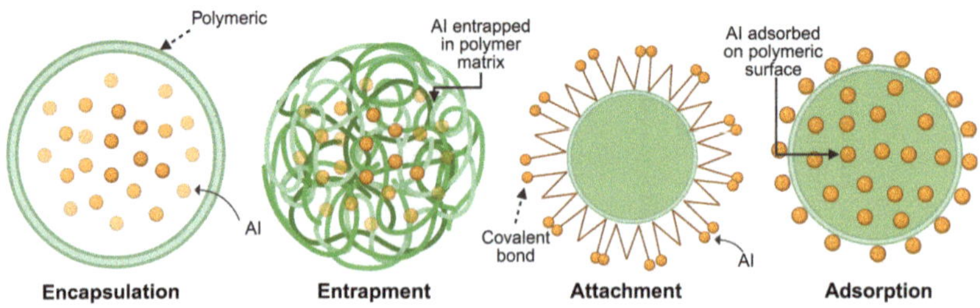

Fig. 8.3 Nanocarriers used in nanopesticides formulations. (Constructed by M.Y. Hernández-Ramírez)

nanocapsules, nanospheres, nanofibers, nanoemulsions, nanomicelles, and nanogels, have been developed, each offering specific advantages depending on their design and application (Chaud et al., 2021; Gahukar & Das, 2020; de Lima et al., 2021). Table 8.2 lists the advances in the development of nanoformulations for pest control.

Encapsulating active ingredients within a protective nanolayer facilitates slow and controlled release, significantly reducing runoff and extending the duration of effectiveness (Ding et al., 2023) (Fig. 8.4). Nanoencapsulation not only minimizes the degradation and evaporation of active ingredients but also enhances their interaction with pests, increasing overall efficiency (Yin et al., 2023). Colloidal systems, such as nanoemulsions, enhance the solubility, stability, and bioavailability of poorly soluble agrochemicals, enabling prolonged release and reducing the quantities required for effective pest control (Anjaneyulu et al., 2024).

Nanogels, known for their high permeability and biocompatibility, are particularly effective for encapsulating both hydrophobic and hydrophilic compounds. They allow precise delivery of active ingredients while minimizing environmental toxicity (Kumar et al., 2019). Similarly, nanomicelles, which are supramolecular assemblies, improve the solubility and stability of active compounds. Meanwhile, nanofibers, with their extensive surface area and high porosity, overcome challenges associated with polymer-based formulations by ensuring a more uniform release (Mondéjar-López et al., 2024).

Innovative nanocarrier systems, such as dual-stimuli-responsive matrices that react to pH and ionic strength, are being developed to optimize release mechanisms, reduce pesticide usage, and minimize environmental impact (Ashitha & Mathew, 2020; Bhaskar et al., 2023). These advancements are paving the way for more sustainable and effective pest control strategies in agriculture (Chaud et al., 2021).

Nanoparticles, classified as inorganic or organic, are particularly valued for their small size, high surface area, and strong adsorption capacity. These properties improve chemical stability, water dispersibility, and pesticide delivery efficiency (Nisha Raj et al., 2021; Yin et al., 2023). Additionally, nanocarriers with functionalized biocompatible groups facilitate rapid and efficient pesticide uptake by plants or pest cells, reducing application rates and environmental risks while enhancing agricultural productivity (Paradva & Kalla, 2023). Recent research has focused on smart-release systems, targeting precise delivery and minimizing pesticide wastage and toxicity. New nanocarrier development is geared towards biodegradability, cost-effectiveness, and low environmental impact (Ashitha & Mathew, 2020).

8.2.5 Mechanisms of Action

Nanopesticides affect plant growth and metabolism depending on their efficiency of absorption through mechanisms such as phagocytosis, pinocytosis, and endocytosis, enabling plants to minimize toxicity (Chaud et al., 2021; da Silva Pontes et al., 2023). The maintenance of cellular flux in plants depends on adjustment responses

Table 8.2 Advances in the development of nanoformulations for pest control

Structured type	Properties/function	Formulation/active ingredient	Concentration	Target pest	Impact	Reference
Nanocapsules	Hollow and solid nanoparticles, wrapping membrane, and oily core. Encapsulate molecules in their central domain (active ingredients). Reduce leaching and volatilization of harmful substances.	*Atrazine/poly (ε-caprolactone)*	0.1-1 mg mL^{-1}	**Weed pests:** *B. pilosa*	80% mortality percentage	Campos et al. (2018); Preisler et al. (2020); Rajkumar et al. (2020); Shoaib et al. (2018)
		Mesoporous nanosilica and silicon dioxide nanoparticles/ emamectin benzoate	0.18 mg L^{-1}	**Insect pests:** *Plutella xylostella*	100% mortality	
		β-Cyclodextringrafted chitosan nanoparticles co-loaded with carvacrol and linalool	0.05 mg/mL	*Corn earworm (Helicoverpa armigera) and Spider mite (Tetranychus urticae)*	80% repellency percentage exceeding	
		Peppermint essential oil in chitosan nanoparticles	75.0 µL/L	*T. castaneum and S. oryzae*	100% mortality	
Nanospheres	Prepared with the pesticide trapped inside, adsorbed uniformly on the surface of the spherical structures. Possess a dense polymer network (amorphous or crystalline). Their structure protects the active ingredient from chemical and enzymatic degradation	*Propiconazole encapsulated/poly lactic acid and polymer poly (lactic-co-glycolic) acid*	100 µM	**Fungus pests:** *F. solani*	Encapsulation efficiency over 42%, antifungal activity	Barrera-Méndez et al. (2019); Giongo et al. (2016); Kan et al. (2022); Wang et al. (2020)
		Poly (-β-hydroxybityrate) suspension/neem based	0.5% and 1.0%	**Insect pests:** *S. frugiperda*	76.67% mortality, antifeedant activity	
		Pyraclostrobin	0.01, 0.25, 0.5,2.5, and 5 mg L^{-1}	**Fungus pests:** *P. ochrochloron*	Fungicidal activity	
			16.7% Cpf		Mortality rates reached 100%	

		Polydopamine modified mesoporous silica/ chlorpyrifos		**Insect pests:** *Larvae (M. separata)*		
Nanogels	Environmentally stable. Can be loaded with volatile substances	*Gelatin-sodium alginate composite*	128, 64, 32, 16, 8, 4, 2, and 1 μg/mL	**Pathogens:** *Staphylococcus aureus*	Inhibited bacterial density by 99.9%	Almeida et al. (2018); Luo et al. (2022); Zhang et al. (2022a)
		Emamectin benzoate nanogel suspensión/ poly(vinyl alcohol)-valine derivatives and lignosulfonate	10 mg L^{-1}	**Insect pests:** *Plutella xylostella*	Long duration of pest control	
		Copper nanoparticle composite in alginate/ CTAC	58 μg mL^{-1}	**Pathogens:** *Pseudomonas syringae*	64% efficiency (Cui et al., 2024; Ma et al., 2022)	
		Nano-gelation particles/ coated cycloxaprid	0.5, 1, 2, 4 and 8 mg/L	**Insect pests:** *Diaphorina citri*	97.6% lethality rate	
Nanoemulsions	Nanopesticide dispersion in water. Surfactant molecules are present at the interface. Coating of agrochemical ingredients	*Emamectin benzoate–sodium lignosulfonate nanoformulation*	100 and 200 ppm	**Insect pests:** *Prodenia litura*	70% fatality rate	Adak et al. (2020); Assalin et al. (2019); Cui et al. (2019); Erlina et al. (2020)
		Thiamethoxam insecticide–polycaprolactone (PCL)–chitosan nanoformulation	901.59 mg L^{-1}	*Raphidocelis subcapitata and C. elegans*	Inhibited growth	
		Piper aduncum extract	0.32% concentration (100% extract)	*Crocidolomia pavonana F. (Lepidoptera: Crambidae) and C. pavonana larvae*	95% mortality	
		Eucalyptus oil				

(continued)

Table 8.2 (continued)

Structured type	Properties/function	Formulation/active ingredient	Concentration	Target pest	Impact	Reference
			0.795 and 4.178 µL/cm^2	*Sitophilus oryzae (L.) and Tribolium castaneum*	Alternative to stored grain pests	
Nanofibers	Controlled delivery systems. Have great length, small diameter, and large specific surface area	*L-Carvone-based thiazolinone–hydrazone/ nanochitosan complexes*	50 µg/mL	**Fungus pests:** *Cercospora arachidicola and Alternaria solani*	58.8~71.0% and 49.0~68.5%, respectively	Czarnobai De Jorge et al. (2022); Farias et al. (2019); Li et al. (2024); Shangguan et al. (2023)
		Cellulose diacetate (CDA)/abamectin or fluopyram	*Not specified*	*Alternaria lineariae*	Inhibits fungal growth	
		Polyhydroxybutyrate/ pheromones S. litura	*50 mg/25 mL, and 1 mL*	**Insect pests:** *Spodoptera litura (Lepidoptera: Noctuidae)*	Trapped and inhibition	
		Nanofiber-based dispensers (PCL)/clove essential oil	*10% of fiber*	*Pear psyllids (Hemiptera: Psyllidae)*	80% mortality	
Nanomicelles	Formed from amphiphilic copolymers in aqueous solution Hydrophobic pesticides are solubilized in the inner region of the micelles	*Chitosan isoleucine polyethylene glycol methyl ether (CIP)/ spinosad*	5 and 20 mg/L	**Fungus pests:** *Fusarium oxysporum*	Higher antifungal property	Dong et al. (2019); Sarkar et al. (2023); Zhang et al. (2020b); Zhou et al. (2024)
		PEG/amphiphilic copolymers (ACPs) a	11.55–38.75 mg/L	**Weed pests:** *Atherigona soccata and Tribolium castaneum*	Encapsulation efficiency 90%, antifungal activity	

	mPEG/camptothecin	25 µg/mL	**Insect pests:** *Brevicoryne brassicae, Tetranychus cinnabarinus, and Bursaphelenchus xylophilus*	75% mortality	
	N-succinyl chitosan (NSCS)/BTL-11	11.1 µg/mL	**Fungus pests:** *Rhizoctonia solani*	81.1–85.9% efficiency	

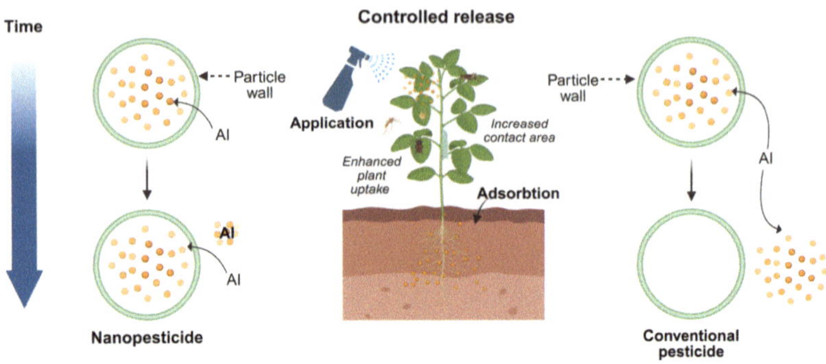

Fig. 8.4 Advantages of nanopesticides in controlled release of active ingredients (AI). (Constructed by M.Y. Hernández-Ramírez)

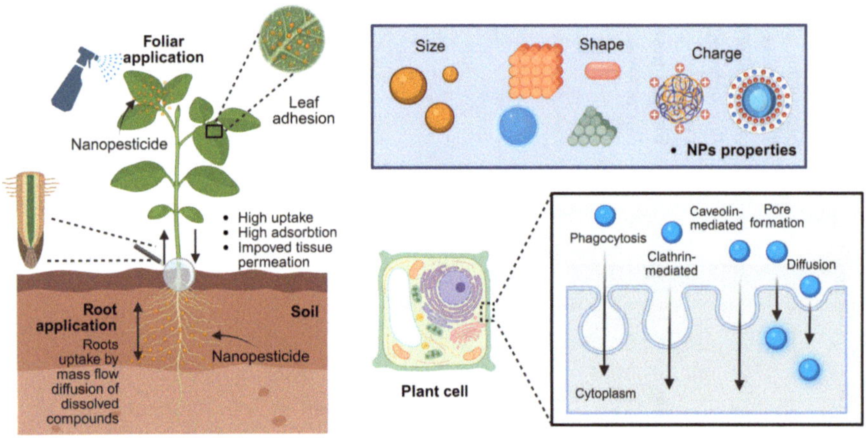

Fig. 8.5 Principal pathways of nanopesticides in plant uptake and physicochemical properties that influence the cellular uptake of nanopesticides through cell membranes. (Constructed by M.Y. Hernández-Ramírez)

induced by nanomaterials (NMs), which are linked to the plant's homeostatic capacity to counteract biotic and abiotic environmental stressors. These responses can include allosteric mechanisms that trigger complex metabolic processes (Li et al., 2023b).

The uptake of nanopesticides occurs when a particle penetrates the plant cell walls after application. The type, composition, concentration, size, shape, and charge of the nanomaterials play a critical role in their absorption and cellular uptake through the cell wall pores (Fig. 8.5) (Fraceto et al., 2020; Kannan et al., 2023). The complex nature of plant cell walls imposes stringent parameters for selecting penetrating nanocarriers in terms of size, morphology, and surface characteristics. For instance, controlled-release matrices and the antimicrobial potential of chitosan nanoparticles are considered innovative strategies for microbial control. Moreover, the selection of the polymer and matrix for controlled pesticide release depends on several factors, including the matrix preparation technique, pesticide type, experimental conditions, and the essential requirement that the polymer is biodegradable (Kannan et al., 2023; Takeshita et al., 2023).

The oxidative stress mechanisms and plant growth inhibition related to nanopesticides involve various pathways where reactive oxygen species (ROS) are natural byproducts of cellular oxidative metabolism (da Silva Pontes et al., 2023; Marslin et al., 2017). However, cellular ROS levels must remain low to maintain cellular homeostasis. Otherwise, high ROS concentrations can be highly toxic to plants, leading to DNA damage, lipid peroxidation, enzyme inhibition, and cell death. While elevated ROS levels are associated with several negative effects on plant cell metabolism, these molecules can also play a role in complex regulatory mechanisms for plant-environment integration (Del Prado-Audelo et al., 2022; Li et al., 2023b). The positive role of ROS in plant regulation depends on the balance between ROS generation and their neutralization by cellular antioxidant agents. For example, metallic nanoparticles used as nanopesticides can induce intracellular ROS generation, suppressing crop pests and pathogens by directly targeting specific sites. Plants exposed to nanopesticides trigger ROS generation, and the induced metabolites act as ROS scavengers (Fraceto et al., 2020).

On the other hand, the application methods of nanopesticides are crucial in determining the plant's uptake efficiency. In root applications, roots specialize in nutrient and water absorption. Thus, nanopesticides can migrate from the epidermal layer to endodermal cells through the apoplastic pathway. For foliar applications, leaves rely on the stomatal pathway as the most likely route for nanopesticide internalization, although the cuticle presents a barrier that hampers substance penetration (Fraceto et al., 2020; Li et al., 2023b).

Nanocarrier systems designed to deliver pesticides, herbicides, and plant growth regulators may be more complex due to the composition of the carrier system. The ability of nanocarriers to protect active ingredients from degradation and adhere to plant surfaces prolongs the contact time between the agrochemical and the plant surface. This feature is often a major factor in the superior efficiency of these nanopesticides compared to the free forms of agrochemicals (da Silva Pontes et al., 2023; Takeshita et al., 2023).

8.2.6 Environmental Impact

The fate and behavior of pesticides during and after their application in the receiving environment are crucial determinants of their potential impact on ecosystems and human health (Panpatte & Jhala, 2019). Generally, nanomaterials such as carbon nanotubes (CNTs) (Al-Esawy et al., 2021; Wang et al., 2022), silver nanoparticles (Ag NPs) (Ashitha & Mathew, 2020; Mhamoud et al., 2019), and zinc oxide nanoparticles (ZnO NPs) have shown low risks when exposed at lower concentrations (El-Ansary et al., 2022; Li et al., 2023a; Malaikozhundan & Vinodhini, 2018;

Pittarate et al., 2021; Shukla et al., 2020). However, the increased application of nanomaterials in agricultural practices or their potential incorporation into food products could elevate exposure levels and disrupt the balance between their benefits and risks (Chaud et al., 2021; Fraceto et al., 2020).

The targeted release of active ingredients in nanopesticides responds to specific stimuli such as pH, light, temperature, or enzymes, enhancing pesticide bioavailability and reducing environmental contamination. By utilizing nanocarriers such as metal-organic frameworks or polymers, nanopesticides can deliver pesticides directly to pests, pathogens, and foliage, improving efficacy and minimizing undesired effects. These formulations release pesticides in response to environmental or biological triggers (Fraceto et al., 2020; Paradva & Kalla, 2023; Takeshita et al., 2023).

Nanopesticides have demonstrated increased effectiveness against various pests, including weeds, insects, and pathogens, by improving delivery and reducing the required pesticide concentration (Chaud et al., 2021). Silver nanoparticles disrupt insect antioxidant and detoxification enzymes, causing oxidative stress and cell death (Amin, 2020; Bapat et al., 2022; Fabiyi et al., 2018; Santos et al., 2021; Shahryari et al., 2020). Chitosan-based and hybrid polymer nanopesticides can alter gene expression, resulting in developmental and reproductive failures in insects (Arjunan et al., 2017; Dong et al., 2021c; Liang et al., 2018; Pontes et al., 2021; Zhou et al., 2024). Metallic nanoparticles can bind to sulfur and phosphorus in proteins and nucleic acids, reducing membrane permeability and causing the denaturation of organelles and enzymes, ultimately leading to cell death (Carley et al., 2020; Dong et al., 2021a).

Mesoporous silica nanopesticides embedded with MoS2 release pesticides in response to low pH, while light-sensitive nanocapsules release pesticides upon UV irradiation, ensuring precise and controlled delivery to target sites (Dong et al., 2021b). Once released, nanoparticles undergo transformations, accumulate in soil, and impact non-target organisms and ecosystems. Properties such as the stability of the carrier influence their environmental behavior, underscoring the need for careful development of nanopesticides, as they can alter soil microbial communities and enzymatic activities (Panpatte & Jhala, 2019).

Similarly, nanopesticides can have significant effects on non-target aquatic organisms (Fig. 8.6). Studies have demonstrated changes in sediment communities, particularly affecting eukaryotic organisms such as protists and algae, indicating potential risks to aquatic biodiversity (Panpatte & Jhala, 2019). Atrazine-loaded nanocapsules were more toxic to some fish species, causing DNA damage and biochemical changes. Silver nanoparticles affected fish reproduction, blood parameters, and survival rates. Copper oxide nanoparticles caused tissue damage and mortality in fish, while zinc oxide nanoparticles reduced protein levels in common carp (Carley et al., 2020; Simonin et al., 2018; Zhang et al., 2020a).

Nanopesticides can influence the degradation rates of other pesticides, potentially leading to prolonged environmental presence and increased exposure risks. The toxicity and bioaccumulation of nanopesticides can vary between species; however,

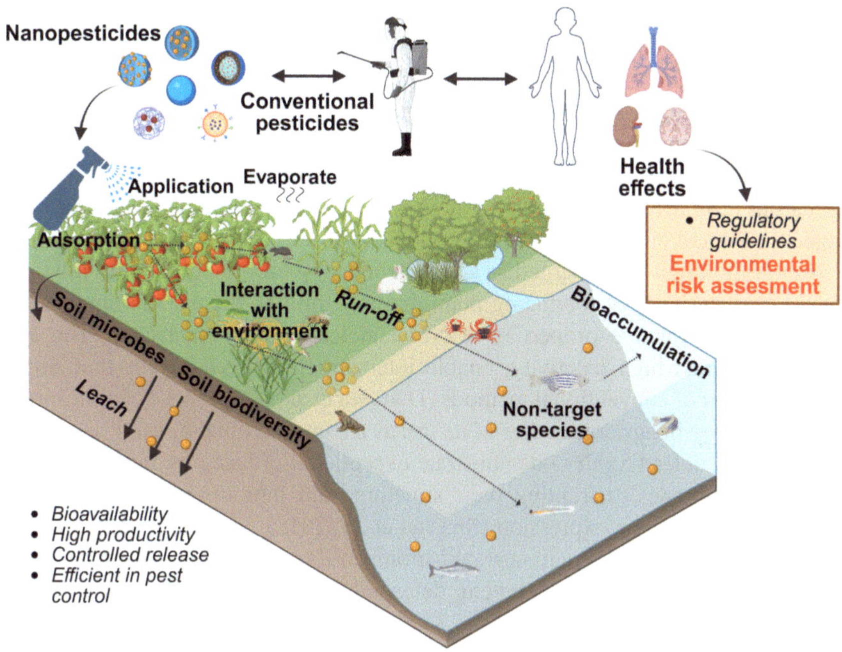

Fig. 8.6 Environmental impact of nanopesticides. (Constructed by M.Y. Hernández-Ramírez)

they may be less toxic and bioaccumulative compared to conventional products, thereby reducing environmental impact (Kah et al., 2016).

It is essential to promote adaptations in risk assessment procedures to address the environmental fate, exposure, and ecotoxicity of nanopesticides. Long-term studies are necessary to better understand the chronic impacts of nanopesticides on ecosystems, particularly under varying agricultural practices and environmental conditions. While nanopesticides offer promising benefits for agricultural productivity, they also present environmental challenges that require comprehensive risk assessments and regulatory adaptations (Chaud et al., 2021; Fraceto et al., 2020; Yin et al., 2023). Their impacts on soil and aquatic ecosystems, as well as their interactions with other agrochemicals, highlight the need for careful management to ensure sustainable use (Fig. 8.6) (Carley et al., 2020; Panpatte & Jhala, 2019).

8.3 Antimicrobial Nanoparticles: Types and Mechanisms

Climate change, combined with the emergence of new pathogens and diseases, highlights the pressing need for solutions to meet the demands of an ever-growing global population. The agricultural sector, heavily dependent on climatic conditions, is particularly vulnerable to these changes. Phytopathogens alone account for annual crop losses of 20–30%, presenting a substantial threat to global food security (Bahadur et al., 2023; Fu et al., 2020).

Traditional crop protection methods rely heavily on the extensive use of chemical pesticides, which often exhibit low efficiency, cause environmental harm to non-target organisms, contribute to pollution, promote resistance development in pathogens, and pose significant risks to human health (Pandey et al., 2018). These limitations have driven the search for alternative and more sustainable approaches to protect crops and ensure agricultural productivity.

Nanotechnology has emerged as a groundbreaking tool in agriculture, providing innovative solutions for combating plant diseases through the use of nanomaterials (NMs). The distinctive properties of NMs, including their nanoscale size, large surface area, and high reactivity, enable interactions with pathogens that surpass the capabilities of conventional methods (Dwivedi et al., 2016). A key application of nanotechnology in agriculture lies in its use as an antimicrobial agent. For instance, silver nanoparticles (AgNPs) demonstrate exceptional efficacy by targeting microbial cell membranes, disrupting their structure, and inhibiting critical biological processes such as gene expression (Sharma et al., 2018). Similarly, polymer-based copper nanoparticles have demonstrated strong antifungal activity against a range of plant pathogens, providing a potent defense against fungal diseases (Chauhan et al., 2020).

Conventional methods often lead to significant waste, with over 90% of applied chemicals failing to reach their intended targets. In contrast, NMs enable the precise delivery of active ingredients, ensuring that pesticides, fungicides, and other treatments are applied directly where they are needed, thereby enhancing efficiency and minimizing waste (Shang et al., 2019). This reduces the required dosage, minimizes environmental impact, and improves overall efficacy (Medina-Pérez et al., 2019).

This section will explore the various types of nanomaterials employed in combating plant pathogens and diseases, emphasizing how their unique properties make them ideal for crop protection. By examining the interactions of metallic nanoparticles, carbon-based NMs, polymeric nanoparticles, and other advanced materials with pathogens, we can uncover their transformative potential in disease management. Furthermore, this section will highlight the integration of these NMs into agricultural systems to tackle critical challenges in crop protection, improve plant health, and contribute to sustainable farming solutions.

8.3.1 Types of Antimicrobial Nanomaterials

Antimicrobial nanomaterials are gaining attention for their potential to boost agricultural productivity by effectively controlling phytopathogens and pests. These materials provide innovative alternatives to traditional approaches, utilizing the unique properties of nanoparticles to achieve potent and targeted antimicrobial activity (Brandelli, 2015; Morsi et al., 2017; Sportelli et al., 2018).

There are various types of nanomaterials that can be used as antimicrobial agents, including those based on metallic nanoparticles, metal oxides, carbon-based materials, and polymers (Morsi et al., 2017). Metallic nanoparticles exhibit antimicrobial properties by interacting with the cell membranes of pathogens, causing structural damage and disrupting essential biological processes (Nisar et al., 2019; Salem et al., 2021). Metal oxides generate reactive oxygen species that can destroy microorganisms through oxidative mechanisms (Rajwade et al., 2020). Carbon-based nanomaterials, with their high surface area and unique structure, can physically disrupt cell membranes or interfere with key metabolic processes (Bhattacharya et al., 2016; Hoseinzadeh et al., 2016). Lastly, polymer-based nanomaterials, known for their biocompatibility and biodegradability, are capable of releasing antimicrobial agents in a controlled manner to maximize their effectiveness (Brandelli, 2015). These properties make each type of nanomaterial suitable for specific applications in crop protection and the management of agricultural diseases (Hoseinzadeh et al., 2016). The following table (Table 8.3) presents an overview of various nanomaterials categorized by type, highlighting their key properties, antimicrobial mechanisms, and applications.

Table 8.3 Classification of nanomaterials by type, their key properties, antimicrobial mechanisms, and agricultural applications

Type	Nanomaterial	Key properties	Antimicrobial mechanism	Applications	Reference
Metallic	Silver (AgNPs)	Highly effective antimicrobial agent Low toxicity and adaptable Antioxidant and anticancer properties	Disrupts bacterial cell membranes Impairs respiration Interacts with proteins and DNA, halting microbial growth	Used in medical coatings, textiles, and agricultural disease management	Brandelli (2015)
	Copper (CuNPs)	Essential trace element in plants Strong antimicrobial activity against bacteria and fungi Low cost	Inhibits bacterial growth (e.g., *S. aureus, B. subtilis, E. coli, K. pneumoniae*) Antifungal activity	Applied in fungicides, pesticides, and as an alternative to AgNPs in agriculture	Kasana et al. (2016); Ul Haq and Ijaz (2019)
Oxide-based	Zinc oxide (ZnO NPs)	Versatile compound with optical, electrical, and photocatalytic properties Nutritional benefits for plants	Exhibits antimicrobial effects by targeting microbial cell walls and oxidative stress pathways	Used in micronutrient fertilizers and for microbial threat management in agricultural settings	Faizan et al. (2020)

(continued)

Table 8.3 (continued)

Type	Nanomaterial	Key properties	Antimicrobial mechanism	Applications	Reference
	Titanium oxide (TiO_2 NPs)	Strong oxidizing potential Non-toxic and chemically stable Enhances photosynthesis in plants	Effective against bacteria, fungi, viruses, and prions Long-term antibacterial effects	Improves crop yields and reduces disease severity; used in sustainable agricultural practices	Nadeem et al. (2018); Rajwade et al. (2020)
Carbon-based	Graphene	Single-layer carbon lattice Exceptional electronic, thermal, and mechanical properties	Functional groups interact with microbial cell walls and proteins Causes oxidative stress	Used in medical devices, agriculture, and environmental applications	Chen et al. (2016); Zhang et al. (2022b)
	Carbon nanotubes (CNTs)	Unique tubular structure (SWCNTs and MWCNTs) High loading capacity Excellent physical properties	Penetrates bacterial membranes Causes structural damage leading to cell death	Drug delivery systems, antimicrobial coatings, and environmental remediation	Al-Esawy et al. (2021); Shoala (2020)
Polymer-based	Chitosan	Biocompatible and biodegradable Derived from crustaceans Proven antimicrobial properties	Inhibits fungal and bacterial pathogens Enhances plant defense mechanisms	Used in agriculture, drug delivery systems, and food preservation	Agnihotri et al. (2004); Brandelli (2015)
	Gelatin	Cost-effective and biodegradable Encapsulates acidic and basic substances	Controlled release of antimicrobial agents through nanocapsules	Drug delivery, tissue engineering, and antimicrobial coatings	
	Nanofibers	Produced via electrospinning High surface area Modifiable with functional groups	Controlled drug release and improved antimicrobial activity Targets *S. aureus* and *E. coli*	Drug delivery, wound healing, and antimicrobial materials	Spasova et al. (2010)

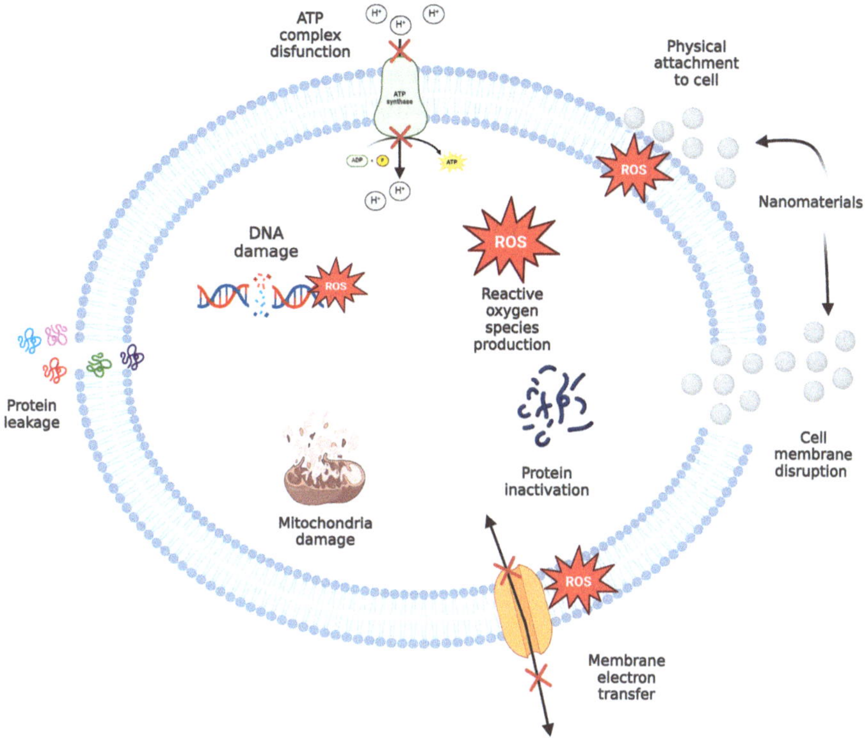

Fig. 8.7 Schematic representation of proposed mechanisms of nanomaterial toxicity against bacteria and fungi. (Constructed by E. Aguirre-García)

8.3.2 *Mechanisms of Antimicrobial Activity*

Nanomaterials exhibit antimicrobial properties through several key mechanisms, making them promising agents against antibiotic-resistant pathogens. These mechanisms include the production of reactive oxygen species (ROS), disruption of cell membranes, and interference with cellular components and processes, leading to cell death (Fig. 8.7) (Dayem et al., 2017).

8.3.2.1 Disruption of Bacterial Cell Membranes

Bacteria are classified as Gram-positive or Gram-negative based on their cell wall structure. Gram-positive bacteria have a thick peptidoglycan layer, while Gram-negative bacteria possess a thinner peptidoglycan layer and an outer membrane of lipopolysaccharides with a periplasmic space. Gram-positive bacteria are generally more resistant to certain nanoparticle mechanisms due to their cell wall composition.

In contrast, Gram-negative bacteria are more susceptible, as their negatively charged lipopolysaccharide molecules strongly attract the positive ions released by many metal nanoparticles, leading to ion accumulation, cellular uptake, and intracellular damage (Slavin et al., 2017). Nanoparticle (NP) antibacterial activity is influenced by factors beyond size, as both smaller and larger NPs can be highly effective. Shape also plays a role, with spherical forms being most common, but other geometries, such as sheets, rods, and plates, contribute to mechanical damage. Physical deformations, like defects in ZnO NPs, can enhance toxicity by generating reactive oxygen species (ROS) under light activation, leading to bacterial membrane penetration (Jiang et al., 2016). Additionally, positively charged NPs disrupt bacterial electron transport chains through electrostatic interactions with negatively charged bacterial cell walls, a process further enhanced under acidic conditions (Ivask et al., 2014).

The antimicrobial mechanisms of carbon nanotubes (CNTs) are believed to arise from physical interactions, such as cell wall disruption and cytoplasmic leakage, as well as chemical effects like oxidative stress and ROS generation (X. Wang et al., 2017). Similarly, graphene oxide (GO) nanoparticles exhibit antimicrobial activity against plant pathogens by damaging microbial membranes, compromising membrane integrity, disrupting electron transport, and increasing ROS production (Mangadlao et al., 2015).

8.3.2.2 Inhibition of Key Metabolic Processes in Microorganisms

ATP synthesis in bacterial cells can be inhibited through multiple mechanisms induced by ROS and NPs. These mechanisms primarily target essential enzymes, proteins and processes involved in ATP production and energy regulation (Slavin et al., 2017).

When bacterial cells like *E. coli* are exposed to ROS generated by nanoparticles such as silver nanoparticles (Ag NPs), key proteins and enzymes essential for ATP production are disrupted. Ag+ ions, for instance, reduce the expression of the S2 protein, a subunit of the 30S ribosome critical for codon-anticodon pairing during translation. This disruption causes ribosome dysfunction and suppresses protein synthesis, including enzymes like succinyl-CoA synthetase, which is vital for the citric acid cycle. As a result, E. coli experiences ATP depletion without leakage into the surrounding medium, impairing cellular energy production. (Yamanaka et al., 2005).

Gold-based nanoparticles (AuNPs) exhibit a dual inhibitory effect on ATP production, particularly in multidrug-resistant Gram-negative bacteria. They disrupt the bacterial membrane potential, inhibiting ATPase activity and reducing ATP levels, while also blocking the binding of ribosomal subunits to tRNA, impairing protein synthesis. Additionally, AuNPs downregulate F-type ATP synthase, a crucial enzyme in oxidative phosphorylation that synthesizes ATP and generates a transmembrane proton gradient essential for molecular transport. By targeting this enzyme, AuNPs cause a significant decline in ATP levels, ultimately leading to metabolic failure (Cui et al., 2012).

8.3.2.3 ROS Generation

Reactive oxygen species (ROS) are highly reactive oxygen-containing molecules generated during cellular metabolism, including free radicals like superoxide and hydroxyl radicals, as well as non-radicals like hydrogen peroxide. While ROS are essential for processes such as cell signaling, survival, differentiation, and inflammatory responses, excessive ROS levels can surpass cellular defense mechanisms, resulting in oxidative stress and causing cell damage or death (Dayem et al., 2017; Slavin et al., 2017).

Mitochondria and NADPH oxidase are the main sources of reactive oxygen species (ROS), producing superoxide radicals that are quickly converted to hydrogen peroxide by superoxide dismutases (SODs). Although hydrogen peroxide is more stable, it can still induce substantial oxidative damage, including DNA denaturation, protein modification, and lipid peroxidation (Dayem et al., 2017).

Nanoparticles are renowned for their capacity to enhance ROS production, with higher concentrations causing a proportional rise in ROS levels. This excessive ROS generation disrupts redox homeostasis, leading to cellular dysfunction and oxidative damage to critical biomolecules. Furthermore, when positively charged nanoparticles interact with negatively charged bacterial cell walls under acidic conditions, the resulting interaction amplifies toxicity, further compromising cellular integrity (Gold et al., 2018; Vazquez-Muñoz et al., 2019).

The bactericidal activity of transition metal nanoparticles (NPs) is primarily attributed to their ability to generate reactive oxygen species (ROS) and their strong affinity for binding to thiol groups (R-SH). Heavy metal ions from non-essential transition metals, such as Ag+ and Cu^{2+}, readily bind to SH groups found in cysteine, disrupting the function of key enzymes and breaking S-S bridges essential for maintaining protein structure. These interactions result in significant damage to the cell's metabolism and physiology, ultimately impairing its survival (Slavin et al., 2017).

8.4 Case Studies and Effectiveness

Nanotechnology is increasingly recognized as a promising approach in agriculture, particularly for pest and disease management (Worrall et al., 2018). Conventional methods often depend on toxic pesticides that present significant risks to human health and the environment (Fu et al., 2020). In contrast, nanotechnology provides innovative solutions by improving the effectiveness and safety of pest control strategies (Jiang et al., 2024).

Nanomaterials (NMs) are being utilized in various ways to manage plant diseases and pests, as mentioned throughout the entire chapter:

- Antimicrobial Agents: Nanoparticles (NPs) such as silver (Ag), copper (Cu), and zinc (Zn) have shown direct antimicrobial properties, effectively targeting plant pathogen (Khanna et al., 2019; My-Thao Nguyen et al., 2021; Shankar et al., 2016).
- Nanocarriers: NMs can serve as carriers for pesticides, allowing for controlled release and targeted delivery, which reduces the amount of active ingredients needed and minimizes environmental impact (Jiang et al., 2024; Worrall et al., 2018).
- Biostimulants: Some NMs act as biostimulants, enhancing plant immunity and resilience against diseases (Fu et al., 2020).

As previously mentioned, various nanomaterials can be used in agriculture as antimicrobial agents. Metal-based nanoparticles, such as AgNPs, ZnONPs, CuNPs, and TiO_2NPs, interact with microbial membranes, showing effectiveness against a wide range of pathogens by inhibiting growth, spore germination, and causing cellular damage (Ocsoy et al., 2013). Carbon-based nanoparticles, including CNTs and graphene, are highly effective against plant pathogens, with enhanced antimicrobial activity when combined with metals like silver or zinc (Wang et al., 2014). Polymer-based nanoparticles, particularly chitosan, are widely studied for their antimicrobial properties and ability to stimulate plant defenses, while nanofibers made from diverse polymers demonstrate significant activity against phytopathogens by disrupting their cell walls and competing for resources (Brandelli, 2015). Further details on these materials and their applications are presented in Table 8.4.

Despite their potential benefits, the commercialization of nanoparticle-based products in agriculture remains limited. Key challenges include the necessity for extensive field trials, a deeper understanding of the interactions between nanomaterials and plant systems, and the resolution of regulatory concerns (Worrall et al., 2018). Future research should prioritize the development of eco-friendly and efficient nanomaterials, investigate their long-term impacts, and establish comprehensive guidelines to ensure their safe and sustainable application in agriculture (Bhattacharyya et al., 2016; Carrillo-Lopez et al., 2024). Establishing clear regulatory guidelines for the use of nano-based pesticides is essential for their widespread adoption in agriculture. These guidelines must address potential risks and ensure safe and effective application practices (Kumar et al., 2019).

Table 8.4 Applications of different nanomaterials in agriculture

Type	Nanomaterial	Pathogen	Host/disease	Effect	Reference
Metallic	Ag NPs	*Botrytis cinerea, Penicillium expansum, aspergillus nigeralternaria* sp.	Multiple diseases in crops such as pome fruits	AgNPs showed an antifungal activity	Fernández et al. (2016)
		Fusarium oxysporum f. sp. *ciceri*	Disease of chickpea	Ag NPs biosynthetized by *Trichoderma* sp. demonstrated significant antifungal activity showing the highest inhibition (95%)	Kaur et al. (2018)
		Bacillus megaterium, Pseudomonas syringae, Xanthomonas oryzae and Burkholderia glumae	Bacterial blight in a variety of crops such as wheat and rice	Damage of Gram-positive bacteria cell membrane due to release of ions from AgNPs	Velmurugan et al. (2012)
	Cu NPs	*Xanthomonas axonopodis*	Bacterial blight of pomegranate	Inhibition of bacteria growth and disruption of cell wall after 48 h	Mondal and Mani (2012)
		Xanthomonas campestris pv. vesicatoria	Tomato bacteria leaf spot disease	Higher bactericidal effect compared to commercial copper products.	Varympopi et al. (2022)
		Xanthomonas oryzae pv oryzae	Rice bacterial leaf blight	Inhibition of bacteria growth and lower increase in lesion area of seedlings compared to control	Datta Majumdar et al. (2021)
Oxide-based	ZnO NPs	*Xanthomonas oryzae pv. oryzae*	Multiple diseases in crops	Inhibition of bacteria growth due to collapsed and ruptured cell membrane with leakage of cytoplasm	Ogunyemi et al. (2019)
	Zn/Fe oxide NPS	*Staphylococcus aureus* and *Escherichia coli*	Multiple diseases in crops	Good bacteriostatic activity on *Staphylococcus aureus*, but to a lesser extent against *E. coli* compared to ZnO alone	Gordon et al. (2011)

(continued)

Table 8.4 (continued)

Type	Nanomaterial	Pathogen	Host/disease	Effect	Reference
	TiO_2 NPs, ZnO NPs	*Pectobacterium betavasculorum, Xanthomonas campestris pv. Beticola, and Pseudomonas syringae pv. aptata*	Bacterial vascular necrosis and root rot disease of beetroot, bacterial leaf spot, and bacterial pocket	Inhibition of bacteria growth, with higher inhibitory effect of ZnO	Siddiqui et al. (2019)
	TiO_2 NPs	*Bipolaris sorokiniana*	Spot blot disease in wheat	Boosted wheat plant resistance against phytopathogen.	Satti et al. (2021)
Carbon-based	Reduced graphene oxide-CuO nanocomposite	*Fusarium oxysporum*	Fusarium wilt disease in tomato, eggplant, pepper, and potato	Smaller CuO nanoparticles enhanced the nanocomposite antifungal activity.	El-Abeid et al. (2020)
Polymer-based	Chitosan	*Xanthomonasoryzae pv. oryzae* and *Xanthomonas oryzae pv. oryzicola*	Rice bacterial leaf blight and leaf streak	Growth inhibition and damage to cell membrane	Li et al. (2013)
	Thiamine-loaded CN-NPs	*Fusarium oxysporum* f. sp. ciceri	Wilt disease of chickpea	Thiamine-loaded CN-NPs boost chickpea growth, defense enzymes, and root formation, reducing cell death.	Muthukrishnan et al. (2019)

8.5 Conclusions

Green synthesized nanotechnology represents a groundbreaking advancement in pest and disease management, offering a sustainable and environmentally friendly alternative to conventional pesticides. These nanomaterials, produced through green synthesis methods, leverage natural, non-toxic precursors, reducing the ecological footprint often associated with traditional chemical and physical nanoparticle synthesis. Their unique properties, such as the ability to generate reactive oxygen species (ROS), disrupt microbial membranes, and enable targeted delivery, make them highly effective in protecting crops and enhancing agricultural productivity while mitigating risks to human health and non-target organisms.

Additionally, this chapter has demonstrated the diverse types and mechanisms of nanopesticides, including organic, inorganic, and hybrid systems, which improve the precision of pesticide delivery while reducing environmental contamination. Case studies further exemplify the real-world effectiveness of these technologies, showcasing their potential to address critical challenges in sustainable agriculture, such as pest resistance, nutrient losses, and environmental toxicity.

However, significant challenges remain. The synthesis of green nanomaterials must be further optimized to enhance scalability, cost-efficiency, and consistency in nanoparticle properties. Long-term studies are necessary to better understand their interactions with non-target organisms, their environmental behavior, and their long-term impacts on ecosystems and soil health. Regulatory frameworks must also evolve to address these specific considerations, balancing innovation with environmental and health safety.

This chapter also connects with earlier discussions on green synthesis, emphasizing that the nanomaterials explored here can be produced through sustainable methods detailed in previous chapters. These connections reinforce the integrated vision of this book, where synthesis, application, and regulation converge to provide a comprehensive perspective on sustainable nanotechnology in agriculture.

Future efforts should focus on interdisciplinary collaboration, continuous innovation in smart and biodegradable nanocarriers, and the establishment of clear policies to ensure safe and effective use. By addressing these areas, green synthesized nanotechnology has the potential to transform pest and disease management, contributing to global food security and environmental preservation.

Acknowledgments M.Y. Hernández-Ramírez and E. Aguirre-García thank to FONFIVE-2024 for the scholarship granted.

Conflicts of Interest The authors declare no conflicts of interest.

References

Adak, T., Barik, N., Patil, N. B., Govindharaj, G. P. P., Gadratagi, B. G., Annamalai, M., Mukherjee, A. K., & Rath, P. C. (2020). Nanoemulsion of eucalyptus oil: An alternative to synthetic pesticides against two major storage insects (Sitophilus oryzae (L.) and Tribolium castaneum (Herbst)) of rice. *Industrial Crops and Products, 143*, 111849. https://doi.org/10.1016/J.INDCROP.2019.111849

Agnihotri, S. A., Mallikarjuna, N. N., & Aminabhavi, T. M. (2004). Recent advances on chitosan-based micro- and nanoparticles in drug delivery. *Journal of Controlled Release, 100*(1), 5–28. https://doi.org/10.1016/j.jconrel.2004.08.010

Ahmed, K. S., Mikhail, W. Z. A., Sobhy, H. M., Radwan, E. M. M., El Din, T. S., & Youssef, A. M. (2019). Effect of lambda-cyhalothrin as nanopesticide on cotton leafworm, Spodoptera littoralis (Boisd.). *Egyptian Journal of Chemistry, 62*(7), 1663–1675. https://doi.org/10.21608/EJCHEM.2019.6871.1581

Al-Esawy, M. T., Alshukri, B. M., & Kahdum, B. J. (2021). Carbon nanotubes as control agents against the Khapra beetle, Trogoderma granarium Everts (Coleoptera: Dermestidae). *Revista Brasileira de Ciencias Agrarias, 16*(4), 1–9. https://doi.org/10.5039/agraria.v16i4a9304

Almeida, R. R., Silva Damasceno, E. T., de Carvalho, S. Y. B., de Carvalho, G. S. G., Gontijo, L. A. P., & de Lima Guimarães, L. G. (2018). Chitosan nanogels condensed to ferulic acid for the essential oil of Lippia origanoides Kunth encapsulation. *Carbohydrate Polymers, 188*, 268–275. https://doi.org/10.1016/j.carbpol.2018.01.103

Amin, H. H. (2020). Biosynthesized silver nanoparticles using Ulva lactuca as a safe synthetic pesticide (in vitro). *Open Agriculture, 5*(1), 291–299. https://doi.org/10.1515/opag-2020-0032

Anjaneyulu, B., Chauhan, V., Mittal, C., & Afshari, M. (2024). Innovative nanocarrier systems: A comprehensive exploration of recent developments in nano-biopesticide formulations. *Journal of Environmental Chemical Engineering, 12*(5), 113693. https://doi.org/10.1016/J.JECE.2024.113693

Arjunan, N., Singaravelu, C. M., Kulanthaivel, J., & Kandasamy, J. (2017). A potential photocatalytic, antimicrobial and anticancer activity of chitosan-copper nanocomposite. *International Journal of Biological Macromolecules, 104*, 1774–1782. https://doi.org/10.1016/j.ijbiomac.2017.03.006

Ashitha, A., & Mathew, J. (2020). Characteristics and types of slow/controlled release of pesticides. In *Controlled release of pesticides for sustainable agriculture* (pp. 141–153). Springer International Publishing. https://doi.org/10.1007/978-3-030-23396-9_6

Assalin, M. R., dos Santos, L. D. L., de Souza, D. R. C., Rosa, M. A., Duarte, R. R. M., Castanha, R. F., Donaire, P. P. R., & Durán, N. (2019). Nanoformulation as a tool for improvement of thiamethoxam encapsulation and evaluation of ecotoxicological impacts. *Energy, Ecology and Environment, 4*(6), 310–317. https://doi.org/10.1007/s40974-019-00138-1

Bahadur, R., Kumar, A., Trivedi, A., Nandeha, N., Agrawal, S., Ahirwar, S. B., & Muthuraman, R. M. (2023). An overview of green nanotechnology as a contemporary instrument for sustainable agriculture. *International Journal of Environment and Climate Change, 13*(12), 704–715. https://doi.org/10.9734/IJECC/2023/V13I123732

Bahattab, O., Khan, I., Bawazeer, S., Rauf, A., Qureshi, M. N., Al-Awthan, Y. S., Muhammad, N., Khan, A., Akram, M., Islam, M. N., & Bin Emran, T. (2021). Synthesis and biological activities of alcohol extract of black cumin seeds (Bunium persicum)-based gold nanoparticles and their catalytic applications. *Green Processing and Synthesis, 10*(1), 440–445. https://doi.org/10.1515/gps-2021-0041

Bapat, M. S., Singh, H., Shukla, S. K., Singh, P. P., Vo, D. V. N., Yadav, A., Goyal, A., Sharma, A., & Kumar, D. (2022). Evaluating green silver nanoparticles as prospective biopesticides: An environmental standpoint. *Chemosphere, 286*, 131761. https://doi.org/10.1016/j.chemosphere.2021.131761

Barrera-Méndez, F., Miranda-Sánchez, D., Sánchez-Rangel, D., Bonilla-Landa, I., Rodríguez-Haas, B., Monribot-Villanueva, J. L., & Olivares-Romero, J. L. (2019). Propiconazole nanoencapsulation in biodegradable polymers to obtain pesticide-controlled delivery systems. *Journal of the Mexican Chemical Society, 63*(1), 50–60. https://doi.org/10.29356/jmcs.v63i1.564

Bathaei, A., & Štreimikienė, D. (2023). Renewable energy and sustainable agriculture: Review of indicators. *Sustainability, 15*(19), 14307. https://doi.org/10.3390/SU151914307

Benelli, G. (2018). Gold nanoparticles – Against parasites and insect vectors. *Acta Tropica, 178*, 73–80. https://doi.org/10.1016/J.ACTATROPICA.2017.10.021

Bhaskar, M., Kumar, A., & Rani, R. (2023). Application of nano formulations in agriculture. *Biocatalysis and Agricultural Biotechnology, 54*, 102934. https://doi.org/10.1016/J.BCAB.2023.102934

Bhattacharya, K., Mukherjee, S. P., Gallud, A., Burkert, S. C., Bistarelli, S., Bellucci, S., Bottini, M., Star, A., & Fadeel, B. (2016). Biological interactions of carbon-based nanomaterials: From coronation to degradation. *Nanomedicine: Nanotechnology, Biology and Medicine, 12*(2), 333–351. https://doi.org/10.1016/J.NANO.2015.11.011

Bhattacharyya, A., Duraisamy, P., Govindarajan, M., Buhroo, A. A., & Prasad, R. (2016). Nano-biofungicides: Emerging trend in insect pest control. In *Advances and applications through fungal nanobiotechnology* (pp. 307–319). Springer. https://doi.org/10.1007/978-3-319-42990-8_15

Bocca, B., Barone, F., Petrucci, F., Benetti, F., Picardo, V., Prota, V., & Amendola, G. (2020). Nanopesticides: Physico-chemical characterization by a combination of advanced analytical techniques. *Food and Chemical Toxicology, 146*, 111816. https://doi.org/10.1016/j.fct.2020.111816

Brandelli, A. (2015). Nanobiotechnology strategies for delivery of antimicrobials in agriculture and food. In *Nanotechnologies in food and agriculture* (pp. 119–139). Springer International Publishing. https://doi.org/10.1007/978-3-319-14024-7_6

Campos, E. V. R., Proença, P. L. F., Oliveira, J. L., Pereira, A. E. S., De Morais Ribeiro, L. N., Fernandes, F. O., Gonçalves, K. C., Polanczyk, R. A., Pasquoto-Stigliani, T., Lima, R., Melville, C. C., Della Vechia, J. F., Andrade, D. J., & Fraceto, L. F. (2018). Carvacrol and linalool co-loaded in β-cyclodextrin-grafted chitosan nanoparticles as sustainable biopesticide aiming pest control. *Scientific Reports, 8*(1), 7623. https://doi.org/10.1038/s41598-018-26043-x

Carley, L. N., Panchagavi, R., Song, X., Davenport, S., Bergemann, C. M., McCumber, A. W., Gunsch, C. K., & Simonin, M. (2020). Long-term effects of copper nanopesticides on soil and sediment community diversity in two outdoor mesocosm experiments. *Environmental Science and Technology, 54*(14), 8878–8889. https://doi.org/10.1021/acs.est.0c00510

Carlisle, L., de Wit, M. M., DeLonge, M. S., Calo, A., Getz, C., Ory, J., Munden-Dixon, K., Galt, R., Melone, B., Knox, R., Iles, A., & Press, D. (2019). Securing the future of US agriculture: The case for investing in new entry sustainable farmers. *Elementa, 7*(1), 17. https://doi.org/10.1525/ELEMENTA.356/112494

Carrillo-Lopez, L. M., Villanueva-Verduzco, C., Villanueva-Sánchez, E., Fajardo-Franco, M. L., Aguilar-Tlatelpa, M., Ventura-Aguilar, R. I., & Soto-Hernández, R. M. (2024). Nanomaterials for plant disease diagnosis and treatment: A review. *Plants, 13*(18), 2634. https://doi.org/10.3390/PLANTS13182634

Chaud, M., Souto, E. B., Zielinska, A., Severino, P., Batain, F., Oliveira-Junior, J., & Alves, T. (2021). Nanopesticides in agriculture: Benefits and challenge in agricultural productivity, toxicological risks to human health and environment. *Toxics, 9*(6), 131. MDPI. https://doi.org/10.3390/toxics9060131

Chauhan, D., Afreen, S., Talreja, N., & Ashfaq, M. (2020). Multifunctional copper polymer-based nanocomposite for environmental and agricultural applications. In *Multifunctional hybrid nanomaterials for sustainable agri-food and ecosystems* (pp. 189–211). Elsevier. https://doi.org/10.1016/B978-0-12-821354-4.00008-X

Chen, J., Sun, L., Cheng, Y., Lu, Z., Shao, K., Li, T., Hu, C., & Han, H. (2016). Graphene oxide-silver nanocomposite: Novel agricultural antifungal agent against Fusarium graminearum for crop disease prevention. *ACS Applied Materials and Interfaces, 8*(36), 24057–24070. https://doi.org/10.1021/ACSAMI.6B05730/SUPPL_FILE/AM6B05730_SI_001.PDF

Chhipa, H. (2017). Nanofertilizers and nanopesticides for agriculture. *Environmental Chemistry Letters, 15*(1), 15–22). Springer Verlag. https://doi.org/10.1007/s10311-016-0600-4

Chinnaperumal, K., Govindasamy, B., Paramasivam, D., Dilipkumar, A., Dhayalan, A., Vadivel, A., Sengodan, K., & Pachiappan, P. (2018). Bio-pesticidal effects of Trichoderma viride formulated titanium dioxide nanoparticle and their physiological and biochemical changes on Helicoverpa armigera (Hub.). *Pesticide Biochemistry and Physiology, 149*, 26–36. https://doi.org/10.1016/J.PESTBP.2018.05.005

Cui, Y., Zhao, Y., Tian, Y., Zhang, W., Lü, X., & Jiang, X. (2012). The molecular mechanism of action of bactericidal gold nanoparticles on Escherichia coli. *Biomaterials, 33*(7), 2327–2333. https://doi.org/10.1016/j.biomaterials.2011.11.057

Cui, J. G., Mo, D. M., Jiang, Y., Gan, C. F., Li, W. G., Wu, A., Li, X. Y., Xiao, J. A., Hu, Q., Yuan, H. Y., Lu, R., & Huang, Y. M. (2019). Fabrication, characterization, and insecticidal activity evaluation of emamectin benzoate-sodium lignosulfonate nanoformulation with pH-responsivity. *Industrial and Engineering Chemistry Research, 58*(43), 19741–19751. https://doi.org/10.1021/acs.iecr.9b03171

Cui, L., Deng, G., Wu, J., Ding, F., Wang, W., Yu, H., Song, Z., Rui, C., Han, H., & Yuan, H. (2024). Fabrication of nanogels to improve the toxicity and persistence of cycloxaprid against Diaphorina citri, the vector of citrus huanglongbing. *Journal of Advanced Research, 73*, 105–116. https://doi.org/10.1016/j.jare.2024.08.036

Czarnobai De Jorge, B., Hummel, H. E., & Gross, J. (2022). Repellent activity of clove essential oil volatiles and development of nanofiber-based dispensers against pear psyllids (Hemiptera: Psyllidae). *Insects, 13*(8), 743. https://doi.org/10.3390/insects13080743

da Silva Pontes, M., de Lima Forini, M. M., Caires, A. R. L., Santiago, E. F., Tripathi, D. K., Kah, M., & Grillo, R. (2023). Nanopesticides—Modes of action and impacts. In *Nano-enabled sustainable and precision agriculture* (pp. 79–103). Academic Press. https://doi.org/10.1016/B978-0-323-91233-4.00001-6

Datta Majumdar, T., Ghosh, C. K., & Mukherjee, A. (2021). Dual role of copper nanoparticles in bacterial leaf blight-infected rice: A therapeutic and metabolic approach. *ACS Agricultural Science and Technology, 1*(3), 160–172. https://doi.org/10.1021/ACSAGSCITECH.0C00064/SUPPL_FILE/AS0C00064_SI_001.PDF

Dayem, A. A., Hossain, M. K., Lee, S. B., Kim, K., Saha, S. K., Yang, G. M., Choi, H. Y., & Cho, S. G. (2017). The role of reactive oxygen species (ROS) in the biological activities of metallic nanoparticles. *International Journal of Molecular Sciences, 18*(1), 120. https://doi.org/10.3390/IJMS18010120

de Lima, P. H. C., Antunes, D. R., Forini, M. M. D. L., Pontes, M. D. S., Mattos, B. D., & Grillo, R. (2021). Recent advances on lignocellulosic-based nanopesticides for agricultural applications. *Frontiers in Nanotechnology, 3*, 809329. Frontiers Media S.A. https://doi.org/10.3389/fnano.2021.809329

Ding, Y., Wang, Q., Zhu, G., Zhang, P., & Rui, Y. (2023). Application and perspectives of nanopesticides in agriculture. *Journal of Nanoparticle Research, 25*(8), 159. Springer Science and Business Media B.V. https://doi.org/10.1007/s11051-023-05811-5

Dong, E., Yang, Z., Zhou, C., Wang, C., Li, S., Ouyang, Q., Kong, L., He, Z., Xie, J., Li, P., & Yang, P. (2019). pH-responsive ultrasonic self-assembly spinosad-loaded nanomicelles and their antifungal activity to Fusarium oxysporum. *Reactive and Functional Polymers, 141*, 123–132. https://doi.org/10.1016/J.REACTFUNCTPOLYM.2019.05.004

Dong, J., Chen, W., Feng, J., Liu, X., Xu, Y., Wang, C., Yang, W., & Du, X. (2021a). Facile, smart, and degradable metal-organic framework nanopesticides gated with FeIII-tannic acid networks in response to seven biological and environmental stimuli. *ACS Applied Materials and Interfaces, 13*(16), 19507–19520. https://doi.org/10.1021/acsami.1c04118

Dong, J., Chen, W., Qin, D., Chen, Y., Li, J., Wang, C., Yu, Y., Feng, J., & Du, X. (2021b). Cyclodextrin polymer-valved MoS2-embedded mesoporous silica nanopesticides toward hierarchical targets via multidimensional stimuli of biological and natural environments. *Journal of Hazardous Materials, 419*, 126404. https://doi.org/10.1016/j.jhazmat.2021.126404

Dong, J., Liu, X., Chen, Y., Yang, W., & Du, X. (2021c). User-safe and efficient chitosan-gated porous carbon nanopesticides and nanoherbicides. *Journal of Colloid and Interface Science, 594*, 20–34. https://doi.org/10.1016/j.jcis.2021.03.001

Dong, W., Ren, Y., & Xue, H. (2024). Fabrication and application of carrier-free and carrier-based nanopesticides in pest management. *Archives of Insect Biochemistry and Physiology, 116*(2), e22124. John Wiley and Sons Ltd. https://doi.org/10.1002/arch.22124

Dönmez, D., Isak, M. A., İzgü, T., & Şimşek, Ö. (2024). Green horizons: Navigating the future of agriculture through sustainable practices. *Sustainability, 16*(8), 3505. https://doi.org/10.3390/SU16083505

Duhan, J. S., Kumar, R., Kumar, N., Kaur, P., Nehra, K., & Duhan, S. (2017). Nanotechnology: The new perspective in precision agriculture. *Biotechnology Reports, 15*, 11–23). Elsevier B.V. https://doi.org/10.1016/j.btre.2017.03.002

Dwivedi, S., Saquib, Q., Al-Khedhairy, A. A., & Musarrat, J. (2016). Understanding the role of nanomaterials in agriculture. In *Microbial inoculants in sustainable agricultural productivity: Vol. 2: Functional applications* (pp. 271–288). Springer India. https://doi.org/10.1007/978-81-322-2644-4_17

El-Abeid, S. E., Ahmed, Y., Daròs, J. A., & Mohamed, M. A. (2020). Reduced graphene oxide nanosheet-decorated copper oxide nanoparticles: A potent antifungal nanocomposite against Fusarium root rot and wilt diseases of tomato and pepper plants. *Nanomaterials, 10*(5), 1001. https://doi.org/10.3390/NANO10051001

El-Ansary, M. S. M., Hamouda, R. A., & Elshamy, M. M. (2022). Using biosynthesized zinc oxide nanoparticles as a pesticide to alleviate the toxicity on banana infested with parasitic-nematode. *Waste and Biomass Valorization, 13*(1), 405–415. https://doi.org/10.1007/s12649-021-01527-6

El-Ramady, H., Abdalla, N., Sári, D., Ferroudj, A., Muthu, A., Prokisch, J., Fawzy, Z. F., Brevik, E. C., & Solberg, S. (2023). Nanofarming: Promising solutions for the future of the global agricultural industry. *Agronomy, 13*(6), 1600. https://doi.org/10.3390/AGRONOMY13061600

El-Saadony, M. T., Saad, A. M., Soliman, S. M., Salem, H. M., Desoky, E. S. M., Babalghith, A. O., El-Tahan, A. M., Ibrahim, O. M., Ebrahim, A. A. M., Abd El-Mageed, T. A., Elrys, A. S., Elbadawi, A. A., El-Tarabily, K. A., & AbuQamar, S. F. (2022). Role of nanoparticles in enhancing crop tolerance to abiotic stress: A comprehensive review. *Frontiers in Plant Science, 13*, 946717. https://doi.org/10.3389/FPLS.2022.946717/BIBTEX

Erlina, L. H., Lina, E. C., Reflinaldon, Djamaan, A., & Arneti. (2020). Insecticidal activity of nanoemulsion of Piper aduncum extract against cabbage head cartepillar Crocidolomia pavonana F. (Lepidoptera: Crambidae). *IOP Conference Series: Earth and Environmental Science, 468*(1), 012001. https://doi.org/10.1088/1755-1315/468/1/012001

Fabiyi, O. A., Olatunji, G. A., & Saadu, A. O. (2018). Suppression of heterodera sacchari in rice with agricultural waste-silver nano particles. *Journal of Solid Waste Technology and Management, 44*(2), 87–91. https://doi.org/10.5276/jswtm.2018.87

Faizan, M., Hayat, S., & Pichtel, J. (2020). Effects of zinc oxide nanoparticles on crop plants: A perspective analysis. In *Sustainable agriculture reviews* (Vol. 41, pp. 83–99). Springer. https://doi.org/10.1007/978-3-030-33996-8_4

Farias, B. V., Pirzada, T., Mathew, R., Sit, T. L., Opperman, C., & Khan, S. A. (2019). Electrospun polymer nanofibers as seed coatings for crop protection. *ACS Sustainable Chemistry and Engineering, 7*(24), 19848–19856. https://doi.org/10.1021/acssuschemeng.9b05200

Fernández, J. G., Fernández-Baldo, M. A., Berni, E., Camí, G., Durán, N., Raba, J., & Sanz, M. I. (2016). Production of silver nanoparticles using yeasts and evaluation of their antifungal activity against phytopathogenic fungi. *Process Biochemistry, 51*(9), 1306–1313. https://doi.org/10.1016/J.PROCBIO.2016.05.021

Foteva, T., & Georgieva, N. (2022). Antimicrobial properties of silica/hydroxypropylcellulose hybrids doped with copper ions. *Journal of Chemical Technology and Metallurgy, 57*, 930–936.

Fraceto, L. F., Grillo, R., de Medeiros, G. A., Scognamiglio, V., Rea, G., & Bartolucci, C. (2016). Nanotechnology in agriculture: Which innovation potential does it have? *Frontiers in Environmental Science, 4*, 186737. https://doi.org/10.3389/FENVS.2016.00020/BIBTEX

Fraceto, L. F., de Castro, V. L. S. S., Grillo, R., Ávila, D., Oliveira, H. C., & Lima, R. (2020). *Nanopesticides: From research and development to mechanisms of action and sustainable use in agriculture*. Springer International Publishing. https://doi.org/10.1007/978-3-030-44873-8

Fu, L., Wang, Z., Dhankher, O. P., & Xing, B. (2020). Nanotechnology as a new sustainable approach for controlling crop diseases and increasing agricultural production. *Journal of Experimental Botany, 71*(2), 507–519. https://doi.org/10.1093/JXB/ERZ314

Gahukar, R. T., & Das, R. K. (2020). Plant-derived nanopesticides for agricultural pest control: Challenges and prospects. *Nanotechnology for Environmental Engineering, 5*(1), 3. Springer Science and Business Media Deutschland GmbH. https://doi.org/10.1007/s41204-020-0066-2

German, R. N., Thompson, C. E., & Benton, T. G. (2017). Relationships among multiple aspects of agriculture's environmental impact and productivity: A meta-analysis to guide sustainable agriculture. *Biological Reviews, 92*(2), 716–738. https://doi.org/10.1111/BRV.12251

Giongo, A. M. M., Vendramim, J. D., & Forim, M. R. (2016). Evaluation of neem-based nanoformulations as alternative to control fall armyworm. *Ciencia e Agrotecnologia, 40*(1), 26–36. https://doi.org/10.1590/S1413-70542016000100002

Gold, K., Slay, B., Knackstedt, M., & Gaharwar, A. K. (2018). Antimicrobial activity of metal and metal-oxide based nanoparticles. *Advanced Therapeutics, 1*(3), 1700033. https://doi.org/10.1002/ADTP.201700033

Gordon, T., Perlstein, B., Houbara, O., Felner, I., Banin, E., & Margel, S. (2011). Synthesis and characterization of zinc/iron oxide composite nanoparticles and their antibacterial properties. *Colloids and Surfaces A: Physicochemical and Engineering Aspects, 374*(1–3), 1–8. https://doi.org/10.1016/j.colsurfa.2010.10.015

Granetto, M., Serpella, L., Fogliatto, S., Re, L., Bianco, C., Vidotto, F., & Tosco, T. (2022). Natural clay and biopolymer-based nanopesticides to control the environmental spread of a soluble herbicide. *Science of the Total Environment, 806*, 151199. https://doi.org/10.1016/J.SCITOTENV.2021.151199

Gregory, Marslin Caroline J., Sheeba Gregory, Franklin. (2017). Nanoparticles Alter Secondary Metabolism in Plants via ROS Burst Frontiers in Plant Science 8. https://doi.org/10.3389/fpls.2017.00832

Gutiérrez-Ramírez, J. A., Betancourt-Galindo, R., Aguirre-Uribe, L. A., Cerna-Chávez, E., Sandoval-Rangel, A., Del Ángel, E. C., Chacón-Hernández, J. C., García-López, J. I., & Hernández-Juárez, A. (2021). Insecticidal effect of zinc oxide and titanium dioxide nanoparticles against Bactericera cockerelli Sulc. (Hemiptera: Triozidae) on tomato solanum lycopersicum. *Agronomy, 11*(8), 1460. https://doi.org/10.3390/agronomy11081460

Hoseinzadeh, E., Makhdoumi, P., Taha, P., Hossini, H., Stelling, J., Kamal, M. A., & Ashraf, G. M. (2016). A review on nano-antimicrobials: Metal nanoparticles, methods and mechanisms. *Current Drug Metabolism, 18*(2), 120–128. https://doi.org/10.2174/1389200217666161201111146

Ivask, A., Elbadawy, A., Kaweeteerawat, C., Boren, D., Fischer, H., Ji, Z., Chang, C. H., Liu, R., Tolaymat, T., Telesca, D., Zink, J. I., Cohen, Y., Holden, P. A., & Godwin, H. A. (2014). Toxicity mechanisms in Escherichia coli vary for silver nanoparticles and differ from ionic silver. *ACS Nano, 8*(1), 374–386. https://doi.org/10.1021/NN4044047/SUPPL_FILE/NN4044047_SI_001.PDF

Jagadheeswari, R., Lakshmi, T., Balusamy, S. R., David, S., & Kumar, S. R. (2020). Biosynthesis of silver nanoparticles using Withania somnifera (L.) Dunal extract and its antibacterial activity against food pathogens. *Annals of Phytomedicine: An International Journal, 9*(1), 195–198. https://doi.org/10.21276/ap.2020.9.1.25

Jia, Y., Klumpp, E., Bol, R., & Amelung, W. (2023). Uptake of metallic nanoparticles containing essential (Cu, Zn and Fe) and non-essential (Ag, Ce and Ti) elements by crops: A meta-analysis. *Critical Reviews in Environmental Science and Technology, 53*(16), 1512–1533). Taylor and Francis Ltd. https://doi.org/10.1080/10643389.2022.2156225

Jiang, Y., Zhang, L., Wen, D., & Ding, Y. (2016). Role of physical and chemical interactions in the antibacterial behavior of ZnO nanoparticles against E. coli. *Materials Science and Engineering: C, 69*, 1361–1366. https://doi.org/10.1016/J.MSEC.2016.08.044

Jiang, X., Yang, F., Jia, W., Jiang, Y., Wu, X., Song, S., Shen, H., & Shen, J. (2024). Nanomaterials and nanotechnology in agricultural pesticide delivery: A review. *Langmuir, 40*(36), 18806–18820. https://doi.org/10.1021/ACS.LANGMUIR.4C01842/ASSET/IMAGES/MEDIUM/LA4C01842_0008.GIF

Kah, M., Weniger, A. K., & Hofmann, T. (2016). Impacts of (nano)formulations on the fate of an insecticide in soil and consequences for environmental exposure assessment. *Environmental Science and Technology, 50*(20), 10960–10967. https://doi.org/10.1021/acs.est.6b02477

Kan, Q., Lu, K., Deng, R., Lv, Z., Wu, W., Gao, S., Dong, S., & Mao, L. (2022). An alkali-triggered polydopamine modified mesoporous silica Nanopesticide for smart delivery of chlorpyrifos with low loss. *ACS Agricultural Science and Technology, 2*(3), 501–511. https://doi.org/10.1021/acsagscitech.1c00269

Kannan, M., Bojan, N., Swaminathan, J., Zicarelli, G., Hemalatha, D., Zhang, Y., Ramesh, M., & Faggio, C. (2023). Nanopesticides in agricultural pest management and their environmental risks: A review. *International journal of Environmental Science and Technology, 20*(9), 10507–10532). Institute for Ionics. https://doi.org/10.1007/s13762-023-04795-y

Kasana, R. C., Panwar, N. R., Kaul, R. K., & Kumar, P. (2016). Copper nanoparticles in agriculture: Biological synthesis and antimicrobial activity. In *Nanoscience in food and agriculture* (pp. 129–143). Springer. https://doi.org/10.1007/978-3-319-48009-1_5

Kaur, P., Thakur, R., Duhan, J. S., & Chaudhury, A. (2018). Management of wilt disease of chickpea in vivo by silver nanoparticles biosynthesized by rhizospheric microflora of chickpea (Cicer arietinum). *Journal of Chemical Technology & Biotechnology, 93*(11), 3233–3243. https://doi.org/10.1002/JCTB.5680

Khan, F., Shariq, M., Asif, M., Siddiqui, M. A., Malan, P., & Ahmad, F. (2022). Green nanotechnology: Plant-mediated nanoparticle synthesis and application. *Nanomaterials, 12*(4), 673. MDPI. https://doi.org/10.3390/nano12040673

Khanna, P., Kaur, A., & Goyal, D. (2019). Algae-based metallic nanoparticles: Synthesis, characterization and applications. *Journal of Microbiological Methods, 163*, 105656. https://doi.org/10.1016/J.MIMET.2019.105656

Kora, A. J. (2022). Copper-based nanopesticides. In *Copper nanostructures: Next-generation of agrochemicals for sustainable agroecosystems* (pp. 133–153). Elsevier. https://doi.org/10.1016/B978-0-12-823833-2.00019-2

Kumar, S., Bhanjana, G., Sharma, A., Sidhu, M. C., & Dilbaghi, N. (2014). Synthesis, characterization and on field evaluation of pesticide loaded sodium alginate nanoparticles. *Carbohydrate Polymers, 101*(1), 1061–1067. https://doi.org/10.1016/j.carbpol.2013.10.025

Kumar, S., Nehra, M., Dilbaghi, N., Marrazza, G., Hassan, A. A., & Kim, K. H. (2019). Nano-based smart pesticide formulations: Emerging opportunities for agriculture. *Journal of Controlled Release, 294*, 131–153. https://doi.org/10.1016/J.JCONREL.2018.12.012

Li, B., Liu, B., Shan, C., Ibrahim, M., Lou, Y., Wang, Y., Xie, G., Li, H. Y., & Sun, G. (2013). Antibacterial activity of two chitosan solutions and their effect on rice bacterial leaf blight and leaf streak. *Pest Management Science, 69*(2), 312–320. https://doi.org/10.1002/PS.3399

Li, C., Han, Y., Gao, T., Zhang, J., Xu, D. X., & Wāng, Y. (2023a). Insecticidal activity of metallic nanopesticides synthesized from natural resources: A review. *Environmental Chemistry Letters, 21*(2), 1141–1176). Springer Science and Business Media Deutschland GmbH. https://doi.org/10.1007/s10311-022-01548-0

Li, X., Chen, Y., Xu, J., Lynch, I., Guo, Z., Xie, C., & Zhang, P. (2023b). Advanced nanopesticides: Advantage and action mechanisms. *Plant Physiology and Biochemistry, 203*, 108051. https://doi.org/10.1016/J.PLAPHY.2023.108051

Li, B., Duan, W., Lin, G., Ma, X., Wen, R., & Zhang, Z. (2024). An effective and promising strategy for plant protection: Synthesis of L-carvone-based thiazolinone–hydrazone/nanochitosan complexes with antifungal activity and sustained releasing performance. *International Journal of Molecular Sciences, 25*(9), 4595. https://doi.org/10.3390/ijms25094595

Liang, W., Yu, A., Wang, G., Zheng, F., Hu, P., Jia, J., & Xu, H. (2018). A novel water-based chitosan-La pesticide nanocarrier enhancing defense responses in rice (Oryza sativa L) growth. *Carbohydrate Polymers, 199*, 437–444. https://doi.org/10.1016/j.carbpol.2018.07.042

Liu, P., Ren, Z., Ding, W., Kong, D., Hermanowicz, S. W., & Huang, Y. (2023). Comparative environmental impact assessment of copper-based nanopesticides and conventional pesticides. *ACS Agricultural Science and Technology, 3*(7), 593–600. https://doi.org/10.1021/acsagscitech.3c00118

Lowry, G. V., Avellan, A., & Gilbertson, L. M. (2019). Opportunities and challenges for nanotechnology in the agri-tech revolution. *Nature Nanotechnology, 14*(6), 517–522. https://doi.org/10.1038/s41565-019-0461-7

Luo, W., Liu, J., Algharib, S. A., & Chen, W. (2022). Antibacterial activity of enrofloxacin loaded gelatin-sodium alginate composite nanogels against intracellular Staphylococcus aureus small colony variants. *Journal of Veterinary Science, 23*(3), e48. https://doi.org/10.4142/jvs.21292

Ma, X., Zhu, X., Qu, S., Cai, L., Ma, G., Fan, G., & Sun, X. (2022). Fabrication of copper nanoparticle composite nanogel for high-efficiency management of Pseudomonas syringae pv. tabaci on tobacco. *Pest Management Science, 78*(5), 2074–2085. https://doi.org/10.1002/ps.6833

Malaikozhundan, B., & Vinodhini, J. (2018). Nanopesticidal effects of Pongamia pinnata leaf extract coated zinc oxide nanoparticle against the Pulse beetle, Callosobruchus maculatus. *Materials Today Communications, 14*, 106–115. https://doi.org/10.1016/j.mtcomm.2017.12.015

Mangadlao, J. D., Santos, C. M., Felipe, M. J. L., De Leon, A. C. C., Rodrigues, D. F., & Advincula, R. C. (2015). On the antibacterial mechanism of graphene oxide (GO) Langmuir–Blodgett films. *Chemical Communications, 51*(14), 2886–2889. https://doi.org/10.1039/C4CC07836E

María, L., Del Prado-Audelo SA., Bernal-Chávez, SC., Gutiérrez-Ruíz, H., Hernández-Parra, IG., Kerdan, JM., Reyna-González, J., Sharifi-Rad, G., Leyva-Gómez. (2022). Stability Phenomena Associated with the Development of Polymer-Based Nanopesticides Oxidative Medicine and Cellular Longevity, (1). 10.1155/omcl.v2022.1 https://doi.org/10.1155/2022/5766199

Maruyama, C. R., Guilger, M., Pascoli, M., Bileshy-José, N., Abhilash, P. C., Fraceto, L. F., & De Lima, R. (2016). Nanoparticles based on chitosan as carriers for the combined herbicides imazapic and imazapyr. *Scientific Reports, 6*, 19768. https://doi.org/10.1038/srep19768

Medina-Pérez, G., Fernández-Luqueño, F., Campos-Montiel, R. G., Sánchez-López, K. B., Afanador-Barajas, L. N., & Prince, L. (2019). Nanotechnology in crop protection: Status and future trends. In *Nano-biopesticides today and future perspectives* (pp. 17–45). Elsevier. https://doi.org/10.1016/B978-0-12-815829-6.00002-4

Mhamoud, M. A., Sherif, B. A. G., & Saleh, S. A. (2019). Field evaluation of the efficacy of silver nanoparticles (AgNP) against mites associated with tomato plants in greenhouses. *Pakistan Journal of Agricultural Sciences, 56*(1), 245–259. https://doi.org/10.21162/PAKJAS/19.8155

Mondal, K. K., & Mani, C. (2012). Investigation of the antibacterial properties of nanocopper against Xanthomonas axonopodis pv. punicae, the incitant of pomegranate bacterial blight. *Annals of Microbiology, 62*(2), 889–893. https://doi.org/10.1007/S13213-011-0382-7/METRICS

Mondéjar-López, M., García-Simarro, M. P., Navarro-Simarro, P., Gómez-Gómez, L., Ahrazem, O., & Niza, E. (2024). A review on the encapsulation of "eco-friendly" compounds in natural polymer-based nanoparticles as next generation nano-agrochemicals for sustainable agriculture and crop management. *International Journal of Biological Macromolecules, 280*, 136030. https://doi.org/10.1016/J.IJBIOMAC.2024.136030

Morsi, R. E., Alsabagh, A. M., Nasr, S. A., & Zaki, M. M. (2017). Multifunctional nanocomposites of chitosan, silver nanoparticles, copper nanoparticles and carbon nanotubes for water treatment: Antimicrobial characteristics. *International Journal of Biological Macromolecules, 97*, 264–269. https://doi.org/10.1016/j.ijbiomac.2017.01.032

Muthukrishnan, S., Murugan, I., & Selvaraj, M. (2019). Chitosan nanoparticles loaded with thiamine stimulate growth and enhances protection against wilt disease in chickpea. *Carbohydrate Polymers, 212*, 169–177. https://doi.org/10.1016/j.carbpol.2019.02.037

Muthusamy, R., Ramkumar, G., Kumarasamy, S., Chi, N. T. L., Al Obaid, S., Alfarraj, S., & Karuppusamy, I. (2023). Synergism and toxicity of iron nanoparticles derived from Trigonella foenum-graecum against pyrethriod treatment in S. litura and H. armigera (Lepidoptera: Noctuidae). *Environmental Research, 231*, 116079. https://doi.org/10.1016/J.ENVRES.2023.116079

My-Thao Nguyen, T., Anh-Thu Nguyen, T., Tuong-Van Pham, N., Ly, Q. V., Thuy-Quynh Tran, T., Thach, T. D., Nguyen, C. L., Banh, K. S., Le, V. D., Nguyen, L. P., Nguyen, D. T., Dang, C. H., & Nguyen, T. D. (2021). Biosynthesis of metallic nanoparticles from waste Passiflora edulis peels for their antibacterial effect and catalytic activity. *Arabian Journal of Chemistry, 14*(4), 103096. https://doi.org/10.1016/J.ARABJC.2021.103096

Nadeem, M., Tungmunnithum, D., Hano, C., Abbasi, B. H., Hashmi, S. S., Ahmad, W., & Zahir, A. (2018). The current trends in the green syntheses of titanium oxide nanoparticles and their applications. *Green Chemistry Letters and Reviews, 11*(4), 492–502. https://doi.org/10.1080/17518253.2018.1538430

Narayanan, M., Devi, P. G., Natarajan, D., Kandasamy, S., Devarayan, K., Alsehli, M., Elfasakhany, A., & Pugazhendhi, A. (2021). Green synthesis and characterization of titanium dioxide nanoparticles using leaf extract of Pouteria campechiana and larvicidal and pupicidal activity on Aedes aegypti. *Environmental Research, 200*, 111333. https://doi.org/10.1016/j.envres.2021.111333

Nisar, P., Ali, N., Rahman, L., Ali, M., & Shinwari, Z. K. (2019). Antimicrobial activities of biologically synthesized metal nanoparticles: An insight into the mechanism of action. *Journal of Biological Inorganic Chemistry, 24*(7), 929–941). Springer Verlag. https://doi.org/10.1007/s00775-019-01717-7

Nisha Raj, S., Anooj, E. S., Rajendran, K., & Vallinayagam, S. (2021). A comprehensive review on regulatory invention of nano pesticides in agricultural nano formulation and food system. *Journal of Molecular Structure, 1239*, 130517. https://doi.org/10.1016/j.molstruc.2021.130517

Ocsoy, I., Paret, M. L., Ocsoy, M. A., Kunwar, S., Chen, T., You, M., & Tan, W. (2013). Nanotechnology in plant disease management: DNA-directed silver nanoparticles on graphene oxide as an antibacterial against Xanthomonas perforans. *ACS Nano, 7*(10), 8972–8980. https://doi.org/10.1021/NN4034794/SUPPL_FILE/NN4034794_SI_001.PDF

Ogunyemi, S. O., Abdallah, Y., Zhang, M., Fouad, H., Hong, X., Ibrahim, E., Masum, M. M. I., Hossain, A., Mo, J., & Li, B. (2019). Green synthesis of zinc oxide nanoparticles using different plant extracts and their antibacterial activity against Xanthomonas oryzae pv. oryzae. *Artificial Cells, Nanomedicine, and Biotechnology, 47*(1), 341–352. https://doi.org/10.1080/21691401.2018.1557671

Pan, X., Nie, D., Guo, X., Xu, S., Zhang, D., Cao, F., & Guan, X. (2023). Effective control of the tomato wilt pathogen using TiO2 nanoparticles as a green nanopesticide. *Environmental Science: Nano, 10*(5), 1441–1452. https://doi.org/10.1039/d3en00059a

Pandey, S., Giri, K., Kumar, R., Mishra, G., & Raja Rishi, R. (2018). Nanopesticides: Opportunities in crop protection and associated environmental risks. *Proceedings of the National Academy of Sciences, India Section B – Biological Sciences, 88*(4), 1287–1308). Springer. https://doi.org/10.1007/s40011-016-0791-2

Pandey, A., Srivastava, S., Aggarwal, N., Srivastava, C., Adholeya, A., & Kochar, M. (2020). Assessment of the pesticidal behaviour of diacyl hydrazine-based ready-to-use nanoformulations. *Chemical and Biological Technologies in Agriculture, 7*(1), 10. https://doi.org/10.1186/s40538-020-0177-9

Panpatte, D. G., & Jhala, Y. K. (2019). *Nanotechnology for agriculture: Crop production & protection*. Springer.

Paradva, K. C., & Kalla, S. (2023). Nanopesticides: A review on current research and future perspective. *ChemistrySelect, 8*(26), e202300756. John Wiley and Sons Inc. https://doi.org/10.1002/slct.202300756

Periakaruppan, R., Romanovski, V., Thirumalaisamy, S. K., Palanimuthu, V., Sampath, M. P., Anilkumar, A., Sivaraj, D. K., Ahamed, N. A. N., Murugesan, S., Chandrasekar, D., & Selvaraj, K. S. V. (2023). Innovations in modern nanotechnology for the sustainable production of agriculture. *ChemEngineering, 7*(4), 61. https://doi.org/10.3390/CHEMENGINEERING7040061

Pho, Q. H., Losic, D., Ostrikov, K., Tran, N. N., & Hessel, V. (2020). Perspectives on plasma-assisted synthesis of N-doped nanoparticles as nanopesticides for pest control in crops. *Reaction Chemistry and Engineering, 5*(8), 1374–1396). Royal Society of Chemistry. https://doi.org/10.1039/d0re00069h

Pittarate, S., Rajula, J., Rahman, A., Vivekanandhan, P., Thungrabeab, M., Mekchay, S., & Krutmuang, P. (2021). Insecticidal effect of zinc oxide nanoparticles against spodoptera frugiperda under laboratory conditions. *Insects, 12*(11), 1017. https://doi.org/10.3390/insects12111017

Pontes, M. S., Antunes, D. R., Oliveira, I. P., Forini, M. M. L., Santos, J. S., Arruda, G. J., Caires, A. R. L., Santiago, E. F., & Grillo, R. (2021). Chitosan/tripolyphosphate nanoformulation carrying paraquat: Insights on its enhanced herbicidal activity. *Environmental Science: Nano, 8*(5), 1336–1351. https://doi.org/10.1039/d0en01128b

Prasad, R., Bhattacharyya, A., & Nguyen, Q. D. (2017). Nanotechnology in sustainable agriculture: Recent developments, challenges, and perspectives. *Frontiers in Microbiology, 8*, 1014. https://doi.org/10.3389/FMICB.2017.01014/BIBTEX

Preisler, A. C., Pereira, A. E. S., Campos, E. V. R., Dalazen, G., Fraceto, L. F., & Oliveira, H. C. (2020). Atrazine nanoencapsulation improves pre-emergence herbicidal activity against Bidens pilosa without enhancing long-term residual effect on Glycine max. *Pest Management Science, 76*(1), 141–149. https://doi.org/10.1002/ps.5482

Priya, M., Venkatesan, R., Deepa, S., Sana, S. S., Arumugam, S., Karami, A. M., Vetcher, A. A., & Kim, S. C. (2023). Green synthesis, characterization, antibacterial, and antifungal activity of copper oxide nanoparticles derived from Morinda citrifolia leaf extract. *Scientific Reports, 13*(1), 18838. https://doi.org/10.1038/s41598-023-46002-5

Rahman, A., Pittarate, S., Perumal, V., Rajula, J., Thungrabeab, M., Mekchay, S., & Krutmuang, P. (2022). Larvicidal and antifeedant effects of copper nano-pesticides against Spodoptera frugiperda (J.E. Smith) and its immunological response. *Insects, 13*(11), 1030. https://doi.org/10.3390/insects13111030

Rai, P. K., Kumar, V., Lee, S. S., Raza, N., Kim, K. H., Ok, Y. S., & Tsang, D. C. W. (2018). Nanoparticle-plant interaction: Implications in energy, environment, and agriculture. *Environment International, 119*, 1–19. https://doi.org/10.1016/J.ENVINT.2018.06.012

Rajkumar, V., Gunasekaran, C., Paul, C. A., & Dharmaraj, J. (2020). Development of encapsulated peppermint essential oil in chitosan nanoparticles: Characterization and biological efficacy against stored-grain pest control. *Pesticide Biochemistry and Physiology, 170*, 104679. https://doi.org/10.1016/j.pestbp.2020.104679

Rajwade, J. M., Chikte, R. G., & Paknikar, K. M. (2020). Nanomaterials: New weapons in a crusade against phytopathogens. *Applied Microbiology and Biotechnology, 104*(4), 1437–1461). Springer. https://doi.org/10.1007/s00253-019-10334-y

Ramkumar, G., Asokan, R., Ramya, S., & Gayathri, G. (2021). Characterization of Trigonella foenum-graecum derived iron nanoparticles and its potential pesticidal activity against Tuta absoluta (Lepidoptera). *Journal of Cluster Science, 32*(5), 1185–1190. https://doi.org/10.1007/s10876-020-01867-8

Rivero-Montejo, S. D. J., Vargas-Hernandez, M., & Torres-Pacheco, I. (2021). Nanoparticles as novel elicitors to improve bioactive compounds in plants. *Agriculture (Switzerland), 11*(2), 1–16). MDPI AG. https://doi.org/10.3390/agriculture11020134

Safaei, M., Taran, M., & Imani, M. M. (2019). Preparation, structural characterization, thermal properties and antifungal activity of alginate-CuO bionanocomposite. *Materials Science and Engineering C, 101*, 323–329. https://doi.org/10.1016/j.msec.2019.03.108

Salem, M. Z. M., EL-Hefny, M., Ali, H. M., Abdel-Megeed, A., El-Settawy, A. A. A., Böhm, M., Mansour, M. M. A., & Salem, A. Z. M. (2021). Plants-derived bioactives: Novel utilization as antimicrobial, antioxidant and phytoreducing agents for the biosynthesis of metallic nanoparticles. *Microbial Pathogenesis, 158*, 105107. https://doi.org/10.1016/J.MICPATH.2021.105107

Sánchez Reyna, P. A. (2016). Síntesis, caracterización y evaluación antimicrobiana de AgNPs, TiO2NPs y Ag/TiO2NPs utilizando como soporte Al2O3 proveniente de brackets cerámico.

Santiteerakul, S., Sopadang, A., Tippayawong, K. Y., & Tamvimol, K. (2020). The role of smart technology in sustainable agriculture: A case study of Wangree Plant Factory. *Sustainability, 12*(11), 4640. https://doi.org/10.3390/SU12114640

Santos, T. S., Silva, T. M., Cardoso, J. C., de Albuquerque-Júnior, R. L. C., Zielinska, A., Souto, E. B., Severino, P., & Mendonça, M. D. C. (2021). Biosynthesis of silver nanoparticles mediated by entomopathogenic fungi: Antimicrobial resistance, nanopesticides, and toxicity. *Antibiotics, 10*(7), 852. MDPI. https://doi.org/10.3390/antibiotics10070852

Sarkar, D. J., Mulu Loha, K., Adak, T., Kaushik, P., Koli, P., Majumder, S., Yadav, D. K., Chowdhury, A. R., Kumari, A., Singh, B. B., Rana, V. S., Kumar, J., & Shakil, N. A. (2023). Amphiphilic copolymer-based pesticide nanoformulations for better crop protection: Advances and future need. *Current Chinese Science, 3*(5), 369–385. https://doi.org/10.2174/2210298103666230809154449

Satti, S. H., Raja, N. I., Javed, B., Akram, A., Mashwani, Z. U. R., Ahmad, M. S., & Ikram, M. (2021). Titanium dioxide nanoparticles elicited agro-morphological and physicochemical modifications in wheat plants to control Bipolaris sorokiniana. *PLoS One, 16*(2), e0246880. https://doi.org/10.1371/JOURNAL.PONE.0246880

Sayed, A. M. M., Kim, S., & Behle, R. W. (2017). Characterisation of silver nanoparticles synthesised by Bacillus thuringiensis as a nanobiopesticide for insect pest control. *Biocontrol Science and Technology, 27*(11), 1308–1326. https://doi.org/10.1080/09583157.2017.1397597

Senbill, H., Hassan, S. M., & Eldesouky, S. E. (2023). Acaricidal and biological activities of titanium dioxide and zinc oxide nanoparticles on the two-spotted spider mite, Tetranychus urticae Koch (Acari: Tetranychidae) and their side effects on the predatory mite, Neoseiulus californicus (Acari: Phytoseiidae). *Journal of Asia-Pacific Entomology, 26*(1), 102027. https://doi.org/10.1016/J.ASPEN.2022.102027

Shafey, A. M. E. (2020). Green synthesis of metal and metal oxide nanoparticles from plant leaf extracts and their applications: A review. *Green Processing and Synthesis, 9*(1), 304–339. https://doi.org/10.1515/GPS-2020-0031/ASSET/GRAPHIC/J_GPS-2020-0031_FIG_007.JPG

Shahryari, F., Rabiei, Z., & Sadighian, S. (2020). Antibacterial activity of synthesized silver nanoparticles by sumac aqueous extract and silver-chitosan nanocomposite against Pseudomonas syringae pv. syringae. *Journal of Plant Pathology, 102*(2), 469–475. https://doi.org/10.1007/s42161-019-00478-1

Shang, Y., Kamrul Hasan, M., Ahammed, G. J., Li, M., Yin, H., & Zhou, J. (2019). Applications of nanotechnology in plant growth and crop protection: A review. *Molecules, 24*(14), 2558. MDPI AG. https://doi.org/10.3390/molecules24142558

Shangguan, W. J., Mei, X. D., Chen, H. P., Hu, S., Xu, C. L., Wang, L., Lv, K. F., Huang, Q. L., Xu, H. L., & Cao, L. D. (2023). Biodegradable electrospun fibers as sustained-release carriers of insect pheromones for field trapping of Spodoptera litura (Lepidoptera: Noctuidae). *Pest Management Science, 79*(12), 4774–4783. https://doi.org/10.1002/ps.7673

Shankar, P. D., Shobana, S., Karuppusamy, I., Pugazhendhi, A., Ramkumar, V. S., Arvindnarayan, S., & Kumar, G. (2016). A review on the biosynthesis of metallic nanoparticles (gold and silver) using bio-components of microalgae: Formation mechanism and applications. *Enzyme and Microbial Technology, 95*, 28–44. https://doi.org/10.1016/J.ENZMICTEC.2016.10.015

Sharifiaghdam, M., Shaabani, E., Asghari, F., & Faridi-Majidi, R. (2022). Chitosan coated metallic nanoparticles with stability, antioxidant, and antibacterial properties: Potential for wound healing application. *Journal of Applied Polymer Science, 139*(10), 51766. https://doi.org/10.1002/app.51766

Sharma, J., Singh, V. K., Kumar, A., Shankarayan, R., & Mallubhotla, S. (2018). Role of silver nanoparticles in treatment of plant diseases. *Microbial Biotechnology, 2*, 435–454). Springer Singapore. https://doi.org/10.1007/978-981-10-7140-9_20

Sharma, A., Sood, K., Kaur, J., & Khatri, M. (2019). Agrochemical loaded biocompatible chitosan nanoparticles for insect pest management. *Biocatalysis and Agricultural Biotechnology, 18*, 101079. https://doi.org/10.1016/J.BCAB.2019.101079

Shoaib, A., Waqas, M., Elabasy, A., Cheng, X., Zhang, Q., & Shi, Z. (2018). Preparation and characterization of emamectin benzoate nanoformulations based on colloidal delivery systems and use in controlling: Plutella xylostella (L.) (Lepidoptera: Plutellidae). *RSC Advances, 8*(28), 15687–15697. https://doi.org/10.1039/c8ra01913d

Shoala, T. (2020). Carbon nanostructures: Detection, controlling plant diseases and mycotoxins. In *Carbon nanomaterials for agri-food and environmental applications* (pp. 261–277). Elsevier. https://doi.org/10.1016/B978-0-12-819786-8.00013-X

Shukla, P. K., Misra, P., & Kole, C. (2016). Uptake, translocation, accumulation, transformation, and generational transmission of nanoparticles in plants. In *Plant nanotechnology: Principles and practices* (pp. 183–218). Springer International Publishing. https://doi.org/10.1007/978-3-319-42154-4_8

Shukla, G., Gaurav, S. S., & Singh, A. (2020). Synthesis of mycogenic zinc oxide nanoparticles and preliminary determination of its efficacy as a larvicide against white grubs (Holotrichia sp.). *International Nano Letters, 10*(2), 131–139. https://doi.org/10.1007/s40089-020-00302-0

Shukla, K., Mishra, V., Singh, J., Varshney, V., Verma, R., & Srivastava, S. (2024). Nanotechnology in sustainable agriculture: A double-edged sword. *Journal of the Science of Food and Agriculture, 104*(10), 5675–5688. https://doi.org/10.1002/JSFA.13342

Siddiqui, Z. A., Khan, M. R., Abd_Allah, E. F., & Parveen, A. (2019). Titanium dioxide and zinc oxide nanoparticles affect some bacterial diseases, and growth and physiological changes of beetroot. *International Journal of Vegetable Science, 25*(5), 409–430. https://doi.org/10.1080/19315260.2018.1523267

Simonin, M., Colman, B. P., Tang, W., Judy, J. D., Anderson, S. M., Bergemann, C. M., Rocca, J. D., Unrine, J. M., Cassar, N., & Bernhardt, E. S. (2018). Plant and microbial responses to repeated Cu(OH)2 nanopesticide exposures under different fertilization levels in an agro-ecosystem. *Frontiers in Microbiology, 9*, 1769. https://doi.org/10.3389/fmicb.2018.01769

Singh, R. P., Handa, R., & Manchanda, G. (2021). Nanoparticles in sustainable agriculture: An emerging opportunity. *Journal of Controlled Release, 329*, 1234–1248). Elsevier B.V. https://doi.org/10.1016/j.jconrel.2020.10.051

Slavin, Y. N., Asnis, J., Häfeli, U. O., & Bach, H. (2017). Metal nanoparticles: Understanding the mechanisms behind antibacterial activity. *Journal of Nanobiotechnology, 15*(1), 1–20. https://doi.org/10.1186/S12951-017-0308-Z/FIGURES/4

Song, S., Wang, Y., Xie, J., Sun, B., Zhou, N., Shen, H., & Shen, J. (2019). Carboxymethyl chitosan modified carbon nanoparticle for controlled emamectin benzoate delivery: Improved solubility, pH-responsive release, and sustainable pest control. *ACS Applied Materials and Interfaces, 11*(37), 34258–34267. https://doi.org/10.1021/acsami.9b12564

Spasova, M., Manolova, N., Naydenov, M., Kuzmanova, J., & Rashkov, I. (2010). Electrospun biohybrid materials for plant biocontrol containing chitosan and Trichoderma viride spores. *Journal of Bioactive and Compatible Polymers, 26*(1), 48–55. https://doi.org/10.1177/0883911510391446

Sportelli, M. C., Izzi, M., Volpe, A., Clemente, M., Picca, R. A., Ancona, A., Lugarà, P. M., Palazzo, G., & Cioffi, N. (2018). The pros and cons of the use of laser ablation synthesis for the production of silver nano-antimicrobials. *Antibiotics, 7*(3), 67. https://doi.org/10.3390/ANTIBIOTICS7030067

Takeshita, V., Munhoz-Garcia, G. V., Pereira, A. E. S., Tornisielo, V. L., & Fraceto, L. F. (2023). Radiometric strategy to track nanopesticides: An important approach to understand the fate, mechanisms of action and toxicity. *TrAC Trends in Analytical Chemistry, 165*, 117156. https://doi.org/10.1016/J.TRAC.2023.117156

Toto, N. A., Elhenawy, H. I., Eltaweil, A. S., El-Ashram, S., El-Samad, L. M., Moussian, B., & El Wakil, A. (2022). Musca domestica (Diptera: Muscidae) as a biological model for the assessment of magnetite nanoparticles toxicity. *Science of the Total Environment, 806*, 151483. https://doi.org/10.1016/J.SCITOTENV.2021.151483

Ul Haq, I., & Ijaz, S. (2019). Use of metallic nanoparticles and nanoformulations as nanofungicides for sustainable disease management in plants. In *Nanotechnology in the life sciences* (pp. 289–316). Springer Science and Business Media B.V. https://doi.org/10.1007/978-3-030-17061-5_12

Usman, M., Farooq, M., Wakeel, A., Nawaz, A., Cheema, S. A., ur Rehman, H., Ashraf, I., & Sanaullah, M. (2020). Nanotechnology in agriculture: Current status, challenges and future opportunities. *Science of the Total Environment, 721*, 137778. https://doi.org/10.1016/J.SCITOTENV.2020.137778

Varympopi, A., Dimopoulou, A., Papafotis, D., Avramidis, P., Sarris, I., Karamanidou, T., Kerou, A. K., Vlachou, A., Vellis, E., Giannopoulos, A., Haralampidis, K., Theologidis, I., Hatzinikolaou, D. G., Tsouknidas, A., & Skandalis, N. (2022). Antibacterial activity of copper nanoparticles against Xanthomonas campestris pv. vesicatoria in tomato plants. *International Journal of Molecular Sciences, 23*, 4080. https://doi.org/10.3390/ijms23084080

Vazquez-Muñoz, R., Meza-Villezcas, A., Fournier, P. G. J., Soria-Castro, E., Juarez-Moreno, K., Gallego-Hernández, A. L., Bogdanchikova, N., Vazquez-Duhalt, R., & Huerta-Saquero, A. (2019). Enhancement of antibiotics antimicrobial activity due to the silver nanoparticles impact on the cell membrane. *PLoS One, 14*(11), e0224904. https://doi.org/10.1371/JOURNAL.PONE.0224904

Velmurugan, P., Lee, S.-M., Iydroose, M., Lee, K.-J., & Oh, B.-T. (2012). Pine cone-mediated green synthesis of silver nanoparticles and their antibacterial activity against agricultural pathogens. *Applied Microbiology and Biotechnology, 97*(1), 361–368. https://doi.org/10.1007/S00253-012-3892-8

Velten, S., Leventon, J., Jager, N., & Newig, J. (2015). What is sustainable agriculture? A systematic review. *Sustainability, 7*(6), 7833–7865. https://doi.org/10.3390/SU7067833

Vidyasagar, Patel, R. R., Singh, S. K., Dehari, D., Nath, G., & Singh, M. (2024). Facile green synthesis of silver nanoparticles derived from the medicinal plant Clerodendrum serratum and its biological activity against Mycobacterium species. *Heliyon, 10*(10), e31116. https://doi.org/10.1016/j.heliyon.2024.e31116

Vivekanandhan, P., Swathy, K., Thomas, A., Kweka, E. J., Rahman, A., Pittarate, S., & Krutmuang, P. (2021). Insecticidal efficacy of microbial-mediated synthesized copper nano-pesticide against insect pests and non-target organisms. *International Journal of Environmental Research and Public Health, 18*(19), 10536. https://doi.org/10.3390/ijerph181910536

Wang, X., Liu, X., Chen, J., Han, H., & Yuan, Z. (2014). Evaluation and mechanism of antifungal effects of carbon nanomaterials in controlling plant fungal pathogen. *Carbon, 68*, 798–806. https://doi.org/10.1016/J.CARBON.2013.11.072

Wang, X., Zhou, Z., & Chen, F. (2017). Surface modification of carbon nanotubes with an enhanced antifungal activity for the control of plant fungal pathogen. *Materials, 10*(12), 1375. https://doi.org/10.3390/MA10121375

Wang, A., Cui, J., Wang, Y., Zhu, H., Li, N., Wang, C., Shen, Y., Liu, P., Cui, B., Sun, C., Zhao, X., Wang, C., Gao, F., Zeng, Z., & Cui, H. (2020). Preparation and characterization of a novel controlled-release nano-delivery system loaded with pyraclostrobin via high-pressure homogenization. *Pest Management Science, 76*(8), 2829–2837. https://doi.org/10.1002/ps.5833

Wang, Y., Tian, J., Wang, Z., Li, C., & Li, X. (2022). Crop-safe pyraclostrobin-loaded multiwalled carbon nanotube delivery systems: Higher fungicidal activity and lower acute toxicity. *ACS Agricultural Science and Technology, 2*(3), 534–545. https://doi.org/10.1021/acsagscitech.1c00293

Worrall, E. A., Hamid, A., Mody, K. T., Mitter, N., & Pappu, H. R. (2018). Nanotechnology for plant disease management. *Agronomy, 8*(12), 285. https://doi.org/10.3390/AGRONOMY8120285

Wu, J., Du, C., Zhang, J., Yang, B., Cuthbertson, A. G. S., & Ali, S. (2022). Synthesis of metarhizium anisopliae–chitosan nanoparticles and their pathogenicity against plutella xylostella (Linnaeus). *Microorganisms, 10*(1), 1. https://doi.org/10.3390/microorganisms10010001

Xu, J., Zhang, K., Cuthbertson, A. G. S., Du, C., & Ali, S. (2020). Toxicity and biological effects of beauveria brongniartii Fe^0 nanoparticles against spodoptera litura (Fabricius). *Insects, 11*(12), 1–15. https://doi.org/10.3390/insects11120895

Yamanaka, M., Hara, K., & Kudo, J. (2005). Bactericidal actions of a silver ion solution on Escherichia coli, studied by energy-filtering transmission electron microscopy and proteomic analysis. *Applied and Environmental Microbiology, 71*(11), 7589–7593. https://doi.org/10.

1128/AEM.71.11.7589-7593.2005/ASSET/D25E0289-4ADE-450E-80C7-F7293385304A/ASSETS/GRAPHIC/ZAM0110560780004.JPEG

Ye, X., Liu, M., Zhao, N., Xiao, C., Xu, H., & Jia, J. (2022). Targeted delivery of emamectin benzoate by functionalized polysuccinimide nanoparticles for the flowering cabbage and controlling Plutella xylostella. *Pest Management Science, 78*(2), 758–769. https://doi.org/10.1002/ps.6689

Yin, J., Su, X., Yan, S., & Shen, J. (2023). Multifunctional nanoparticles and nanopesticides in agricultural application. *Nanomaterials, 13*(7), 1255. MDPI. https://doi.org/10.3390/nano13071255

Zhang, X., Xu, Z., Qian, X., Lin, D., Zeng, T., Filser, J., Li, L., & Kah, M. (2020a). Assessing the impacts of Cu(OH)2 nanopesticide and ionic copper on the soil enzyme activity and bacterial community. *Journal of Agricultural and Food Chemistry, 68*(11), 3372–3381. https://doi.org/10.1021/acs.jafc.9b06325

Zhang, Z. J., Shang, X. F., Yang, L., Shi, Y. B., Liu, Y. Q., Li, J. C., Yang, G. Z., & Yang, C. J. (2020b). Engineering of peglayted camptothecin into nanomicelles and supramolecular hydrogels for pesticide combination control. *Frontiers in Chemistry, 7*, 922. https://doi.org/10.3389/fchem.2019.00922

Zhang, D. X., Wang, R., Cao, H., Luo, J., Jing, T. F., Li, B. X., Mu, W., Liu, F., & Hou, Y. (2022a). Emamectin benzoate nanogel suspension constructed from poly(vinyl alcohol)-valine derivatives and lignosulfonate enhanced insecticidal efficacy. *Colloids and Surfaces B: Biointerfaces, 209*, 112166. https://doi.org/10.1016/J.COLSURFB.2021.112166

Zhang, X., Cao, H., Wang, J., Li, F., & Zhao, J. (2022b). Graphene oxide exhibits antifungal activity against Bipolaris sorokiniana in vitro and in vivo. *Microorganisms, 10*(10), 1994. https://doi.org/10.3390/MICROORGANISMS10101994

Zhao, M., Zhou, H., Hao, L., Chen, H., & Zhou, X. (2021). Natural rosin modified carboxymethyl cellulose delivery system with lowered toxicity for long-term pest control. *Carbohydrate Polymers, 259*, 117749. https://doi.org/10.1016/j.carbpol.2021.117749

Zhou, Q., Xia, Z., Zhang, Y., Sun, Z., Zeng, W., Zhang, N., Yuan, C., Gong, C., Zhou, Y., & Xue, W. (2024). Design of a delivery vehicle chitosan-based self-assembling: Controlled release, high hydrophobicity, and safe treatment of plant fungal diseases. *Journal of Nanobiotechnology, 22*(1), 121. https://doi.org/10.1186/s12951-024-02386-8

Chapter 9
Sustainable Nanofertilizers and Plant Nutrition

Sara Khalaji, Sima Panahirad, and Gholamreza Gohari

9.1 Overview of Nanofertilizers

9.1.1 Definition and Significance in Sustainable Agriculture

Nano-fertilizer is a new insight in the agricultural industry based on nanotechnology, in which the fertilizers are engineered in nanoparticle materials with targeted delivery and controlled release features that improve the nutrient uptake and increase crop yield. Moreover, a reduction in application frequency and environmental impacts in comparison to chemical fertilizers will be obtained (Raliya et al., 2017). Nano-fertilizers enhance the solubility, mobility, and nutrient availability in soil, which is a challenge in conventional fertilizers and an approach in sustainable agriculture (Solanki et al., 2015).

Food security is a global concern due to rapid population growth and climate change. Climate change increases the risks for food security all around the world (IPPC, 2014). Observations of the effects of climate change on crop production indicate that it has already negatively affected wheat and maize yields (Moore and Lobell et al., 2015). Furthermore, the world population is projected to reach 9.7 billion people in 2050 and 10.4 billion by 2100 (Fig. 9.1) (Valavanidis,

S. Khalaji
Department of Horticultural Sciences, Faculty of Agriculture, University of Tehran, Tehran, Iran

S. Panahirad
Department of Landscape Engineering, Faculty of Agriculture, University of Tabriz, Tabriz, Iran

G. Gohari (✉)
Department of Horticultural Sciences, Faculty of Agriculture, University of Maragheh, Maragheh, Iran
e-mail: gohari.gh@maragheh.ac.ir

K. Esquivel Escalante et al. (eds.), *Green Nanomaterials*, Nanotechnology in the Life Sciences, https://doi.org/10.1007/978-3-032-04492-1_9

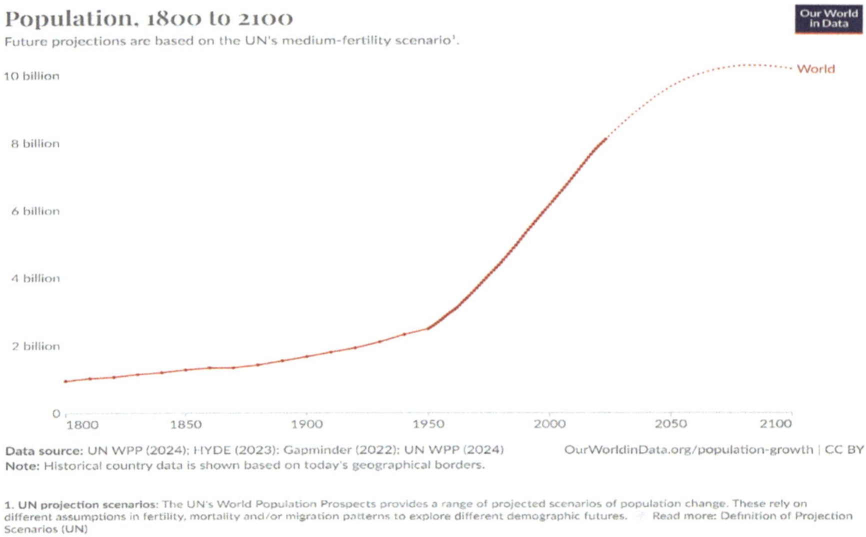

Fig. 9.1 World population growth estimation by 2100 (https://ourworldindata.org/grapher/population-long-run-with-projections)

2023; UN WPP, 2022). This growth creates a 60% more food demand by 2025 (FAO, 2022), while the urbanization process is on the growth and will limit the arable land. Hence, an urgent approach that promises solutions to overcome these challenges is needed, which involves innovation in agricultural technology, sustainable agricultural methods, and global cooperation (FAO, 2022). The invention of nano-fertilizers, which is the synthesis of materials in nanometers (1–100 nm), could serve as a potential solution in sustainable agriculture because scientific research has highlighted significant capacity for nanofertilizer application in the food security crisis.

For instance, nanofertilizers increase the availability of essential nutrients that improve the growth of plants and stress tolerance, like drought or poor nutrient conditions in the soil (Verma et al., 2022; Gade et al., 2023). Nanofertilizers help reduce concerns about the overuse of conventional agricultural chemicals by reducing the frequency and quantity of classical chemical fertilizers. This helps maintain healthy soil in the long term, which is essential for the sustainability of agriculture with a growing population and increasing demand for food. Due to their safe and efficient application, nano-fertilizers hold great promise in the pursuit of future food security and therefore form the basis for the development of sustainable agricultural systems.

9.1.2 Contrast with Traditional Fertilizers

The application of traditional fertilizers is vital in agriculture because they provide the essential nutrients for plants, such as nitrogen, phosphorus, and potassium, whereas the negative points, like inefficiency in nutrient uptake, environmental pollution, leaching, runoff, and volatilization, are serious and hazardous. For instance, more than 50% of nitrogen in urea, a very common N fertilizer, is lost to the atmosphere before there is any chance of its uptake by plant roots, leading to excess groundwater contamination and eutrophication (Chen et al., 2020). Overuse of conventional fertilizers, such as this, jeopardizes nutrient soil imbalance, which has an impact on plant well-being and the general ecosystem (Marschner et al., 2020). On the other hand, the application of nano-fertilizers is a more reliable option due to their advantageous properties to overcome the problems caused by conventional fertilizers. One of these properties is called slow and controlled nutrient release that sustains nutrient availability for a longer period and lets the plants absorb them based on their need, which limits the application's frequency and hence less nutrient loss (Kah et al., 2021). Moreover, nano-fertilizers are prepared in nanoscale particles; therefore, the interaction surface area, solubility, nutrient uptake, and, as a result, crop yield will be increased with a reduction in environmental impacts (Raliya et al., 2021).

Nanofertilizers are also more efficient in terms of the precise use of nutrients. The small size and improved properties of nanoparticles allow for higher interaction with plant root systems, therefore improving nutrient bioavailability and mobility within the soil (Adisa et al., 2021). This precision reduces not just the amount of fertilizer applied but also minimizes problems starting to emerge with nutrient runoff and leaching. For instance, a common problem is when phosphorus from conventional fertilizers becomes immobilized in the soil and hence unavailable for plants. However, phosphorus in nanofertilizers would stay more available because of the increased solubility (Sharma et al., 2022a). While nanofertilizers have tremendous potential benefits, technology remains in infancy. Long-term environmental and human health risks are most likely, considering the fact that the environmental fate of nanoparticles has been poorly characterized so far, according to Rico et al. (2020). Several studies have suggested that nanoparticles might be deposited in the soil or aquatic system with a possibility for toxicity or other adverse effects on ecosystems (Kah et al., 2021). Conclusively, deeper studies to ensure safe and efficient use of nanofertilizers in the agricultural sector should be done. The application of conventional fertilizers, though central to modern agriculture, has engendered inefficiencies and environmental consequences, hence inspiring the quest for better and more environmentally friendly options. Therefore, nanofertilizers become exceptionally promising alternatives due to their enhanced nutrition delivery and environmental friendliness (Fig. 9.2). However, further research and development need to be conducted in relation to their complete potential and risks involved.

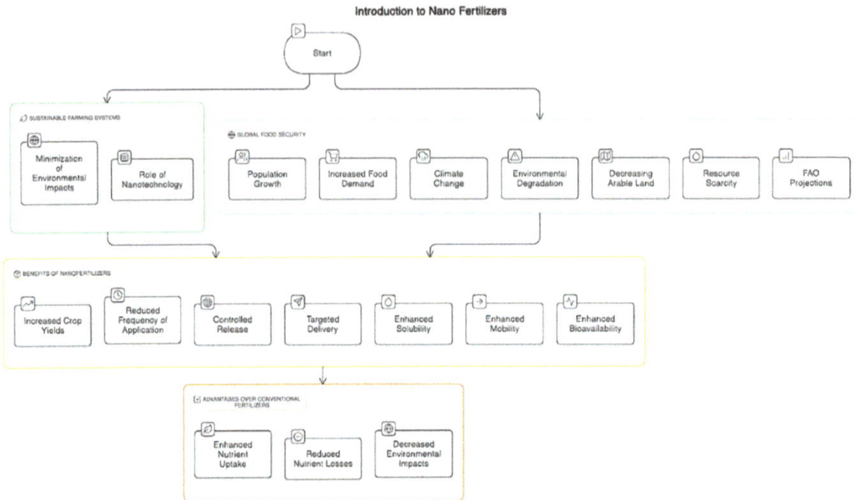

Fig. 9.2 An introduction to nano fertilizers

9.2 The Role of Green NanoFertilizers in Agriculture

9.2.1 Introduction to Green Nano Fertilizers

Green nanofertilizers are fantastic due to their innovative structure. They stir up significant change in agriculture by dramatically reducing environmental impacts. They are developed in a way that while nutrients are effectively delivered, harmful side effects often associated with their chemical counterparts are avoided. These green nanofertilizers use nanotechnology to enhance the efficiency in the use of nutrients and the prevention of waste. These are environment-friendly since plant fertilizers are constituted of biodegradable nanomaterials that ensure the slow, controlled release of basic nutrients (N, P, K) for plant growth requirements (Kah et al., 2021). With the imminent reality of the growth in population anticipated globally to reach 10 billion by 2050, food demand is expected to increase, further intensifying pressure on agriculture in terms of increased outputs out of depleting resources (FAO, 2022). Green nanofertilizers also form a perfect means to alleviate the situation by enhancing nutrient use and lessening the ecological hit (Kah et al., 2021).

9.2.2 *Importance of Sustainability and Circular Economy in Agriculture*

Green nanofertilizers were designed after the raised concerns about the environmental cost and economic losses that came with conventional farming practices. Lots of impacts are all about environmental degradation, like soil destruction, water pollution, and nitrogen aerosol pollution as a result of the overuse of conventional fertilizers (Liu & Lal, 2021). Green nanofertilizers, hence, strive to correct and solve the issue by providing substantial nutrient absorption and requiring little fertilizer application to achieve the maximum crop yields-all products generated from materials that are safe to human health and safe to the ecological balance, owing to green chemistry principles (Usman et al., 2022a, b). Green nanofertilizers should be put in terms of ECV (ecological and economic viability). Nanofertilizers, keeping their size small enough with increased surface area, help to energize the plant root system and thereby increase nutrient bioavailability and efficient nutrient uptake rate as an advantage (Raliya et al., 2021), while this site specificity of nutrient delivery inhibits excess fertilizer application and thus input costs for the farmer, with no reduction or even an increase of crop yields (Usman et al., 2022a, b).

9.2.3 *Green Synthesis of Nanofertilizers*

Environmental problems are creating restrictions on the use of hazardous chemicals. Therefore, there is a great need for the development of safer alternatives in the field of sustainable agriculture. Furthermore, it has the ethos of minimal environmental impact using renewable and biodegradable sources under mild reaction conditions (Sharma et al., 2022a; Khan et al., 2023). It involves choosing a proper biological agent, most commonly plant extracts, bacteria, fungi, and algae, which may serve as a natural reducing agent apart from being a stabilizing agent. These biological materials contain a variety of metabolites, such as alkaloids, proteins, phenolics, and flavonoids, that help in the reduction process of the metal salt into nanoparticles. These are usually reductive agents of metal ions, such as zinc, iron, and copper, to their nanoparticle form; these are then stabilized by organic compounds present within the extract. Not only is it efficient in nutrient delivery and controlled nutrient release but also an eco-friendly product with less damage to the environment (Khan et al., 2023).

Based on a study conducted in 2023, ZnO nanoparticles, which were derived from plant extracts, were applied to maize crops under drought stress. The results showed significant improvement in water-use efficiency, chlorophyll content, and antioxidant enzyme activity, enhancing growth and productivity under stress conditions (Singh et al., 2023). By preventing oxidative damage in this process, the nanoparticles considerably enhanced the endurance and survival potential of the crops under drought conditions. The present study underlines the possibilities of green nanotechnology to act as an alternative to conventional fertilizers and, therefore, as an ecologically sound solution for the abiotic stress challenge (Singh et al., 2023).

9.3 Green Nanofertilizers for Fertilizer Development

9.3.1 Description of Green Nanomaterials Used (Bio-Based Nanoparticles, Plant-Based Extract)

Green nanomaterials are placed in a better position to act as an alternative to improve sustainability. They are mostly synthesized from natural materials, such as plant-based extracts or biobased nanoparticles. The most significant advantages of green nanomaterials are their biodegradability and nontoxicity, with low environmental impact compared to chemical fertilizers (Sharma et al., 2023a).

9.3.1.1 Bio-Based Nanoparticles

Bio-based nanoparticles can be formed from natural polymers or organic materials like chitosan, cellulose, or starch. Nanoparticles from these materials have found applications in agriculture for nutrient delivery, pest control, and soil health improvement. As an example, chitosan nanoparticles hold a special position due to their antimicrobial activities, which help in reducing plant diseases and promoting nutrient uptake. These biobased nanoparticles are also biodegradable, thus limiting the concerns about long-term environmental contaminants (Sharma et al., 2023a). Other examples of green nanomaterials involve nanoparticles based on cellulose, finding broad applications in slow-release fertilizers. Such nanoparticles ensure that crops receive critical nutrients, including nitrogen and phosphorus, in controlled ways to prevent losses either through leaching or volatilization, hence enhancing nutrient use efficiency. In addition, they ensure that soil structure and fertility are retained and become a vital part of sustainable agricultural developments (Usman et al., 2022a, b).

9.3.1.2 Plant-Based Extract

Plant extracts are crucial sources of green nanoparticle synthesis. The plant extracts could be sourced from plant leaves, fruits, and seeds that serve as reducing/stabilizing agents during the synthesis of nanomaterials. The green synthesis holds beckons because it is very simple and cost-effective, unlike conventional methods that produce hazardous byproducts. For example, the silver nanoparticles synthesized using *Azadirachta indica* (neem) extract have been shown to be highly effective in controlling various plant pathogens while being completely environmentally non-toxic (Rico et al., 2021).

9.3.2 *Methods of Synthesizing Nanofertilizers Using Green Chemistry*

Green chemistry has been formulated to develop a holistic synthesis of nanofertilizers that opens eco-friendly and environmentally responsible doors to produce those products that are eco-sustainable and do not employ pressurizing hazardous chemicals in the framework of the environment conservation mode. Some methods are presented here, which explain how synthetic nanofertilizers will be made using the principles of green chemistry, mainly for the period 2014–2020:

1. *Plant-Based Synthesis*

Most commonly, plant extracts play the role of reductive and stabilizing agents in synthesizing nanofertilizers within green methods; the efficiency of this unique technique lies largely in the fact that it avoids the use of hazardous chemicals, thereby providing a toxic route of is nature-friendly for nanoparticle synthesis. Within the extracts of plants, the natural biomolecules neutralize particles of metal ions, like alkaloids, flavonoids, terpenoids, and phenolic acids. These extracts of certain plants, such as *Moringa oleifera* and *Azadirachta indica* work well in synthesizing nanoparticle forms of zinc and silver, respectively. These nanoparticles can be incorporated into fertilizers to deliver nutrients effectively to plants (Sharma et al., 2023a; Rico et al., 2021). The green synthesis process using plants as reducers and stabilizers is an environmentally friendly and sustainable method for nanoparticle production. Green synthesis is based on the extraction of flavonoids, alkaloids, and phenolic acids from plant parts such as leaves, seeds, or bark, followed by mixing them with solutions to which salts of metals such as silver nitrate or zinc acetate are added. The bioactive compounds function as reducing agents for the metal ions, reducing into nanoparticles at the same time, passively stabilizing them against aggregation. Later, these nanoparticles can be characterized for their size, shape, and stability using techniques such as UV–visible spectrophotometry, electron microscopy, and zeta potential analysis. These various advantages make it clear that plant-based nanofertilizers may be a possible stepping stone toward sustainable agriculture. Currently, while given their simple and scalable synthesis, low environmental impact in comparison with other chemical methods of synthesis, and their quite high potential yield for the improved growth of crop plants, the greener process of making well-performing nanofertilizers from agricultural products demonstrates some feasibility (Sharma et al., 2023a; Rico et al., 2021).

2. *Microbial Synthesis*

Another key approach toward green synthesis is with the help of microorganisms like bacteria and fungi. In general, microorganisms can act as "biofactories" in the synthesis of metal nanoparticles through the reduction of metal ions to their nano forms in enzymatic activity. This approach finds excellent application in synthesizing metal-based nanofertilizers like silver, gold, and copper nanoparticles. Microbial enzymes such as nitrate reductase enable the conversion of metal ions into stable nanoparticles. This approach is considered eco-friendly and efficient due to the ease of culturing microorganisms and the ambient conditions under which they operate (Bairwa et al., 2023; Belaiche et al., 2021).

3. *Biopolymer-Based Synthesis*

Chitosan, starch, and cellulose are other biopolymers commonly used for the green synthesis of nanofertilizers. These natural polymers not only act as reducing agents but also stabilize the nanoparticles by preventing agglomeration. More importantly, chitosan is a favorite because of its biodegradability, antimicrobial properties, and ability to enhance plant nutrient use efficiency. Nanofertilizers made of biopolymers provide a slow-release pattern for the plants to tap the nutrition over an extended time period. This consequently enhances the efficiency of the application with an additional benefit of reducing environmental impact (Usman et al., 2022a, b; Avellan et al., 2021).

4. *Extracellular and Intracellular Biosynthesis*

Another process of biosynthesis includes the involvement of microbial cultures through extracellular and intracellular methods. Under extracellular biosynthesis, metal ions are reduced to nanoparticles by enzymes produced by microorganisms, which are secreted into the surrounding medium. In intracellular biosynthesis, the metal ions are transported inside the microbial cells, where they are reduced by enzymes inside the cell. Both methods produce nanoparticles with defined shapes and sizes, which are critical for the precision delivery of nutrients in agricultural systems (Belaiche et al., 2021; Bairwa et al., 2023).

5. *Cell-Free Approaches*

A method variant of microbial synthesis is a cell-free approach; the supernatant or culture medium of microbial cells is utilized for nanoparticle synthesis. This technique makes use of enzymes and biomolecules present within the medium to act both as reducing and capping agents, hence ensuring an efficient nanoparticle synthesis without the use of whole organisms. It is a faster and more scalable method suitable for agricultural applications, such as developing nanoparticle-based fertilizers (Rico et al., 2021; Sharma et al., 2023a).

9.3.3 Case Studies of Successful Green Synthesis Processes

Green synthesis of nanoparticles is a revolutionary approach in nanotechnology, combining sustainability with functional novelty. This method often takes natural resources like plant extracts, whose synthesis has a lesser environmental impact and is devoid of the use of hazardous chemicals common in conventional nanoparticle synthesis. The author hereby presents some cases as illustrations of the success of green synthesis techniques in recent decades.

- *Acalypha indica Antibacterial Application*: The synthesis of AgNPs prepared from the extracts of Acalypha indica leaves was revealed by Chopra et al. (2022). The bioreduction process turned these nanoparticles into metals with exceptional activity. Among other findings, they showed antibacterial potential and rows of promising activities in the strains of *Escherichia coli* and *Staphylococcus aureus*. These results conclusively demonstrate the great potential of plant-based nanoparticles of Acalypha for design as a green nanoantibiotic agent for application against multidrug-resistant strains.
- *Neem (Azadirachta indica) for Dual-Action Nanoparticles*: Ahmed et al. (2016) used neem leaf extract in the biosynthesis of AgNPs, which had both antimicrobial and anticancer properties. These nanoparticles showed significant action on pathogens and cytotoxic effects on HeLa cancer cells, thereby proving the incredible potential of neem in the healthcare nano-technology domain.
- *Ginger (Zingiber officinale) for Antifungal Applications*: AgNPs were biosynthesized using a ginger extract that caused notable antifungal activity against *Candida albicans* and consequently, with probable various applications in antifungal therapy, especially against diseases resistant to conventional medications Mubaraki et al., (2023), Liu et al., (2022).
- *Mango (Mangifera indica) for Antioxidant and Anti-Inflammatory Uses*: AgNPs prepared using mango leaf extracts demonstrated exceptional antioxidant and anti-inflammatory activities, introducing their potential therapeutic uses, including treatment against oxidative stress and inflammatory diseases (Alissa, 2025).
- *Tulsi (Ocimum sanctum) for Water Disinfection*: Biosynthesized AgNPs using *Ocimum sanctum* (tulsi) leaf extracts displayed broad-spectrum antimicrobial activity and were highly effective in water disinfection systems for practical solutions to clean drinking water in rural and urban setups (Kumar et al., 2018a).
- *Tea Leaves for Antibacterial Enhancement*: AgNPs biosynthesized using the tea leaves extracts demonstrated distinctive antibacterial activity against *Staphylococcus aureus* and might have the capacity to be used as a coating on medical equipment and packaging materials to avoid bacterial contamination.
- *Aloe Vera for Wound Healing*: AgNPs achieved by adding *Aloe vera* gel had significant antibacterial properties, a worthy wound-healing capability, and biocompatibility. As a result, Aloe vera-based nanoparticles could be used in a delivery system in biomedical topical treatment (Gangwar et al., 2025).
- *Skimmia laureola for Broad-Spectrum Antimicrobial*: Leaf extracts of *Skimmia laureola* were exploited for AgNPs biosynthesizing, leading to AgNPs with

inhibitory effects against a variety of pathogens like *Pseudomonas aeruginosa* and *Escherichia coli*, underlining their utility in healthcare and industrial applications (Shahzadi et al., 2025).

- *Eclipta alba for Environmental Remediation*: *Eclipta alba*-biosynthesized AgNPs showed great efficiency in the wastewater treatment and therefore with great potential in applications in environmental remediation along with their antimicrobial activities (Sharma et al., 2023a).

All these case studies further prove that green synthesis methods have vast prospective for sustainable nanoparticle production. Moreover, the range of plant materials used indicates the versatility and adaptability of the approach across numerous fields, including healthcare, agriculture, and environmental science. Further research is essential to step up these methods to industrial applications without compromising on cost efficiency and environmental integrity.

9.3.4 Development of Eco-Friendly Nanofertilizers

A new frontier in sustainable agriculture has been obtained through the development of eco-friendly nanofertilizers with regard to improve nutrient use efficiency and reduce environmental impact. Incidentally, green methods-synthesized nanofertilizers enable site-specific nutrient delivery with reduced nutrient losses, thus leading to sustainable crop production. Consequently, the emphasis of green synthesis techniques for nanofertilizers is on plant extracts. For instance, zinc oxide nanoparticles (ZnONPs) synthesized using *Azadirachta indica* (neem) leaf extracts demonstrated higher nutrition bioavailability and plant growth as well as lessening chemical use (Kumar et al., 2021a). This approach ensures environmental safety with an enhancement in soil fertility and crop yield. Likewise, progress in encapsulation methods has additionally made potential biodegradable nanofertilizers with slow-release properties. Several polymer-based nanocarriers loaded with P and N caused a prolonged nutrient release pattern, minimized leaching, and enhanced nutrient-use efficiency (Ali et al., 2022a). The novel advancements could address the critical challenges of soil nutrient deficiency and water pollution.

Besides, nanofertilizers combination with microbial inoculants exhibits potential for regenerative agriculture. Regarding silica nanoparticles functionalized with nitrogen-fixing bacteria such as *Azospirillum brasilense* caused higher root development and drought tolerance in wheat crops, hence proving beneficial for mitigating climate-induced stress (Patel et al., 2023). All these advances reveal that nanofertilizers with an ecological approach, combined with advanced green synthesis and biological integration, could be capable of altering agricultural practices through providing higher productivity with reduced environmental impact.

9.4 Mechanisms of Nutrient Delivery in Nanofertilizers

Nanofertilizers are considered a novel progress in agriculture in order to enhance the efficiency of nutrient delivery systems. In this regard, green nanofertilizers that are prepared through environmentally friendly procedures have proved relatively beneficial effects thanks to their very low environmental impact and greater efficiency in nutrient uptake by plants. ZnONPs that are synthesized using plant-mediated methods increased Zn bioavailability in wheat plants, very effective in improving crop yields and micronutrient uptakes due to the nanoscale size that ensures efficient penetration into the root cells for rapid absorption and translocation (Sharma et al., 2021a). The interaction of functionalized silica nanoparticles with plant systems modulated root morphology and nutrient transporter activities, inducing root growth and enhancing critical nutrients uptake, such as N and P, by plants (Ali et al., 2022a). On top, controlled release mechanisms promote the efficiency of green nanofertilizers. For instance, nitrogen nanoparticles coated with polymers exploited a steady nutrient supply and decreased environmental losses through leaching and volatilization (Patel et al., 2023). All highlight green nanofertilizers roles in improving agricultural sustainability and nutrient management.

9.4.1 How Green Nanofertilizers Enhance Nutrient Uptake

At the moment, green nanofertilizers are considered as a revolution in modern agriculture thanks to their role in improving nutrient availability and uptake, besides promoting environmentally sustainable farming practices. These nanofertilizers provide a bioavailable form of nutrients to certify precise targeting at the cellular level for enhanced plant growth, development, and health. An example is green-synthesized ZnONPs by *Moringa oleifera* extract that demonstrated significant potential in order to increase Zn uptake and availability in plants, as an essential micronutrient for plant growth, than conventional chemical fertilizers (Sharma et al., 2021a). This property, as mentioned, is due to the nanosize that helps them easily pass through barriers present in soil, effectively penetrating the plant root system, thus better translocating to other plant tissues and moving freely within the vascular systems of the plant, and then after, very superior to conventional fertilizers (Ali et al., 2019). On top, employing plant extracts like *Moringa oleifera* in the synthesis process offers eco-friendly and low-cost alternatives to conventional physical or chemical synthesis methods, ensuring the obtained nanoparticles' biodegradability and their lower environmental concerns (Dubey et al., 2020).

Nanofertilizers have been reported to improve agricultural productivity by increasing key micronutrient uptake. For example, green-synthesized ZnONPs increased seedling growth, root elongation, and tolerance to stress conditions, consequently with the capability to modify plant physiological responses (Kumar et al., 2018a). All propose the great potential of nanotechnology in revolutionizing nutrient management strategies through decreasing nutrient losses and improving plants' access to elements under different environmental conditions. Besides, ZnONPs ability to cooperate with cellular membranes and transport across plant vascular pathways causes a unique mechanism for improving crop performance in areas with Zn deficiency (Kumar et al., 2019), proving the importance of green-synthesized nanofertilizers as an excellent and environmentally friendly alternative to classical soil amendments.

9.4.2 Interaction Between Green Nanomaterials and Plant Systems

The interaction of green-synthesized nanomaterials in the form of nanofertilizers with plant systems is a complex process that involves interactions at multiple pathways. The biocompatibility of green nanomaterials in modulating root morphology and enhancing nutrient transport occurs via ion channels. The green synthesized silica nanoparticles stimulated root elongation in maize through an improvement in nitrate (Ali et al., 2022a). Moreover, they increased root exudate production that are essential for inducing a symbiosis with the soil microbiota.

9.4.3 How Green Nanofertilizers Improve the Efficiency of Nutrient Uptake

Green nanofertilizers could enhance nutrient-use efficiency through facilitating controlled release and minimizing their leaching or volatilization losses. In this regard, polymer-coated N nanoparticles reduce N runoff and provide a constant flow of nutrients to the crops for much longer periods. The higher biocompatibility of nanofertilizers causes better bioavailability and functionality in using the nutrients in plants (Patel et al., 2022).

9.5 Nanofertilizers and Crop Yields and Quality

9.5.1 Analysis of Crop Yield Improvements Compared to Traditional and Conventional Fertilizers

Nanofertilizers cause higher crop yield as compared to conventional fertilizers. For example, green-synthesized ZnONPs enhanced the wheat yield by 25% as compared to chemical zinc sulfate fertilizer (Kumar et al., 2021a). All over again, due to better use efficiency and reduced losses in nutrients, polymer-coated N nanofertilizers increased the rice productivity by 30% (Patel et al., 2022). In vegetables, potassium nanoparticle formulations increased the tomato and spinach yields via improving nutrient uptake and reducing environmental runoff (Ali et al., 2023). In fruit crops, silver nanoparticle-based fertilizers enhanced the fruit set and weight and reduced synthetic chemical inputs application (Sharma et al., 2021a).

9.5.2 *Impact on Crop Quality*

Green nanofertilizers have additionally remarkable impacts on crop quality, regarding the nutritional value, shelf life, and marketability. The application of iron oxide nanoparticles improved the iron content of spinach leaves by 40%; consequently, they were nutritionally superior (Singh et al., 2022). Furthermore, calcium-enriched nanofertilizers increased apple postharvest quality and shelf life through decreasing tissue deterioration and increasing tolerance to microbial diseases (Choudhary et al., 2023). Likewise, silica nanoparticles application in cereal crops enhanced grain protein content that making its application a sustainable way to handle nutritional deficiencies in staple crops (Patel et al., 2023). Green nanofertilizers improved antioxidant levels and market appeal of leafy vegetables (Kumar et al., 2021a).

9.5.3 *Effects on Storage, Shelf Life, and Marketability of Crops*

Green nanofertilizers improve plant quality, leading to enhanced storage and shelf life through decreasing oxidative stress and microbial decay. For instance, nanocopper fertilizers reduced spoilage and increased the shelf life by up to 15 days with complete texture and flavor in mangoes as well as storability (Ali et al., 2023). Nano-silica application improved tomato firmness and decreased water loss, leading to higher marketability and lower post-harvest losses (Sharma et al., 2021a). All highlight that nanofertilizers have a dual role: increasing both the yield and quality of crops, leading to a sustainable and efficient future in agriculture.

9.6 Environmental and Ecological Impact of Nanofertilizers

Nanofertilizers are developing as a revolutionary approach in agriculture, primarily through contributing to environmental sustainability and ecological balance. Nanofertilizers help in overcoming environmental challenges and improving agricultural practices. The following sections give a brief explanation of how they will affect in this regard.

9.6.1 Lower Greenhouse Gas Emissions and Energy Efficiency

Nanofertilizers, such as nano-urea and NPK-coated formulations, have controlled-release mechanisms. This property leads to an important decrease in greenhouse gas emissions (e.g., nitrous oxide) generally associated with conventional fertilizers. Nano-urea application increased crop productivity by 8% and reduced over-application of chemical fertilizers, resulting in lower emissions and resource use. Furthermore, nanosize and high efficiency lessen total input quantities and consequently cause savings in transportation and storage energy costs (Manikandan & Subramanian, 2020; Ha et al., 2020).

9.6.2 Positive Effects on Soil Health and Biodiversity

Nanofertilizers efficiently provide nutrients and improve soil health through supporting beneficial microbial populations and enhancing soil structure. Nanofertilizers could promote plant-microbe symbiosis, increase nutrient availability, and, for that reason, contribute to a much healthier soil ecosystem through reducing eutrophication and groundwater contamination as major environmental issues (Prasad et al., 2020; Chemical and Biological Technologies in Agriculture, 2021).

9.6.3 Potential for Recycling and Reuse

Nanofertilizers, through using resource-efficient synthesis methods, contribute to circular agriculture. In fact, bio-based or green synthesized nanofertilizers have the potential for recycling and reuse through the development of integrated systems with waste materials or renewable inputs, consistent with the greater objective of a reduction in agricultural wastes and enhancement of sustainability (Chebli et al., 2021).

9.6.4 Challenges and Future Directions

Besides the advantages of nanofertilizers, some challenges, including their high production costs, possible risks of nanomaterials to non-target organisms, and regulatory uncertainties, still exist, leading to the need for more research in this area. Future research has to focus on the mentioned challenges and concerns through improving the size of green nanofertilizer production, understanding the long-term ecological impacts, and developing comprehensive regulatory frameworks that would assure their safe adoption.

References

Adisa, I. O., Pullagurala, V. L. R., Rawat, S., Adisa, O. A., & Elmer, W. H. (2021). Nanofertilizers and the sustainable development goals: Challenges and perspectives. *Frontiers in Plant Science, 12*, 658473. https://doi.org/10.3389/fpls.2021.658473

Ahmed, S., Ahmad, M., Swami, B.L. and Ikram, S., 2016. Green synthesis of silver nanoparticles using Azadirachta indica aqueous leaf extract. Journal of radiation research and applied sciences, 9(1), pp.1–7.

Ali, M., et al. (2019). Mechanisms of zinc oxide nanoparticle uptake in plant roots and its role in improving nutrient use efficiency. *Plant and Soil Journal, 33*(2), 215–230.

Ali, A., Ahmad, T., & Khan, A. (2022a). Biodegradable polymer-coated nanofertilizers for sustainable agriculture: Slow-release behavior and enhanced nutrient use efficiency. *Journal of Agricultural Science and Technology, 24*(2), 145–160.

Ali, A., Singh, R., & Kumar, P. (2023). Role of nanofertilizers in enhancing the productivity and quality of vegetable crops. *Journal of Sustainable Agriculture, 18*(2), 155–167.

Alissa, M. (2025). Green Synthesis of Silver Nanoparticles using Mangifera indica Leaves and Their Antibacterial Efficacy Against Multidrug-Resistant Bacteria. Journal of Herbal Medicine, p.101039.

Bairwa, P., Kumar, N., Devra, V., & Abd-Elsalam, K. A. (2023). Nano-biofertilizers synthesis and applications in agroecosystems. *Agrochemicals, 2*(1), 118–134.

Belaiche, Y., Khelef, A., Laouini, S. E., Bouafia, A., Tedjani, M. L., & Barhoum, A. (2021). Green synthesis and characterization of silver/silver oxide nanoparticles using aqueous leaves extract of *Artemisia Herba-Alba. Revista Romana de Materiale, 51*, 342–352.

Chen, J., Lü, S., Zhang, Z., Zhao, X., Li, X., Ning, P., & Liu, M. (2020). Environmentally friendly fertilizers: A review of materials used and their effects on the environment. *Science of the Total Environment, 711*, 134810. https://doi.org/10.1016/j.scitotenv.2019.134810

Chopra, H., Bibi, S., Singh, I., Hasan, M.M., Khan, M.S., Yousafi, Q., Baig, A.A., Rahman, M.M., Islam, F., Emran, T.B. and Cavalu, S. (2022). Green metallic nanoparticles: biosynthesis to applications. Frontiers in Bioengineering and Biotechnology, 10, p.874742.

Choudhary, M., Sharma, N., & Gupta, S. (2023). Enhancing postharvest quality of apples using calcium-enriched nanofertilizers. *Postharvest Biology and Technology, 105*(4), 67–80.

Dubey, S., et al. (2020). Green synthesis methods for nanoparticles using plant extracts and their agricultural applications. *Journal of Environmental Nanotechnology, 12*(1), 67–80.

FAO. (2022). *The future of food and agriculture: Trends and challenges*. Food and Agriculture Organization of the United Nations.

Gade, A., et al. (2023). Nanofertilizers: The next generation of agrochemicals for long-term impact on sustainability in farming systems. *Agrochemicals, 2*(2), 257–278.

Gangwar, R., Rao, K.T., Khatun, S., Rengan, A.K., Subrahmanyam, C. and Vanjari, S.R.K., 2025. Aloe vera-driven green synthesis of silver nanoparticles: a facile approach for superior antibacterial activity and enhanced wound-healing. Applied Nanoscience, 15(3), pp.1–11.

Ha, C. H., Vu, N. K., & Kim, S. J. (2020). Effects of nanofertilizers on crop production and their environmental implications. *Frontiers in Environmental Science, 8*, 89.

IPCC. (2014). Impacts, adaptation, and vulnerability. Part A: Global and sectoral aspects. Contribution of Working Group II to the Fifth Assessment Report of the Intergovernmental Panel on Climate Change. C.B. Field, V.R. Barros, D.J. Dokken, K.J. Mach, M.D. Mastrandrea, T.E. Bilir, M. Chatterjee, K.L. Ebi, Y.O. Estrada, R.C. Genova, B. Girma, E.S. Kissel, A.N. Levy, S. MacCracken, P.R. Mastrandrea & L.L. White, eds. Cambridge, UK, and New York, USA, Cambridge University Press.

Kah, M., Tufenkji, N., & White, J. C. (2021). Nanotechnology for precision agriculture: Opportunities and challenges. *Nature Nanotechnology, 16*(1), 98–103. https://doi.org/10.1038/s41565-020-00750-4

Liu, Y., Kang, S., Li, K., Chen, J., Bae, B., Hwang, I., Ahn, E.Y., Park, Y., Chun, K.H. and Lee, J. (2022). Ecofriendly and enhanced biogenic synthesis of silver nanoparticles using deep eutectic solvent-based green tea extracts. Journal of Cleaner Production, 379, p.134655.

Khan, M. A., Iqbal, M., & Rehman, M. U. (2023). Advances in green synthesis of nanofertilizers: A sustainable approach for crop productivity. *Journal of Environmental Management, 324*, 115049.

Kumar, R., et al. (2018a). Physiological effects of ZnO nanoparticles on zinc uptake, growth, and stress resistance in crops. *Crop Physiology Journal, 30*(4), 202–217.

Kumar, A., Kumar, S., & Rajput, M. (2021a). Recent advances in the green synthesis of biogenic nanoparticles and their applications: A review. *Journal of Cleaner Production, 318*, 128473. https://doi.org/10.1016/j.jclepro.2021.128473

Liu, R., & Lal, R. (2021). Potentials of engineered nanoparticles as fertilizers for increasing agronomic productions. *Science of the Total Environment, 825*, 153866. https://doi.org/10.1016/j.scitotenv.2021.153866

Manikandan, S., & Subramanian, K. (2020). Role of nanofertilizers in sustainable agriculture: Recent advancements. *Microbial Cell Factories, 19*(12), 1–13.

Marschner, P., Crowley, D. E., & Rengel, Z. (2020). Rhizosphere interactions between microorganisms and plants govern iron and phosphorus acquisition along the root axis–model and research methods. *Soil Biology and Biochemistry, 149*, 107936. https://doi.org/10.1016/j.soilbio.2020.107936

Moore, F.C. and Lobell, D.B. (2015). The fingerprint of climate trends on European crop yields. Proceedings of the National Academy of sciences, 112(9), pp.2670–2675.

Mubaraki, M.A., Mustafa, K., Fozia, F., Aslam, M., Ahmad, I. and Ahmad, N. (2023). Antimicrobial, antiasthmatic and cytotoxic activities of silver nanoparticles synthesized by green method using Zingiber officinale extract. Crystals, 13(2), p.333.

Patel, M., Chauhan, H., & Yadav, N. (2022). Advances in polymer-coated nanofertilizers: Applications in cereal and fruit crops. *Agricultural Nanotechnology Journal, 9*(1), 112–125.

Patel, M., Yadav, N., & Chauhan, H. (2023). Controlled-release nitrogen nanoparticles: Reducing environmental impact and enhancing crop yields. *Agriculture and Nanotechnology, 8*(2), 102–118.

Prasad, R., Bhattacharyya, A., & Nguyen, Q. D. (2020). Nanotechnology in sustainable agriculture: Recent developments and future prospects. *Frontiers in Microbiology, 11*, 1–13.

Raliya, R., et al. (2017). Nanofertilizers for precision and sustainable agriculture: Current state and future perspectives. *Journal of Agricultural and Food Chemistry, 66*, 6487–6503.

Raliya, R., Saharan, V., Dimkpa, C., & Biswas, P. (2021). Nanofertilizer for precision and sustainable agriculture: Current state and future perspectives. *Journal of Agricultural and Food Chemistry, 69*(45), 13120–13130. https://doi.org/10.1021/acs.jafc.1c03300

Rico, C. M., Majumdar, S., Duarte-Gardea, M., Peralta-Videa, J. R., & Gardea-Torresdey, J. L. (2020). Interaction of nanoparticles with edible plants and their possible implications in the food chain. *Journal of Agricultural and Food Chemistry, 59*(8), 3485–3498. https://doi.org/10.1021/jf104517j

Rico, C. M., Barrios, A. C., & Peralta-Videa, J. R. (2021). Nanotechnology in agriculture: Advancements in nanomaterial applications and implications. *Frontiers in Plant Science, 12*, 637594. https://doi.org/10.3389/fpls.2021.637594

Sharma, P., et al. (2021a). Green synthesis of zinc oxide nanoparticles from *Moringa oleifera* extract and their role in enhancing zinc uptake in wheat. *Agricultural Nanotechnology Journal, 25*(3), 451–463.

Sharma, R., Agrawal, N., & Satya, S. (2022a). Phosphorus nanofertilizers for sustainable agriculture: A critical review. *Environmental Science & Technology, 56*(10), 6217–6230. https://doi.org/10.1021/acs.est.1c08436

Sharma, R., Agrawal, N., & Satya, S. (2023a). Green nanotechnology in agriculture: Development and implications of biobased nanomaterials. *Environmental Science & Technology, 57*(2), 324–339. https://doi.org/10.1021/acs.est.2c06154

Shahzadi, S., Fatima, S., Shafiq, Z. and Janjua, M.R.S.A. (2025). A review on green synthesis of silver nanoparticles (SNPs) using plant extracts: a multifaceted approach in photocatalysis, environmental remediation, and biomedicine. RSC advances, 15(5), pp.3858–3903.

Singh, T., Gupta, R., & Mehra, A. (2022). Improving the nutritional content of leafy vegetables using green nanofertilizers. *Functional Foods and Agriculture, 16*(2), 205–218.

Singh, P., Sharma, N., & Kaur, J. (2023). Green-synthesized zinc oxide nanoparticles enhance drought stress tolerance in maize by modulating physiological and biochemical attributes. *Environmental Science and Pollution Research, 30*(5), 6318–6329.

Solanki, P., Bhargava, A., Chhipa, H., Jain, N. and Panwar, J. (2015). Nano-fertilizers and their smart delivery system. In Nanotechnologies in food and agriculture (pp. 81–101). Cham: Springer International Publishing.

United Nations Publication. World Population Prospects (2022). Summary of Results; UN DESA/POP/2022/TR/NO. 3 (UN, 2022); United Nations Publication: New York, NY, USA.

Usman, M., Farooq, M., Wakeel, A., Nawaz, A., & Siddique, K. H. M. (2022a). Role of biobased nanomaterials in enhancing agricultural productivity and sustainability. *Journal of Agricultural and Food Chemistry, 70*(4), 1050–1060. https://doi.org/10.1021/acs.jafc.1c06512

Usman, M., Farooq, M., Wakeel, A., Nawaz, A., Alam Cheema, S., & Siddique, K. H. M. (2022b). Nanotechnology in agriculture: Current status, challenges, and future opportunities. *Science of the Total Environment, 825*, 153967. https://doi.org/10.1016/j.scitotenv.2021.153967

Valavanidis, A. (2023). Feeding a Growing World Population. Critical rethinking of lessons from the "Green Revolution" in the 1960s as a warning for feeding a hungry planet, 1, pp.1–47.

Verma, K. K., et al. (2022). Nanofertilizer possibilities for healthy soil, water, and food in future: An overview. *Frontiers in Plant Science, 13*, 865048.

Verma, K.K., Song, X.P., Degu, H.D., Guo, D.J., Joshi, A., Huang, H.R., Xu, L., Singh, M., Huang, D.L., Rajput, V.D. and Li, Y.R. (2023). Recent advances in nitrogen and nano-nitrogen fertilizers for sustainable crop production: a mini-review. Chemical and Biological Technologies in Agriculture, 10(1), p.111.

Chapter 10
Post-Harvest Applications of Green-Synthesized Nanomaterials

Deepoo Meena, Rakesh Kumar Gupta, and Prem Prakash Srivastav

10.1 Introduction

Nanoparticles (NPs) that are between 1 and 100 nm in size serve as a link between atomic or molecular structures and bulk materials (Joudeh & Linke, 2022; Gupta et al., 2024a). Their tiny size, wide surface area with free dangling bonds, and increased reactivity compared to their bulk counterparts provide them with exceptional and intriguing features. Though the methods are still unknown, scientists have known since the eighteenth century that living entities may decrease metal precursors (Khan et al., 2017). Researchers have turned to biological approaches as a result of the development of effective green synthesis that uses natural reducing, capping, and stabilizing agents instead of costly, hazardous chemicals that consume a lot of energy. The Earth's atmosphere is deteriorating due to rapid industrialization, urbanization, and population growth, and a vast number of undesirable and dangerous compounds are being emitted. Learning about the mysteries found in nature and its natural components, which result in improvements in the synthesis processes of nanoparticles, is long overdue. Additionally, NPs are frequently used in locations where people come into contact with them, and there is a rising need to create synthesis methods that do not include harsh, hazardous chemicals. Therefore, a potential substitute for chemical and physical approaches is the green/biological production of NPs (Gour & Jain, 2019; Ullah et al., 2024a, b; Gupta et al., 2024).

"Why are green-synthesized nanoparticles (GSNP) so interesting and gaining importance nowadays?" is the first question pertaining to the development of green nanomaterials. Biologically manufactured nanoparticles are favored over physicochemically created nanomaterials due to their distinct features. NPs can be created via physicochemical techniques (Ealia & Saravanakumar, 2017). These techniques

D. Meena · R. K. Gupta (✉) · P. P. Srivastav
Agricultural and Food Engineering Department, Indian Institute of Technology Kharagpur, Kharagpur, India

K. Esquivel Escalante et al. (eds.), *Green Nanomaterials*, Nanotechnology in the Life Sciences, https://doi.org/10.1007/978-3-032-04492-1_10

are costly and fraught with issues, such as the use of hazardous solvents, the production of dangerous byproducts, and surface structure imperfections. Because of their composition's ambiguity and unpredictability, chemical procedures typically involve several chemical species or molecules that may enhance particle reactivity and toxicity as well as endanger human health and the environment. Green synthesis yields different particles than physicochemical methods do. Green synthesis, a bottom-up method, is comparable to chemical reduction in that it uses an extract of a natural product, such fruit or tree/crop leaves, to synthesize metal or metal oxide nanoparticles instead of a costly chemical reducing agent. There is enormous potential for biological organisms to produce NPs (Pechyen et al., 2024).

Metal precursors can be reduced biogenically to their corresponding nanoparticles (NPs) in a sustainable, cost-effective, chemically contaminant-free, and mass-producible manner. Furthermore, costly metal salts found in waste streams, such as gold and silver, can be recycled thanks to the biological generation of NPs. The price of these metals fluctuates, and their deposits are limited. Green NPs with the required characteristics may be obtained. The biological components that readily stabilize NPs—mostly proteins, enzymes, sugars, and even entire cells—allow NPs to interact with other biomolecules, which enhances their interactions with microbes and boosts their antimicrobial effectiveness. The NPs' biological production makes it simple to separate them from the reaction medium or to use centrifugation to increase their concentration. The antibacterial activity of biogenic silver nanoparticles was 20 times greater than that of chemically generated NPs. The added value of the biological material itself determines which plant extracts are used to make NPs. *Spirulina platensis* algal cells were selected due to their reducing agent characteristics as well as their medicinal and nutraceutical qualities (Ullah et al., 2024a, b). In the reaction procedures, unicellular bacteria and extracts of multicellular eukaryotes transform metal precursors into desired NP sizes and forms. Furthermore, living entities have the stabilizing and capping molecules needed to prevent the aggregation/agglomeration process and act as growth terminators. The size and form of NPs are influenced by the type of biological entity and its concentrations in conjunction with organic reducing agents. Furthermore, growth media characteristics, including pH, temperature, salt content, and exposure duration, have a significant impact on the size and form of NPs (Uzair et al., 2020). For the creation of nanomaterials, metal precursors are bio-reduced either in vitro or in vivo. However, the primary functions of enzymes, proteins, carbohydrates, and phytochemicals such as terpenoids, flavonoids, phenolics, and cofactors are stabilizing and reducing. It has been documented that bacteria, yeast, fungi, algae, and plants may produce NPs in vivo. For in vitro synthesis, which entails purifying bio-reducing agents and carefully combining them with an aqueous solution of the appropriate metal precursor, biological extracts are typically utilized (Uzair et al., 2020). The process happens on its own at room temperature. However, occasionally further heating and stirring are required. Plants or their extracts appear to be the most effective biological entities among those listed above since they are readily available, can produce NPs in large quantities, and, in contrast to certain microbial extracts, their waste products are environmentally benign.

Although physicochemical methods have been used extensively in nanotechnology research, green synthesis has not been widely used to synthesize nanoparticles (NPs). Only a small number of studies have sought to clarify green synthesis and its possible uses in the food business. Green-synthesized nanoparticles alter sustainability in numerous sectors by improving food safety, quality, and shelf life. Green-synthesized nanoparticles can help industry meet sustainable development goals by addressing environmental and health issues caused by traditional procedures. Their use signals a turning point toward more sustainable, efficient, and consumer-friendly industrial methods, demonstrating responsible innovation. This chapter explains the ideas, methods, and applications of green-synthesized nanoparticles, emphasizing their role in post-harvest operations (Das et al., 2009). This chapter introduces green synthesis methods—from microbial to plant-mediated—and their novel applications in food preservation, packaging, and safety to demonstrate the transformative potential of these nanoparticles to promote sustainable food and packaging practices. This chapter will also examine green-nanoparticle technology's practical and theoretical uses, including case studies and applications. These green technologies reduce food waste, synthetic chemical use, and environmental impact, boosting the economy and environment. The analysis also examines regulatory, consumer, and scalability problems related to large-scale green nanomaterials use. This research highlights the potential of green-synthesized nanomaterials to modernize post-harvest techniques and create a more sustainable and efficient food supply chain. This chapter hopes to help academics, industry experts, and politicians understand how green nanotechnology might improve agriculture and food processing processes. It will also emphasize multidisciplinary interactions between chemists, biologists, and environmental scientists to advance green nanotechnology.

10.2 Green Synthesis Process of Nanoparticle

Green nanotechnology is broadly characterized as the utilization of biological methods, such as bacteria, fungi, or plants, for the creation of nanomaterials (or nanoparticles) via different biotechnological processes. The produced nanoparticles are devoid of harmful substances and are environmentally benign. We shall examine many biological pathways for the creation of nanoparticles in depth.

10.2.1 Plant-Based Green Nanoparticle Generation

The utilization of plants for the synthesis of silver nanoparticles has garnered interest in recent years due to its quick, eco-friendly, non-pathogenic, and cost-effective protocol, which offers a streamlined approach for the biosynthetic process. The reduction and stabilization of silver ions through a combination of biomolecules, including proteins, amino acids, polysaccharides, terpenes, alkaloids, phenolics, saponins, and vitamins present in plant extracts, offers a cost-effective and straightforward method for synthesizing silver nanoparticles. The primary benefits of utilizing plant extracts for the synthesis of silver nanoparticles are their accessibility, safety, low toxicity in most instances, a diverse array of metabolites that facilitate the reduction of silver ions, and expedited synthesis compared to microbial methods. Photochemicals render plant-assisted reduction the primary mechanism regarded for this activity. The elevated concentrations of steroids, saponins, carbohydrates, and flavonoids function as reducing agents and phytoconstituents, serving as capping agents that confer stability to the silver nanoparticles. Various nanoparticles have been synthesized using green synthetic methods (Sur et al., 2018; Liu et al., 2018; Devi & Ahmaruzzaman, 2018; Begum & Ahmaruzzaman, 2018; Medina Cruz et al., 2018; Jin et al., 2018; Olajire et al., 2017; Gao et al., 2020; Alghuthaymi et al., 2021; Mohseni et al., 2020; Gopalakrishnan et al., 2023; Mandal et al., 2024).

10.2.2 Algal-Based Green Nanoparticle Generation

Numerous studies have documented the use of algae for nanoparticle production. Algae are a vast category of species that may be generally defined as oxygen-producing, photosynthetic, unicellular, or multicellular organisms (Pipliya et al., 2025). *Spirulina platensis*, a blue-green alga, has been utilized in the manufacture of protein-mediated gold nanoparticles, yielding uniformly sized nanoparticles with an average dimension of roughly 5 nm. These were employed for antibacterial experiments targeting *Bacillus subtilis* and *Staphylococcus aureus* (Suganya et al., 2015; Yedoti & Supraja, 2024). *Sargassum wightii*, a brown seaweed, has been documented to produce stable, well-dispersed gold nanoparticles measuring 8–12 nm, as verified by UV–visible spectroscopy, transmission electron microscopy, and X-ray diffraction analysis (Singaravelu et al., 2007; Rizki & Klaypradit, 2023). In a similar manner, silver nanoparticles were produced with *Sargassum wightii* and subsequently evaluated for their antibacterial efficacy (Govindaraju et al., 2009; Hamouda & Aljohani, 2024). Gold nanoparticles, EPS-gold, and silica-gold bionanocomposites have been created using diatoms, specifically *Navicula atomus* and *Diadesmis gallica*, as part of various algae-mediated green nanoparticle synthesis techniques (Uzair et al., 2020; Chan et al., 2022; Michael et al., 2022; Alprol et al., 2023a, b; Jalali et al., 2024; Patel et al., 2024).

10.2.3 Yeast-Based Green Nanoparticle Generation

Yeast, a unicellular eukaryotic microbe, is mostly utilized to make semiconductors. *Candida glabrata* formed 20 Å spherical CdS quantum crystallites, while *Schizosaccharomyces pombe* produced 1–1.5 nm wurtzite hexagonal crystals in mid-log phase (Dameron et al., 1989). *Torulopsis* sp. vacuoles generated 2–5 nm PbS nanocrystallites, which were utilized to make a perfect diode. Silver nanoparticles are synthesized extracellularly using silver-tolerant yeast strain MKY3. When challenged with 1 mM soluble silver in its log phase, the strain generated 2–5 nm silver nanoparticles, which were separated by differential thawing. Optical absorption, transmission electron microscopy, X-ray diffraction, and photoelectron spectroscopy verified silver nanoparticle production (Kowshik et al., 2002a, b; Roychoudhury, 2020; Kumari et al., 2023). Salem (2022) biosynthesized AgNPs from *Saccharomyces cerevisiae* cells in a clean, environmentally friendly method. Biosynthesized AgNPs were identified by UV–Vis, XRD, FTIR, TEM, and DLS. Ag particles were spherical and 3–60 nm in size. FTIR research on Ag nanoparticle-reducing and stabilizing NPs reveals functional groups matching yeast cell metabolites. AgNPs biosynthesized were evaluated against pathogens for antibacterial activity. AgNPs exhibited antimicrobial activity against pathogens (*E. coli*, *P. aeruginosa*, *B. subtilis*, *S. aureus*, and *C. albicans*), with MICs of 12.5 μg/mL for *S. aureus* and *C. albicans* and 25 μg/mL for *E. coli*, *P. aeruginosa*, and *B. subtilis*. Finally, baker's yeast-derived AgNPs may be safe microbe-fighting agents. Biosynthesized AgNPs from baker's yeast cells are antibacterial and useful in disease biocontrol and therapy.

10.2.4 Actinomycetes-Based Green Nanoparticle Generation

Actinomycetes, which belong to the phylum Actinobacteria, share many traits with fungi and are the source of at least two-thirds of all presently used antibacterial chemicals. However, there is still much to learn about the use of actinomycetes in the production of nanoparticles. Researchers have used a novel extremophilic actinomycete, *Thermomonospora* sp., to create monodisperse spherical gold nanoparticles with an average size of 8 nm. These particles were stable for over 6 months. Using Fourier-Transform Infrared Spectroscopic (FTIR) research, it was verified that the proteins that stabilize the nanoparticles were present on their surface. *Rhodococcus* sp., an alkalotolerant actinomycete, was reported to exhibit intracellular accumulation of gold nanoparticles ranging in size from 5 to 15 nm. The particles were seen to aggregate on the cell wall and cell membrane as Ag (Ahmad et al., 2003a, b; Kabiri et al., 2023; Nagda et al., 2024). Similarly, it has been reported that a variety of actinomycetes, including *Streptomyces* sp., *Thermoactinomyces* sp., and *Rhodococcus* sp., have generated silver nanoparticles. The produced nanoparticles were successfully evaluated for their antibacterial properties against a range of diseases and used in a variety of biomedical applications (Alani et al., 2012; Deepa et al., 2013; Otari et al., 2012; Golinska et al., 2014; Jain et al., 2022; Prashanth, 2023).

10.2.5 Fungi-Based Green Nanoparticle Generation

Fungi are utilized to create metallic nanoparticles because they have tolerance, metal bioaccumulation, strong binding capacity, and intracellular intake, like bacteria. Various fungi such as *Fusarium*, *Colletotrichum*, and *Phaenerocheate chrysosporium* are employed to make nanoparticles (Mukherjee et al., 2002: Shankar et al., 2003; Vigneshwaran et al., 2006; Kisimba et al., 2023; Patel et al., 2022; Ranjan, 2023; Nain et al., 2024). Fungi grow faster and are easier to handle and create in a lab than bacteria, making them better for nanoparticle creation. Fungal mycelial mesh can endure bioreactor flow pressure and agitation. Fungi release vast amounts of enzymes to degrade silver ions and generate metallic nanoparticles. Extracellular enzymes like naphthoquinones and anthraquinones help in reduction (Ahmad et al., 2003a, b; Ramani et al., 2003; Kabiri et al., 2023; Nagda et al., 2024). The NADPH-dependent nitrate reductase and shuttle quinine extracellular mechanism may generate nanoparticles in *F. oxysporum*. The first fungus-mediated synthesis of metal nanoparticles, AgNPs with a diameter of 25 ± 12 nm, was conducted in the early twentieth century utilizing *Verticillium*. *Verticillium luteoalbum* generated intracellular gold nanoparticles with pH-dependent shape and size. Extracellular, fungus-mediated nanoparticle biosynthesis produced gold, silver, and silver-gold nanoparticles from *Volvariella volvacea* extract (Mukherjee et al., 2001a, b; Gericke & Pinches, 2006a, b; Philip, 2009; Kisimba et al., 2023; Patel et al., 2022; Ranjan, 2023; Nain et al., 2024).

10.2.6 Bacteria-Based Green Nanoparticle Generation

Bacteria create diverse inorganic materials intra- or extracellularly, making them good biofactories for silver and gold nanoparticle production. Silver is recognized for biocidal qualities. Some silver-resistant bacteria can collect silver on their cell walls to 25% of their dry weight biomass, suggesting its applicability in commercial silver ore recovery (Paulkumar et al., 2013; Pooley, 1982; Prabhu & Poulose, 2012; Tsekhmistrenko et al., 2020). *Pseudomonas stutzeri* AG259 from a silver mine synthesized silver nanoparticles for the first time. The enzyme nitrate reductase turns nitrate into nitrite, the most common silver biosynthetic process. By employing alpha-nicotinamide adenine dinucleotide phosphate reduced form (NADPH)-dependent nitrate reductase, bacteria may synthesize silver in vitro without downstream processing (Chen & Schluesener, 2008; Tsekhmistrenko et al., 2020). A study found that intracellular *E. coli* DH5α synthesized gold nanoparticles from chloroauric acid, resulting in spherical, triangular, and quasihexagonal shapes, deposited on the cell surface for direct electrochemistry of proteins, including hemoglobin (Du et al., 2007). *Geobacter metallireducens* GS-15, a nonmagnetotactic bacterium from Potomac River sediments, was used to synthesize magnetite nanoparticles 10–50 nm in anaerobic conditions. Crystalline ferromagnetic $Co3O_4$ nanoparticles 5–7 nm were extracellularly produced by metal-tolerant marine bacteria *Brevibacterium casei* using cobalt acetate as precursor (Lovley et al., 1987; Kumar et al., 2008; Qamar & Ahmad, 2021; Lahiri et al., 2021; Mughal et al., 2021; Jeevanandam et al., 2022; Kumar et al., 2023). Wei and Zheng (2024) used the L-cysteine desulfurization method to create ZnS nanoparticles using the bacteria *Lysinibacillus* sp. The ZnS generated by bacteria has a cubic crystal structure called sphalerite, according to X-ray diffraction (XRD) study. The nanoparticles in the size range of 3–5 nm are visible in TEM pictures. The polycrystalline nature of the ZnS produced by SH74 is confirmed by SAED analysis. According to FTIR research, the bacterial ZnS nanoparticles' surface contains organic molecules.

10.3 Application of GSNP-Based Materials in Post-Harvest Processing

GSNP has emerged as a transformative agent in the food industry, offering sustainable and innovative solutions across multiple domains, including food preservation, packaging, safety monitoring, and fortification. Their antimicrobial properties are pivotal in extending the shelf life of perishable foods by inhibiting microbial growth and preventing spoilage. For example, silver and zinc oxide nanoparticles synthesized through green methods are incorporated into food coatings and packaging materials to combat bacteria and fungi (Bahrulolum et al., 2021). Possible application of GSNP-based materials in post-harvest processing is shown in Fig. 10.1. In intelligent packaging, these nanoparticles enable real-time food quality monitoring through sensors that detect spoilage indicators such as gases or pH changes, providing visual or digital alerts (Nikolic et al., 2021). Such innovations are particularly useful for fresh produce like fruits and vegetables, where ethylene gas and moisture levels must be controlled to reduce post-harvest losses (Nikolic et al., 2021). Green nanoparticles also enhance food packaging by improving barrier properties against oxygen, moisture, and UV radiation, thereby preserving food quality. Biodegradable and sustainable packaging materials, incorporating nanoparticles with biopolymers like chitosan, are gaining popularity as eco-friendly alternatives to conventional plastics (Omerović et al., 2021). In food fortification, green synthesis nanoparticles play a crucial role by encapsulating bioactive compounds, such as vitamins and minerals, to improve their stability and bioavailability (Kumari & Chauhan, 2022). This ensures effective delivery of nutrients in functional foods and beverages. Additionally, green-synthesized nanoparticles derived from plant extracts or agricultural waste promote a circular economy, utilizing renewable resources while minimizing environmental impact. Their applications also extend to wastewater treatment and food processing, where they aid in reducing contaminants and enhancing product quality. However, challenges such as scalability, regulatory approval, and consumer acceptance must be addressed to maximize their potential

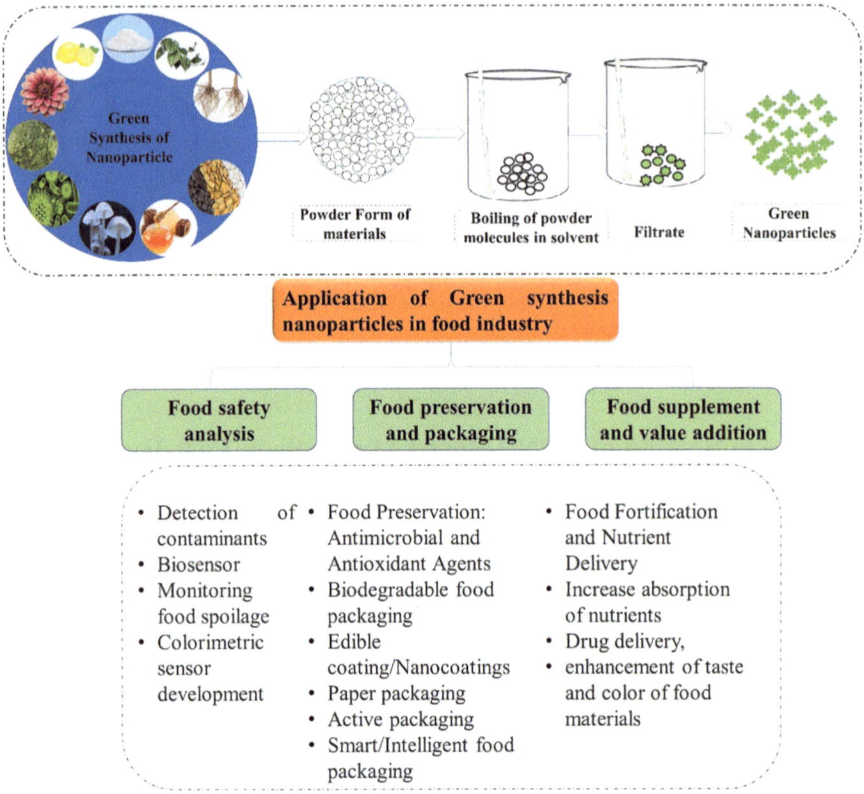

Fig. 10.1 Production and application of green-synthesized nanoparticles

(Polat et al., 2019). Application of GSNP in different areas of post-harvest processing is given in Table 10.1.

10.3.1 Application in the Production of Food Packaging

Green synthesis nanoparticles have revolutionized food packaging, particularly for perishable products like fruits and vegetables, by enhancing functionality, sustainability, and shelf-life preservation. These applications can be broadly categorized into active packaging, intelligent packaging, nanocoatings, and sustainable solutions. Active packaging integrates green-synthesized nanoparticles such as silver, zinc oxide, and titanium dioxide to inhibit microbial growth, ensuring food safety and reducing spoilage. For fruits and vegetables, these nanoparticles play a critical role in minimizing microbial decay and maintaining freshness during storage and transportation. Additionally, nanoparticles function as oxygen scavengers, reducing oxidative spoilage, and as moisture regulators to maintain optimal humidity levels,

Table 10.1 Application of GSNP in post-harvest processing

GSNP	Sources	Type of Film	Application	References
AgNPs	*Gliricidia sepium*	Alginate-based films	Extend shelf life of foods	Martinez-Molina et al. (2022)
Selenium NPs (Se-NPs)	Cinnamon (*Cinnamomum zeylanicum*) bark extract (CIE)	Cht-NPs		Alghuthaymi et al. (2021)
SeNPs	Pomegranate peels (PPE	Chitosan nanoparticles (NCT)	Edible coatings for citrus green mold protection	Salem et al. (2022)
Iron oxide (IO) Nanoparticles (NPs)	Supplemented with ginger and garlic extracts	Chitosan (CH)	Coatings for enhancing straw-berry shelf-life	Sani et al. (2024)
Titanium dioxide (TiO_2) nanoparticles and	Mousami peel extract (MPE)	Sodium alginate (SA)	Edible coatings on post-harvest qual-ity of peach fruits	Khan et al. (2023)
Zinc oxide nanoparticles (ZnO-NPs)	Aloe vera (AV) and green synthesized	Sodium carboxymethyl cellulose (CMC)	Coating	Salama et al. (2024)
ZnO nanoparticles	*Citrus sinensis* peel extract (SNP)	Sodium carboxymethyl cellulose	Nanocoatings on fresh strawberries	Gao et al. (2020)
ZnO nanoparticles (ZnO-NPs)	*Coriandrum sativum* extract	CMC/Xa com-posite coating	Coated tomato	Yashni et al. (2019)
TiO_2 NPs	*Psidium Guajava* leaves	Chitosan-TiO_2	Antimicrobial packaging	Shafie et al. (2024)
Ag	Pomegranate and Kinnow fruit peel	Cellulose wrap-per for pack-aged bread	Shelf-life exten-sion of packaged bread	Gopalakrishnan et al. (2023)
Nanocellulose	Orange peel	Poly (vinyl alcohol) PVA-based film		Espinosa et al. (2022)
Cellulose nanocrystals		PVA-based films	Mango fruit packaging	Dey et al. (2021)
ZnONP	Grapefruit seed	Cellulose nanofiber-based nanocomposite films	Active food packaging	Roy et al. (2021)
ZnO	Pineapple peel	Starch nanocomposite films	Active food packaging	Hassan Basri et al. (2020)
AgNP	Mangrove extract of *Bruguiera gymnorrhiza*	Fish-gelatin-based edible film	Edible film	Nurdiani et al. (2024)

(continued)

Table 10.1 (continued)

GSNP	Sources	Type of Film	Application	References
TiO_2	*Psidium guajava* leaves	Biofilm of chitosan	Smart packaging	Shafie et al. (2024)
AgNp	Tea waste	Carbon paste electrode (CPE)	Bio-sensing glucose	Pektaş et al. (2024)
AgNP	Tailed pepper seed extract		Novel self-ratiometric fluorescent nanosensor	El Hamd et al. (2024)
Iron oxide NP	Petal extracts of *Hibiscus rosa-sinensis*		Fortifying wheat biscuits	Razack et al. (2024)
ZnO NP	Aqueous extract of guava (*Psidium guajava*) leaves		Shelf life and sensory quality of minced fish	Mahato et al. (2024)

preserving texture and preventing dehydration or mold formation. In intelligent packaging, green nanoparticles facilitate real-time monitoring of food quality. Nano-biosensors can detect spoilage indicators such as ethylene gas, a ripening agent released by fruits, or other volatile compounds associated with deterioration. Colorimetric indicators using pH- or temperature-sensitive nanomaterials provide visual signals of changes in storage conditions, helping consumers identify product freshness. Advanced tracking systems enhanced with nanotechnology, such as nanoparticle-based QR codes and RFID tags, bolster supply chain transparency, ensuring that fruits and vegetables reach consumers in optimal condition (Souza & Fernando, 2016; Nikolic et al., 2021; Garavand et al., 2022; del Rosario Herrera-Rivera et al., 2024). Nanocoatings derived from green synthesis improve the barrier properties of packaging materials, protecting fruits and vegetables from oxygen, moisture, and UV light exposure. Silica or clay-based nanoparticles create thin, protective films that reduce respiration rates in fresh produce, extending their shelf life. These coatings are not only functional but also environmentally friendly, reducing reliance on synthetic plastic films while maintaining product quality. Furthermore, biodegradable packaging solutions incorporating green nanoparticles and biopolymers such as chitosan or polylactic acid offer a sustainable alternative for packaging fresh produce. These nanocomposites are lightweight, durable, and designed to decompose naturally, addressing the pressing global issue of plastic waste. For fruits and vegetables, green nanoparticles also enable innovative treatments that directly enhance post-harvest quality. Nanoemulsions incorporating bioactive compounds can be applied as edible coatings, providing antimicrobial and antioxidant protection while being safe for consumption. Such applications are particularly effective in reducing post-harvest losses, which are a major concern in the fresh produce supply chain (Andriani & Handayani, 2023; Khan et al., 2023). By extending shelf life, preserving quality, and supporting sustainability, green synthesis nanoparticles offer transformative solutions for packaging fruits, vegetables, and other perishable foods, ensuring safer and more efficient food systems for the future.

10.3.1.1 Bio-Based and Biodegradable Food Packaging Films

Green-synthesized nanoparticles improve functionality and environmental compatibility of biodegradable packaging. Nowadays, various types of natural polymers have been used for the production of nanoparticles such as starch, protein, cellulose, and chitosan (Gupta et al., 2023, 2024c; Shrivastava et al., 2024). Biopolymers containing starch nanoparticles have higher mechanical strength and barrier characteristics without sacrificing biodegradability (Bahrulolum et al., 2021). Mohseni et al. (2020) produced AgNPs from pomegranate seeds (Green synthesis). Ag-corn starch nanocomposite was produced. Found that nanocomposite Ag-starch enhanced the thermoplastic film tensile strength and extensibility. Adding StNPs and AgNPs to maize starch films created nanocomposites with good mechanical characteristics. This study found that adding nanocomposites with specified ratios to films increased their mechanical characteristics and physical resistance. Kumar et al. (2020) created chitosan–gelatin nanocomposite hybrid films with green ZnO nanoparticles. The films' antimicrobial activity was tested using Gram-negative (*Escherichia coli*) and Gram-positive (*Staphylococcus aureus*) bacteria. ZnO NPs were precipitated using *Cassia fistula* fruit extract employing green synthesis. The hybrid films reinforced with ZnO NPs increased thermal stability, EAB, and compactness. Compared to the control (chitosan-gelatin hybrid) films, the 2% and 4% ZnO NP films had smooth, compact, and diverse surface morphology. Disc diffusion testing indicated that the nanocomposite film has strong *E. coli* antibacterial activity. Developed hybrid nanocomposite films might provide biodegradable postharvest packaging for fresh fruits and vegetables. Zahiri Oghani et al. (2021) green-synthesized ZnO nanoparticles using Nettle leaf extract to improve the antibacterial capabilities of chitosan (CS) and carboxymethyl cellulose-based nonstoichiometric polyelectrolyte complex. Slow dropwise addition of CMC to chitosan produced anionic or cationic particles in the 35.5–69.7 nm size range. The transparency and opacity of Chitosan: CMC (75:25) are optimal, while increasing CMC increases nanocomposite water absorption and ash percentage. The composite without nano ZnO had the highest antibacterial activity for *Escherichia coli*, whereas the nanocomposite with nano ZnO (PCC3/ZnO) in *Staphylococcus aureus* medium had the best outcome. El-Naggar et al. (2022) created biodegradable antimicrobial films from cellulose acetate/polycaprolactone and green copper nanoparticles (CuNPs) utilizing isolated fungus *Aspergillus terreus* AH20. The polymers-based film (cellulose acetate and polycaprolactone) containing CuNPs had a rough surface due to nanoparticles being deposited as packed particles or thin layers on the outer surface. The films made with CuNPs have better mechanical characteristics and air permeability than those made without. All developed films were tested for antimicrobial and cytotoxic properties. Antimicrobial results showed that CuNP-loaded films were more effective against

Escherichia coli, *Pseudomonas aeruginosa*, *Staphylococcus aureus*, *Bacillus subtilis*, *Candid. albicans*, *Aspergillus niger*, *A. terreus*, *Penicillium expansum*, and *Fusarium oxysporum*. Cytotoxicity findings showed that all films in the current investigation had no cytotoxicity on Wi38 normal cell line, confirming their safety. These findings show that the myco-synthesized CuNP-loaded biodegradable film is an effective hydrophobic film ideal for food biodegradable packaging.

10.3.1.2 Edible Coatings on Food

Green nanoparticles are improving edible packaging, a growing food technology industry. Eco-friendly nanoparticles like curcumin or other plant extracts are incorporated onto edible films for many uses. They lengthen product shelf life and provide nutritional or health advantages, meeting customer demand for health-oriented products (Ali et al., 2025). Green-synthesized nanoparticles strengthen edible films' structural integrity and moisture barrier, making them more durable and useful for culinary applications. This unique usage of green nanoparticles improves edible packaging's usefulness and appeal to ecologically aware consumers (Andriani & Handayani, 2023; Khan et al., 2023). Biopolymer films and coatings are widely employed to enhance fresh vegetable shelf life. These films and coatings perform like synthetic coatings and protective packaging (Kumar et al., 2019). Food-approved nanoparticles enable edible coatings. Nanoemulsions, polymer nanoparticles, nanofibers, SLNs, NLCs, and inorganic–organic polymer nanocomposites make up these coatings (Bahrami et al., 2020). Chitosan is a popular edible film component. Natural polymer chitosan is (1,4)-linked 2-amino-deoxy-b-D-glucan. Many studies have revealed that chitosan is non-toxic, biodegradable, biofunctional, biocompatible, and antimicrobial. Chitosan is also a great material for edible packaging since it forms translucent films with strong mechanical qualities (Singh et al., 2015). Chitosan and nanoparticles work well together. Chitosan coated with Fe3O4 nanoparticles inhibits *E. coli*, *B. subtilis*, *Candida albicans*, *Aspergillus niger*, and *Fusarium solani*, according to Nehra et al. (2018). Shrifian-Esfahni et al. (2015) observed that chitosan–Fe_3O_4 nanoparticle films inhibited *E. coli* and *P. aeruginosa* in vitro. SPIONs made of magnetite (Fe_3O_4) can be utilized for antibacterial applications because the FDA has deemed them to be biocompatible (BC) (Lattuada & Hatton, 2007). Gao et al. (2020) produced ZnO nanoparticles from *Citrus sinensis* peel extract (SNP), described and compared them to commercial CNP, and put them as nanocoatings on fresh strawberries to test their preservation effect. SNP has a comparable crystal structure to CNP but greater crystallinity and particle size. SNP was similar to CNP in cytotoxicity against HUVECs below 25 mg/L. SNP exhibited a considerably lower minimum bactericidal concentration (MBC) against *E. coli* than CNP, but both had equal MICs against *E. coli* and *S. aureus*. CNP has higher antifungal activity than SNP against *Botrytis cinereal* isolated from mildewed strawberries at concentrations below 400 mg/L. After covering fresh strawberries with carboxymethylcellulose sodium (CMC) containing SNP and CNP, weight loss, rotting, and storage quality improved. Thus, green-prepared

SNP offers significant food packaging potential. Using plant extract from *Nymphae odorata*, Gudimalla et al. (2021) created sodium alginate films doped with green-manufactured silver nanoparticles (Ag NPs) and investigated the antibacterial activity of both the doped films and the green-synthesized Ag NPs. The results showed that a very low concentration of Ag NPs inhibited the complete bacterial activity of *Escherichia coli* and *Staphylococcus aureus*, and the impact was also seen in the films. The produced Ag NPs and films' effectiveness as antibacterial agents was validated by the findings. In another work, Srikhao et al. (2021) examined the impact of green synthesis Ag NPs with oregano essential oil (OEO) on polyvinyl alcohol PVA/starch film. According to their findings, the films with NPs and 5 weight percent OEO had the lowest inhibitory concentrations against *S. aureus* and *E. coli*. Additionally, adding Ag NPs with oregano essential oil improved the films' mechanical qualities and water resistance. This work demonstrated that the effectiveness of bioactive nanocomposite film for food packaging applications was increased by adding green-generated Ag NPs and OEO. Nurdiani et al. (2024) examined the effects of green-synthesized silver nanoparticles utilizing Mangrove extract of *Bruguiera gymnorrhiza* on *Pangasius* sp. skin gelatin consumable film physicochemical properties. A higher concentration of AgNPs made the fish-gelatin-based edible film darker. The inclusion of AgNPs did not significantly affect the thickness (145–216 μm), tensile strength (14.58–19.72 MPa), elongation (21.86–54.19%), or water vapor transmission (30.91–42.55 $g/m^2/day$) of the fish-gelatin-based edible film. Color is the only influence of AgNPs on gelatin-based edible film physicochemical parameters. Thus, the edible film can be improved. Mandal et al. (2024) optimized Chitosan-based films with green-synthesized TiO_2 nanoparticles (CT) and *Averrhoa carambola* extract (CP) at varying concentrations for antioxidant-enhanced tensile, moisture–gas barrier, and retention. Antimicrobial profiling showed good inhibition against gram-positive bacteria (*E. faecalis* and *S. aureus*), gram-negative bacteria (*P. aeruginosa* and *E. coli*), and a fungus (*C. tropicalis*). Chitosan was best with 0.06% TiO_2 NP and 6% carambola extract. The biodegradable sheets might be used as food packaging or edible covering to postpone harvest maturation.

10.3.1.3 Active/Smart/Intelligent Food Packaging

Consumers now need active packaging. Active food packaging combines packaging and product innovation to improve food safety and shelf life (Gupta et al., 2022, 2024d). Stalling microbial growth, lipid oxidation, and moisture loss degrading responses yields superior outcomes than typical food packaging. Active packaging can absorb food-related substances from food, the environment, the pack headspace, or release preservatives, antioxidants, and flavorings into the food or environment. An active ingredient can be directly incorporated into a polymer matrix, packaged, or labeled (Garavand et al., 2022; del Rosario Herrera-Rivera et al., 2024). Active packaging solutions, including antimicrobial and antioxidant packaging, carbon dioxide emission, and oxygen/CO_2 absorbers are popular. Active food packaging methods are particularly promising for antioxidant and antibacterial packaging (Janani et al., 2020). New smart packaging technologies can boost food traceability, safety, and quality. This technology allows food packaging systems to track changes in the external and interior surroundings and packaged food to notify customers, merchants, and producers of the system's status. Throughout the "lifecycle of the product," smart packaging checks freshness and informs customers. Additionally, this packaging can detect and convey food qualities, telling consumers about product quality and safety. Smart packaging may evaluate active packaging systems. In summary, smart packaging tracks, senses, communicates, records, monitors, and uses scientific reasoning to enhance shelf life, improve quality, and offer product information. Smart features like detection, communication, recording, and identification distinguish intelligent packaging from active packaging (Baneshi et al., 2024). Intelligent packaging with radio frequency identification/sensors, time–temperature indicators, freshness indicators, etc., is promising to improve information transmission in the product distribution chain and consumers' homes.

Accordingly, several systematic studies have shown the great potential of these unique nanostructures as nanofillers/additives to enhance the thermal, mechanical, physical, UV barrier, and water barrier properties of food packaging, as well as their potential to fabricate active packaging with antibacterial/antioxidant properties and smart packaging with food freshness/spoila detection (Garavand et al., 2022; del Rosario Herrera-Rivera et al., 2024). A thorough literature review shows that green/biosynthesized CDs (Carbon quantum dots) have become increasingly used to make bio-nanocomposite films/coatings with multiple functions (pH sensing, UV-shielding, antioxidant, antibacterial, etc.) (Zhao et al., 2022; Riahi et al., 2022). Green produced CDs are growing in popularity; however, maintaining their physicochemical features (e.g., fluorescence quantum yield) is difficult. This difficulty has affected their application in food packaging but also enabled the creation of cheaper and biocompatible nanocomposite films/coatings for smart food packaging (Moradi et al., 2021; Ezati et al., 2022; Zhao et al., 2020; Riahi et al., 2022; Wang et al., 2022; Fu et al., 2022; Kilic et al., 2022). Zhao et al. (2020) examined how green-manufactured banana CDs and aqueous chitosan (CH) solution affected soy milk shelf life and stability. This study tested soy milk shelf life at room temperature with varied concentrations of chitosan solution ("0.00%, 0.08%, 0.12%, 0.16%, and 0.20%") and banana-based CDs ("4%, 6%, and 8%"). The optimal amount of green produced by CDs inhibited *Bacillus subtilis*, *Staphylococcus aureus*, and *Escherichia coli* growth, according to the results. Soy milk samples with 0.16% chitosan and 8% CDs had a longer shelf life and reduced total bacterial count after 4 days at room temperature. When the CD concentration grew from 0% to 10%, bacterial growth fell from 100% to 0%. *B. subtilis*, *S. aureus*, and *E. coli* thrive in solid media like liquid media, although colony numbers decrease as CD concentration increases. Few bacteria grew on solid plates with 8% CDs (61, 11, 8 for *B. subtilis*, *S. aureus*, and *E. coli*). Most bacteria colonies were absent on plates with 10% CD, demonstrating room-temperature soymilk sample appearance variations. The control sample precipitated after 2 days of storage, but the soy milk samples were more stable. After 4 days of storage, the control and 0.16% CH samples showed color changes, but the 0.16% CH + 8% CDs sample did not show any color change or precipitation. The results indicate that CDs combined with an aqueous chitosan solution can extend the shelf life and improve the stability of soymilk, making it a potential composite combination and alternative to heat treatment. In another study, Koshy et al. (2021) green-synthesized nitrogen-doped CDs using soy protein isolate (SPI, a natural source) and incorporated them in a starch matrix containing *Clitoria ternatea* flower (CTE) anthocyanin to create an intelligent and cost-effective biopolymer matrix to monitor the freshness of packed pork samples at ambient conditions. CDs lowered water sensitivity and boosted film mechanical strength, according to the scientists. Green manufactured CDs boosted starch film thermal stability. Thermal stability rises due to intermolecular hydrogen bonding between the CD functional groups and the matrix. The optimized starch/CD/CTE films were used to verify the freshness of stored pork since pH indicator films change color with pH.

10.3.2 Food Quality Analysis

Nanoparticles are commonly employed in food analysis. Magnetic nanoparticles are used as electrode modifiers and sample preconcentrators in detection methods (Xu et al., 2014). One approach uses magnetic microparticles coated with protein A, a cell wall component of various *Staphylococcus aureus* strains. This can accurately identify the allergen Ara h3, a heat-stable protein that makes up 19% of peanut extract protein, in food at 0.2 mg/kg. The procedure is repeatable and reproducible (Speroni et al., 2010). Another example is glucose biosensors with magnetic nanoparticles. The biosensors transform target analytes into electrochemically detectable products using immobilized oxidase. This simplifies glucose testing in meals (Cao et al., 2012). Magnetite nanoparticles detect silver, copper, lead, cadmium, and mercury in food (Hernández-Hernández et al., 2017). Yang et al. (2011) found that SPIONs enhance food bacterium preconcentration. PCR target microorganisms are concentrated utilizing submicron superparamagnetic anion exchangers. Moving the PCR detection limit from 104–105 CFU mL^{-1} to 102 CFU mL^{-1} was achieved. *Escherichia coli* and *Agrobacterium tumefaciens* were bacterial models. Chen et al. (2015) extracted pathogenic bacteria DNA using silica-coated SPIONs. This approach avoids sample preincubation and food matrix impact. Pektaş et al. (2024) developed a glucose amperometric biosensor employing green synthesis waste tea-based silver nanoparticles. To do this, a carbon paste electrode (CPE) was modified with green-synthesized (WT-AgNPs) and cross-linked to immobilize glucose oxidase enzyme. The glucose measurement was done by comparing hydrogen peroxide reduction at +0.4 V to Ag/AgCl. The linear working range was chosen as 0.10–1.0 μM. The biosensor also measured glucose in commercial fruit juice. The biosensor has a low detection limit, high repeatability, selectivity, and nearly extended shelf life. To measure vanillin in adult and baby meals and human plasma, El Hamd et al. (2024) developed a self-ratiometric fluorescent nanosensor using green-synthesized silver nanoparticles (Ag-NPs). A microwave-assisted method employing tailed pepper seed extract to synthesize Ag-NPs in under 1 minute is suggested. When stimulated at 265 nm, the produced Ag-NPs fluoresced strongly with an intense emission band at 360 nm and a shoulder peak at 430 nm. When vanillin interacts with Ag-NPs, their fluorescence peak at 360 nm drops concentration-dependently and shifts to 385 nm. Vanillin has little effect on the shoulder fluorescence peak at 430 nm. Therefore, a novel Ag-NP self-ratiometric probe was constructed and validated for vanillin measurement utilizing the 385 nm peak and the 430 nm shoulder peak as built-in reference peaks. The improved vanillin measurement method achieved a 9.0 ng/mL detection limit and 0.05–8.0 μg/mL linear range without pre-derivatization or expensive equipment. High recovery and accuracy were achieved while measuring vanillin in adult and baby milk formula, biscuits, and human plasma samples.

10.3.3 Sensory and Quality Enhancement

Green synthetic nanoparticles are redefining sensory and quality enhancement in the food sector by improving product aesthetics and functional performance, including color, flavor, and texture. Titanium dioxide nanoparticles lighten dairy goods (Rathore & Pandey, 2024), whereas silica nanoparticles make powdered meals more fluid and anti-caking (Kwon et al., 2024). Antimicrobial nanoparticles like zinc oxide and silver enhance shelf life and keep perishables fresh. This increases food taste and quality and safety, which boosts consumer satisfaction (Yiblet et al., 2024). Thus, green-synthesized nanoparticles in food items improve customer appeal while meeting sustainability and safety norms, signifying a move toward new, consumer-focused food innovations (Gour & Jain, 2019; Ullah et al., 2024a, b). Mahato et al. (2024) examined how chemically and greenly generated zinc oxide nanoparticles (ZnO-NPs) affect fish flesh shelf life and taste. ZnO-NPs were manufactured using colloidal chemistry (CZnO-NPs) and green synthesis (GZnO-NPs) techniques and evaluated for morphology in this work. Synthetic ZnO-NPs, ZnO, and ZnA were employed to preserve and fortify *Pangasius hypophthalmus* meat at 20 mg/kg Zn. Fish samples were evaluated for sensory attributes (color and odor), physicochemical quality (pH and total volatile base nitrogen), oxidative changes (thiobarbituric acid-reactive substances and peroxide value), and microbial loads at 0, 3, and 6 days of 4 °C storage. The synthesized CZnO-NPs fortified raw fish had superior color and odor and maintained a pH non-conducive to microbial development during storage than the control, ZnO, and ZnA-fortified samples. GZnO-NPs occasionally outperformed the control, ZnO, and ZnA samples, but mostly did not outperform CZnO-NPs. CZnO-NPs in chilled fish had a lower oxidative status and total volatile base nitrogen than other treatments. The ZnO-NP-fortified fish contained the fewest live bacteria, coliforms, Staphylococcus, and Vibrio. Thus, adding generated CZnO-NPs to fish in storage may minimize microbiological deterioration and lipid peroxidation. ZnO-NPs were biosynthesized by lactobacilli strains by El-Sayed et al. (2021). Integrated biogenesis ZnO-NPs in yogurt increased its shelf life for 4 weeks without affecting taste. Fortified yogurt's microbiological population decreased more than control during storage. Chemically, fortified yogurt had more dry matter, protein, and ash than the control. The results revealed that low biosynthesized ZnO-NPs cause integrated yogurt characteristics. Additionally, adding biosynthesized ZnO-NPs to yogurt may be an excellent source of zinc for elderly people or vegans at risk of zinc shortage.

10.3.4 Nanoparticles as Ingredients in Foods

Green nanoparticles improve food safety and shelf life. Their antibacterial qualities prevent rotting organisms and harmful germs from growing, extending freshness and minimizing waste. Nutritional supplements are being transformed by green nanoparticles to boost nutritional and functional ingredient bioavailability. Vitamins, minerals, and antioxidants are encapsulated in nanoparticles to prevent digestive breakdown and provide controlled release (Altemimi et al., 2024). This method boosts supplement nutrition, improving health outcomes. Additionally, these nanoparticles are used to create functional meals with health advantages. Due to its increased surface area and bioavailability, nano-curcumin, generated from turmeric and processed by green synthesis, has better anti-inflammatory and antioxidant effects than its non-nano counterpart (Gharibzahedi et al., 2023). Iron's main biological role is oxygen transport as part of the heme nucleus in hemoglobin and myoglobin. Iron is involved in about 200 enzyme systems that are required for cellular energy usage and DNA, RNA, and protein production (Luo et al., 2012). Neurotransmitter metabolism, vitamin D activation, collagen metabolism, and cholesterol catabolism include iron. Animals, including humans, need iron to operate properly. Dave and Gao (2009) say ferritin is an intracellular protein found in the liver and immune cells. When needed, it releases inactive iron from the body. Kumari and Chauhan (2021) found that high-bioavailability iron enrichment compounds ($FeSO_4$) are chemically reactive. Foods get rotten, taste metallic, and change color due to these processes. $FeSO_4$ is replaced by water-insoluble $FePO_4$, $Fe_4(P_2O_7)_3$, or iron ions due to these issues. Despite not affecting food qualities, these additives are weakly absorbed. Following von Moos et al. (2017) and Perfecto et al. (2017), IONPs are an alternative to typical iron supplements that prevent iron-deficiency illnesses utilizing $FePO_4$ NPs. In addition to iron oxides and phosphates, Hilty et al. (2011) employed Ca and Mg complexes. Humans need these minerals; thus, adding them to the diet is safe. By adding calcium and magnesium to Fe_2O_3, i. d.a., solubility $>85\%$ was achieved, indicating good bioavailability. Caking was absent during storage at varied humidity settings. Asaikkutti et al. (2016) produced green Mn3O4 NPs for freshwater prawn *Macrobrachium rosenbergii* diets. Glutamic pyruvate transaminase, antioxidant enzymatic activity, and glutamic oxaloacetate transaminase were greatly enhanced by green-produced MnO NPs. *M. rosenbergii's* ultimate weight, growth, and survival improved by Mn_3O_4 NPs.

Conclusion: Green generated Mn_2O_4 NPs were safe and effective food supplements for freshwater prawn *Macrobrachium rosenbergii*. This study suggests it might complement aquatic creatures' diets. Using dietary curcumin-assisted green-synthesized zinc oxide nanoparticles (CUR-ZnONPs), Yazdani et al. (2023) examined Nile tilapia growth performance, haemato-biochemical profile, antioxidant capability, immunological response, carcass composition, and Zn bioaccumulation. Principal component analysis showed that CUR-ZnONP-fed fish have considerably higher growth performance, immunological responses, and antioxidant status than control diet fish. Overall, dietary CUR-ZnONPs, especially at 10 mg/kg, can increase Nile tilapia development, antioxidant status, and immunological markers. Razack et al. (2020) used *Hibiscus rosa-sinensis* flower extracts to synthesize iron oxide nanoparticles for wheat cookies. Fortified biscuits had more iron than controls, according to inductively coupled plasma optical emission spectroscopy. Microbial investigation for 30 days suggested enriched biscuits might last longer. In summary, the first publication on iron oxide nanoparticles advised using them as food fortifiers and treating malnourished, iron-deficient, and anemic individuals.

10.4 Economic and Environmental Benefits of Nanoparticles in the Food Industry

Nanotechnology has revolutionized the food business, bringing economic and environmental benefits. Nanoparticles' high surface-area-to-volume ratio and greater functionality have enabled food quality, safety, and sustainability advancements. Functional foods use nanoparticles to increase nutritional and active ingredient absorption. This invention meets the increased need for health-related items, boosting their value. To preserve vitamin and omega-3 fatty acid efficacy throughout processing and storage, nanoencapsulation is used (Pandohee et al., 2023). These premium products may boost profits and market distinction for manufacturers. Nanocomposites have improved food packaging performance. These materials have greater mechanical strength, oxygen and moisture barriers, and antibacterial characteristics. Clay nanocomposites in plastic films minimize gas permeability, prolonging perishable product shelf life and reducing food waste. Antimicrobial nanocoatings reduce spoilage and improve food safety by inhibiting bacterial growth on packaging. These innovations save producers and merchants money and ensure customer quality consumers (Andriani & Handayani, 2023; Khan et al., 2023). Nanotechnology in the food business has considerable environmental benefits, especially in waste, pollution, and sustainability. Nanosensors can detect deterioration or contamination in real time in food. Preventing the disposal of safe but mislabeled items lowers food waste. These sensors are increasingly used in intelligent packaging solutions to give merchants and customers meaningful data. Nanocomposites in packaging materials help the environment. Traditional plastic packaging is practical but non-biodegradable, posing environmental issues.

However, biodegradable bio-nanocomposites generated from renewable resources have comparable capabilities (Garavand et al., 2022; del Rosario Herrera-Rivera et al., 2024). These materials reduce petroleum-based plastics, reducing pollution and supporting global carbon reduction initiatives. Nanotechnology is vital to food production improvement. Precision agriculture with nanopesticides and nanofertilizers reduces chemical runoff into waterways, conserving aquatic ecosystems and soil health. Nanopesticides provide active components directly to target pests, decreasing environmental consequences. Nano-enabled water filtration systems purify irrigation and processing water, alleviating agricultural water scarcity. Food packaging becomes more recyclable and energy-efficient with nanotechnology. High-performance coatings with metal oxide nanoparticles promote recyclability without losing barrier characteristics. These coatings save energy in manufacturing and recycling, saving the industry energy. Food safety and quality assurance are greatly improved by nanotechnology. Nanostructures and nanosensors can detect infections and chemical residues at low concentrations. Precision guarantees that only safe products reach consumers, preserving public health and manufacturer brand integrity. Silver nanoparticles on food contact surfaces reduce foodborne disease by inhibiting bacterial growth. Nanotechnology-enabled traceability solutions improve food safety. Food goods may be tracked throughout the supply chain using nano-enabled tracking sensors in packaging. Perishable commodities benefit from real-time temperature, humidity, and other data for efficient storage and delivery (Andriani & Handayani, 2023; Khan et al., 2023). Nanotechnology in the food business has clear benefits, yet governmental monitoring and public perception remain issues. Nanoparticle concerns, including bioaccumulation and toxicity, must be carefully assessed to protect consumers. The European Food Safety Authority (EFSA) and the US Food and Drug Administration (FDA) have frameworks for assessing nanomaterial safety, although technological advances require ongoing modifications (EFSA, 2021). Nanotechnology adoption also depends on public perception. Building customer trust requires transparent communication about its advantages and safety. Education and stakeholder involvement may dispel myths and demonstrate nanotechnology's role in food sustainability and efficiency.

10.5 Regulation on the Use of Nanomaterials in Food Industry

The manipulation of materials at the nanoscale, or nanotechnology, has significantly advanced a number of sectors, including the food sector. Because of their special qualities, which include their high surface-to-volume ratio and new functions, nanoparticles are sought after for a variety of uses, including food packaging, nutritional content enhancement, and shelf-life extension. These developments, however, also bring up safety issues because of possible toxicity, bioaccumulation, and biological system interactions. The characteristics that make nanoparticles appealing for use in food applications include increased reactivity, changed solubility, and improved functional capabilities. These same characteristics, however, provide difficulties for risk assessment and a need for sophisticated analytical techniques. Concerns have been raised by scientific research on possible health hazards linked to nanoparticles: silver and titanium dioxide are two examples of nanoparticles that might cause inflammatory reactions in biological systems (EFSA, 2021). Nanoparticles may interact with DNA, according to genotoxicity studies, which might have long-term health consequences. Certain nanoparticles build up in tissues, which raises questions regarding long-term exposure. Strict regulation is required to protect the environment and public health. This report integrates the most recent scientific and technological data to analyze the worldwide regulatory landscape for the use of nanoparticles in the food sector.

United States

Applications of nanotechnology in food are governed by the US Food and Drug Administration (FDA) within preexisting frameworks. On a case-by-case basis, the FDA assesses the safety of food components and packaging materials that use nanotechnology. The FDA advises manufacturers to seek advice before launching goods based on nanotechnology. The FDA produced recommendations in 2014 titled "Assessing the Effects of Significant Manufacturing Process Changes, Including Emerging Technologies, on the Safety and Regulatory Status of Food Ingredients and Food Contact Substances." This report emphasizes how important it is to evaluate the safety effects of major process modifications, such as the application of nanotechnology (FDA, 2014). For compounds that come into contact with food, the FDA mandates a comprehensive safety assessment that takes into account characteristics such as particle size, shape, and propensity for migration into food. Critics contend that present laws do not sufficiently address the long-term toxicity and bioaccumulation issues connected with nanoparticles, despite the FDA's strong framework.

European Union

Nanotechnology in food is approached cautiously by the European Union. A key player in risk evaluations and the creation of regulations pertaining to the use of nanoparticles is the European Food Safety Authority (EFSA). Nanomaterials that are utilized in food fall under the "novel food" category and need pre-market approval. Manufacturers are required to submit comprehensive safety information, including toxicological analyses (EFSA, 2021). In 2022, the European Union outlawed titanium dioxide as a food additive after the EFSA revealed that it was genotoxic. The ruling reaffirms the precautionary principle‘s importance to the EU (European Commission, 2022). The European Parliament (2011) states that Regulation (EU) No 1169/2011 requires that food containing artificial nanoparticles be clearly labeled with the term "nano" in the ingredient list to ensure consumer knowledge. Better ways to describe nanoparticles are needed, according to recent EFSA guidelines, which place special emphasis on factors such as particle dispersion, aggregation, and interactions with biological systems (EFSA, 2021).

Australia and New Zealand

The application of nanotechnology in food is governed by Food Standards Australia New Zealand (FSANZ) in both Australia and New Zealand. For nanoparticles in food and materials that come into contact with it, FSANZ mandates thorough risk evaluations that concentrate on toxicological information and possible exposure (FSANZ, 2018). To maintain openness and dispel myths around nanotechnology, FSANZ regularly interacts with the public. In order to exchange scientific insights and standardize regulatory standards, FSANZ works with foreign institutions (FSANZ, 2018).

India

Since advances in nanotechnology provide enormous possibilities for food preservation, packaging, and safety, regulations governing nanoparticles in the Indian food business are steadily changing. The main regulatory agency in charge of food safety is the Food Safety and Standards Authority of India (FSSAI), although as of right now, it does not have any rules specifically pertaining to nanoparticles. Rather, general food safety rules and the current regulations under the "Food Safety and Standards (Packaging and Labeling) Regulations, 2018" are used. Establishing safety regulations and encouraging research into nanotechnology uses for food and agriculture are other goals of the Department of Biotechnology (DBT) and the Bureau of Indian Standards (BIS). The lack of explicit rules, inadequate testing facilities, and low public knowledge of food nanoparticles are among the difficulties. Nonetheless, initiatives are being made to create testing procedures, conform to global guidelines such as those set forth by the European Food Safety Authority (EFSA), and implement label transparency. The laws governing food nanotechnology in India are changing. Guidelines focusing on safety and research have been released by the Food Safety and Standards Authority of India (FSSAI). Assessing the safety of nanoparticles in food additives and packaging materials is the main objective of the FSSAI's framework (FSSAI, 2019). Studies on the toxicological and environmental effects of nanoparticles are being carried out by Indian research institutes, offering a solid scientific foundation for regulation. Significant obstacles still exist in the form of inadequate infrastructure for improved safety evaluations and a lack of comprehensive rules tailored to nanoparticles.

Global Harmonization Efforts

Global standards for food nanotechnology are being standardized by international bodies like the *Codex Alimentarius* Commission. These initiatives seek to promote global trade and provide uniform safety evaluation procedures. When it comes to regulating nanotechnology in food, the Codex Alimentarius Commission stresses the significance of risk assessment and openness. Countries creating their regulatory frameworks might use their recommendations as a guide. To solve shared issues, including analytical constraints and public concerns, nations are urged to exchange scientific data and best practices in regulation (Codex Alimentarius, 2020). The general public's perspective of food nanotechnology differs greatly. Some customers voice worries about safety and ethical ramifications, while others see the potential advantages. In order to build consumer trust, regulators must guarantee open and honest communication regarding the advantages and safety of nanotechnology. The ethical ramifications of uneven access to breakthroughs created by nanotechnology must be taken into account by policymakers, especially in low-income areas.

10.6 Limitations and Future Challenges

GSNPs provide an environmentally beneficial substitute for their chemically synthesized equivalents; nonetheless, there are several obstacles and restrictions on their use in the food sector and large-scale manufacturing. Because plant extracts, microbial cultures, or other biological agents utilized as reducing and stabilizing agents might change in composition based on ambient and seasonal conditions, one of the main obstacles is the variability of synthesis processes. Because of this heterogeneity, the size, shape, and functional characteristics of the nanoparticles vary, which makes repeatability and standardization difficult for applications in the food business. Furthermore, the scalability and production costs of green synthesis techniques are limited due to their tendency to produce lower nanoparticle concentrations and longer reaction times. Lack of infrastructure and technology to allow large-scale synthesis further limits the scalability of GSNP manufacturing. Exact control over nanoparticle characteristics is necessary for industrial-scale manufacturing, but existing techniques frequently rely on labor-intensive, hard-to-scale batch procedures. To guarantee efficiency and consistency, a substantial capital investment and process optimization would be required when switching to continuous processes. A further technological obstacle in large-scale manufacturing is guaranteeing uniform dispersion of GSNPs in food items, as aggregation and uneven distribution might jeopardize the efficacy and quality of the final product.

Another major obstacle is achieving the appropriate stability of GSNPs in food matrices. Because of their extreme sensitivity to environmental factors, including pH, temperature, and ionic strength, these nanoparticles can aggregate or stop working in intricate food systems. Antimicrobial nanoparticles, for example, may become less efficient at prolonging shelf life or improving food safety when added to settings that are heavy in fat or acid. Furthermore, a major bottleneck still exists in the purification of GSNPs to eliminate biological residues. The economic viability of large-scale manufacturing is further limited by the energy and cost-intensive nature of processes like centrifugation and filtering, which are necessary for creating high-purity nanoparticles. From a regulatory standpoint, the absence of precise rules and regulations hinders the food industry's use of GSNPs. To assess any dangers related to nanoparticles, such as toxicity, bioaccumulation, and environmental impact, regulatory agencies like the European Food Safety Authority (EFSA) and the US Food and Drug Administration (FDA) need thorough safety data. Compliance efforts are made more difficult by the lack of widely recognized procedures for evaluating the safety of GSNPs, which delays their commercial use (EFSA, 2021). Additionally, a major obstacle still exists in the public's perception of nanotechnology. Despite the "green" designation, market adoption has been hampered by customer reluctance due to worries about possible undiscovered health effects linked to nanoparticles in food items. The incorporation of GSNPs into current food systems such as food processing and packaging. To handle formulations based on nanoparticles, conventional production equipment frequently has to be modified, requiring additional capital expenditure. For instance, adding GSNPs to active packaging materials to improve barrier functions or give antimicrobial qualities may necessitate modifications to extrusion or coating methods, raising the cost and complexity of operations. Furthermore, it is difficult to sustain the environmentally beneficial potential of GSNPs during the course of manufacturing. Although the synthesis process is naturally more environmentally friendly than chemical procedures, the environmental advantages may be outweighed by energy-intensive downstream steps like stabilization and functionalization or the employment of chemical stabilizers. Lastly, little is known about how GSNPs affect the environment, especially how they affect ecosystems over the long run. Aquatic and terrestrial ecosystems may be at danger when these nanoparticles penetrate the food chain and are eventually released into the environment through waste streams or container breakdown. To solve these issues and guarantee that the "green" designation accurately represents sustainable practices, thorough life cycle evaluations and environmental impact studies are required. It will take multidisciplinary research, scalable technology investments, and the creation of strong regulatory frameworks to overcome these constraints. Building trust, encouraging their use in the food sector, and facilitating the shift to large-scale manufacturing will all require open and honest communication with customers that highlights the advantages and safety of GSNPs.

10.7 Conclusion and Summary

Green-synthesized nanomaterials have the potential to completely transform post-harvest food systems by offering creative ways to improve food safety, quality, and shelf life. By decreasing waste, prolonging food freshness, and enhancing overall food safety, their many uses—such as active packaging, edible coatings, and food analysis—align with global sustainability goals. These developments are essential as the food sector looks for environmentally friendly technology to satisfy growing customer demands and address environmental issues. However, a number of restrictions prevent green-synthesized nanomaterials from being widely used. Since existing techniques frequently result in inconsistent nanoparticle characteristics and need tuning for consistent manufacturing, scalability is still a major hurdle. Furthermore, their implementation in the food industry is complicated by the absence of widely recognized regulatory frameworks and thorough safety recommendations. Despite their environmentally friendly synthesis, consumer concern about the usage of nanoparticles highlights the value of openness and education in fostering acceptance and trust. Future research must concentrate on developing green synthesis methods to improve yield, scalability, and consistency while guaranteeing environmental advantages in order to overcome these obstacles. To create strong safety standards and laws that encourage innovation without endangering public health, cooperation between researchers, legislators, and industry stakeholders is crucial. Additionally, customer acceptability and confidence may be increased by public awareness efforts that emphasize the advantages and security of green-synthesized nanomaterials. Green-synthesized nanoparticles have the potential to be a key component of sustainable food technology by tackling these issues. Their capacity to blend technical innovation with environmental responsibility makes them an essential part of building a more robust and effective food business that can fulfill future demands while reducing its ecological impact.

Conflict of Interest The authors declare no conflict of interest.

References

Ahmad, A., Mukherjee, P., Senapati, S., Mandal, D., Khan, M. I., Kumar, R., & Sastry, M. (2003a). Extracellular biosynthesis of silver nanoparticles using the fungus Fusarium oxysporum. *Colloids and Surfaces B: Biointerfaces, 28*(4), 313–318.

Ahmad, A., Senapati, S., Khan, M. I., Kumar, R., & Sastry, M. (2003b). Extracellular biosynthesis of monodisperse gold nanoparticles by a novel extremophilic actinomycete, Thermomonospora sp. *Langmuir, 19*(8), 3550–3553.

Alani, F., Moo-Young, M., & Anderson, W. (2012). Biosynthesis of silver nanoparticles by a new strain of Streptomyces sp. compared with Aspergillusfumigatus. *World Journal of Microbiology and Biotechnology, 28*(3), 1081–1086.

Alghuthaymi, M. A., Diab, A. M., Elzahy, A. F., Mazrou, K. E., Tayel, A. A., & Moussa, S. H. (2021). Green biosynthesized selenium nanoparticles by cinnamon extract and their antimicrobial activity and application as edible coatings with nano-chitosan. *Journal of Food Quality, 2021*(1), 6670709.

Ali, E. A., Mohammed, D. M., Abd El Gawad, F., Orabi, M. A., Gupta, R. K., & Srivastav, P. P. (2025). Valorization of food processing waste byproducts for essential oil production and their application in food system. *Waste Management Bulletin, 3*, 100200.

Alprol, A. E., Mansour, A. T., El-Beltagi, H. S., & Ashour, M. (2023a). Algal extracts for green synthesis of zinc oxide nanoparticles: Promising approach for algae bioremediation. *Materials, 16*(7), 2819.

Alprol, A. E., Mansour, A. T., Abdelwahab, A. M., & Ashour, M. (2023b). Advances in green synthesis of metal oxide nanoparticles by marine algae for wastewater treatment by adsorption and photocatalysis techniques. *Catalysts, 13*(5), 888.

Altemimi, A. B., Farag, H. A. M., Salih, T. H., Awlqadr, F. H., Al-Manhel, A. J. A., Vieira, I. R. S., & Conte-Junior, C. A. (2024). Application of nanoparticles in human nutrition: A review. *Nutrients, 16*(5), 636.

Andriani, V., & Handayani, N. A. (2023). Recent technology of edible coating production: A review. *Materials Today: Proceedings, 87*, 200–206.

Asaikkutti, A., Bhavan, P. S., Vimala, K., Karthik, M., & Cheruparambath, P. (2016). Dietary supplementation of green synthesized manganese-oxide nanoparticles and its effect on growth performance, muscle composition and digestive enzyme activities of the giant freshwater prawn Macrobrachium rosenbergii. *Journal of Trace Elements in Medicine and Biology, 35*, 7–17.

Bahrami, A., Delshadi, R., Assadpour, E., Jafari, S. M., & Williams, L. (2020). Antimicrobial-loaded nanocarriers for food packaging applications. *Advances in Colloid and Interface Science, 278*, 102140.

Bahrulolum, H., Nooraei, S., Javanshir, N., Tarrahimofrad, H., Mirbagheri, V., Easton, A., & Ahmadian, G. (2021). Green synthesis of metal nanoparticles using microorganisms and their application in the agrifood sector. *Journal of Nanobiotechnology, 19*, 86.

Baneshi, M., Aryee, A. N., English, M., & Mkandawire, M. (2024). Designing plant-based smart food packaging solutions for prolonging the consumable life of perishable foods. *Food Chemistry Advances*, 100769.

Begum, S., & Ahmaruzzaman, M. (2018). Green synthesis of SnO2 nanoparticles loaded on activated carbon and its application as photocalayst in the degradation of alizarin red S dye. *Materials Today Proceedings, 5*(1), 2314–2320.

Cao, M., Li, Z., Wang, J., Ge, W., Yue, T., Li, R., Colvin, V. L., & Yu, W. W. (2012). Food related applications of magnetic iron oxide nanoparticles: Enzyme immobilization, protein purification, and food analysis. *Trends in Food Science and Technology, 27*, 47–56.

Chan, S. S., Low, S. S., Chew, K. W., Ling, T. C., Rinklebe, J., Juan, J. C., et al. (2022). Prospects and environmental sustainability of phyconanotechnology: A review on algae-mediated metal nanoparticles synthesis and mechanism. *Environmental Research, 212*, 113140.

Chen, X., & Schluesener, H. J. (2008). Nanosilver: A nanoproduct in medical application. *Toxicology Letters, 176*(1), 1–2.

Chen, X., Wu, X., Gan, M., Xu, F., He, L., Yang, D., Xu, H., Shah, N. P., & Wei, H. (2015). Rapid detection of Staphylococcus aureus in dairy and meat foods by combination of capture with silica-coated magnetic nanoparticles and thermophilic helicase-dependent isothermal amplification. *Journal of Dairy Science, 98*, 1563–1570.

Codex Alimentarius. (2020). *Guidelines on food safety and nanotechnology*. https://www.fao.org/fao-who-codexalimentarius/en/

Dameron, C. T., Reese, R. N., Mehra, R. K., Kortan, A. R., Carroll, P. J., Steigerwald, M. L., Brus, L. E., & Winge, D. R. (1989). Biosynthesis of cadmium sulphide quantum semiconductor crystallites. *Nature, 338*(6216), 596. References 23.

Das, M., Saxena, N., & Dwivedi, P. D. (2009). Emerging trends of nanoparticles application in food technology: Safety paradigms. *Nanotoxicology, 3*(1), 10–18.

Dave, S. R., & Gao, X. (2009). Monodisperse magnetic nanoparticles for biodetection, imaging, and drug delivery: A versatile and evolving technology. *Wiley Interdisciplinary Reviews: Nanomedicine and Nanobiotechnology., 1*, 583–609.

Deepa, S., Kanimozhi, K., & Panneerselvam, A. (2013). Antimicrobial activity of extracellularly synthesized silver nanoparticles from marine derived actinomycetes. *International Journal of Current Microbiology and Applied Sciences, 2*(2), 223–230.

del Rosario Herrera-Rivera, M., Torres-Arellanes, S. P., Cortés-Martínez, C. I., Navarro-Ibarra, D. C., Hernández-Sánchez, L., Solis-Pomar, F., et al. (2024). Nanotechnology in food packaging materials: Role and application of nanoparticles. *RSC Advances, 14*(30), 21832–21858.

Devi, T. B., & Ahmaruzzaman, M. (2018). Facile preparation of Copper nanaoparticles using Coccinia grandis fruit extract and its application towards the reduction of toxic nitro compound. *Materials Today Proceedings, 5*(1), 2098–2104.

Dey, D., Dharini, V., Selvam, S. P., Sadiku, E. R., Kumar, M. M., Jayaramudu, J., & Gupta, U. N. (2021). Physical, antifungal, and biodegradable properties of cellulose nanocrystals and chitosan nanoparticles for food packaging application. *Materials Today: Proceedings, 38*, 860–869.

Du, L., Jiang, H., Liu, X., & Wang, E. (2007). Biosynthesis of gold nanoparticles assisted by Escherichia coli DH5α and its application on direct electrochemistry of hemoglobin. *Electrochemistry Communications, 9*(5), 1165–1170.

Ealia, S. A. M., & Saravanakumar, M. P. (2017, November). A review on the classification, characterisation, synthesis of nanoparticles and their application. In *IOP conference series: Materials science and engineering* (Vol. 263, No. 3, p. 032019). IOP Publishing.

EFSA. (2021). *Guidance on nanotechnology in food chain (EFSA Journal)*. https://www.efsa.europa.eu/en/efsajournal/pub/6768

El Hamd, M. A., El-Maghrabey, M., Almawash, S., El-Shaheny, R., & Magdy, G. (2024). Self-ratiometric fluorescence approach based on plant extract-assisted synthesized silver nanoparticles for the determination of vanillin. *Microchimica Acta, 191*(1), 16.

El-Naggar, M. E., Hasanin, M., & Hashem, A. H. (2022). Eco-friendly synthesis of superhydrophobic antimicrobial film based on cellulose acetate/polycaprolactone loaded with the green biosynthesized copper nanoparticles for food packaging application. *Journal of Polymers and the Environment, 30*(5), 1820–1832.

El-Sayed, H. S., El-Sayed, S. M., & Youssef, A. M. (2021). Novel approach for biosynthesizing of zinc oxide nanoparticles using lactobacillus gasseri and their influence on microbiological, chemical, sensory properties of integrated yogurt. *Food Chemistry, 365*, 130513.

Espinosa, E., Rincón, E., Morcillo-Martín, R., Rabasco-Vílchez, L., & Rodríguez, A. (2022). Orange peel waste biorefinery in multi-component cascade approach: Polyphenolic compounds and nanocellulose for food packaging. *Industrial Crops and Products, 187*, 115413.

European Commission. (2022). *Decision on banning titanium dioxide*. https://ec.europa.eu/newsroom/sante/items/732079/en

European Parliament. (2011). *Regulation (EU) No 1169/2011 on food information*. https://eur-lex.europa.eu/legal-content/EN/TXT/?uri=CELEX%3A32011R1169

Ezati, P., Roy, S., & Rhim, J.-W. (2022). Pectin/gelatin-based bioactive composite films reinforced with sulfur functionalized carbon dots. *Colloids and Surfaces A: Physicochemical and Engineering Aspects, 636*, 128123.

FDA. (2014). *Assessing the effects of significant manufacturing process changes, including emerging technologies, on the safety and regulatory status of food ingredients and food contact substances*. U.S. Food and Drug Administration. https://www.fda.gov/science-research/nanotechnology-programs-fda/nanotechnology-guidance-documents

FSANZ. (2018). *Risk assessment of nanotechnology*. https://www.foodstandards.gov.au/

FSSAI. (2019). *Guidelines on nanotechnology in food*. https://pmc.ncbi.nlm.nih.gov/articles/PMC10320728/

Fu, B., Liu, Q., Liu, M., Chen, X., Lin, H., Zheng, Z., Zhu, J., Dai, C., Dong, X., & Yang, D.-P. (2022). Carbon dots enhanced gelatin/chitosan bio-nanocomposite packaging film for perishable foods. *Chinese Chemical Letters, 33*, 4577–4582.

Gao, Y., Xu, D., Ren, D., Zeng, K., & Wu, X. (2020). Green synthesis of zinc oxide nanoparticles using Citrus sinensis peel extract and application to strawberry preservation: A comparison study. *Lwt, 126*, 109297.

Garavand, F., Cacciotti, I., Vahedikia, N., Rehman, A., Tarhan, Ö., Akbari-Alavijeh, S., et al. (2022). A comprehensive review on the nanocomposites loaded with chitosan nanoparticles for food packaging. *Critical Reviews in Food Science and Nutrition, 62*(5), 1383–1416.

Gericke, M., & Pinches, A. (2006a). Biological synthesis of metal nanoparticles. *Hydrometallurgy, 83*(1–4), 132–140.

Gericke, M., & Pinches, A. (2006b). Microbial production of gold nanoparticles. *Gold Bulletin, 39*(1), 22–28.

Gharibzahedi, S. M. T., Moghadam, M., Amft, J., Tolun, A., Hasabnis, G., & Altintas, Z. (2023). Recent advances in dietary sources, health benefits, emerging encapsulation methods, food fortification, and new sensor-based monitoring of vitamin B12: A critical review. *Molecules, 28*(22), 7469.

Golinska, P., Wypij, M., Ingle, A. P., Gupta, I., Dahm, H., & Rai, M. (2014). Biogenic synthesis of metal nanoparticles from actinomycetes: Biomedical applications and cytotoxicity. *Applied Microbiology and Biotechnology, 98*(19), 8083–8097.

Gopalakrishnan, K., Chandel, M., Gupta, V., Kaur, K., Patel, A., Kaur, K., et al. (2023). Valorisation of fruit peel bioactive into green synthesized silver nanoparticles to modify cellulose wrapper for shelf-life extension of packaged bread. *Food Research International, 164*, 112321.

Gour, A., & Jain, N. (2019). Advances in green synthesis of nanoparticles. *Artificial Cells, Nanomedicine, and Biotechnology, 47*, 844–851.

Govindaraju, K., Kiruthiga, V., Kumar, V. G., & Singaravelu, G. (2009). Extracellular synthesis of silver nanoparticles by a marine alga, Sargassum wightii Grevilli and their antibacterial effects. *Journal of Nanoscience and Nanotechnology, 9*(9), 5497–5501.

Gudimalla, A., Jose, J., Varghese, R. J., & Thomas, S. (2021). Green synthesis of silver nanoparticles using *Nymphae odorata* extract incorporated films and antimicrobial activity. *Journal of Polymers and the Environment, 29*, 1412–1423.

Gupta, R. K., Guha, P., & Srivastav, P. P. (2022). Natural polymers in bio-degradable/edible film: A review on environmental concerns, cold plasma technology and nanotechnology application on food packaging-A recent trends. *Food Chemistry Advances, 1*, 100135.

Gupta, R. K., Guha, P., & Srivastav, P. P. (2023). Effect of high voltage dielectric barrier discharge (DBD) atmospheric cold plasma treatment on physicochemical and functional properties of taro (Colocasia esculenta) starch. *International Journal of Biological Macromolecules, 253*, 126772.

Gupta, R. K., Guha, P., & Srivastav, P. P. (2024a). Investigating the toxicological effects of nanomaterials in food packaging associated with human health and the environment. *Journal of Hazardous Materials Letters, 5*, 100125.

Gupta, R. K., Abd El Gawad, F., Ali, E. A., Karunanithi, S., Yugiani, P., & Srivastav, P. P. (2024). Nanotechnology: Current applications and future scope in food packaging systems. *Measurement: Food, 13*, 100131.

Gupta, R. K., Guha, P., & Srivastav, P. P. (2024c). Insights into the physicochemical, crystalline, multistructural, thermal and digestibility behavior of taro starch (Colocasia esculenta) modified by cold plasma treatment with high voltage dielectric barrier discharge (HVDBD): An alternative technique for chemical modification of starch and environmental sustainability. *Journal of Cleaner Production, 461*, 142710.

Gupta, R. K., Guha, P., & Srivastav, P. P. (2024d). Dielectric barrier discharge plasma: A green and novel method to change the structure of taro peel starch and improve the physicochemical properties of taro peel starch films. *Plasma Processes and Polymers, 21*(9), 2400047.

Hamouda, R. A., & Aljohani, E. S. (2024). Assessment of silver nanoparticles derived from brown algae sargassum vulgare: Insight into antioxidants, anticancer, antibacterial and hepatoprotective effect. *Marine Drugs, 22*(4), 154.

Hassan Basri, H., Talib, R. A., Sukor, R., Othman, S. H., & Ariffin, H. (2020). Effect of synthesis temperature on the size of ZnO nanoparticles derived from pineapple peel extract and antibacterial activity of ZnO–starch nanocomposite films. *Nanomaterials, 10*(6), 1061.

Hernández-Hernández, A. A., Álvarez-Romero, G. A., Contreras-López, E., Aguilar-Arteaga, K., & Castañeda-Ovando, A. (2017). Food analysis by microextraction methods based on the use of magnetic nanoparticles as supports: Recent advances. *Food Analytical Methods, 10*, 2974–2993.

Hilty, F. M., Knijnenburg, J. T., Teleki, A., Krumeich, F., Hurrell, R. F., Pratsinis, S. E., & Zimmermann, M. B. (2011). Incorporation of Mg and Ca into nanostructured Fe2O3 improves Fe solubility in dilute acid and sensory characteristics in foods. *Journal of Food Science, 76*, N2–N10.

Jain, S., Gupta, I., Walia, P., & Swami, S. (2022). Application of actinobacteria in agriculture, nanotechnology, and bioremediation. In *Actinobacteria-diversity, applications and medical aspects*. IntechOpen.

Jalali, T., Ghanavati, F., & Osfouri, S. (2024). Green synthesis of ZnO nanoparticles using marine brown algae (Cystoseira) extract comprising sol–gel, and combustion techniques based on dye-sensitized solar cells application. *International Journal of Modern Physics B, 38*(14), 2450178.

Janani, N., Zare, E. N., Salimi, F., & Makvandi, P. (2020). Antibacterial tragacanth gum-based nanocomposite films carrying ascorbic acid antioxidant for bioactive food packaging. *Carbohydrate Polymers, 247*, 116678.

Jeevanandam, J., Kiew, S. F., Boakye-Ansah, S., Lau, S. Y., Barhoum, A., Danquah, M. K., & Rodrigues, J. (2022). Green approaches for the synthesis of metal and metal oxide nanoparticles using microbial and plant extracts. *Nanoscale, 14*(7), 2534–2571.

Jin, X., Liu, Y., Tan, J., Owens, G., & Chen, Z. (2018). Removal of Cr (VI) from aqueous solutions via reduction and absorption by green synthesized iron nanoparticles. *Journal of Cleaner Production, 176*, 929–936.

Joudeh, N., & Linke, D. (2022). Nanoparticle classification, physicochemical properties, characterization, and applications: A comprehensive review for biologists. *Journal of Nanobiotechnology, 20*(1), 262.

Kabiri, F., Aghaei, S. S., Pourbabaee, A. A., Soleimani, M., & Komeili Movahhed, T. (2023). Antibiofilm and cytotoxic potential of extracellular biosynthesized gold nanoparticles using actinobacteria Amycolatopsis sp. KMN. *Preparative Biochemistry & Biotechnology, 53*(3), 265–278.

Khan, I., Saeed, K., & Khan, I. (2017). Nanoparticles: Properties, applications and toxicities. *Arabian Journal of Chemistry*. https://doi.org/10.1016/J.ARABJC.2017.05.011

Khan, O. A., Zaidi, S., Islam, R. U., Naseem, S., & Junaid, P. M. (2023). Enhanced shelf-life of peach fruit in alginate based edible coating loaded with TiO2 nanoparticles. *Progress in Organic Coatings, 182*, 107688.

Kilic, B., Dogan, V., Kilic, V., & Kahyaoglu, L. N. (2022). Colorimetric food spoilage monitoring with carbon dot and UV light reinforced fish gelatin films using a smartphone application. *International Journal of Biological Macromolecules, 209*, 1562–1572.

Kisimba, K., Krishnan, A., Faya, M., Byanga, K., Kasumbwe, K., Vijayakumar, K., & Prasad, R. (2023). Synthesis of metallic nanoparticles based on green chemistry and their medical biochemical applications: Synthesis of metallic nanoparticles. *Journal of Renewable Materials, 11*(6), 2575–2591.

Koshy, R. R., Koshy, J. T., Mary, S. K., Sadanandan, S., Jisha, S., & Pothan, L. A. (2021). Preparation of pH sensitive film based on starch/carbon nano dots incorporating anthocyanin for monitoring spoilage of pork. *Food Control, 126*, 108039.

Kowshik, M., Deshmukh, N., Vogel, W., Urban, J., Kulkarni, S. K., & Paknikar, K. M. (2002a). Microbial synthesis of semiconductor CdS nanoparticles, their characterization, and their use in the fabrication of an ideal diode. *Biotechnology and Bioengineering, 78*(5), 583–588.

Kowshik, M., Ashtaputre, S., Kharrazi, S., Vogel, W., Urban, J., Kulkarni, S. K., & Paknikar, K. M. (2002b). Extracellular synthesis of silver nanoparticles by a silver-tolerant yeast strain MKY3. *Nanotechnology, 14*(1), 95.

Kumar, U., Shete, A., Harle, A. S., Kasyutich, O., Schwarzacher, W., Pundle, A., & Poddar, P. (2008). Extracellular bacterial synthesis of protein-functionalized ferromagnetic Co3O4 nanocrystals and imaging of self-organization of bacterial cells under stress after exposure to metal ions. *Chemistry of Materials, 20*(4), 1484–1491.

Kumar, S., Ye, F., Dobretsov, S., & Dutta, J. (2019). Chitosan nanocomposite coatings for food, paints, and water treatment applications. *Applied Sciences, 9*, 2409.

Kumar, S., Mudai, A., Roy, B., Basumatary, I. B., Mukherjee, A., & Dutta, J. (2020). Biodegradable hybrid nanocomposite of chitosan/gelatin and green synthesized zinc oxide nanoparticles for food packaging. *Foods, 9*(9), 1143.

Kumar, V., Kaushik, N. K., Tiwari, S. K., Singh, D., & Singh, B. (2023). Green synthesis of iron nanoparticles: Sources and multifarious biotechnological applications. *International Journal of Biological Macromolecules, 253*, 127017.

Kumari, A., & Chauhan, A. K. (2021). Iron nanoparticles as a promising compound for food fortification in iron deficiency anemia: A review. *Journal of Food Science and Technology, 59*, 3319–3335.

Kumari, A., & Chauhan, A. K. (2022). Iron nanoparticles as a promising compound for food fortification in iron deficiency anemia: A review. *Journal of Food Science and Technology, 59*(9), 3319–3335.

Kumari, K. A., Mangatayaru, K. G., & Reddy, G. B. (2023). Fungal and yeast-mediated biosynthesis of metal nanoparticles: Characterization and bio applications. In *Fungal cell factories for sustainable nanomaterials productions and agricultural applications* (pp. 309–336). Elsevier.

Kwon, R. Y., Youn, S. M., & Choi, S. J. (2024). Oral excretion kinetics of food-additive silicon dioxides and their effect on in vivo macrophage activation. *International Journal of Molecular Sciences, 25*(3), 1614.

Lahiri, D., Nag, M., Sheikh, H. I., Sarkar, T., Edinur, H. A., Pati, S., & Ray, R. R. (2021). Microbiologically-synthesized nanoparticles and their role in silencing the biofilm signaling cascade. *Frontiers in Microbiology, 12*, 636588.

Lattuada, M., & Hatton, T. A. (2007). Functionalization of monodisperse magnetic nanoparticles. *Langmuir, 23*, 2158–2168.

Liu, Y., Jin, X., & Chen, Z. (2018). The formation of iron nanoparticles by Eucalyptus leaf extract and used to remove Cr (VI). *Science of the Total Environment, 627*, 470–479.

Lovley, D. R., Stolz, J. F., Nord, G. L., & Phillips, E. J. (1987). Anaerobic production of magnetite by a dissimilatory iron-reducing microorganism. *Nature, 330*(6145), 252–254.

Luo, Y.-B., Yu, Q.-W., Yuan, B.-F., & Feng, Y.-Q. (2012). Fast microextraction of phthalate acid esters from beverage, environmental water and perfume samples by magnetic multi-walled carbon nanotubes. *Talanta, 90*, 123–131.

Mahato, A., Chatterjee, P. N., Sarkar, S., Sen, A. R., Pal, A., Roy, S., & Patra, A. K. (2024). Effects of chemically and green synthesized zinc oxide nanoparticles on shelf life and sensory quality of minced fish (Pangasius hypophthalmus). *Foods, 13*(17), 2810.

Mandal, C., Sunooj, K. V., Bosco, S. J. D., Raghu, S., Reddy, K. K. K., Navaf, M., & George, J. (2024). Synergistic interaction of green-synthesized titanium dioxide nanoparticles and Averrhoa carambola extract in chitosan films. *International Journal of Biological Macromolecules*, 138539.

Martinez-Molina, E. C., Freile-Pelegrin, Y., Ovando-Chacon, S. L., Gutierrez-Miceli, F. A., Ruiz-Cabrera, M. Á., Grajales-Lagunes, A., et al. (2022). Development and characterization of alginate-based edible film from Sargassum fluitans incorporated with silver nanoparticles obtained by green synthesis. *Journal of Food Measurement and Characterization, 16*, 1–11.

Medina Cruz, D., Mi, G., & Webster, T. J. (2018). Synthesis and characterization of biogenic selenium nanoparticles with antimicrobial properties made by Staphylococcus aureus, methicillinresistant Staphylococcus aureus (MRSA), Escherichia coli, and Pseudomonas aeruginosa. *Journal of Biomedical Materials Research Part A, 106*, 1400.

Michael, A., Singh, A., Roy, A., & Islam, M. R. (2022). [Retracted] fungal-and algal-derived synthesis of various nanoparticles and their applications. *Bioinorganic Chemistry and Applications, 2022*(1), 3142674.

Mohseni, M. S., Khalilzadeh, M. A., Mohseni, M., Hargalani, F. Z., Getso, M. I., Raissi, V., & Raiesi, O. (2020). Green synthesis of Ag nanoparticles from pomegranate seeds extract and synthesis of Ag-Starch nanocomposite and characterization of mechanical properties of the films. *Biocatalysis and Agricultural Biotechnology, 25*, 101569.

Moradi, M., Molaei, R., Kousheh, S. A., Guimarães, J. T., & McClements, D. J. (2021). Carbon dots synthesized from microorganisms and food by-products: Active and smart food packaging applications. *Critical Reviews in Food Science and Nutrition, 63*, 1943–1959.

Mughal, B., Zaidi, S. Z. J., Zhang, X., & Hassan, S. U. (2021). Biogenic nanoparticles: Synthesis, characterisation and applications. *Applied Sciences, 11*(6), 2598.

Mukherjee, P., Ahmad, A., Mandal, D., Senapati, S., Sainkar, S. R., Khan, M. I., Parishcha, R., Ajaykumar, P. V., Alam, M., Kumar, R., & Sastry, M. (2001a). Fungus-mediated synthesis of silver nanoparticles and their immobilization in the mycelial matrix: A novel biological approach to nanoparticle synthesis. *Nano Letters, 1*(10), 515–519.

Mukherjee, P., Ahmad, A., Mandal, D., Senapati, S., Sainkar, S. R., Khan, M. I., Ramani, R., Parischa, R., Ajayakumar, P. V., Alam, M., & Sastry, M. (2001b). Bioreduction of $AuCl4-$ ions by the fungus, Verticillium sp. and surface trapping of the gold nanoparticles formed. *Angewandte Chemie, International Edition, 40*(19), 3585–3588.

Mukherjee, P., Senapati, S., Mandal, D., Ahmad, A., Khan, M. I., Kumar, R., & Sastry, M. (2002). Extracellular synthesis of gold nanoparticles by the fungus Fusarium oxysporum. *Chembiochem, 3*(5), 461–463.

Nagda, G., Rai, N., Jaya, S., Bhalothia, C., & Singh, N. A. (2024). Nanoparticles synthesis using extremophilic microbes and their potential agricultural applications. In *Extremophiles for sustainable agriculture and soil health improvement* (pp. 455–483). Springer Nature Switzerland.

Nain, S., Kumari, S., Singh, B., & Singh, N. (2024). Green synthesis, characterization methods, and applications of nanoparticles: An overview. *African Journal of Biomedical Research, 27*(4S), 420–437.

Nehra, P., Chauhan, R., Garg, N., & Verma, K. (2018). Antibacterial and antifungal activity of chitosan coated iron oxide nanoparticles. *British Journal of Biomedical Science, 75*, 13–18.

Nikolic, M. V., Vasiljevic, Z. Z., Auger, S., & Vidic, J. (2021). Metal oxide nanoparticles for safe active and intelligent food packaging. *Trends in Food Science & Technology, 116*, 655–668.

Nurdiani, R., Prihanto, A. A., Firdaus, M., Aini, F. N., Nabilah, F. A., Talib, R. A., & Huda, N. (2024). Physicochemical characteristics of Pangasius sp. skin-gelatin-based-edible film enriched with silver nanoparticles. *F1000Research, 12*, 160.

Olajire, A. A., Kareem, A., & Olaleke, A. (2017). Green synthesis of bimetallic Pt@ Cu nanostructures for catalytic oxidative desulfurization of model oil. *Journal of Nanostructure in Chemistry, 7*(2), 159–170.

Omerović, N., Djisalov, M., Živojević, K., Mladenović, M., Vunduk, J., Milenković, I., et al. (2021). Antimicrobial nanoparticles and biodegradable polymer composites for active food packaging applications. *Comprehensive Reviews in Food Science and Food Safety, 20*(3), 2428–2454.

Otari, S. V., Patil, R. M., Nadaf, N. H., Ghosh, S. J., & Pawar, S. H. (2012). Green biosynthesis of silver nanoparticles from an actinobacteria Rhodococcus sp. *Materials Letters, 72*, 92–94.

Pandohee, J., Timilsena, Y. P., Shankaraswamy, J., & That, L. F. L. N. (2023). Techniques and processes involved in nanoencapsulation of omega-3,-6, and-9 fatty acids. In *Handbook of nanoencapsulation* (pp. 167–188). CRC Press.

Patel, M. P., Bharadia, P. D., Bodiwala, K. B., Mansuri, M. M., & Patel, J. (2022). Green synthesis of nanoparticles in oligonucleotide drug delivery system. In *Gene delivery* (pp. 51–64). CRC Press.

Patel, D., Patel, B., Yadav, V. K., Sudhakar, M. P., Alharbi, S. A., Salmen, S. H., et al. (2024). Silver nanoparticles synthesized from marine algae Spatoglossum asperum: Antioxidant properties and seed germination enhancement. *Journal of Hazardous Materials Advances, 16*, 100478.

Paulkumar, K., Rajeshkumar, S., Gnanajobitha, G., Vanaja, M., Malarkodi, C., & Annadurai, G. (2013). Biosynthesis of silver chloride nanoparticles using Bacillus subtilis MTCC 3053 and assessment of its antifungal activity. *ISRN Nanomaterials, 2013*, 1.

Pechyen, C., Tangnorawich, B., Toommee, S., Marks, R., & Parcharoen, Y. (2024). Green synthesis of metal nanoparticles, characterization, and biosensing applications. *Sensors International, 5*, 100287.

Pektaş, S. Ü., Keskin, M., Bodur, O. C., & Arslan, F. (2024). Green synthesis of silver nanoparticles and designing a new amperometric biosensor to determine glucose levels. *Journal of Food Composition and Analysis, 129*, 106133.

Perfecto, A., Elgy, C., Valsami-Jones, E., Sharp, P., Hilty, F., & Fairweather-Tait, S. (2017). Mechanisms of iron uptake from ferric phosphate nanoparticles in human intestinal Caco-2 cells. *Nutrients, 9*(4), 359.

Philip, D. (2009). Biosynthesis of Au, Ag and Au–Ag nanoparticles using edible mushroom extract. *Spectrochimica Acta Part A, Molecular and Biomolecular Spectroscopy, 73*(2), 374–381.

Pipliya, S., Kumar, S., Gupta, R. K., Das, R. S., Meena, D., Srivastav, P. P., Tiwari, B. K., & Garcia-Vaquero, M. (2025). The future of algal proteins: Innovations in extraction and modifications, functional properties, and sustainable food applications. *Future Foods*, 100549.

Polat, S., Ağçam, E., Dündar, B., & Akyildiz, A. (2019). Nanoparticles in food packaging: Opportunities and challenges. In *Health and safety aspects of food processing technologies* (pp. 577–611). Springer International Publishing.

Pooley, F. D. (1982). Bacteria accumulate silver during leaching of sulphide ore minerals. *Nature, 296*(5858), 642–643.

Prabhu, S., & Poulose, E. K. (2012). Silver nanoparticles: Mechanism of antimicrobial action, synthesis, medical applications and toxicity effects. *International Nano Letters, 2*(1), 32.

Prashanth, K. M. (2023). *Biosynthesis of nanoparticles from actinomycetes and their biomedical applications* (Doctoral dissertation). Kuvempu University.

Qamar, S. U. R., & Ahmad, J. N. (2021). Nanoparticles: Mechanism of biosynthesis using plant extracts, bacteria, fungi, and their applications. *Journal of Molecular Liquids, 334*, 116040.

Ranjan, D. (2023). Nanomaterials originated from microbes for the removal of toxic pollutants from water. In *Environmental applications of microbial nanotechnology* (pp. 347–363). Elsevier.

Rathore, S., & Pandey, A. K. (2024). Applications of nanotechnology in food processing. In *Advanced research methods in food processing technologies* (pp. 401–417). Apple Academic Press.

Razack, S. A., Suresh, A., Sriram, S., Ramakrishnan, G., Sadanandham, S., Veerasamy, M., et al. (2020). Green synthesis of iron oxide nanoparticles using Hibiscus rosa-sinensis for fortifying wheat biscuits. *SN Applied Sciences, 2*(5), 898.

Riahi, Z., Rhim, J.-W., Bagheri, R., Pircheraghi, G., & Lotfali, E. (2022). Carboxymethyl cellulose-based functional film integrated with chitosan-based carbon quantum dots for active food packaging applications. *Progress in Organic Coating, 166*, 106794.

Rizki, I. N., & Klaypradit, W. (2023). Utilization of marine organisms for the green synthesis of silver and gold nanoparticles and their applications: A review. *Sustainable Chemistry and Pharmacy, 31*, 100888.

Roy, S., Kim, H. C., Panicker, P. S., Rhim, J. W., & Kim, J. (2021). Cellulose nanofiber-based nanocomposite films reinforced with zinc oxide nanorods and grapefruit seed extract. *Nanomaterials, 11*(4), 877.

Roychoudhury, A. (2020). Yeast-mediated green synthesis of nanoparticles for biological applications. *Indian Journal of Pharmaceutical and Biological Research, 8*(03), 26–31.

Salama, H. E., Khattab, M. A., Sabaa, M. W., & Aziz, M. S. A. (2024). Xanthan/carboxymethyl cellulose-based edible coatings enriched with greenly synthesized ZnO-NPs for active packaging applications. *Cellulose, 31*(13), 8189–8202.

Salem, S. S. (2022). Baker's yeast-mediated silver nanoparticles: Characterisation and antimicrobial biogenic tool for suppressing pathogenic microbes. *BioNanoScience, 12*(4), 1220–1229.

Salem, M. F., Abd-Elraoof, W. A., Tayel, A. A., Alzuaibr, F. M., & Abonama, O. M. (2022). Antifungal application of biosynthesized selenium nanoparticles with pomegranate peels and nanochitosan as edible coatings for citrus green mold protection. *Journal of Nanobiotechnology, 20*(1), 182.

Sani, A., Hassan, D., Chanihoon, G. Q., Melo Máximo, D. V., & Sánchez-Rodríguez, E. P. (2024). Green chemically synthesized iron oxide nanoparticles–chitosan coatings for enhancing strawberry shelf-life. *Polymers, 16*(23), 3239.

Shafie, N. F. A., Zulkifli, N. S., Ali, R. R., Tarmizi, Z. I., & Faizo, S. M. (2024). Green synthesis of titanium dioxide nanoparticles using extraction of psidium guajava for smart packaging application. *Journal of Research in Nanoscience and Nanotechnology, 11*(1), 16–23.

Shankar, S. S., Ahmad, A., Pasricha, R., & Sastry, M. (2003). Bioreduction of chloroaurate ions by geranium leaves and its endophytic fungus yields gold nanoparticles of different shapes. *Journal of Materials Chemistry, 13*(7), 1822–1826.

Shrifian-Esfahni, A., Salehi, M. T., Nasr-Esfahni, M., & Ekramian, E. (2015). Chitosan-modified superparamgnetic iron oxide nanoparticles: Design, fabrication, characterization and antibacterial activity. *Chemik, 69*, 19–32.

Shrivastava, A., Gupta, R. K., & Srivastav, P. P. (2024). Exploring novel frontiers of advancements in purple yam (Dioscorea alata L.) starch extraction, modification, characterization, applications in food and other industries. *Measurement: Food, 15*, 100196.

Singaravelu, G., Arockiamary, J. S., Kumar, V. G., & Govindaraju, K. (2007). A novel extracellular synthesis of monodisperse gold nanoparticles using marine alga, Sargassum wightii Greville. *Colloids and Surfaces B: Biointerfaces, 57*(1), 97–101.

Singh, T. P., Chatli, M. K., & Sahoo, J. (2015). Development of chitosan based edible films: Process optimization using response surface methodology. *Journal of Food Science and Technology, 52*, 2530–2543.

Souza, V. G. L., & Fernando, A. L. (2016). Nanoparticles in food packaging: Biodegradability and potential migration to food—A review. *Food Packaging and Shelf Life, 8*, 63–70.

Speroni, F., Elviri, L., Careri, M., & Mangia, A. (2010). Magnetic particles functionalized with PAMAM-dendrimers and antibodies: A new system for an ELISA method able to detect Ara H3/4 peanut allergen in foods. In *Proceedings of the analytical and bioanalytical chemistry* (Vol. 397, pp. 3035–3042). Springer: Berlin/Heidelberg, Germany.

Srikhao, N., Kasemsiri, P., Ounkaew, A., Lorwanishpaisarn, N., Okhawilai, M., Pongsa, U., Hiziroglu, S., & Chindaprasirt, P. (2021). Bioactive nanocomposite film based on cassava starch/polyvinyl alcohol containing green synthesized silver nanoparticles. *Journal of Polymers and the Environment, 29*, 672–684.

Suganya, K. U., Govindaraju, K., Kumar, V. G., Dhas, T. S., Karthick, V., Singaravelu, G., & Elanchezhiyan, M. (2015). Blue green alga mediated synthesis of gold nanoparticles and its antibacterial efficacy against Gram positive organisms. *Materials Science and Engineering: C, 47*, 351–356.

Sur, U. K., Ankamwar, B., Karmakar, S., Halder, A., & Das, P. (2018). Green synthesis of silver nanoparticles using the plant extract of Shikakai and Reetha. *Materials Today Proceedings, 5*(1), 2321–2329.

Tsekhmistrenko, S. I., Bityutskyy, V. S., Tsekhmistrenko, O. S., Horalskyi, L. P., Tymoshok, N. O., & Spivak, M. Y. (2020). Bacterial synthesis of nanoparticles: A green approach. *Biosystems Diversity, 28*(1), 9–17.

Ullah, I., Ullah, A., Al-Qaaneh, A., Aman, K., Bekhit, M., Ullah, I., Zengin, G., & Farid, A. (2024a). Green synthesis of silver and copper nanoparticles using Parthenium hysterophorus: Antibacterial, antioxidant, and anticancer properties. *ChemistrySelect, 9*, e202401573.

Ullah, I., Rauf, A., Khalil, A. A., Luqman, M., Islam, M. R., Hemeg, H. A., et al. (2024b). Peganum harmala L. extract-based Gold (Au) and Silver (Ag) nanoparticles (NPs): Green synthesis, characterization, and assessment of antibacterial and antifungal properties. *Food Science & Nutrition, 12*, 4459–4472.

Uzair, B., Liaqat, A., Iqbal, H., Menaa, B., Razzaq, A., Thiripuranathar, G., et al. (2020). Green and cost-effective synthesis of metallic nanoparticles by algae: Safe methods for translational medicine. *Bioengineering, 7*(4), 129.

Vigneshwaran, N., Kathe, A. A., Varadarajan, P. V., Nachane, R. P., & Balasubramanya, R. H. (2006). Biomimetics of silver nanoparticles by white rot fungus, Phaenerochaete chrysosporium. *Colloids and Surfaces B: Biointerfaces, 53*(1), 55–59.

Von Moos, L. M., Schneider, M., Hilty, F. M., Hilbe, M., Arnold, M., Ziegler, N., Mato, D. S., Winkler, H., Tarik, M., Ludwig, C., et al. (2017). Iron phosphate nanoparticles for food fortification: Biological effects in rats and human cell lines. *Nanotoxicology, 11*, 496–506.

Wang, L., Liu, X., Qi, P., Sun, J., Jiang, S., Li, H., Gu, X., & Zhang, S. (2022). Enhancing the thermostability, UV shielding and antimicrobial activity of transparent chitosan film by carbon quantum dots containing N/P. *Carbohydrate Polymers, 278*, 118957.

Wei, S., & Zheng, Q. (2024). Biosynthesis and characterization of zinc sulphide nanoparticles produced by the bacterium Lysinibacillus sp. SH74. *Ceramics International, 50*(2), 2637–2642.

Xu, J., Sun, J., Wang, Y., Sheng, J., Wang, F., & Sun, M. (2014). Application of iron magnetic nanoparticles in protein immobilization. *Molecules, 19*, 11465–11486.

Yashni, G., Al-Gheethi, A. A., Mohamed, R. M. S. R., & Hashim, M. A. (2019). Green synthesis of ZnO nanoparticles by Coriandrum sativum leaf extract: structural and optical properties. *Desalination and Water Treatment, 167*, 245–257.

Yang, K., Jenkins, D. M., & Su, W. W. (2011). Rapid concentration of bacteria using submicron magnetic anion exchangers for improving PCR-based multiplex pathogen detection. *Journal of Microbiological Methods, 86*, 69–77.

Yazdani, Z., Mehrgan, M. S., Khayatzadeh, J., Shekarabi, S. P. H., & Tabrizi, M. H. (2023). Dietary green-synthesized curcumin-mediated zinc oxide nanoparticles promote growth performance, haemato-biochemical profile, antioxidant status, immunity, and carcass quality in Nile tilapia (Oreochromis niloticus). *Aquaculture Reports, 32*, 101717.

Yedoti, V., & Supraja, N. (2024). A review on algal mediated synthesized metallic nanoparticles: An eco-friendly approach for sustainable nanotechnology. *Current Journal of Applied Science and Technology, 43*(6), 1–10.

Yiblet, Y., Abdu, I., & Belew, B. (2024). Comprehensive literature review on metal nanoparticle for enhanced shelf life of mango fruit. *The Scientific World Journal, 2024*(1), 4782328.

Zahiri Oghani, F., Tahvildari, K., & Nozari, M. (2021). Novel antibacterial food packaging based on chitosan loaded ZnO nano particles prepared by green synthesis from Nettle leaf extract. *Journal of Inorganic and Organometallic Polymers and Materials, 31*, 43–54.

Zhao, L., Zhang, M., Wang, H., & Devahastin, S. (2020). Effect of carbon dots in combination with aqueous chitosan solution on shelf life and stability of soy milk. *International Journal of Food Microbiology, 326*, 108650.

Zhao, L., Zhang, M., Mujumdar, A. S., & Wang, H. (2022). Application of carbon dots in food preservation: A critical review for packaging enhancers and food preservatives. *Critical Reviews in Food Science and Nutrition, 63*(24), 6738–6756.

Chapter 11
Environmental and Ecotoxicological Impact

Mohammad Harun-Ur-Rashid, Abu Bin Imran, and Md. Abu Bin Hasan Susan

11.1 Introduction

Nanomaterials are at the vanguard of scientific innovation. Unique physicochemical properties of nanomaterials, which inter alia include high surface-to-volume ratio, tunable reactivity, and distinctive optical and electrical properties, have propelled their advances across a wide range of disciplines. These are indispensable in modern photonics, as demonstrated by stimuli-responsiveness of smart materials (Harun-Ur-Rashid et al., 2009a; Takeoka et al., 2013), tunability of soft photonic crystals by structurally colored gels (Harun-ur-Rashid et al., 2009b; Huda et al., 2010), and achievement of structural color without regard to angle in colloidal amorphous arrays (Harun-Ur-Rashid et al., 2010; Harun-Ur-Rashid & Imran, 2019). Polyrotaxane actuators, a pioneering development in material science, represent another leap in designing functional and adaptive systems (Imran et al., 2019). Nanomaterials have also revolutionized wastewater treatment. Polyphenol oxidases and nanomaterials are promoting sustainable practices for food packaging by enhancing food safety and extending shelf life (Harun-Ur-Rashid et al., 2023a,

M. Harun-Ur-Rashid
Department of Chemistry, International University of Business Agriculture and Technology (IUBAT), Dhaka, Bangladesh
e-mail: mrashid@iubat.edu

A. B. Imran
Department of Chemistry, Bangladesh University of Engineering and Technology, Dhaka, Bangladesh
e-mail: abimran@chem.buet.ac.bd

Md. Abu Bin H. Susan (✉)
Department of Chemistry, University of Dhaka, Dhaka, Bangladesh

Dhaka University Nanotechnology Centre (DUNC), University of Dhaka, Dhaka, Bangladesh
e-mail: susan@du.ac.bd

K. Esquivel Escalante et al. (eds.), *Green Nanomaterials*, Nanotechnology in the Life Sciences, https://doi.org/10.1007/978-3-032-04492-1_11

2024d). In parallel, green polymer nanocomposites are addressing industry needs while prioritizing environmental sustainability (Harun-Ur-Rashid et al., 2023c). Nanotechnology is also transforming the medical and healthcare sectors. Rapid diagnostic methods, such as nano-biosensors for COVID-19 detection, highlight its relevance in addressing global health crises (Harun-Ur-Rashid et al., 2022). Hybrid nanostructures have been pivotal in pioneering modern therapeutic advances (Harun-Ur-Rashid & Jahan, 2024). Furthermore, breakthroughs in vaccinology and bio-inspired nanomaterials for medical devices illustrate how nanotechnology is reshaping clinical applications (Harun-Ur-Rashid et al., 2023d, e). Nanomaterials also play an essential role in sustainable energy systems. Engineered nanomaterials for energy conversion and stretchable hydrogels for energy storage exemplify their potential to revolutionize the energy sector (Rezaul Karim et al., 2020; Islam et al., 2024a, b, c). Fire-resistant polymeric foams and nanocomposites continue to enhance the safety and versatility of advanced materials (Harun-Ur-Rashid et al., 2023b).

Despite these benefits, nanomaterials also raise concerns regarding toxicity and environmental impact. Risk assessments of pesticides and heavy metals in agriculture emphasize the significance of assessing potential hazards to the environment and human health (Huda et al., 2024). Furthermore, molecular techniques offer innovative approaches for environmental monitoring and safety (Selim et al., 2024; Harun-Ur-Rashid et al., 2023f, 2024a, b; Hassan et al., 2024). In material engineering, nanotechnology has driven trends that focus on performance enhancement, sustainability, and resilience (Karim et al., 2023; Harun-Ur-Rashid & Imran, 2024; Harun-Ur-Rashid et al., 2024c; Jahan et al., 2024). There have been increasing concerns about their lifecycle, including environmental release and impacts on ecosystems and non-target species. Bioaccumulation and ecological imbalances can harm plants, microorganisms, and aquatic organisms. Nanoparticles like TiO_2 and silver induce oxidative stress in aquatic species and disrupt soil microbial communities vital for nutrient cycling. These risks, along with potential human health impacts via air and water exposure, demand further research to mitigate adverse effects. This chapter systematically examines the environmental and ecotoxicological impact of nanomaterials, with a focus on their fate, transport, and interactions across soil, water, and air while emphasizing their physicochemical properties and environmental transformations. It evaluates ecological risks to non-target species in terrestrial and aquatic ecosystems, highlighting exposure pathways and biological impacts. Strategies for risk assessment and mitigation are discussed within regulatory frameworks to promote sustainable use. This chapter also explores emerging green nanotechnology trends, advocating for circular economy principles to minimize environmental impacts and enhance resource efficiency. By addressing these aspects, it provides insights into the dual role of nanomaterials as solutions and challenges, guiding strategies that balance their benefits with ecological and human health considerations.

11.2 Fate and Transport of Nanomaterials in the Environment

11.2.1 Sources and Pathways

Nanomaterials are introduced into the environment through diverse sources, primarily from industrial, consumer, agricultural, and medical applications. Industrial processes, including manufacturing and waste disposal, release significant amounts of engineered nanomaterials (ENMs). Nanomaterials in electronics, coatings, and construction materials contribute to unintentional discharges during synthesis, application, or disposal. Similarly, consumer products such as cosmetics, textiles, and cleaning agents release nanoparticles through routine use and eventual degradation (Donia & Carbone, 2019). In agriculture, nanotechnology is increasingly applied in fertilizers, pesticides, and soil amendments to enhance crop productivity. However, these nano-agrochemicals can accumulate in soil and water and result in long-term ecological effects (Okeke et al., 2022). Medical applications, including drug delivery systems, diagnostic agents, and antimicrobial coatings, also contribute to environmental loads. Discharge from healthcare facilities and improper disposal of nano-enabled medical devices introduces nanomaterials into wastewater.

11.2.2 Environmental Transformations

Nanomaterials undergo various transformations in the environment that affect their behavior, reactivity, and potential toxicity. These transformations can be categorized as physical, chemical, or biological.

11.2.2.1 Physical Transformations

Physical changes such as aggregation, sedimentation, and dissolution significantly influence the mobility and bioavailability of nanoparticles. Aggregation often occurs in aquatic systems where they interact with dissolved organic matter, forming larger clusters that may settle to the bottom (Peng et al., 2017). The size, density, and surface properties of the nanomaterials all affect sedimentation rates.

11.2.2.2 Chemical Transformations

Chemical processes, including oxidation, reduction, and surface functionalization, alter properties of nanomaterials. For example, silver nanoparticles can oxidize to silver ions, which are more bioavailable and potentially more toxic (Zhang et al.,

2018). Such transformations depend on environmental factors like ionic strength, pH, and the presence of oxidizing agents.

11.2.2.3 Biological Transformations

Microorganisms play an important role in the transformation of nanomaterials. Microbial reduction of metal oxides can alter their chemical structure, changing their toxicity and reactivity (Uddin et al., 2020). Biological transformations often involve enzyme-mediated processes that enhance the degradation or stabilization of nanoparticles. Environmental factors such as pH, temperature, and salinity further modulate these transformations. Acidic conditions may accelerate dissolution, while temperature variations influence reaction kinetics. These factors collectively determine the fate of nanomaterials in natural systems.

11.2.3 Mobility and Persistence

Nanomaterials exhibit diverse transport behaviors across air, soil, and water systems, depending on their physicochemical properties and environmental interactions.

11.2.3.1 Transport in Air

Airborne nanoparticles, released from industrial emissions or combustion processes, can travel long distances before deposition. Their mobility is influenced by aerodynamic diameter, atmospheric conditions, and interactions with other airborne particles.

11.2.3.2 Transport in Soil

In terrestrial environments, the interaction of nanomaterials with organic matter and soil particles affects their mobility. Strong adsorption onto soil particles limits transport but increases persistence. Conversely, weak interactions enhance leaching into groundwater.

11.2.3.3 Transport in Water

In aquatic systems, the mobility of nanomaterials is governed by hydrodynamic flow, aggregation behavior, and interactions with dissolved substances. For example, coated nanoparticles exhibit enhanced stability and mobility compared to their uncoated counterparts.

Factors influencing the persistence of nanomaterials include size, shape, surface coating, and functionalization. Smaller particles with hydrophilic coatings tend to remain suspended longer, whereas hydrophobic coatings promote aggregation and sedimentation. Understanding these factors is critical for predicting the long-term environmental presence of nanomaterials.

11.2.4 Modeling Fate and Transport

Predicting the environmental behavior of nanomaterials requires robust computational models that integrate experimental data and theoretical approaches. These models help simulate the dispersion, accumulation, and transformation of nanomaterials across environmental compartments.

11.2.4.1 Mechanistic Models

Mechanistic models incorporate physical and chemical processes, such as aggregation, sedimentation, and dissolution, to predict the transport and fate of nanomaterials (Malekjani & Jafari, 2021).

11.2.4.2 Multimedia Models

Multimedia models evaluate the distribution of nanomaterials across air, water, soil, and biota. These models use partitioning coefficients and mass balance equations to estimate concentrations in different environmental compartments (Tang et al., 2024).

11.2.4.3 Reactive Transport Models

Reactive transport models consider both transport dynamics and chemical reactions, providing insights into how transformations affect mobility and persistence. For example, models for silver nanoparticles include oxidation and dissolution processes (Katzourakis & Chrysikopoulos, 2021).

11.2.4.4 Machine Learning Approaches

Machine learning approaches help predict nanomaterial behaviors under varying environmental conditions. These integrate large datasets to identify patterns and predict outcomes more accurately (Ahmad et al., 2021). By combining experimental studies with computational predictions, these models facilitate a comprehensive

Table 11.1 Nanomaterial transformations in different environmental compartments

Nanomaterial type	Environmental transformation processes	Persistence level	Major outcomes
Silver nanoparticles	Oxidation, dissolution, aggregation	Medium to high	Formation of silver ions, increased bioavailability, potential toxicity to aquatic organisms
TiO_2	Photocatalytic activity, aggregation, sedimentation	High	Enhanced reactivity under UV light, reduced mobility in aquatic environments
Carbon nanotubes	Functionalization, aggregation, microbial degradation	Medium	Enhanced sedimentation, partial degradation by microorganisms, reduced bioavailability
ZnO nanoparticles	Dissolution, photooxidation, surface passivation	Low to medium	Release of zinc ions, increased toxicity, reduced reactivity due to surface passivation
Iron oxide nanoparticles	Oxidation, reduction, aggregation	High	Formation of stable aggregates, reduced mobility, long-term persistence in soil and aquatic systems
Gold nanoparticles	Aggregation, surface functionalization, adsorption of organic matter	Very high	Strong persistence due to inert nature, limited bioavailability, reduced toxicity
Polymer-coated nanoparticles	Surface degradation, photolysis, microbial interaction	Low to medium	Breakdown of coatings, altered transport properties, potential release of core material

understanding of fate, enabling informed risk assessments of nanomaterials and guiding sustainable nanotechnology practices.

Table 11.1 analyzes nanomaterial transformations in different environmental compartments.

11.3 Toxicological Effects on Non-target Species

11.3.1 Aquatic Ecosystems

The presence of nanomaterials in aquatic environments significantly impacts diverse organisms, including algae, zooplankton, fish, and benthic species. Algae, as primary producers, are highly susceptible to nanomaterials that interfere with photosynthesis and growth. For instance, silver nanoparticles inhibit algal photosystem II activity, reducing oxygen production and impacting higher trophic levels (Lekamge et al., 2020).

Zooplankton, crucial in aquatic food webs, face bioaccumulation and oxidative stress originating from exposure to nanomaterials such as TiO_2 nanoparticles. These impacts can impair reproduction and mobility, cascading through the ecosystem (Kumar & Pramanick, 2022). Fish, often exposed to waterborne nanoparticles, experience gill damage, behavioral changes, and bioaccumulation, especially in organs like the liver and kidneys.

Benthic organisms, which interact with sediment-bound nanomaterials, face risks from the aggregation and deposition of nanoparticles. For instance, studies on benthic invertebrates in marine environments have highlighted reduced metabolic activity and reproductive success due to long-term exposure (Rajala et al., 2016).

Case studies across freshwater and marine systems provide further evidence. In a controlled study, lakes treated with engineered nanomaterials exhibited disruptions in algal bloom cycles and reduced zooplankton populations, demonstrating ecosystem-wide effects. Similarly, marine environments near industrial discharge zones showed altered benthic community structures due to nanoparticle contamination (Esposito et al., 2021).

11.3.2 Terrestrial Ecosystems

Nanomaterials impact terrestrial ecosystems through their interactions with soil microorganisms, plants, and invertebrates. Soil microorganisms, critical for nutrient cycling, are particularly sensitive to metallic nanoparticles like zinc oxide (ZnO), which can inhibit enzymatic activity and alter microbial diversity (Rajput et al., 2018). These disruptions can reduce soil fertility and affect plant growth.

Plants absorb nanoparticles through roots, leading to bioaccumulation in edible tissues. While some nanoparticles promote growth, others cause toxicity. For instance, silver nanoparticles disrupt root elongation and nutrient uptake in agricultural crops.

Terrestrial invertebrates, such as earthworms and insects, are critical to soil health and pollination. Exposure to nanoparticles like carbon nanotubes has been shown to impair reproduction and alter burrowing behaviors in earthworms, affecting soil aeration and organic matter decomposition (Hlavkova et al., 2020).

Bioaccumulation of nanomaterials in terrestrial food webs raises concerns about trophic transfer and long-term ecological impacts. Studies have demonstrated that nanoparticles ingested by primary consumers, such as insects, can move up the food chain, impacting higher-order predators (Wu et al., 2023).

11.3.3 Human Health Concerns

Human exposure to nanomaterials occurs through environmental pathways, including air, water, and food. Inhalation is a significant route of exposure, especially for airborne nanoparticles released during industrial processes or from consumer products. Nanoparticles such as carbon black and silica can penetrate deep into the lungs, causing inflammation and oxidative stress (Forest et al., 2021).

Oral exposure through contaminated water and food introduces nanoparticles into the gastrointestinal tract. TiO_2 nanoparticles can disrupt gut microbiota balance and increase intestinal permeability (Chen et al., 2019). Dermal exposure, while less common, occurs through contact with nanoparticle-containing cosmetics and textiles, potentially leading to skin irritation and transdermal absorption (Crosera et al., 2009).

Emerging concerns include long-term exposure effects, such as chronic inflammation and potential carcinogenicity, which remain insufficiently understood. Current research also highlights gaps in understanding cumulative exposure risks from multiple pathways and mixed nanomaterials (Yokel & MacPhail, 2011).

11.3.4 Mechanisms of Toxicity

Nanomaterials may become toxic following several mechanisms, like genotoxicity, cellular damage, and oxidative stress. Oxidative stress arises from reactive oxygen species (ROS) generated from interactions of nanoparticles with cells. Silver and ZnO nanoparticles cause mitochondrial dysfunction by inducing excessive ROS production (Sengul & Asmatulu, 2020).

Genotoxicity occurs when nanoparticles interact with DNA, leading to strand breaks, mutations, or chromosomal aberrations. Metallic nanoparticles such as cobalt and nickel have been implicated in disrupting cellular replication processes and inducing genotoxic effects in mammalian cells (Datkhile et al., 2021).

Nanoparticles can also disrupt cellular membranes and interact with macromolecules. They are embedded in lipid bilayers, altering membrane integrity and affecting ion channels (Abarca-Cabrera et al., 2021). These interactions can impair cellular signaling and lead to apoptosis or necrosis, depending on the exposure level and nanoparticle type. Table 11.2 systematically organizes information to provide a clear overview of nanomaterial-induced toxicity.

11.4 Risk Assessment and Management

11.4.1 Risk Assessment Frameworks

Effective risk assessment for nanomaterials requires a systematic approach comprising identification of hazards, assessment of exposure, and characterization of risks. Identification of hazards evaluates the intrinsic properties, such as the potential of nanomaterials to cause toxicity through oxidative stress, genotoxicity, or bioaccumulation. For example, metallic nanoparticles like silver (Park et al., 2009) and ZnO (Yu et al., 2013) are known to induce ROS, to cause cellular damage in aquatic and terrestrial organisms. Assessment of exposure quantifies the degree and routes of exposure in environmental and human systems. This involves evaluating air, water, soil, and food contamination pathways, considering variables like particle size, coating, and aggregation behavior (Scott-Fordsmand et al., 2017).

Table 11.2 Summary of nanomaterial-induced toxicity across ecosystems

Ecosystem type	Affected species	Mechanism of toxicity	Reported impacts
Aquatic	Algae	Oxidative stress, disruption of photosystem II	Reduced oxygen production, inhibited growth
	Zooplankton	Bioaccumulation, oxidative stress	Impaired mobility, reduced reproductive success
	Fish	Gill damage, bioaccumulation, genotoxicity	Liver and kidney damage, behavioral changes
	Benthic organisms	Sedimentation, reduced metabolic activity	Altered community structures, reduced reproductive success
Terrestrial	Soil microorganisms	Enzymatic inhibition, oxidative stress	Altered microbial diversity, reduced nutrient cycling
	Plants	Disruption of nutrient uptake, genotoxicity	Stunted root elongation, reduced crop yields
	Invertebrates (e.g., earthworms, insects)	Bioaccumulation, reproductive impairment	Altered burrowing behavior, reduced soil aeration
Human health	Lung cells (via inhalation)	Oxidative stress, cellular inflammation	Chronic respiratory inflammation, potential fibrosis
	Gut microbiota (via ingestion)	Disruption of microbiota balance, genotoxicity	Increased intestinal permeability, altered digestion
	Skin (via dermal exposure)	Membrane disruption, oxidative stress	Skin irritation, potential transdermal absorption

Characterization of risks integrates estimates of possible negative consequences by combining exposure and hazard data. This is critical for formulating safety thresholds and informing regulatory measures (Hassellöv et al., 2008). International guidelines and regulations provide foundational frameworks for risk assessment. For instance, the Working Party on Manufactured Nanomaterials of OECD and the regulation of Registration, Evaluation, Authorization, and Restriction of Chemicals (REACH) of the European Union emphasize the need for standard testing methodologies tailored to nanomaterials (Doak et al., 2023). Despite progress, these frameworks often face challenges in capturing the unique behaviors of nanomaterials, such as their dynamic transformations and nanoscale reactivity.

11.4.2 Risk Management Strategies

Mitigating the risks associated with nanomaterials involves both technical and policy-driven approaches. Innovative strategies in nanomaterial design can minimize environmental and health risks. For instance, surface modifications, such as biocompatible coatings, reduce toxicity while enhancing stability in environmental systems (Schubert & Chanana, 2018). Other approaches include developing biodegradable nanomaterials that degrade into harmless byproducts under natural conditions.

Safe disposal practices are essential to prevent uncontrolled release. Techniques like advanced filtration, chemical neutralization, and incineration have been employed to capture and deactivate nanomaterials from industrial and medical waste streams (Dhall et al., 2024). Governments and organizations worldwide have introduced policies to curb nanomaterial emissions. These include emission limits for industrial processes, stricter disposal regulations, and mandatory environmental impact assessments for new nanomaterial products (Carlander & Skentelbery, 2021). Integrating such policies into national and international environmental protection frameworks is critical for reducing environmental release.

11.4.3 Challenges and Uncertainties

Despite advancements, several challenges and uncertainties persist in risk assessment and management. It is yet unclear how nanomaterials may affect ecosystems and human health in the long run. For instance, the cumulative effects of chronic low-dose exposure or the potential for nanomaterials to evolve into more toxic forms over time are under-researched. The diversity of nanomaterials poses significant difficulties in standardizing risk assessment methods. Nanomaterials differ widely in size, shape, surface chemistry, and functionalization, requiring tailored evaluation techniques. Current guidelines often lack the flexibility to address these variations (Kroll et al., 2009). Bridging gaps between scientific research and policy implementation is another significant hurdle. The rapidly evolving field of nanotechnology often outpaces regulatory updates, leading to inconsistencies in enforcement (Roco, 2005).

11.4.4 Future Directions in Risk Management

Future risk management strategies must prioritize sustainability and public engagement while leveraging advances in science and technology. Green nanotechnology principles, such as designing for environmental compatibility and reducing resource usage, should guide the development of nanomaterials. For example, using renewable raw materials and minimizing waste during synthesis can significantly reduce the environmental footprint. Effective communication strategies are essential to foster public understanding of nanotechnology risks and benefits. Public awareness campaigns, educational programs, and transparent risk communication can build trust and ensure informed decision-making. Dynamic regulatory frameworks that incorporate real-time monitoring and predictive modeling are vital for managing emerging risks. Machine learning tools and artificial intelligence can enhance the prediction of nanomaterial behaviors and inform adaptive regulations.

Table 11.3 systematically compares global frameworks for assessing nanomaterial risks, emphasizing their relevance, strengths, and challenges.

Table 11.3 Comparison of risk assessment frameworks for nanomaterials

Framework/guideline	Key features	Strengths	Limitations	References
Guidelines of OECD for testing of chemicals	Focused on standardized testing methods for nanomaterial toxicity and ecotoxicity	Provides internationally accepted protocols, facilitating regulatory harmonization	Limited flexibility to address emerging nanomaterials with novel properties	Doak et al. (2012)
REACH	Requires comprehensive data submission for nanomaterial safety, including exposure and hazard assessments	Promotes extensive data collection and ensures accountability for chemical producers	High compliance costs and complexity, particularly for small and medium-sized enterprises	Schwirn et al. (2014)
ISO/TR 13121: nanotechnologies—nanomaterial risk evaluation	Focuses on qualitative risk evaluation and provides a structured framework for risk communication	Encourages risk-benefit analysis and integration of stakeholder perspectives	Lacks quantitative approaches for detailed risk characterization	https://www.iso.org/obp/ui/#iso:std:iso:tr:13121:ed-1:v1:en (accessed on 18 November 2024)
US EPA framework for nanomaterials	Includes hazard identification, dose-response assessment, and exposure assessment for nanomaterials	Addresses environmental and human health impacts with a strong focus on exposure pathways	Limited scope for international applicability; data gaps in long-term ecological effects	https://www.epa.gov/chemical-research/research-nanomaterials (accessed on 18 November 2024)
Safe by design approach	Incorporates risk reduction at the design stage, emphasizing green chemistry and sustainability	Proactive strategy to minimize risks during development; aligns with green nanotechnology principles	Difficult to implement for existing nanomaterials; requires significant upfront investment in research	Corsi et al. (2023)

11.5 Conclusion

The intersection of nanotechnology with environmental sustainability presents a dual challenge and opportunity. This chapter has elucidated the multifaceted impacts of nanomaterials on ecological and human systems, emphasizing their transformative potential and inherent risks. Through an in-depth analysis of their environmental fate, transport mechanisms, and ecotoxicological effects, we identified critical pathways where nanomaterials interact with air, water, soil, and biological systems.

These interactions often yield significant benefits but may also result in bioaccumulation, toxicity, and ecological imbalances. This chapter also highlighted the urgent need for robust risk assessment frameworks and sustainable design principles. By adopting green nanotechnology approaches and aligning with circular economy strategies, it is possible to mitigate environmental hazards while maximizing the utility of these innovative materials. Advanced modeling techniques, regulatory frameworks, and stakeholder engagement are indispensable for bridging the gaps in our current understanding and ensuring responsible integration of nanotechnology into various sectors. Moving forward, prioritizing interdisciplinary research and proactive policymaking will be crucial for balancing the benefits of nanomaterials with their ecological and human health considerations. This balanced approach will not only enhance the efficacy and safety of nanotechnology applications but also open the door to a more resilient and sustainable future.

Acknowledgments The authors acknowledge the Bose Centre for Advanced Research in Natural Sciences for financial support on a related Project.

Conflict of Interest There are no conflicts of interest to declare.

References

Abarca-Cabrera, L., Fraga-García, P., & Berensmeier, S. (2021). Bio-nano interactions: Binding proteins, polysaccharides, lipids and nucleic acids onto magnetic nanoparticles. *Biomaterials Research, 25*(1), 12. https://doi.org/10.1186/s40824-021-00212-y

Ahmad, F., Mahmood, A., & Muhmood, T. (2021). Machine learning-integrated omics for the risk and safety assessment of nanomaterials. *Biomaterials Science, 9*(5), 1598–1608. https://doi.org/10.1039/D0BM01672A

Carlander, D., & Skentelbery, C. (2021). EU regulations and nanotechnology innovation. In J. R. Lead, S. H. Doak, & M. J. D. Clift (Eds.), *Nanotoxicology in humans and the environment* (pp. 229–248). Springer International Publishing. https://doi.org/10.1007/978-3-030-79808-6_8

Chen, Z., Han, S., Zhou, D., Zhou, S., & Jia, G. (2019). Effects of oral exposure to titanium dioxide nanoparticles on gut microbiota and gut-associated metabolism in vivo. *Nanoscale, 11*(46), 22398–22412. https://doi.org/10.1039/C9NR07580A

Corsi, I., Venditti, I., Trotta, F., & Punta, C. (2023). Environmental safety of nanotechnologies: The eco-design of manufactured nanomaterials for environmental remediation. *Science of the Total Environment, 864*, 161181. https://doi.org/10.1016/j.scitotenv.2022.161181

Crosera, M., Bovenzi, M., Maina, G., Adami, G., Zanette, C., Florio, C., & Filon Larese, F. (2009). Nanoparticle dermal absorption and toxicity: A review of the literature. *International Archives of Occupational and Environmental Health, 82*(9), 1043–1055. https://doi.org/10.1007/s00420-009-0458-x

Datkhile, K. D., Patil, S. R., Durgawale, P. P., Patil, M. N., Hinge, D. D., Jagdale, N. J., Deshmukh, V. N., & More, A. L. (2021). Biogenic synthesis of gold nanoparticles using Argemone mexicana L. and their cytotoxic and genotoxic effects on human colon cancer cell line (HCT-15). *Journal of Genetic Engineering and Biotechnology, 19*(1), 9. https://doi.org/10.1186/s43141-020-00113-y

Dhall, S., Nigam, A., Harshavardhan, M., Mukherjee, A., & Srivastava, P. (2024). A comprehensive overview of methods involved in nanomaterial production and waste disposal from research labs and industries and existing regulatory guidelines for handling engineered nanomaterials. *Environmental Chemistry and Ecotoxicology, 6*, 269–282. https://doi.org/10.1016/j.enceco.2024.06.002

Doak, S. H., Manshian, B., Jenkins, G. J. S., & Singh, N. (2012). In vitro genotoxicity testing strategy for nanomaterials and the adaptation of current OECD guidelines. *Mutation Research/Genetic Toxicology and Environmental Mutagenesis, 745*(1), 104–111. https://doi.org/10.1016/j.mrgentox.2011.09.013

Doak, S. H., Andreoli, C., Burgum, M. J., Chaudhry, Q., Bleeker, E. A. J., Bossa, C., Domenech, J., Drobne, D., Fessard, V., Jeliazkova, N., Longhin, E., Rundén-Pran, E., Stępnik, M., El Yamani, N., Catalán, J., & Dusinska, M. (2023). Current status and future challenges of genotoxicity OECD Test Guidelines for nanomaterials: A workshop report. *Mutagenesis, 38*(4), 183–191. https://doi.org/10.1093/mutage/gead017

Donia, D. T., & Carbone, M. (2019). Fate of the nanoparticles in environmental cycles. *International journal of Environmental Science and Technology, 16*(1), 583–600. https://doi.org/10.1007/s13762-018-1960-z

Esposito, M. C., Corsi, I., Russo, G. L., Punta, C., Tosti, E., & Gallo, A. (2021). The era of nanomaterials: A safe solution or a risk for marine environmental pollution? *Biomolecules, 11*(3). https://doi.org/10.3390/biom11030441

Forest, V., Pourchez, J., Pélissier, C., Audignon Durand, S., Vergnon, J.-M., & Fontana, L. (2021). Relationship between occupational exposure to airborne nanoparticles, nanoparticle lung burden and lung diseases. *Toxics, 9*(9). https://doi.org/10.3390/toxics9090204

Harun-Ur-Rashid, M., & Imran, A. B. (2019). Superabsorbent hydrogels from carboxymethyl cellulose. In I. H. Mondal (Ed.), *Carboxymethyl cellulose, vol 1. Synthesis and characterization* (Biochemistry research trends) (pp. 159–182). Nova Science Publishers.

Harun-Ur-Rashid, M., & Imran, A. B. (2024). Emerging trends in engineering polymers: A paradigm shift in material engineering. *Recent Progress in Materials, 06*(03), 024. https://doi.org/10.21926/rpm.2403024

Harun-Ur-Rashid, M., & Jahan, I. (2024). Innovative hybrid nanostructures: Pioneering advances in modern therapy. *Frontiers in Nanotechnology, 6*. https://doi.org/10.3389/fnano.2024.1458894

Harun-Ur-Rashid, M., Imran, A. B., Seki, T., Takeoka, Y., Ishii, M., & Nakamura, H. (2009a). Template synthesis for stimuli-responsive angle independent structural colored smart materials. *Transactions of the Materials Research Society of Japan, 34*(2), 333–337. https://doi.org/10.14723/tmrsj.34.333

Harun-ur-Rashid, M., Seki, T., & Takeoka, Y. (2009b). Structural colored gels for tunable soft photonic crystals. *The Chemical Record, 9*(2), 87–105. https://doi.org/10.1002/tcr.20169

Harun-Ur-Rashid, M., Bin Imran, A., Seki, T., Ishii, M., Nakamura, H., & Takeoka, Y. (2010). Angle-independent structural color in colloidal amorphous arrays. *ChemPhysChem, 11*(3), 579–583. https://doi.org/10.1002/cphc.200900869

Harun-Ur-Rashid, M., Foyez, T., Jahan, I., Pal, K., & Imran, A. B. (2022). Rapid diagnosis of COVID-19 via nano-biosensor-implemented biomedical utilization: A systematic review. *RSC Advances, 12*(15), 9445–9465. https://doi.org/10.1039/D2RA01293Fb

Harun-Ur-Rashid, M., Imran, A. B., & Susan, M. (2023a). Prospective nanomaterials for food packaging and safety. In N. B. Singh, M. A. B. H. Susan, & R. G. Chaudhary (Eds.), *Emerging applications of nanomaterials* (Vol. 141, pp. 327–352). Materials Research Foundations. https://doi.org/10.21741/9781644902295-13

Harun-Ur-Rashid, M., Imran, A. B., & Susan, M. A. B. H. (2023b). Fire-resistant polymeric foams and their applications. In *Polymeric foams: Applications of polymeric foams (volume 2)* (ACS symposium series) (Vol. 1440, pp. 97–121). American Chemical Society. https://doi.org/10.1021/bk-2023-1440.ch005

Harun-Ur-Rashid, M., Imran, A. B., & Susan, M. A. B. H. (2023c). Green polymer nanocomposites in automotive and packaging industries. *Current Pharmaceutical Biotechnology, 24*(1), 145–163. https://doi.org/10.2174/1389201023666220506111027

Harun-Ur-Rashid, M., Jahan, I., Foyez, T., & Imran, A. B. (2023d). Bio-inspired nanomaterials for micro/nanodevices: A new era in biomedical applications. *Micromachines, 14*(9). https://doi.org/10.3390/mi14091786

Harun-Ur-Rashid, M., Jahan, I., & Imran, A. B. (2023e). Recent advancement of nanostructured materials for clinical challenges in vaccinology. In K. Pal (Ed.), *Nanovaccinology: Clinical application of nanostructured materials research to translational medicine* (pp. 135–160). Springer International Publishing. https://doi.org/10.1007/978-3-031-35395-6_8

Harun-Ur-Rashid, M., Pal, K., & Imran, A. B. (2023f). Hybrid nanocomposite fabrication of nanocatalyst with enhanced and stable photocatalytic activity. *Topics in Catalysis*. https://doi.org/10.1007/s11244-023-01809-4

Harun-Ur-Rashid, M., Imran, A. B., & Susan, M. A. B. H. (2024a). Emerging 3D printed polymers and composites for water quality preservation. In *Reference module in materials science and materials engineering*. Elsevier. https://doi.org/10.1016/B978-0-323-95486-0.00010-7

Harun-Ur-Rashid, M., Imran, A. B., & Susan, M. A. B. H. (2024b). Magnetic polymer nanocomposites for removal of dyes and metals from wastewater. In *Reference module in materials science and materials engineering*. Elsevier. https://doi.org/10.1016/B978-0-323-95486-0.00008-9

Harun-Ur-Rashid, M., Jahan, I., Islam, M. J., Kumer, A., Huda, M. N., Imran, A. B., Gouadria, S., & Alsalhi, S. A. (2024c). Global advances and smart innovations in supramolecular polymers. *Journal of Molecular Structure, 1304*, 137665. https://doi.org/10.1016/j.molstruc.2024.137665

Harun-Ur-Rashid, M., Naidu Krishna, S. B., Golla, N., & Bin Imran, A. (2024d). Polyphenol oxidases: The future toward global sustainability. In P. Verma, K. Agrawal, & M. P. Shah (Eds.), *Function, wastewater remediation, and biosensors* (pp. 309–344). De Gruyter. https://doi.org/10.1515/9783111033525-012

Hassan, M. K., Karim, M. T., Biswas, P., Howlader, D., Harun-Ur-Rashid, M., & Kumer, A. (2024). Computational investigation for tetragonal crystals of Zn(GaS2)2, Zn(GaSe2)2, and Zn(GaTe2)2 photocatalysts for wastewater treatment: First principle approaches. *Advanced Journal of Chemistry-Section B: Natural Products and Medical Chemistry*, 46–66. https://doi.org/10.48309/ajcb.2024.421434.1204

Hassellöv, M., Readman, J. W., Ranville, J. F., & Tiede, K. (2008). Nanoparticle analysis and characterization methodologies in environmental risk assessment of engineered nanoparticles. *Ecotoxicology, 17*(5), 344–361. https://doi.org/10.1007/s10646-008-0225-x

Hlavkova, D., Beklova, M., Kopel, P., & Havelkova, B. (2020). Effects of silver nanoparticles and ions exposure on the soil invertebrates Folsomia candida and Enchytraeus crypticus. *Bulletin of Environmental Contamination and Toxicology, 105*(2), 244–249. https://doi.org/10.1007/s00128-020-02909-7

Huda, M. N., Seki, T., Suzuki, H., Kabir, A. N. M. H., Rashid, M. H., & Takeoka, Y. (2010). Characteristics of high-density poly(N-isopropylacrylamide) (PNIPA) brushes on silicon surface by atom transfer radical polymerization. *Transactions of the Materials Research Society of Japan, 35*(4), 845–848. https://doi.org/10.14723/tmrsj.35.845

Huda, M. N., Harun-Ur-Rashid, M., Hosen, A., Akter, M., Islam, M. M., Emon, S. Z., Rahman, A., Jashim, Z. B., Shahrukh, S., & Ismail, M. (2024). A potential toxicological risk assessment of heavy metals and pesticides in irrigated rice cultivars near industrial areas of Dhaka, Bangladesh. *Environmental Monitoring and Assessment, 196*(9), 794. https://doi.org/10.1007/s10661-024-12927-1

Imran, A. B., Harun-Ur-Rashid, M., & Takeoka, Y. (2019). Polyrotaxane actuators. In K. Asaka & H. Okuzaki (Eds.), *Soft actuators: Materials, modeling, applications, and future perspectives* (pp. 81–147). Springer Singapore. https://doi.org/10.1007/978-981-13-6850-9_6

Islam, D. M. J., Sayara, N., Talukder, T., & Harun-Ur-Rashid, M. (2024a). Trapping light, revealing properties: Laser trapping as a powerful tool for photoluminescence spectroscopy. *Chemical Physics Impact, 9*, 100764. https://doi.org/10.1016/j.chphi.2024.100764

Islam, M. J., Pal, K., Abdullah, M. M., & Harun-Ur-Rashid, M. (2024b). Threshold bandgap of lead halide hybrid perovskites for making stable single crystal by NIR laser trapping. *Journal of Molecular Structure, 1301*, 137322. https://doi.org/10.1016/j.molstruc.2023.137322

Islam, M. J., Pal, K., Harun-Ur-Rashid, M., Kumar, A., Hossain, M. F., Asthana, N., Rajendran, S., Vallinayagam, S., & Malik, A. (2024c). Optical maneuvering of photofunctioning hybrid perovskite for future photonics potential application. *Journal of Molecular Liquids, 408*, 125343. https://doi.org/10.1016/j.molliq.2024.125343

Jahan, I., Islam, M. A., Harun-Ur-Rashid, M., & Sultana, G. N. N. (2024). Cancer prevention at the microscopic level with the potent power of micronutrients. *Heliyon, 10*(21). https://doi.org/10.1016/j.heliyon.2024.e39680

Karim, M. R., Harun-Ur-Rashid, M., & Imran, A. B. (2023). Effect of sizes of vinyl modified narrow-dispersed silica cross-linker on the mechanical properties of acrylamide based hydrogel. *Scientific Reports, 13*(1), 5089. https://doi.org/10.1038/s41598-023-32185-4

Katzourakis, V. E., & Chrysikopoulos, C. V. (2021). Modeling the transport of aggregating nanoparticles in porous media. *Water Resources Research, 57*(1), e2020WR027946. https://doi.org/10.1029/2020WR027946

Kroll, A., Pillukat, M. H., Hahn, D., & Schnekenburger, J. (2009). Current in vitro methods in nanoparticle risk assessment: Limitations and challenges. *European Journal of Pharmaceutics and Biopharmaceutics, 72*(2), 370–377. https://doi.org/10.1016/j.ejpb.2008.08.009

Kumar, V., & Pramanick, B. (2022). Impact of nanoparticles on the dynamics of a Crowley–Martin type phytoplankton–zooplankton interaction model. *Results in Control and Optimization, 8*, 100139. https://doi.org/10.1016/j.rico.2022.100139

Lekamge, S., Miranda, A. F., Abraham, A., Ball, A. S., Shukla, R., & Nugegoda, D. (2020). The toxicity of coated silver nanoparticles to the alga Raphidocelis subcapitata. *SN Applied Sciences, 2*(4), 596. https://doi.org/10.1007/s42452-020-2430-z

Malekjani, N., & Jafari, S. M. (2021). Modeling the release of food bioactive ingredients from carriers/nanocarriers by the empirical, semiempirical, and mechanistic models. *Comprehensive Reviews in Food Science and Food Safety, 20*(1), 3–47. https://doi.org/10.1111/1541-4337.12660

Okeke, E. S., Ezeorba, T. P. C., Mao, G., Chen, Y., Feng, W., & Wu, X. (2022). Nano-enabled agrochemicals/materials: Potential human health impact, risk assessment, management strategies and future prospects. *Environmental Pollution, 295*, 118722. https://doi.org/10.1016/j.envpol.2021.118722

Park, H.-J., Kim, J. Y., Kim, J., Lee, J.-H., Hahn, J.-S., Gu, M. B., & Yoon, J. (2009). Silver-ion-mediated reactive oxygen species generation affecting bactericidal activity. *Water Research, 43*(4), 1027–1032. https://doi.org/10.1016/j.watres.2008.12.002

Peng, C., Zhang, W., Gao, H., Li, Y., Tong, X., Li, K., Zhu, X., Wang, Y., & Chen, Y. (2017). Behavior and potential impacts of metal-based engineered nanoparticles in aquatic environments. *Nanomaterials, 7*(1). https://doi.org/10.3390/nano7010021

Rajala, J. E., Mäenpää, K., Vehniäinen, E.-R., Väisänen, A., Scott-Fordsmand, J. J., Akkanen, J., & Kukkonen, J. V. K. (2016). Toxicity testing of silver nanoparticles in artificial and natural sediments using the benthic organism Lumbriculus variegatus. *Archives of Environmental Contamination and Toxicology, 71*(3), 405–414. https://doi.org/10.1007/s00244-016-0294-4

Rajput, V. D., Minkina, T., Sushkova, S., Tsitsuashvili, V., Mandzhieva, S., Gorovtsov, A., Nevidomskyaya, D., & Gromakova, N. (2018). Effect of nanoparticles on crops and soil microbial communities. *Journal of Soils and Sediments, 18*(6), 2179–2187. https://doi.org/10.1007/s11368-017-1793-2

Rezaul Karim, M., Harun-Ur-Rashid, M., & Imran, A. B. (2020). Highly stretchable hydrogel using vinyl modified narrow dispersed silica particles as cross-linker. *ChemistrySelect, 5*(34), 10556–10561. https://doi.org/10.1002/slct.202003044

Roco, M. C. (2005). The emergence and policy implications of converging new technologies integrated from the nanoscale. *Journal of Nanoparticle Research, 7*(2), 129–143. https://doi.org/10.1007/s11051-005-3733-0

Schubert, J., & Chanana, M. (2018). Coating matters: Review on colloidal stability of nanoparticles with biocompatible coatings in biological media, living cells and organisms. *Current Medicinal Chemistry, 25*(35), 4553–4586. https://doi.org/10.2174/0929867325666180601101859

Schwirn, K., Tietjen, L., & Beer, I. (2014). Why are nanomaterials different and how can they be appropriately regulated under REACH? *Environmental Sciences Europe, 26*(1), 4. https://doi.org/10.1186/2190-4715-26-4

Scott-Fordsmand, J. J., Peijnenburg, W. J. G. M., Semenzin, E., Nowack, B., Hunt, N., Hristozov, D., Marcomini, A., Irfan, M. A., Jiménez, A. S., Landsiedel, R., Tran, L., Oomen, A. G., Bos, P. M. J., & Hund-Rinke, K. (2017). Environmental risk assessment strategy for nanomaterials. *International Journal of Environmental Research and Public Health, 14*(10). https://doi.org/10.3390/ijerph14101251

Selim, S., Harun-Ur-Rashid, M., Jahan, I., & Mostafa, E. M. (2024). Culture-independent molecular techniques for bacterial detection in bivalves. *Egyptian Journal of Aquatic Research*. https://doi.org/10.1016/j.ejar.2024.11.002

Sengul, A. B., & Asmatulu, E. (2020). Toxicity of metal and metal oxide nanoparticles: A review. *Environmental Chemistry Letters, 18*(5), 1659–1683. https://doi.org/10.1007/s10311-020-01033-6

Takeoka, Y., Yoshioka, S., Teshima, M., Takano, A., Harun-Ur-Rashid, M., & Seki, T. (2013). Structurally coloured secondary particles composed of black and white colloidal particles. *Scientific Reports, 3*(1), 2371. https://doi.org/10.1038/srep02371

Tang, W., Zhang, X., Hong, H., Chen, J., Zhao, Q., & Wu, F. (2024). Computational nanotoxicology models for environmental risk assessment of engineered nanomaterials. *Nanomaterials, 14*(2). https://doi.org/10.3390/nano14020155

Uddin, M. N., Desai, F., & Asmatulu, E. (2020). Engineered nanomaterials in the environment: Bioaccumulation, biomagnification and biotransformation. *Environmental Chemistry Letters, 18*(4), 1073–1083. https://doi.org/10.1007/s10311-019-00947-0

Wu, J., Sun, J., Bosker, T., Vijver, M. G., & Peijnenburg, W. J. G. M. (2023). Toxicokinetics and particle number-based trophic transfer of a metallic nanoparticle mixture in a terrestrial food chain. *Environmental Science & Technology, 57*(7), 2792–2803. https://doi.org/10.1021/acs.est.2c07660

Yokel, R. A., & MacPhail, R. C. (2011). Engineered nanomaterials: Exposures, hazards, and risk prevention. *Journal of Occupational Medicine and Toxicology, 6*(1), 7. https://doi.org/10.1186/1745-6673-6-7

Yu, K.-N., Yoon, T.-J., Minai-Tehrani, A., Kim, J.-E., Park, S. J., Jeong, M. S., Ha, S.-W., Lee, J.-K., Kim, J. S., & Cho, M.-H. (2013). Zinc oxide nanoparticle induced autophagic cell death and mitochondrial damage via reactive oxygen species generation. *Toxicology In Vitro, 27*(4), 1187–1195. https://doi.org/10.1016/j.tiv.2013.02.010

Zhang, W., Xiao, B., & Fang, T. (2018). Chemical transformation of silver nanoparticles in aquatic environments: Mechanism, morphology and toxicity. *Chemosphere, 191*, 324–334. https://doi.org/10.1016/j.chemosphere.2017.10.016

Chapter 12
Ensuring Safe Nanotechnology: A Biosafety and Regulatory Perspective

Himani Pandey and Rakesh K. Pandey

12.1 Introduction

Materials with dimensions ranging from 1 to 100 nm possess distinct physical, chemical, and biological properties due to their high surface area-to-volume ratio, differentiating them from bulk materials (Alberto et al., 2022; Baig, 2023). These properties make nanomaterials necessary in various applications, including medicine, electronics, energy, and environmental remediation. Nanomaterials represent a significant advancement in materials science with wide-ranging applications. However, their unique properties also demand careful consideration of their safety and environmental impact (Yang et al., 2021). Ongoing research and regulations are crucial to harnessing the full potential of nanomaterials while minimizing the risks. Due to their small size and high reactivity, nanomaterials can potentially be toxic to living organisms (Zhang et al., 2014; Lewinski et al., 2008). They can penetrate biological membranes, interact with cellular components, and induce oxidative stress (Donaldson et al., 2013; Kus-Liskiewicz et al., 2021; Ma et al., 2024; Summer et al., 2024). For example, specific metal oxide nanoparticles can generate reactive oxygen species that damage cells (Summer et al., 2024). They can also interact with environmental systems in complex ways, affecting soil, water, and air quality (Zhao et al., 2021). Their long-term environmental impacts, including bioaccumulation and persistence, are active research areas (Wahab et al., 2024). Understanding and mitigating the toxic effects of nanomaterials is crucial for their safe application, particularly in medicine and consumer products (Elsaesser & Howard, 2012). There

H. Pandey
Department of Biotechnology, Mahatma Gandhi Central University, Motihari, Bihar, India

R. K. Pandey (✉)
Department of Chemistry, Mahatma Gandhi Central University, Motihari, Bihar, India
e-mail: rakeshpandey@mgcub.ac.in

K. Esquivel Escalante et al. (eds.), *Green Nanomaterials*, Nanotechnology in the Life Sciences, https://doi.org/10.1007/978-3-032-04492-1_12

is a need for standardized testing and regulation to ensure the safe development and use of nanomaterials.

Nanomaterials can be classified by origin, structure, and composition. They include natural particles like inorganic ash, soot, sulfur, and mineral particles found in the environment and nanoparticles produced by bacteria and yeasts (Griffin et al., 2017). Engineered nanomaterials, such as gold and silver nanoparticles, are the most common. Structurally, nanomaterials are described as zero-dimensional (e.g., quantum dots), one-dimensional (e.g., nanotubes), two-dimensional (e.g., graphene), or three-dimensional (e.g., nanoporous materials) (Byakodi et al., 2022). Their composition varies and includes metallic types, such as gold and silver (Bouche et al., 2020); semiconductors, like TiO_2 and ZnO (Cho et al., 2011; Jovanović, 2015); polymeric materials, such as polyaniline, poly(lactide-co-glycolide) or PLGA, lipid-based structures, liposomes, and porphyrin aggregates based nanostructures (Eker et al., 2024; Voigt et al., 2014; Pandey et al., 2009, 2012). These nanomaterials demonstrate remarkable size-dependent effects, such as quantum confinement, which enhances their electronic, optical, and magnetic properties. Furthermore, their high surface reactivity makes them highly suitable for applications in catalysis, drug delivery, and sensors (Pavlicek et al., 2023; Soldado et al., 2022; Qureshi et al., 2023). Their optical properties, like SPR, improve imaging and photothermal therapies (Choi et al., 2020; Li et al., 2022).

Chemically, nanoparticles excel in catalytic activity and surface functionalization for targeted drug delivery (Seke et al., 2024). Mechanically, they enhance strength, flexibility, and toughness like carbon nanotubes (Ferrier & Honeychurch, 2021). Magnetic and electrical properties, including superparamagnetism and high conductivity, find use in biomedical and electronic devices (Mehak et al., 2023; Yang et al., 2021). Additionally, their thermal stability and conductivity aid heat management (Ferrier & Honeychurch, 2021; Coetzee et al., 2020). Gold, silver, and other biodegradable nanoparticles are suitable for medical applications. Silver nanoparticles, for example, provide antimicrobial properties for coatings and medical devices (Sakthi Devi et al., 2022; Pasparakis, 2022; Hancharova et al., 2024). Table 12.1 summarizes common nanomaterials' classification, applications, and potential risks.

This chapter mainly explores the emerging concerns of nanotechnology surrounding their safety, environmental impact, and regulatory considerations. Figure 12.1a shows the applications and hazards associated with nanotechnology. By presenting an integrative perspective, this chapter serves as a foundation for understanding the potential and limitations of nanomaterials in driving innovation across industries. Numerous studies have identified deficiencies in existing legislation and regulatory guidelines, emphasizing the necessity for comprehensive, universal safety protocols to further the application of metal nanomaterials in healthcare (Kus-Liskiewicz et al., 2021).

Table 12.1 Classification of various types of nanomaterials, their applications, and potential risks

Nanomaterial type	Example	Common applications	Potential risks
Carbon-based			
Fullerenes (Ershova et al., 2016)	C60, C70	Drug delivery, antioxidants, solar cells	Possible cytotoxicity, DNA damage
Carbon nanotubes (CNTs) (Donaldson et al., 2013)	Single-walled (SWCNT), multi-walled (MWCNT)	Composite materials, conductive films, biosensors	Pulmonary fibrosis, asbestos-like effects, biopersistence in lungs
Graphene and graphene oxide (Gurcan et al., 2020)	Monolayer graphene, reduced graphene oxide (rGO)	Flexible electronics, batteries, supercapacitors, biomedical coatings	Oxidative stress, potential genotoxicity, cellular uptake issues
Carbon quantum dots (Das et al., 2018)	Nitrogen-doped carbon dots	Bioimaging, drug delivery, photodynamic therapy	Generally considered low risk, but potential bioaccumulation needs evaluation
Metal-based			
Silver nanoparticles (AgNPs) (Zhao et al., 2021)	Colloidal silver, coated AgNPs	Antibacterial coatings, wound dressings, textiles	Toxicity to aquatic organisms, disrupts bacterial flora
Gold nanoparticles (AuNPs) (Cai et al., 2022)	Citrate-capped AuNPs	Cancer therapy, biosensors, catalysis	Generally low toxicity but may cause immune responses
Copper nanoparticles (Ingle et al., 2014)	CuNPs, CuO	Conductive inks, antimicrobial agents	High cytotoxicity, can induce oxidative stress
Aluminum nanoparticles (Kermani et al., 2018)	AlNPs	Aerospace coatings, fuel additives	Potential neurotoxicity, environmental persistence
Metal oxide-based			
Titanium dioxide (TiO_2) (Jovanović, 2015)	Nanorods, spherical TiO_2	Sunscreens, photocatalysts, self-cleaning surfaces	Phototoxicity, dermal absorption risk
Zinc oxide (ZnO) (Cho et al., 2011)	ZnO nanoparticles	UV-blocking materials, paints, cosmetics	Potential lung inflammation, harmful to marine life
Iron oxide (Fe_2O_3, Fe_3O_4) (Valdiglesias et al., 2015)	Magnetite, hematite	MRI contrast agents, targeted drug delivery	Generally biocompatible, minimal toxicity reported
Tungsten oxide (WO_3) (Chinde & Grover, 2017)	WO_3 nanowires	Electrochromic devices, energy storage	Possible oxidative stress in cells
Polymeric			
Dendrimers (Jain et al., 2010)	PAMAM, PPI	Drug delivery, gene therapy, antimicrobial coatings	Cytotoxicity due to surface charge, biodegradability concerns
Polymeric nanoparticles (Bobo et al., 2016; Voigt	PLA, PLGA, PCL, PANI	Controlled drug release, optical properties,	Safe if biodegradable, but long-term effects of

(continued)

Table 12.1 (continued)

Nanomaterial type	Example	Common applications	Potential risks
et al., 2014; Pandey et al., 2009)		nanocoatings, biodegradable implants	synthetic polymers are unknown
Nanocapsules (Le Roux et al., 2017)	Lipid-based nanocapsules	Targeted drug delivery, encapsulation of nutraceuticals	Lipid degradation products can cause inflammation

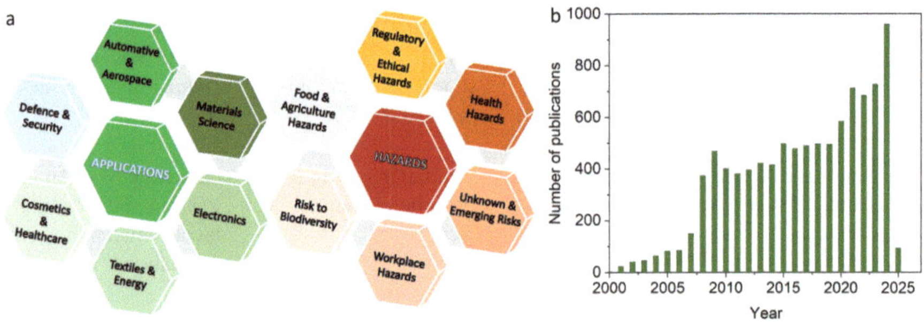

Fig. 12.1 (**a**) A schematic figure showing the applications and hazards associated with nanotechnology. (**b**) Articles appearing in PubMed search for the keyword "nanotechnology safety" during 2000–2025. (Constructed by Rakesh K. Pandey)

12.2 History, Development of Regulations

Even before understanding nanotechnology, ancient civilizations unknowingly used nanomaterials. For instance, artisans used gold and silver nanoparticles in stained glass windows in medieval Europe, giving them vibrant colors (Szunerits & Boukherroub, 2012). Key advancements came with the invention of tools like electron microscopes (SEM, TEM), scanning tunneling microscope (STM), and atomic force microscope (AFM), allowing scientists to observe and manipulate materials at the atomic level (Miyahara et al., 2017). As the use of nanomaterials grew, concerns about safety and environmental impact led governments to start considering regulations. Initial guidelines were broad, given the limited understanding of nanomaterial risks. Agencies like the US Environmental Protection Agency (EPA) and the European Union (EU) commenced more specific regulations, targeting industries like cosmetics and electronics (https://www.epa.gov/tsca-inventory).

For instance, the EU's REACH (Registration, Evaluation, Authorization, and Restriction of Chemicals) regulation required companies to provide safety data for nanomaterials (https://euon.echa.europa.eu/general-information). Today, regulatory bodies like the International Organization for Standardization (ISO) and the US Food and Drug Administration (FDA) have developed specialized guidelines for the safe development, testing, and handling of nanomaterials. Regulations continue to evolve as research provides new insights into nanomaterial safety and environmental effects. A PubMed search using the keyword "nanotechnology safety" reveals a consistent increase in the number of publications from 2000 to 2025 (Fig. 12.1b). This trend highlights the ongoing research efforts to implement precautionary measures to safely use nanomaterials.

12.3 Current Regulations on Nanomaterials

The regulation of nanomaterials varies significantly across regions, reflecting differing priorities, resources, and approaches to risk assessment. In the USA, it is regulated under the Toxic Substances Control Act (TSCA), which is administered by the EPA and focuses on regulating chemicals used in industrial, commercial, and consumer applications, excluding those specifically regulated by other laws, like food, drugs, and cosmetics (https://www.epa.gov/tsca-inventory). The FDA oversees nanomaterials in food, cosmetics, and medical products (Bobo et al., 2016). EPA requires the reporting of nanomaterials use, manufacture, and disposal. Its mission is to protect human health and the environment by ensuring clean air, land, and water, reducing environmental risks, and enforcing laws fairly. It achieves this by developing regulations, funding grants, conducting research, forming partnerships, educating the public, and providing information to promote environmental management and safety (https://www.epa.gov/aboutepa/our-mission-and-what-we-do). The European Chemicals Agency (ECHA) collaborates with EU member states, the European Commission, NGOs, industry associations, and international organizations like the OECD (Organization for Economic Co-operation and Development) to implement EU chemicals legislation for nanomaterials (https://www.oecd.org). Nanomaterials are regulated under REACH and CLP (Classification, Labelling, and Packaging) as substances, following the provisions of both regulations. In 2011, the European Commission provided a recommended definition for nanomaterials to ensure consistency across European regulations, including REACH and CLP (https://echa.europa.eu/regulations/nanomaterials; Armstrong et al., 2021). EU's REACH entered into force on 1st June 2007. It sets procedures for collecting and assessing substance properties and hazards (https://echa.europa.eu/regulations/reach/understanding-reach). Another law related to nanomaterials under ECHA is the Biocidal Products Regulation (BPR) (https://echa.europa.eu/regulations/biocidal-products-regulation/understanding-bpr). It has very specific provisions for nanomaterials. The provisions apply to products and substances meeting BPR criteria based on the Commission's nanomaterials definition recommendation.

In India, there are no dedicated nanotechnology safety laws yet, and nanomaterials fall under general environmental and chemical safety regulations. However, The Bureau of Indian Standards (BIS) has drafted guidelines for nanotechnology (https://www.bis.gov.in/the-bureau/about-bis). It was established under the BIS Act 2016 and is India's national standards body. It ensures standardization, quality certification, and reliability. This benefits the economy by promoting safety, reducing health risks, supporting exports, and managing product variety through certification and testing.

Worksafe Australia oversees the regulation of nanomaterials within occupational environments, including carbon nanotubes (Koiranen et al., 2017; https://www.worksafe.com.au/). Concurrently, the Australian Competition and Consumer Commission (ACCC) regulates consumer products containing nanomaterials that are not governed by other regulatory entities.

In Canada, Canadian departments and agencies, including Health Canada and Environment and Climate Change Canada, focus on the safe use and risk assessment of nanotechnology and nanoplastics (Allan et al., 2021). Their regulatory approach emphasizes risk mitigation, product inventory, and biological impact assessment, considering lifecycle and exposure. The Chemical Management Plan (CMP) oversees nanomaterial regulation, aligning with OECD safety recommendations and various Canadian laws. Canada collaborates closely with the USA on nanomaterial regulation through the Canada-US Regulatory Cooperation Council, developing a common classification scheme.

The regulatory frameworks for nanomaterials in Japan, China, and South Korea are evolving to address the unique challenges posed by these materials. Japan employs a Chemical Substances Control Law (CSCL), which regulates nanomaterials as part of its broader chemical management framework (Matsumoto et al., 2015). Japan's Ministry of Health, Labor, and Welfare (MHLW) has compiled hazard information on existing chemicals to evaluate potential initial risks to human health. It requires notification for new chemical substances, including nanomaterials, and evaluates them for environmental and health risks. Exemptions may apply to low-volume and small-scale uses. In China, the New Chemical Substance Notification (NCSN), which the Ministry of Ecology and Environment oversees, requires the pre-market registration of new chemical substances, including nanomaterials (Ministry of Ecology and Environment, 2021). It governs the environmental management of new chemical substances in China, covering production, import, and use activities. It excludes unprocessed exports, specific products (e.g., food, cosmetics), and most radioactive substances unless used industrially or designed to release chemicals. China regulates nanomaterials at both the substance and product levels. Depending on a product's position in the supply chain, various orders from the state administration of work safety may apply. The regulatory system is complex due to the involvement of multiple ministries in chemical management. In South Korea, the K-REACH (Act on the Registration and Evaluation of Chemicals) mandates registration, hazard evaluation, and risk assessment of chemical substances, including nanomaterials (Ha et al., 2016). It includes provisions for annual reporting and registration of new and existing substances. It also encourage collaboration with industries to manage nano-safety effectively. The government plans to control domestic policies related to nano-safety management, indicating a proactive approach to assessing and managing risks associated with nanomaterials.

These regulations aim to ensure nanomaterials safe use, development, and management while addressing their unique characteristics. It is important to note that these regulatory frameworks continually grow as each country adapts to the rapid advancements in nanotechnology. Staying informed about the latest regulations and guidelines is crucial for compliance and safe handling of nanomaterials.

12.4 Comparative Analysis: Asian Versus Western Regulatory Frameworks

Scope and Coverage Asian frameworks, such as CSCL, NCSN, and K-REACH, often integrate nanomaterials into broader chemical management laws, whereas Western frameworks (e.g., EU REACH) include specific guidelines for nanomaterials. The USA (via TSCA) has provisions for nanoscale substances, but lacks specific legislation focused solely on nanomaterials (Park & Yeo, 2016).

Risk Assessment Western counterparts emphasize detailed risk assessment, lifecycle analysis, and sustainability for nanomaterials. Asian frameworks are increasingly adopting these practices, but may focus more on environmental hazards rather than comprehensive lifecycle management (Schwirn et al., 2020).

Compliance and Reporting Both Asian and Western systems require pre-market notifications and registrations, but Western frameworks often have more stringent data requirements, such as mandatory toxicological studies. Asia's approach is more flexible, with simplified notification pathways for smaller-scale applications.

Innovation and Regulation Balance Western frameworks often prioritize precautionary principles, which can sometimes slow down innovation. Asian frameworks tend to balance regulation and promote industrial growth, which may lead to faster adoption but potentially safety gaps.

In conclusion, while Asian and Western regulatory frameworks share similarities in managing nanomaterials, differences lie in their emphasis on lifecycle analysis, data requirements, and the balance between innovation and regulation. Strengthening international collaboration and harmonizing standards can bridge gaps, ensuring that safety and innovation thrive in nanotechnology development (Isigonis et al., 2020).

12.5 Safety Protocols and Guidelines for Nanomaterials

Due to the unique properties and potential risks of nanomaterials, safety protocols and guidelines are essential to protect workers, consumers, and the environment (Powell et al., 2022). Some standard safety practices to follow when working with nanomaterials include the following:

(i) Workplace Safety Risk assessment in the workplace involves identifying hazards by assessing the physicochemical properties of nanomaterials, such as size, shape, and toxicity, and considering exposure pathways like inhalation, dermal contact, ingestion, or environmental release. Engineering controls are essential, including using ventilation systems like fume hoods or exhaust systems to reduce airborne nanoparticles and enclosing processes to prevent the release of nanomaterials into the workspace. Personal protective equipment (PPE) such as masks or respirators designed for nanomaterials, gloves, and protective clothing to prevent skin contact, and safety goggles to guard against accidental splashes are critical. Workplace practices should minimize aerosol-generating activities and include regular cleaning of work areas using wet cleaning or HEPA-filtered vacuums, avoiding dry sweeping, which can spread nanomaterials. Workers must be trained in the safe handling of nanomaterials and emergency procedures (Tsai et al., 2012).

(ii) Product Safety For product safety, it is important to label products containing nanomaterials clearly, indicating their presence and potential risks. Testing should be conducted to determine the toxicity and environmental impact tailored to the nanoscale properties of materials. Nanomaterials should be stored in sealed, labelled containers in well-ventilated areas away from ignition sources (Woskie, 2010).

(iii) Environmental Safety Implementing measures to reduce the release of nanomaterials into the air, water, and soil during production and disposal is essential for environmental safety. Nanomaterial-containing waste should be treated as hazardous unless proven otherwise, using appropriate waste management techniques such as incineration or containment.

(iv) Research and Development Guidelines Research and development should follow international safety standards from organizations like ISO, which provide guidelines for terminology, risk assessment, and the safe use of nanotechnology, and the OECD, which offers testing strategies for evaluating the safety of manufactured nanomaterials. Detailed documentation of all experiments involving nanomaterials, including quantities used, exposure controls, and incident reports, must be maintained (https://www.oecd.org).

(v) Consumer Safety Consumer safety regulations for products containing nanomaterials, such as cosmetics, food additives, and medicines, should ensure compliance with standards and educate consumers about the potential benefits and risks of these materials. Emergency protocols must establish clear procedures for handling spills, exposure incidents, and contamination, ensuring easy access to safety data sheets specific to the nanomaterials used in the facility and providing first aid training and emergency response plans tailored to nanoparticle exposure. Challenges in working with nanomaterials include the lack of universally accepted standards for testing and safety, rapid technological advancements outpacing regulatory updates, and difficulty in measuring exposure to nanomaterials due to their small size and unique properties. By adhering to these protocols and guidelines, the safe use of nanomaterials can be ensured while minimizing risks to health and the environment (Alberto et al., 2022).

(vi) Emergency Protocols Establish clear procedures for handling spills, exposure incidents, and contamination. Ensure easy access to safety data sheets specific to nanomaterials used in the facility. Provide first aid training and emergency response plans tailored to nanoparticle exposure. There could be various challenges that one can face while working on nanomaterials due to the lack of universally accepted standards for testing and safety. Rapid technological advancements outpace regulatory updates and difficulty in measuring exposure to nanomaterials due to their small size and unique properties. By adhering to these protocols and guidelines, the safe use of nanomaterials can be ensured while minimizing risks to health and the environment (Alberto et al., 2022).

12.6 International Standards and Compliance

ISO is a global group comprising national standards organizations (ISO member bodies) that regulate and standardize nanomaterials globally. By developing and publishing technical standards, ISO ensures uniformity and safety in nanomaterials' production, use, and disposal (https://www.iso.org; Caputo et al., 2024). ISO works closely with the International Electrotechnical Commission (IEC) on standards related to electrical and electronic systems. Some key ISO Standards Relevant to Nanomaterials are the following.

(i) ISO/TR 11360 (2010) Establishes a framework for classifying and categorizing nanomaterials based on their size, structure, and properties. Helps researchers and regulators maintain consistency in terminology (https://www.iso.org).

(ii) ISO 10808 (2010) Focuses on characterizing workplace exposure to airborne nanomaterials. Provides measurement protocols to protect workers from potential health hazards (https://www.iso.org).

(iii) ISO 29701 (2010) Provides guidelines for detecting endotoxins in nanomaterials using the Limulus amebocyte lysate (LAL) test. It ensures the safety of nanomaterials used in in-vitro systems by identifying potential endotoxin contamination, which can cause harmful biological reactions. This addresses challenges posed by nanomaterials unique properties and is essential for biomedical, pharmaceutical, and research applications (https://www.iso.org).

(iv) ISO/TS 80004 Series Defines a standardized vocabulary for nanotechnology, covering terms related to nanoparticles, nanoscale materials, and their functionalities. It ensures clear communication, supports innovation, and helps global collaboration and regulation in nanotechnology (https://www.iso.org).

(v) ISO 19007 (2018) Provides guidance on using the MTS (micronucleus test) in vitro to evaluate the potential genotoxicity of nanomaterials. It focuses on identifying DNA damage caused by exposure to nanomaterials in cell cultures, helping assess their safety for health and environmental applications. This standard is essential for understanding the risks associated with nanotechnology products. ISO standards are essential for harmonizing nanotechnology practices globally, facilitating trade, ensuring safety, and promoting innovation (https://www.iso.org).

12.7 Organization for Economic Co-operation and Development (OECD) Guidelines

OECD is an important player in developing regulatory frameworks for nanomaterials, focusing on environmental health and safety (https://www.oecd.org). It is an international organization established in 1961 to promote policies that improve the economic and social well-being of people worldwide (Revised Introduction to the OECD Guidelines for Testing of Chemicals, Section 3, 2006). OECD provides a platform for governments to work together to address economic growth, trade, education, and sustainable development challenges. It currently has 38 member countries, most of which are high-income economies. OECD work on nanomaterials focuses on ensuring their safe development and use by providing guidelines, fostering research, and facilitating international cooperation to address their potential risks (Revised Introduction to the OECD Guidelines for Testing of Chemicals, Section 3, 2006).

12.8 Some Key Initiatives and Contributions

(i) Testing Guidelines OECD has developed extensive testing protocols for assessing nanomaterials' physicochemical, toxicological, and ecotoxicological properties, for example, OECD TG 318 (2017) focuses on evaluating the dispersion stability of nanomaterials in environmental conditions (https://www.oecd.org).

(ii) Working Party on Manufactured Nanomaterials (WPMN) Launched in 2006, this initiative facilitates international collaboration on nanomaterial safety. It conducts intergovernmental case studies on materials like titanium dioxide and carbon nanotubes (https://ised-isde.canada.ca/site/canadian-nanotechnologies-industry/en/oecd-working-party-manufactured-nanomaterials).

(iii) Harmonization and Data Sharing OECD promotes data sharing among member countries to avoid duplication in testing efforts. Their standardized approaches improve the reliability of risk assessments globally.

(iv) ProSafe and Safe-by-Design (SbD) Initiatives It describes a structured methodology for integrating safety considerations into industrial innovation processes, focusing on risk assessment and uncertainty reduction. The Key components include workflows for SbD implementation, a Safety Dossier, a Safety Profile, and a harmonized inventory of SbD protocols and procedures (Hohener et al., 2016).

12.9 Challenges in Global Harmonization

Achieving global regulatory harmonization for nanomaterials presents several challenges and barriers due to the unique properties of nanomaterials, the diverse regulatory environments across countries, and the complexity of the science involved. For example, different countries have varying approaches to nanomaterial regulation. The EU's REACH regulation emphasizes comprehensive lifecycle analysis, while countries like China and the USA focus on pre-market risk assessments (Dolez, 2015).

In addition to the above challenge, the regulation of nanomaterials encounters several other obstacles. One major issue is the absence of a universally accepted definition, as some regulations classify them based on size (1–100 nm), while others consider properties like reactivity or functionality (Bleeker et al., 2013; Ma et al., 2024). Resource gaps further complicate regulatory enforcement, as developing nations often lack the infrastructure, expertise, and funding to implement robust frameworks (Elzein, 2024). Additionally, the rapid evolution of nanotechnology makes it difficult for regulations to remain relevant, with emerging materials like 2D materials and quantum dots not fitting existing categories (Baig, 2023; Yirak & Chaujar, 2024). Industry concerns also play a role, as companies may resist harmonization efforts due to increased compliance costs and potential delays in product approvals (Ma et al., 2024). Furthermore, data-sharing barriers arise from confidentiality concerns over exclusive information, hindering global regulatory collaboration. Lastly, public perception significantly impacts regulatory implementation, as misinformation or a lack of awareness about nanotechnology can lead to resistance, making it challenging to enforce necessary policies effectively (Arabeyyat et al., 2022; Festus-Ikhuoria et al., 2023).

12.10 Conclusion

The various properties of nanomaterials make them potentially hazardous if not handled carefully. Improper handling can lead to occupational exposure, environmental contamination, and health risks, including respiratory, dermal, or systemic effects (Ma et al., 2024). Therefore, strict safety protocols, including using protective equipment, proper ventilation systems, and following regulatory guidelines, are essential when working with nanomaterials. This ensures their safe use in various applications, from medicine to electronics. The regulation of nanomaterials is essential to ensure their safe development and use while promoting innovation. However, regional differences in regulatory frameworks present significant challenges to achieving global harmonization. Companies face increased costs and complexity in adhering to multiple regulatory standards, while inconsistent risk assessments can lead to environmental and health safety gaps. Moreover, scientific uncertainties worsen these challenges, including data gaps on nanomaterials' long-term impacts. The EU, the USA, and Asian countries have established their own approaches; the absence of standardized definitions, testing methods, and risk assessment protocols creates barriers to international trade and compliance. Collaborative efforts, such as those led by the OECD, are crucial in bridging these gaps, promoting consistent guidelines, and addressing scientific uncertainties.

A globally harmonized regulatory framework will enhance safety and environmental protection, supporting the sustainable growth of nanotechnology industries worldwide and enabling uniform international trade (Caputo et al., 2024; Cho et al., 2011). To achieve this, it is crucial to provide developing nations with the necessary resources and expertise for the equitable implementation of standards. Additionally, regulatory frameworks must evolve alongside advancements in nanotechnology to effectively address new risks. Clear communication between regulators, industries, and the public is essential for building trust and promoting compliance. Future efforts should focus on developing unified international guidelines and advancements in testing methodologies tailored to nanomaterials to address their unique properties and potential risks. Continuous monitoring and adaptive regulatory frameworks will be crucial for keeping pace with the rapid evolution of nanotechnology.

Acknowledgments This work has been financially supported by the WOS-A research grant of DST (DST/WOS-A/LS-50/2020) to Dr. Himani Pandey.

Conflict of Interest The authors have no conflicts of interest to declare that are relevant to the content of this chapter.

References

Alberto, A. R., Matos, C., Carmona-Aparicio, G., & Iten, M. (2022). Nanomaterials, a new challenge in the workplace. *Advances in Experimental Medicine and Biology, 1357*, 379–402.

Allan, J., Belz, S., Hoeveler, A., Hugas, M., Okuda, H., Patri, A., Rauscher, H., Silva, P., Slikker, W., Sokull-Kluettgen, B., Tong, W., & Anklam, E. (2021). Regulatory landscape of nanotechnology and nanoplastics from a global perspective. *Regulatory Toxicology and Pharmacology, 122*, 104885.

Arabeyyat, Z. H., Jamaliah, M. M., & Khalaf, M. A. (2022). Public awareness of nanotechnology and its implications for health in Jordan. *Sustainability, 14*, 5786.

Armstrong, V., Karyakina, N. A., Nordheim, E., Arnold, I., & Krewski, D. (2021). Overview of REACH: Issues involved in the registration of metals. *Neurotoxicology, 83*, 186–198.

Baig, N. (2023). Two-dimensional nanomaterials: A critical review of recent progress, properties, applications, and future directions. *Composites Part A: Applied Science and Manufacturing, 165*, 107362.

Bleeker, E. A., de Jong, W. H., Geertsma, R. E., Groenewold, M., Heugens, E. H., Koers-Jacquemijns, M., van de Meent, D., Popma, J. R., Rietveld, A. G., Wijnhoven, S. W., Cassee, F. R., & Oomen, A. G. (2013). Considerations on the EU definition of a nanomaterial: Science to support policy making. *Regulatory Toxicology and Pharmacology, 65*, 119–125.

Bobo, D., Robinson, K. J., Islam, J., Thurecht, K. J., & Corrie, S. R. (2016). Nanoparticle-based medicines: A review of FDA-approved materials and clinical trials to date. *Pharmaceutical Research, 33*, 2373–2387.

Bouche, M., Hsu, J. C., Dong, Y. C., Kim, J., Taing, K., & Cormode, D. P. (2020). Recent advances in molecular imaging with gold nanoparticles. *Bioconjugate Chemistry, 31*, 303–314.

Byakodi, M., Shrikrishna, N. S., Sharma, R., Bhansali, S., Mishra, Y., Kaushik, A., & Gandhi, S. (2022). Emerging 0D, 1D, 2D, and 3D nanostructures for efficient point-of-care biosensing. *Biosensors and Bioelectronics: X, 12*, 100284.

Cai, F., Li, S., Huang, H., Iqbal, J., Wang, C., & Jiang, X. (2022). Green synthesis of gold nanoparticles for immune response regulation: Mechanisms, applications, and perspectives. *Journal of Biomedical Materials Research Part A, 110*, 424–442.

Caputo, F., Favre, G., Borchard, G., Calzolai, L., Fisicaro, P., Frejafon, E., Gunday-Tureli, N., Koltsov, D., Minelli, C., Nelson, B. C., Parot, J., Prina-Mello, A., Zou, S., & Ouf, F. X. (2024). Toward an international standardisation roadmap for nanomedicine. *Drug Delivery and Translational Research, 14*, 2578–2588.

Chinde, S., & Grover, P. (2017). Toxicological assessment of nano and micron-sized tungsten oxide after 28 days repeated oral administration to Wistar rats. *Mutation Research/Genetic Toxicology and Environmental Mutagenesis, 819*, 1–13.

Cho, W.-S., Duffin, R., Howie, S. E. M., Scotton, C. J., Wallace, W. A. H., MacNee, W., Bradley, M., Megson, I. L., & Donaldson, K. (2011). Progressive severe lung injury by zinc oxide nanoparticles; the role of Zn2+ dissolution inside lysosomes. *Particle and Fibre Toxicology, 8*, 27.

Choi, J. H., Lee, J. H., Son, J., & Choi, J. W. (2020). Noble metal-assisted surface plasmon resonance immunosensors. *Sensors (Basel), 20*, 1003.

Coetzee, D., Venkataraman, M., Militky, J., & Petru, M. (2020). Influence of nanoparticles on thermal and electrical conductivity of composites. *Polymers (Basel), 12*, 742.

Das, R., Bandyopadhyay, R., & Pramanik, P. (2018). Carbon quantum dots from natural resource: A review. *Materials Today Chemistry, 8*, 96–109.

Dolez, P. (2015). *Nanoengineering: Global approaches to health and safety issues*. Elsevier Science.

Donaldson, K., Poland, C. A., Murphy, F. A., MacFarlane, M., Chernova, T., & Schinwald, A. (2013). Pulmonary toxicity of carbon nanotubes and asbestos – Similarities and differences. *Advanced Drug Delivery Reviews, 65*, 2078–2086.

Eker, F., Duman, H., Akdasci, E., Bolat, E., Saritas, S., Karav, S., & Witkowska, A. M. (2024). A comprehensive review of nanoparticles: From classification to application and toxicity. *Molecules, 29*, 3482.

Elsaesser, A., & Howard, C. V. (2012). Toxicology of nanoparticles. *Advanced Drug Delivery Reviews, 64*, 129–137.

Elzein, B. (2024). Nano revolution: "Tiny tech, big impact: How nanotechnology is driving SDGs progress". *Heliyon, 10*, e31393.

Ershova, E. S., Sergeeva, V. A., Chausheva, A. I., Zheglo, D. G., Nikitina, V. A., Smirnova, T. D., Kameneva, L. V., Porokhovnik, L. N., Kutsev, S. I., Troshin, P. A., Voronov, I. I., Khakina, E. A., Veiko, N. N., & Kostyuk, S. V. (2016). Toxic and DNA damaging effects of a functionalized fullerene in human embryonic lung fibroblasts. *Mutation Research/Genetic Toxicology and Environmental Mutagenesis, 805*, 46–57.

Ferrier, D. C., & Honeychurch, K. C. (2021). Carbon nanotube (CNT)-based biosensors. *Biosensors (Basel), 11*, 486.

Festus-Ikhuoria, I. C., Constance, O. N., Adekola, A. R., & Kayode, O. O. (2023). Nanotechnology in consumer products: A review of applications and safety considerations. *World Journal of Advanced Research and Reviews, 21*, 2050–2059.

Griffin, S., Masood, M. I., Nasim, M. J., Sarfraz, M., Ebokaiwe, A. P., Schafer, K. H., Keck, C. M., & Jacob, C. (2017). Natural nanoparticles: A particular matter inspired by nature. *Antioxidants (Basel), 7*, 3.

Gurcan, C., Taheri, H., Bianco, A., Delogu, L. G., & Yilmazer, A. (2020). A closer look at the genotoxicity of graphene based materials. *Journal of Physics: Materials, 3*, 014007.

Ha, S., Seidle, T., & Lim, K. M. (2016). Act on the Registration and Evaluation of Chemicals (K-REACH) and replacement, reduction or refinement best practices. *Environmental Health and Toxicology, 31*, e2016026.

Hancharova, M., Halicka-Stepien, K., Dupla, A., Lesiak, A., Soloducho, J., & Cabaj, J. (2024). Antimicrobial activity of metal-based nanoparticles: A mini-review. *Biometals, 37*, 773–801.

Hohener, K., Hock, J., & Lehman, H. C. (2016). ProSafe safe-by-design (SbD) implementation concept. ProSafe. In *Institute of Physical Chemistry "Ilie Murgulescu" Romania.*

https://echa.europa.eu/regulations/biocidal-products-regulation/understanding-bpr. Accessed 19 Dec.

https://echa.europa.eu/regulations/nanomaterials

https://echa.europa.eu/regulations/reach/understanding-reach

https://euon.echa.europa.eu/general-information

https://ised-isde.canada.ca/site/canadian-nanotechnologies-industry/en/oecd-working-party-manufactured-nanomaterials. Accessed 21 Sept

https://www.bis.gov.in/the-bureau/about-bis

https://www.epa.gov/aboutepa/our-mission-and-what-we-do. Accessed 1 May

https://www.epa.gov/tsca-inventory. Accessed 25 Jan.

https://www.iso.org

https://www.oecd.org

https://www.worksafe.com.au/

Ingle, A. P., Duran, N., & Rai, M. (2014). Bioactivity, mechanism of action, and cytotoxicity of copper-based nanoparticles: A review. *Applied Microbiology and Biotechnology, 98*, 1001–1009.

Isigonis, P., Afantitis, A., Antunes, D., Bartonova, A., Beitollahi, A., Bohmer, N., Bouman, E., Chaudhry, Q., Cimpan, M. R., Cimpan, E., Doak, S., Dupin, D., Fedrigo, D., Fessard, V., Gromelski, M., Gutleb, A. C., Halappanavar, S., Hoet, P., Jeliazkova, N., Jomini, S., Lindner, S., Linkov, I., Longhin, E. M., Lynch, I., Malsch, I., Marcomini, A., Mariussen, E., de la Fuente, J. M., Melagraki, G., Murphy, F., Neaves, M., Packroff, R., Pfuhler, S., Puzyn, T., Rahman, Q., Pran, E. R., Semenzin, E., Serchi, T., Steinbach, C., Trump, B., Vrcek, I. V., Warheit, D., Wiesner, M. R., Willighagen, E., & Dusinska, M. (2020). Risk governance of emerging technologies demonstrated in terms of its applicability to nanomaterials. *Small, 16*, e2003303.

Jain, K., Kesharwani, P., Gupta, U., & Jain, N. K. (2010). Dendrimer toxicity: Let's meet the challenge. *International Journal of Pharmaceutics, 394*, 122–142.

Jovanović, B. (2015). Review of titanium dioxide nanoparticle phototoxicity: Developing a phototoxicity ratio to correct the endpoint values of toxicity tests. *Environmental Toxicology and Chemistry, 34*, 1070–1077.

Kermani, Z. R., Haghighi, S. S., Hajihosseinali, S., Fashami, A. Z., Akbaritouch, T., Akhtari, K., Shahpasand, K., & Falahati, M. (2018). Aluminium oxide nanoparticles induce structural changes in tau and cytotoxicity of the neuroblastoma cell line. *International Journal of Biological Macromolecules, 120*, 1140–1148.

Koiranen, T., Nevalainen, T., Virkki-Hatakka, T., Aalto, H., Murashko, K., Backfolk, K., Kraslawski, A., & Pyrhönen, J. (2017). The risk assessment of potentially hazardous carbon nanomaterials for small scale operations. *Applied Materials Today, 7*, 104–111.

Kus-Liskiewicz, M., Fickers, P., & Ben Tahar, I. (2021). Biocompatibility and cytotoxicity of gold nanoparticles: Recent advances in methodologies and regulations. *International Journal of Molecular Sciences, 22*, 10952.

Le Roux, G., Moche, H., Nieto, A., Benoit, J.-P., Nesslany, F., & Lagarce, F. (2017). Cytotoxicity and genotoxicity of lipid nanocapsules. *Toxicology In Vitro, 41*, 189–199.

Lewinski, N., Colvin, V., & Drezek, R. (2008). Cytotoxicity of nanoparticles. *Small, 4*, 26–49.

Li, Y., Chen, C., Liu, F., & Liu, J. (2022). Engineered lanthanide-doped upconversion nanoparticles for biosensing and bioimaging application. *Mikrochimica Acta, 189*, 109.

Ma, X., Tian, Y., Yang, R., Wang, H., Allahou, L. W., Chang, J., Williams, G., Knowles, J. C., & Poma, A. (2024). Nanotechnology in healthcare, and its safety and environmental risks. *Journal of Nanobiotechnology, 22*, 715.

Matsumoto, M., Kobayashi, K., Takahashi, M., Hirata-Koizumi, M., Ono, A., & Hirose, A. (2015). Summary information of human health hazard assessment of existing chemical substances (I). *Bulletin of National Institute of Health Sciences, 133*, 42–47.

Mehak, Thummer, R. P., & Pandey, L. M. (2023). Surface modified iron-oxide based engineered nanomaterials for hyperthermia therapy of cancer cells. *Biotechnology & Genetic Engineering Reviews, 39*, 1187–1233.

Ministry of Ecology and Environment. (2021). Measures for the registration of new chemical substances for environmental management. In *Ministerial Order No. 12*, edited by Ministry of Ecology and Environment, China.

Miyahara, Y., Roy-Gobeil, A., & Grutter, P. (2017). Quantum state readout of individual quantum dots by electrostatic force detection. *Nanotechnology, 28*, 064001.

Pandey, R. K., Suchand Sandeep, C. S., Philip, R., & Lakshminarayanan, V. (2009). Enhanced optical nonlinearity of polyaniline–Porphyrin nanocomposite. *The Journal of Physical Chemistry C, 113*, 8630–8634.

Pandey, R. K., Chitgupi, U., & Lakshminarayanan, V. (2012). Porphyrin aggregates in the form of nanofibers and their unusual aggregation induced emission. *Journal of Porphyrins and Phthalocyanines, 16*, 1055–1058.

Park, H.-G., & Yeo, M.-K. (2016). Nanomaterial regulatory policy for human health and environment. *Molecular & Cellular Toxicology, 12*, 223–236.

Pasparakis, G. (2022). Recent developments in the use of gold and silver nanoparticles in biomedicine. *Wiley Interdisciplinary Reviews. Nanomedicine and Nanobiotechnology, 14*, e1817.

Pavlicek, A., Neubauer, S., Zafiu, C., Huber-Humer, M., Ehmoser, E. K., & Part, F. (2023). The use and detection of quantum dots as nanotracers in environmental fate studies of engineered nanoparticles. *Environmental Pollution, 317*, 120461.

Powell, L. G., Gillies, S., Fernandes, T. F., Murphy, F., Giubilato, E., Cazzagon, V., Hristozov, D., Pizzol, L., Blosi, M., Costa, A. L., Prina-Mello, A., Bouwmeester, H., Sarimveis, H., Janer, G., & Stone, V. (2022). Developing Integrated Approaches for Testing and Assessment (IATAs) in order to support nanomaterial safety. *Nanotoxicology, 16*, 484–499.

Qureshi, A., Shaikh, T., & Niazi, J. H. (2023). Semiconductor quantum dots in photoelectrochemical sensors from fabrication to biosensing applications. *Analyst, 148*, 1633–1652.

Revised Introduction to the OECD Guidelines for Testing of Chemicals, Section 3. (2006).

Sakthi Devi, R., Girigoswami, A., Siddharth, M., & Girigoswami, K. (2022). Applications of gold and silver nanoparticles in theranostics. *Applied Biochemistry and Biotechnology, 194*, 4187–4219.

Schwirn, K., Voelker, D., Galert, W., Quik, J., & Tietjen, L. (2020). Environmental risk assessment of nanomaterials in the light of new obligations under the REACH regulation: Which challenges remain and how to approach them? *Integrated Environmental Assessment and Management, 16*, 706–717.

Seke, M., Zivkovic, M., & Stankovic, A. (2024). Versatile applications of fullerenol nanoparticles. *International Journal of Pharmaceutics, 660*, 124313.

Soldado, A., Barrio, L. C., Diaz-Gonzalez, M., de la Escosura-Muniz, A., & Costa-Fernandez, J. M. (2022). Advances in quantum dots as diagnostic tools. *Advances in Clinical Chemistry, 107*, 1–40.

Summer, M., Ashraf, R., Ali, S., Bach, H., Noor, S., Noor, Q., Riaz, S., & Khan, R. R. M. (2024). Inflammatory response of nanoparticles: Mechanisms, consequences, and strategies for mitigation. *Chemosphere, 363*, 142826.

Szunerits, S., & Boukherroub, R. (2012). Sensing using localised surface plasmon resonance sensors. *Chemical Communications (Cambridge), 48*, 8999–9010.

Tsai, C. S., Echevarria-Vega, M. E., Sotiriou, G. A., Santeufemio, C., Schmidt, D., Demokritou, P., & Ellenbecker, M. (2012). Evaluation of environmental filtration control of engineered nanoparticles using the Harvard Versatile Engineered Nanomaterial Generation System (VENGES). *Journal of Nanoparticle Research, 14*, 812.

Valdiglesias, V., Kiliç, G., Costa, C., Fernández-Bertólez, N., Pásaro, E., Teixeira, J. P., & Laffon, B. (2015). Effects of iron oxide nanoparticles: Cytotoxicity, genotoxicity, developmental toxicity, and neurotoxicity. *Environmental and Molecular Mutagenesis, 56*, 125–148.

Voigt, N., Henrich-Noack, P., Kockentiedt, S., Hintz, W., Tomas, J., & Sabel, B. A. (2014). Toxicity of polymeric nanoparticles in vivo and in vitro. *Journal of Nanoparticle Research, 16*, 2379.

Wahab, A., Muhammad, M., Ullah, S., Abdi, G., Shah, G. M., Zaman, W., & Ayaz, A. (2024). Agriculture and environmental management through nanotechnology: Eco-friendly nanomaterial synthesis for soil-plant systems, food safety, and sustainability. *Science of the Total Environment, 926*, 171862.

Woskie, S. (2010). Workplace practices for engineered nanomaterial manufacturers. *Wiley Interdisciplinary Reviews. Nanomedicine and Nanobiotechnology, 2*, 685–692.

Yang, W., Wang, L., Mettenbrink, E. M., DeAngelis, P. L., & Wilhelm, S. (2021). Nanoparticle toxicology. *Annual Review of Pharmacology and Toxicology, 61*, 269–289.

Yirak, M. G., & Chaujar, R. (2024). Fundamental of emerging nanomaterials. In Y. S. Song, L. R. Thoutam, S. Tayal, S. B. Rahi, & T. S. Arun Samuel (Eds.), *Handbook of emerging materials for semiconductor industry*. Springer Nature Singapore.

Zhang, M., Jin, J., Chang, Y. N., Chang, X., & Xing, G. (2014). Toxicological properties of nanomaterials. *Journal of Nanoscience and Nanotechnology, 14*, 717–729.

Zhao, J., Wang, X., Hoang, S. A., Bolan, N. S., Kirkham, M. B., Liu, J., Xia, X., & Li, Y. (2021). Silver nanoparticles in aquatic sediments: Occurrence, chemical transformations, toxicity, and analytical methods. *Journal of Hazardous Materials, 418*, 126368.

Chapter 13
Economic and Sustainability Aspects

Adriana Hernández-Rangel, Enrique Samaniego-Benítez, Edgar Iván Hernández-Carbajal, and Alicia Rosales-Pérez

13.1 Introduction

A fundamental principle of sustainable agriculture is reducing the use of agrochemicals to protect the environment and conserve biodiversity. Establishing sustainable agriculture practices with low-input systems is essential, as these approaches reduce production costs while increasing net returns (Araújo et al., 2023). In this way, nanotechnology emerges as a promising solution to enhance the efficiency of agricultural inputs (Tauseef & Uddin, 2024). Improving agricultural productivity, particularly by controlling harmful insect populations, is a crucial factor for the economic stability of many countries worldwide. While nanomaterials (NMs) have demonstrated benefits for the agricultural sector, only a limited number of nano-products are currently available on the commercial market. Agrochemical companies primarily focus on maximizing crop yield and economic gains, often paying less attention to environmental toxicity (Carvalho, 2017; Xu et al., 2021).

To address this, cost-effective and environmentally friendly synthesis methods, such as green synthesis using plant-based materials or biodegradable agricultural waste, have been proposed (Jeevanandam et al., 2022). The implications of employing nanoparticles (NPs) in environmental contexts must prioritize safe and sustainable use, as research has previously highlighted the potential negative effects

A. Hernández-Rangel · E. I. Hernández-Carbajal
Instituto Politécnico Nacional, ESIQIE, Mexico City, Mexico
e-mail: adhernandezra@ipn.mx

E. Samaniego-Benítez
CONAHCyT, CICATA-Legaria/Instituto Politécnico Nacional, Mexico City, Mexico
e-mail: jose.samaniego@secihti.mx

A. Rosales-Pérez (✉)
Universidad Tecnologica Metropolitana de Aguascalientes, Aguascalientes, Mexico
e-mail: brosales@utma.edu.mx

K. Esquivel Escalante et al. (eds.), *Green Nanomaterials*, Nanotechnology in the Life Sciences, https://doi.org/10.1007/978-3-032-04492-1_13

of chemically synthesized NPs on plants (Zhang et al., 2021). The green synthesis of NPs offers an economical, biocompatible, and eco-friendly alternative. This approach ensures reduced toxicity, improved stability, and ease of replication (Acharya et al., 2019). Green-synthesized nanoparticles (GNPs) are often derived from various plant parts, utilizing phytochemical compounds like phenolics, terpenoids, polysaccharides, and flavonoids to drive oxidation-reduction processes (Vijayaram et al., 2024).

The chemical composition, size, surface functional groups, coating type, and concentration of NPs play a crucial role in determining their interactions within plant cells. These factors, along with the uptake of NPs, impact plant physiology, often result in positive effects on growth and development. Incorporating NMs into agricultural practices holds significant potential for fostering sustainability, enhancing food security, and improving both crop production and nutritional quality. In particular, GNPs have shown the capability to promote plant growth and yield by improving nutrient availability and increasing tolerance to environmental stress (Kale et al., 2024).

Advances in science and technology hold significant potential to increase the added value of current production systems (Vijayaram et al., 2024). A substantial rise in agricultural productivity can be achieved by leveraging modern knowledge in nanotechnology to develop effective systems for plant nutrition and protection (Semenova et al., 2024). The need for advanced, efficient methodologies to drive agricultural intensification has become critical, particularly in light of global population growth. Among the most promising strategies for enhancing crop yields is the use of GNPs in fertilizers. These GNPs enable controlled release and targeted delivery of nanoscale components, improving plant productivity while minimizing environmental pollutants (Mary Isabella Sonali et al., 2022).

To achieve sustainable global growth of the agro-industrial sector, it is essential to design an efficient plant nutrition and protection system that minimizes environmental damage (Araújo et al., 2023). This underscores the urgent need for non-toxic, eco-friendly NPs production technologies. In recent years, a variety of safe, reproducible, cost-effective, and scalable green synthesis methods for NPs have been developed, driven by the "safety-by-design" concept (Hano & Abbasi, 2022).

This chapter explores the sustainable aspect of GNPs, detailing their generation, manufacturing processes, and application in soil and plant systems. It examines their environmental and product-related benefits and limitations, while also considering their economic implications for agricultural production. By focusing on the synergy between sustainability and economic viability, this chapter highlights the potential of GNPs for safe and effective applications in agriculture.

13.2 Cost-Benefit Analysis of Green-Synthesized Nanomaterials

Green synthesis of NMs using microbial and plant sources presents many advantages over physical and chemical methods normally used. With suitable conditions, stable NMs can be generated without the need for high temperatures, high pressures, or the use of toxic solvents and reducing agents (Shah et al., 2015; Ying et al., 2022). To date, no cost-benefit analysis has been conducted for the commercial implementation of microbial and plant synthesis routes for GNPs; however, a few studies related to nano-fertilizers and nano-pesticides in general, whether green routes are used for their obtention or not, are available. In general, it is stated that the advantages of NMs over regular compounds used in agriculture compensate for their costs. Nanofertilizers such as ZnO-NPs and Fe_2O_3-NPs have demonstrated crop yield improvements of 10–30%, offering substantial economic benefits (Kah et al., 2018). For example, ZnO NPs and CuO NPs require application rates of only 0.1–0.15 kg/ha, whereas common micronutrient fertilizers often need 3–5 kg/ha. Nanofertilizers are estimated to significantly boost agricultural revenues, adding $133.2, $66.0, and $86.4 (US dollars) per hectare for corn (*Zea mays*), wheat (*Triticum*), and soybeans (*Glycine max*), respectively (Su et al., 2022). Yiming Su et al. (2022) emphasized that the relative efficiency of nanofertilizers compared to traditional fertilizers is a key factor in revenue generation. Low-efficiency nanofertilizers fail to produce positive revenue. For moderately efficient nanofertilizers, a co-dosing strategy with conventional fertilizers is recommended, with a ratio of 49 kg/ha of conventional fertilizers to 24.5 kg/ha of nanofertilizers, to achieve higher profits. Full replacement of conventional fertilizers requires highly efficient nanofertilizers; however, achieving such levels of efficiency remains a significant challenge. This efficiency is a result of NMs properties, such as size, shape, surface charge and area, and solubility (Thangavelu et al., 2024). These characteristics are related to the synthesis methods and optimal conditions for NPs obtention. Green synthesis methods for nanoparticle production face significant challenges in reaching industrial scalability, primarily due to inefficiencies in achieving true nanoscale dimensions during scale-up. In contrast, chemical and physical synthesis routes remain the preferred choice for industrial applications, as they consistently offer better control over particle size and production efficiency.

13.2.1 Sustainability Principles on GNPs

GNPs synthesis methods provide several benefits compared to conventional methods, including the following:

(a) Use of natural and non-toxic precursors
(b) Reduction of energy consumption

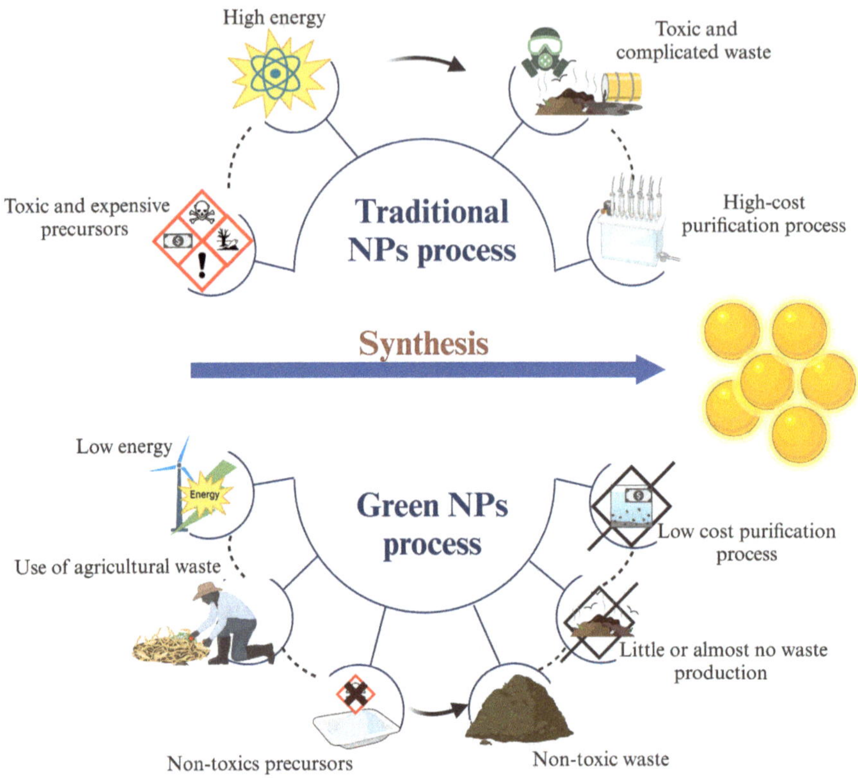

Fig. 13.1 Characteristics and differences between traditional processes and green routes for the synthesis of NPs. (Constructed by E. Samaniego-Benítez and A. Rosales-Pérez)

(c) Waste minimization
(d) Safety and non-toxicity

In conventional metal NPs synthesis, precursors and reducing agents such as metal nitrates ($M(NO_3)_3$), sodium borohydride ($NaBH_4$), and atrazine ($C_8H_{14}CIN_5$) are commonly used. However, these substances generate toxic by-products that pose environmental risks during synthesis, handling, and disposal. In green synthesis, natural reducing agents are used which do not generate toxic by-products and facilitate their manipulation during the process (see Fig. 13.1). One of the most used precursors for the synthesis of GNP's are those derived from different plants such as leaf extracts (*Aloe vera*, *Azadirachta indica*, etc.) (Anju et al., 2021; Alharbi & Alsubhi, 2022), seed extracts (*Cassia tora*, *Cakile maritima*, etc.) (Suhail & Prathibha, 2022; Elazab et al., 2024) or whole plant extracts (*Morus alba L*, *Mentha piperita*, etc.) (Razavi et al., 2020; Hussain et al., 2023). Plan extracts derived from various parts of the plants have demonstrated the ability to reduce metallic NPs thanks to their content of secondary metabolites, including alkaloids, flavonoids, phenolic acid, and terpenoids (Iravani, 2011). Similarly, fungi have proven to be

Table 13.1 Raw materials and synthesis temperature of different metallic GNPs and metallic oxides by green synthesis

Green nanoparticles	Raw material	Synthesis temperature (° C)	Ref
Silver (Ag)	Banana peel	~25	Ohiduzzaman et al. (2024)
Cerium oxide (CeO_2)	Banana peel	85	Manojkumar et al. (2024)
Copper oxide (CuO)	Camellia sinensis (Tea)	85	Kumar et al. (2024)
Ruthenium (Ru)	Aloe vera	~25	Begum et al. (2025)
Zinc oxide (ZnO)/selenium oxide (SeO_2)	Pomegranate peel (*Punica granatum L.*) extract	30	Hashem et al. (2023)
Iron oxide (Fe_3O_4)	Seaweed (*Sargassum crassifolium*) extract	~25	Budlayan et al. (2020)
Titanium oxide (TiO_2)	Onion (*Allium cepa L.*) and garlic (*Allium sativum*) peel	~25	Ali et al. (2023)
Titanium oxide (TiO_2)	*Calendula officinalis* extract	30	Zhao et al. (2024)

effective reducing agents in green synthesis, showcasing strong capabilities for NPs production (Guilger-Casagrande & de Lima, 2019). The mechanism of NPs synthesis by fungi follows two different routes: intracellular or extracellular. The difference between the two routes is where the metal precursors are added: in the case of the intracellular route, they are internalized in the biomass, while in the extracellular route, they are introduced into the aqueous filtrate that contains only the fungal biomolecules (Costa Silva et al., 2017; Molnár et al., 2018). In both processes, the reagents and by-products are non-toxic, and the synthesized NPs can be recovered by simple methods such as filtration or centrifugation.

A major drawback of traditional methods for NPs synthesis is the use of high-energy processes, such as thermal decomposition, mechanical milling, and sputtering (Baig et al., 2021; Altammar, 2023). This disadvantage not only affects the possible large-scale production of NPs by traditional methods, but also has an impact on the environment by generating a high carbon footprint if they are to be applied industrially. Most GNPs synthesis processes are conducted at low temperatures and with low energy consumption, significantly reducing the carbon footprint associated with the process (Khorrami et al., 2019; Phan et al., 2020; Carbajal Arízaga et al., 2022). Table 13.1 reports different NPs obtained by green synthesis using different base materials; in all of them, the process was chemical reduction at low temperatures (<90 °C). It is observed that both metallic and metallic oxide GNPs can be obtained using low temperatures through the green synthesis route, which saves the synthesis process compared to other techniques.

Green synthesis of NPs produces significantly less waste compared to traditional methods by utilizing natural materials like plant extracts, agricultural residues, or fruit peels as reducing and stabilizing agents. By utilizing renewable and biodegradable resources, it not only minimizes environmental impact but also promotes more efficient management of organic waste, converting it into valuable materials like catalysts, sensors, or antimicrobial agents. For instance, fruit waste like banana or orange peel serves as both a reducing and stabilizing agent, eliminating the need for extra chemicals and further minimizing waste generation. Finally, NPs synthesized by green routes tend to be more biocompatible compared to those synthesized by traditional methods. This is due to several factors inherent to the green approach, which uses natural materials as reducing agents. Some authors have reported greater biocompatibility of pure metal NPs and metal oxides synthesized by the green route compared to traditional methods (Shukla et al., 2005; Mousavi-Khattat et al., 2018; Dowlath et al., 2021; Banjara et al., 2024). Similarly, some research has shown that NPs synthesized by green routes induce less inflammatory responses and less cellular damage compared to traditional ones, increasing their potential for biomedical applications (Osonga et al., 2016; Boruah et al., 2021; Das et al., 2022).

13.2.2 Benefit of GNPs over Agriculture

The green synthesis of NPs is an environmentally friendly, cost-effective, and safe approach that employs beneficial microorganisms, such as bacterial, fungal, and plant cell cultures (Khan et al., 2022). These environmentally clean synthesis methods emphasize simplicity, cost-effectiveness, minimal time requirements, and the avoidance of toxic solvents. Techniques like microwave irradiation and sonochemistry have firmly established themselves as environmentally friendly methods (Bartolucci et al., 2020). The use of non-toxic reagents and solvents is crucial in green synthesis, as aqueous solvents typically require less environmental treatment than organic ones. Guided by the 12 Green Chemistry Principles (GCPs), this approach informs the design of nanoscale products, the development of NMs synthesis methods, and their applications. It emphasizes the use of lower concentrations, reduced temperatures, minimal energy consumption, renewable inputs, and a comprehensive consideration of the material life cycle (Martínez-Barbosa & Figueroa-Pizano, 2024).

Among biological sources, plants are favored for NPs synthesis by their simplicity, accessibility, and environmental safety. Unlike cell cultures, plants eliminate the need for time-consuming maintenance and provide a rich source of metabolites and pharmacological constituents. These act as natural agents in the synthesis of the NPs, making them an ideal choice (Alshammari et al., 2023).

GNPs are gaining considerable attention for their reliability, sustainability, and eco-friendliness characteristics. Green synthesis offers a critical solution for minimizing the harmful effects of traditional NPs synthesis methods, often used in laboratories and industry (Singh et al., 2018). The properties of NPs, including morphology, size, activity, and stability, are influenced by diverse secondary metabolites produced by different plants. The synthesis of NPs from plant extracts occurs

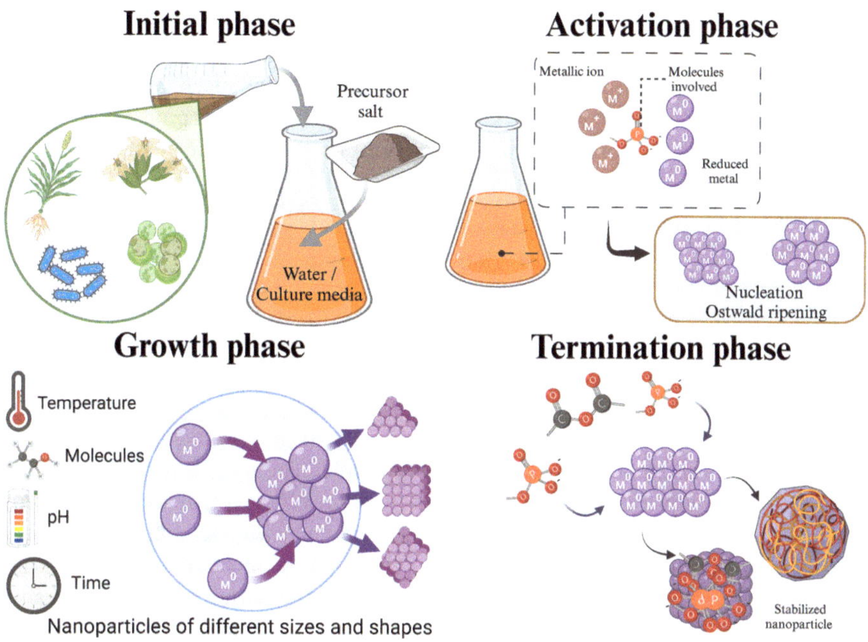

Fig. 13.2 Phases of the synthesis of GNPs. (Constructed by A. Rosales-Pérez)

in three key phases: activation phase, where bio-reduction of metal ions o salts and nucleation of reduced metal ions occur, the growth phase, where the smaller particles combine into larger ones through Ostwald ripening; and the termination phase, where the NPs attain their definitive shape (see Fig. 13.2) (Baraketi & Khwaldia, 2024).

Although numerous physical and chemical methods exist for fabricating metallic NPs, nanobiotechnology offers a greener, cost-effective, and environmentally friendly alternative by utilizing microorganisms. This approach addresses the limitations of conventional methods, making biological synthesis more favorable in terms of cost, environmental impact, and scalability for large-scale production (Noman et al., 2020). Bioinspired manganese (Mn) and manganese oxide (MnO) NPs can be synthesized through biological sources such as plants, bacteria, fungi, and algae. These methods are characterized by their low cost, eco-friendliness, and high feasibility (Zhang et al., 2023).

It is worth noting that registering any new active substance incurs significant costs. However, the lower production costs of GNPs nano-formulations may partially offset these expenses. While industrial-scale production of GNPs has not yet been realized, the absence of chemical compounds and the minimal energy requirements of these processes are expected to reduce costs further. Utilizing recyclable waste materials in biosynthesized NMs could also enhance cost-effectiveness, contributing to sustainable development goals (SDG 12) by adding value to waste materials (Bartolucci et al., 2020).

GNPs have demonstrated several advantages in agriculture. For example, their use as nano-fertilizers promotes soil microorganism activity without altering soil physicochemical properties, an important consideration for sustainable agriculture. Their stability is enhanced by plant-derived biomolecules such as proteins, enzymes, and sugars, which also exhibit strong antioxidant properties (Jiang et al., 2022).

GNPs have greatly influenced agriculture by facilitating the creation of fertilizers, pesticides, and antifungal solutions. The effects of GNPs in agriculture depend on their concentration, physicochemical properties, plant species, and the treatments are determinant methods.

One of the toxic NPs is the iron-based NPs where the GNPs decrease the effect and increase the growth (Haydar et al., 2022), and photosynthetic parameters (Soliemanzadeh & Fekri, 2021), or present antifungal properties (Elbasuney et al., 2022).

A major advantage of GNPs is their capability to overcome seed dormancy, a major agricultural challenge that reduces crop yield. The use of Manganese GNPs (Mn GNPs) demonstrated high effectiveness over the efficiency of seed germination (Shinde et al., 2020; Sharma et al., 2022). Another way to enhance the seed germination and other parameters is the GNPs used as bimetallic NPs, where the effectiveness is enhanced. For example, bimetallic MnO_x/FeO_x (de França Bettencourt et al., 2020), ZnO and TiO_2 GNPs (Hossain et al., 2019), Ag/Zn alloy GNPs (Ehsan et al., 2021), and Cu and Fe GNPs (Dey et al., 2023).

High concentrations and prolonged exposure to NPs have demonstrated negative effects on plants, including inhibited seed germination, reduced root lengths, decreased photosynthetic parameters, and enzymatic changes. Some of these NPs are the Ag NPs (Bapat et al., 2022) and Cu NPs (Rajput et al., 2018; Yang et al., 2020). However, to mitigate these negative effects, green synthesis promotes positive outcomes, such as the immobilization of chromium (Cr) in soil employing Cu GNPs (Noman et al., 2020), enhancing plant survival under salt stress using Ag GNPs (Wahid et al., 2020), or decrease the toxicity of NPs and enhance crop production, (Castillo-Henríquez et al., 2020). Another alternative solution to minimize cytotoxic effects is the use of Chitosan a polysaccharide with significant benefits in agriculture (Saad et al., 2022), or as a composite with toxic NPs like Cu NPs (Tarakanov et al., 2023) or Zn NPs (Choudhary et al., 2019).

In this context, the GNPs are emerging as effective delivery systems for pesticides, fertilizers, herbicides, and growth regulators. Their controlled-release mechanisms ensure prolonged activity, improving efficacy while reducing the required dosage and application frequency of agrochemicals (herbicides and pesticides) (Martínez-Barbosa & Figueroa-Pizano, 2024). An example of this approach is biopesticides, which align with the GCPs as opposed to synthetic pesticides. GCPs emphasize renewable resources, minimal energy use, prevention and reduction of waste, and the creation of non-toxic end products. Biopesticides adhere to these criteria by being environmentally safe, non-harmful to non-target organisms, and effective in small quantities.

The application of green chemistry principles ensures safety, effectiveness, and economic viability, while fostering sustainable technologies (Slimani & Rais, 2024).

For instance, silver GNPs (Ag GNPs) synthesized via green methods have shown reduced toxicity, enhanced photosynthesis, and improved biocompatibility due to functional groups facilitating dispersion (Zhang et al., 2021). Plant-based synthesis of Ag GNPs is particularly attractive due to its simplicity, single-step process, non-pathogenic nature, cost-effectiveness, and eco-friendliness (Khan et al., 2022).

Polysaccharides, abundant in agri-food products, are effective stabilizers during NP synthesis. Their multiple binding sites enable attachment to metal surfaces, forming an organic-inorganic network that prevents aggregation and chemical alteration, a process referred to as static stabilization (Loza et al., 2019). Similarly, agro-food waste rich in phytochemicals or antioxidants, such as flavonoids and polyphenols, can improve the antioxidant and antimicrobial properties of synthesized NPs (Baraketi & Khwaldia, 2024).

Green synthesis also utilizes organic waste as a resource for NPs production, enhancing economic feasibility. Organic materials from food, agricultural, or lignocellulosic waste act as biological reducing agents in this process. Plant-based waste, in particular, offers high reducing agent content, boosting NPs production and delivering economic benefit (Ndaba et al., 2022).

13.2.3 Limitations of the GNPs Manufacturing

Biological or green synthesis, also known as biogenic reduction, provides an eco-friendly and cost-efficient method for transforming metal precursors into NPs. This method avoids chemical contaminants, making it especially ideal for medical and biological applications where the purity of NPs is essential (Salem & Fouda, 2021). These advantages can be harnessed to develop new green pathways for synthesizing GNPs with enhanced properties for use across various scientific fields. In contrast, physicochemical methods are resource-intensive and present significant challenges, such as the reliance on toxic solvents, the production of hazardous by-products, and imperfections in NPs' surface structures, all of which complicate the production of these materials (Hussain et al., 2016; Abinaya et al., 2021).

After reviewing the different aspects related to the production and manufacturing of GNPs, the challenges associated are the same as those for typical NPs synthesis methods (Hussain et al., 2016). However, despite the advantages of green synthesis, current methods are hindered by several obstacles, including the safe and appropriate selection of solvents, optimization of process parameters, nanomaterial cytotoxicity concerns, limitations in bulk production, challenges in controlling NM morphology, tedious maintenance requirements, and gaps in knowledge. As a result, the green synthesis of NMs remains predominantly confined to the laboratory phase, limiting its scalability and widespread application.

One example of green synthesis involves the use of microorganisms; however, this approach has significant drawbacks, including complex steps such as microbial sampling, isolation, culturing, and storage. Furthermore, recovering NPs produced by this method necessitates extensive downstream processing (Bahrulolum et al.,

Table 13.2 Summary of disadvantages of biological/green bottom-up methods for synthesizing NPs

Bottom-up biological methods	Pros	Cons
Bacteria	Bacteria growth is quick Ability of the bacteria to adapt to extreme conditions Easy to cultivate	Safety risk due to bacteria contamination of side effects of using them High aseptic conditions
Fungi	Scalability approach Fungi are more efficient than bacteria	NPs size control is not possible Safety risk due to fungi used
Plant and plant products	Ecofriendly synthesis of NPs Non pathogenicity	Selectivity of the plants' metabolite Energy needed for the reaction
Electrochemical deposition[a]	Fast, simple and cost-effective Unique method that controls size and morphology	Scalability due to electrochemical equipment feasibility
Microwave[a]	Highly effective for NPs preparation Simple Rapid heat and reaction rate operation	Low crystallization profile
Supercritical fluids[a]	Low cost and easy method High-purity NPs Sustainable method due to the recovery of the gases used No organic solvents are required (high eco-friendly method)	High operation conditions (pressure and temperature)
Ultrasound technique[a]	Eco-friendly, green, fast, and easy method to produce NPs Several forms can be obtained	Specialized equipment required Sonochemical reduction rate depends on ultrasonic potency Non-scalable yet

[a]These methods can be considered into the common bottom-up methods quoted before, but with the usage of green chemistries, they can be fit into the biological methods classification

2021). In contrast, plant-based green synthesis lowers preparation cost and removes the need for specialized techniques and preparations in the synthesis of nanocomposites. However, a significant limitation of this method is the challenge of identifying the specific reactive components in plant extracts, as they contain a wide variety of organic compounds (Perveen et al., 2021).

Despite these challenges, the environmental friendliness, biocompatibility, and sensitivity of green-synthesized NMs make them highly applicable in environmental remediation, biomedical science, and consumer industries. To support their large-scale implementation, future research should focus on elucidating the mechanisms behind synthesis processes, discovering additional biological and chemical agents for synthesis, improving the feasibility of green synthesis in an industrial context, and optimizing the factors that impact the synthesis process (Ahmed et al., 2022).

The synthesis method of GNPs popular in the last years is the bottom-up method, which is based on the formation NPs from smaller molecules or small particles. However, there exist limitations that are summarized in Table 13.2 (Jamkhande et al., 2019). Equipment feasibility and synthesis conditions such as innocuity, temperature, and pressure affect the reproducibility, scalability, and final application of the GNPs.

13.2.4 Comparison Between NPs and GNPs Cost and Productivity

The synthesis of NMs poses potential risks to both the environment and society. However, green nanotechnology, which employs biomolecules and living organisms as capping agents, presents a promising solution to reducing the generation of toxic by-products (Islam et al., 2022). Traditional physical methods of NPs synthesis require expensive equipment, high temperatures and pressures, and significant space for machine setup, making them resource intensive. Chemical methods, on the other hand, involve toxic reagents that can harm both the environment and the individuals handling them. Moreover, research indicates that residues of these toxic chemicals may remain in the nanoparticles, raising concerns about their safety in applications, particularly in the medical field (Tirumala et al., 2021).

Biosynthesis of GNPs utilizes microorganisms and plants to synthesize NPs with biomedical applications. This environmentally friendly, cost-effective, biocompatible, and safe approach aligns with the GCPs. Green synthesis methods involve a variety of biological sources, including plants, bacteria, fungi, and algae, to facilitate large-scale GNPs production while reducing reliance on costly and toxic chemicals (Agarwal et al., 2017). Collectively referred to as biogenic nanoparticles, these GNPs represent a sustainable alternative for various applications (Singh et al., 2018). Figure 13.3 illustrates the methodologies employed in this green synthesis pathway.

The advancement of green synthesis over traditional physical and chemical methods lies in its environmental sustainability, cost-effectiveness, and scalability for large-scale NPs production. Unlike conventional approaches, it eliminates the need for high temperatures and pressures, excessive energy, or harmful chemicals, making it a more practical and eco-friendly alternative (Rafique et al., 2017).

The three key principles of green synthesis for NPs preparation involve choosing a solvent medium (preferably water), utilizing an environmentally friendly reducing agent, and employing a non-toxic stabilizing material for the GNPs. To enhance energy efficiency, synthesis should occur under near-ambient temperature and pressure conditions and at a neutral pH. Biological systems are particularly well-suited for achieving these sustainable and natural chemistry conditions (Rauwel et al., 2015).

Green synthesis of NPs using various biological entities effectively mitigates many of the adverse effects linked to traditional physical and chemical methods. This approach facilitates the biosynthesis of NPs under mild pH, pressure, and temperature conditions, eliminating the need for toxic substances as well as external reducing, capping, and stabilizing agents. Conventional methods for NPs synthesis

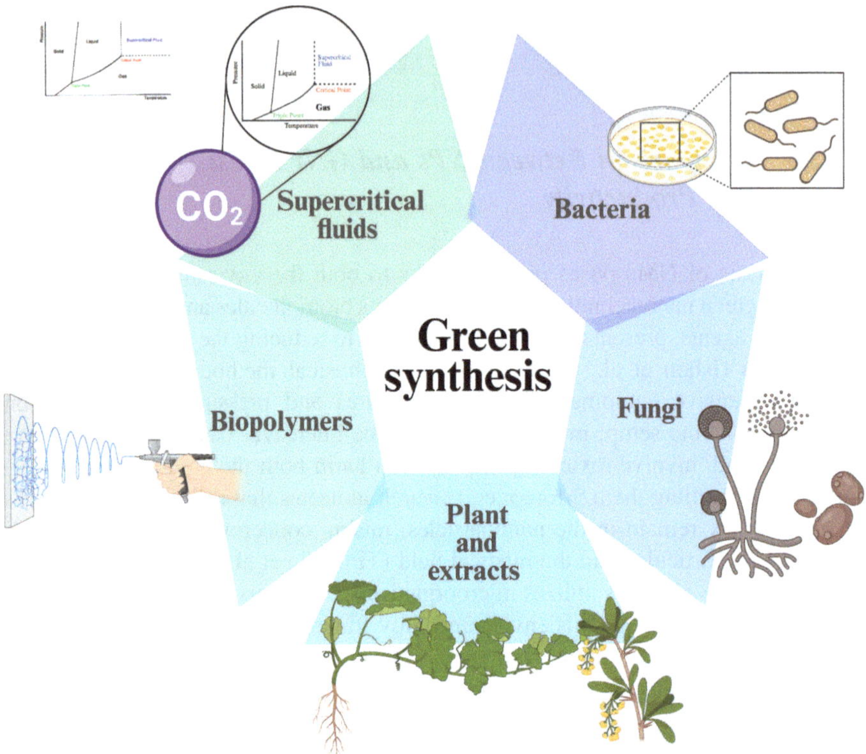

Fig. 13.3 GNPs green synthesis approach summary. (Constructed by A. Rosales-Pérez and E. Hernández-Carbajal)

often involve high operational costs, toxicity, and energy inefficiency, contributing to significant environmental issues. In contrast, green synthesis provides a sustainable and eco-friendly alternative to these challenges.

The primary challenges hindering green synthesis processes can be summarized as follows: First, optimization of the synthesis process is necessary to produce NPs with specific sizes and shapes, as these characteristics significantly influence their biological activities. Second, gaining insight into the role of each compound I the bio-fabrication process necessitates a chemical analysis of the biological biomass filtrate. Third, scaling up NP production for commercial applications demands further research and development. Additionally, the mechanisms underlying NP fabrication need more detailed explanation and exploration. To overcome these challenges, the green synthesis of NMs requires collaboration among basic science, chemical engineering, and industrial sectors to innovate and develop new commercial materials (Salem & Fouda, 2021).

Researchers are actively working to develop simple, efficient, and reliable green chemistry processes for producing NMs. It is crucial to establish more dependable and sustainable methods for synthesizing GNPs. Key factors to consider include economic feasibility, environmental sustainability, social adaptability, and the availability of local resources. To keep nanotechnology-based products affordable for consumers, industries must carefully balance the adoption of eco-friendly green processes with their long-term viability (Saif et al., 2016).

Life cycle assessment (LCA) and technoeconomic analysis (TEA) have been employed for certain classes of NMs and their applications. However, there remains a critical need to deepen our understanding of the drivers, trade-offs, and co-benefits linked to the economic and environmental performance of different NMs production pathways. These insights can serve as essential guidance for designing and optimizing processes to produce greener and more cost-effective NMs (Lan et al., 2024).

13.2.5 Life Cycle Assessment

Life cycle assessments (LCA) of NMs aim to evaluate and understand their environmental impact across their entire life cycle, from manufacturing to disposal (Salieri et al., 2018). The synthesis method used to produce NMs significantly influences their life cycle impact, along with the type of NPs being synthesized (Mitrano et al., 2015; Gallagher et al., 2017). However, the wide variety of production methods and NMs makes it challenging to provide comprehensive data for all cases. In the case of iron oxide NPs, Asifur Rahman et al. (Rahman et al., 2022) conducted a study comparing seven different synthesis methods, encompassing chemical, physical, and biological approaches, including bacterial reduction. Their findings indicated that biological synthesis had the highest environmental impact across all categories, mainly due to the substantial energy requirements for preparing bacterial culture media and the continuous operation of large-scale fermenters. Another drawback of the biological method was the relatively large average particle size produced on a large scale, measured at 40 nm, the highest among all the tested methods. Similarly, Fan Wu et al. (2019) carried out an LCA for TiO_2 NPs, comparing seven synthesis methods, including biological approaches. Their study showed that the bacterial culture media accounted for over 91% of the environmental impact, largely due to limited production data on the key components of the media, such as meat extract and peptone. Additionally, the study assumed a 100% production yield, with complete conversion of the precursor $TiO(OH)_2$ into NPs. However, experimental data indicated that producing 1 kg of TiO_2 NPs would require approximately 2500 kg of bacterial culture media. Currently, the biological synthesis of GNPs remains at the laboratory scale and faces significant challenges for industrial application. To scale up production, improvements are needed in energy efficiency and conversion processes. The use of bioreactors could enhance and better control bacterial growth, increasing production yield. Furthermore, the energy and materials required for purifying NPs and managing biological waste must be incorporated into LCAs to achieve a genuine "green" synthesis approach (Wu et al., 2019; Rahman et al., 2022). While biological synthesis offers the potential for sustainable production, further advancements are necessary to improve efficiency and reduce environmental burdens. Green synthesis methods must evolve to compete with mainstream chemical and physical synthesis techniques effectively.

13.3 Materials and Processing of GNPs

The increasing use of NPs has emphasized the need to improve synthesis for environmental and economic viability, particularly for industrial scaling. NPs synthesis is categorized into top-down and bottom-up methods. Top-down methods involve breaking down bulk materials into nanoscale particles, offering advantages like simplicity, quick production, and industrial scalability. However, they encounter challenges, including size variability, contamination, high energy consumption, and difficulties in producing particles smaller than 20 nm. In contrast, bottom-up methods build NPs from atoms or molecules, allowing precise control over size and shape, energy efficiency, and suitability for specialized applications. However, they are often limited by the high cost of precursors, complex process conditions, and the generation of chemical by-products, which complicates large-scale implementation. Balancing these methods′ advantages and limitations is essential for advancing NPs synthesis (Saravanan et al., 2022).

13.3.1 Manufacturing of GNPs at Large Scale

Global interest in advancing the production of various NMs is steadily increasing, fueled by their broad applications across industries like cosmetics, medicine, agriculture, and environmental remediation. However, scaling up NPs production via green synthesis from the laboratory to commercial scale remains the most challenging step. There are technical, engineering, economic, and environmental limitations affecting large-scale production.

Achieving high reproducibility in the synthesis of GNPs represents the greatest challenge, especially when trying to scale up this approach. Focusing on plants and fungi are exposed to a wide range of environmental stresses, both abiotic and biotic, which differ across regions of the world. Abiotic stresses include extreme temperatures, heavy metal contamination, floods, droughts, and nutrient imbalances, while biotic stress arises from herbivore attacks, and infections by fungi, bacteria, and parasites (Gull et al., 2019). Likewise, global environmental changes such as unusual rainfall, droughts, and high temperatures have considerably altered ecosystems, decreasing crop yields and hindering the growth of certain plants and fungi. In addition to environmental variations, the application of man-made chemicals, such as fertilizers to enhance plant growth and pesticides or herbicides to control pests such as weeds, insects, or rodents, further influences the conditions plants and fungi face (Zuanazzi et al., 2020); and even environmental pollution varies from region to region. In countries of the European Union, some banned pesticides are still used in countries of the American or Asian continents due to the difference in regulations in each country (Gianessi, 2013).

Environmental and non-environmental factors significantly affect the production of primary and secondary metabolites in plants and the growth of fungi. These influences can either enhance or reduce metabolic production, depending on regional conditions, leading to considerable variations in the metabolic profiles between batches. Even plants of the same species grown in different locations exhibit differences in metabolite concentration within their extracts. Moreover, plants from the same region can vary in extract quality due to the differences in genetic makeup and the age of the plant parts as the time of extraction (Hussain et al., 2012). These variations directly impact the quality, morphology, and properties of the NPs obtained.

To date, there is limited understanding of the reaction mechanisms and bioactive components in plant extracts involved in green synthesis. Several authors have reported that alcohol, amine, and acid groups play a crucial role in reducing metal particles, while others suggested that flavonoid compounds are mainly responsible ones (Shankar et al., 2004, 2005; Terenteva et al., 2015). This is a major issue when trying to scale up, since if one attempts to produce batches of GNPs consistently but some percentage of the base material lacks the bioactive metabolite (either due to variations in the age of the plants used or the region of origin), an incomplete reaction would occur, which would significantly affect the yield and homogeneity of the GNPs. The number of potential reducing agents present in extracts complicates research into the mechanism of nanoparticle formation. One of the initial steps required to scale up GNPs production through green routes involves conducting studies with pure solutions of active components found in plant extracts. These studies aim to provide a deeper understanding of the effectiveness of these compounds in green synthesis.

Another complication in scaling up nanoparticle production by green synthesis is the lack of standardized physicochemical parameters. As is known, extracts from different parts of plants or the use of different fungi exhibit different properties, but even for the same part of a plant or the use of the same fungus, variation in physicochemical properties/experimental conditions such as pH, extraction or process temperature, reaction time, and ratio of extract to metal precursor can affect particle size, shape and morphology.

Regarding the engineering challenges to scale up nanoparticle synthesis through green routes, the two main ones are the lack of appropriate equipment and the lack of process performance efficiency. The problem with synthesizing NPs (both by conventional and green methodologies) is the dramatic variation in characteristics and properties when using tanks of different volume capacities. Therefore, the use of automated flow or batch reactors is sought to optimize. However, this remains a challenge to date (Saldanha et al., 2017; Gao et al., 2020; Besenhard et al., 2023). Yield is a way to determine the productivity, profitability, and efficiency of synthesized NPs. Most of the existing literature emphasizes the quality of starting reagents, such as the metal compound and the plant extract, but often overlooks the efficiency of the process yield. Since yield is directly linked to profitability, industries such as pharmaceuticals, paint, or construction are unlikely to invest their money and resources without clear evidence of high process efficiency. Therefore, it is essential to conduct studies that not only detail the use and quantity of precursors for NPs synthesis through green routes but also report on the process yield. Such comprehensive research would encourage industries to invest in these methods, ultimately promoting the large-scale synthesis of NPs from plant-based sources.

13.3.2 End of Life of GNPs

Materials produced at the nanoscale are now widely utilized in different fields, including electronics, polymers, energy, medical devices, textiles, and cosmetics. NPs can be released into the environment at various stages including manufacturing, transportation, product usage, and final waste disposal (Bundschuh et al., 2018). Once released, NPs may retain their original form or undergo transformations through interactions with their surroundings. Physical, chemical, and biological changes occur when NPs come into contact with air, water, or soil (Abbas et al., 2020). These interactions and transformations are affected by the size of the NPs and the physicochemical properties of both the particles and the environmental medium (Zhang et al., 2018). NPs are known to aggregate in various environments, resulting in increased particle size and a corresponding reduction in reactivity, a characteristic attributed to their nanoscale dimensions. Depending on the size of the aggregated NPs, they may remain suspended, settle as sediment, or dissolve (Abbas et al., 2020; Ahmed et al., 2021). However, aggregation does not entirely prevent the bioavailability of NPs or their adverse interactions with living organisms, including algae, microbial communities, plants, food crops, animals, and humans. In the case of green-synthesized NPs, evidence suggests that environmentally friendly synthesis methods can partially reduce their toxicity.

13.3.2.1 Toxicity of Green Nanoparticles

The toxicity of NPs has been well established in prokaryotic and eukaryotic cells by means of reactive oxygen species (ROS) generation, genotoxicity, etc. (Ganguly et al., 2018). However, NPs toxicity cannot be generalized, since it is dependent on factors such as type, size, capping agent, route of synthesis, and doses used (Ganguly et al., 2018; Rana et al., 2020). Even though green synthesized NPs are known to be the safest and result in less toxicity than the chemically produced ones, they can be harmful at certain doses (Rana et al., 2020). For example, AuNPs produced using *F. oxysporum* showed toxic effects when tested in vivo (Yahyaei et al., 2019), the liver, kidney, and testis were used to evaluate structural changes after 3 days of NPs administration. The results showed a dose-dependent toxicity with mild changes in the structure of the tissues tested. In another study (Ramachandran et al., 2018), the toxicity of biologically synthesized Ag NPs and Au NPs was compared employing as a model organism zebrafish. The findings revealed that grill and liver tissues treated with Ag NPs exhibited morphological changes, including cell membrane damage, irregular cell outlines, pyknotic nuclei, and complete disruption of gill and liver cells. In contrast, Au NPs treatment induced such changes only in liver tissues. Ag NPs-treated liver cells showed significantly higher levels of ROS generation compared to the control, whereas Au NPs treatment resulted in lower ROS levels than the control. Furthermore, blood cells treated with Ag NPs demonstrated micronuclei formation and nuclear abnormalities, effects that were absent in Au NPs-treated blood cells. Other investigations have compared the toxic effects of chemically and biologically produced Ag NPs to the clam *Mercenaria mercenaria* and the silver carp fish (*Hypophthalmichthys molitrix*) (Jassim et al., 2022; Younas et al., 2022). The results clearly showed a decreased adverse effect when green routes for GNPs obtention were used. Nonetheless, all types of NPs induce toxic effects in the species tested as a function of Ag concentration. For *Mercenaria mercenaria* the lethal concentration 50 (LC50) was 248.8 mg L^{-1} for commercial Ag NP and around 100 mg L^{-1} for chemical and biological Ag NPs. In the case of silver carp fish, LC50 values were 700 and 1300 mg L^{-1} for chemical Ag NPs and biological Ag NPs, respectively. The authors claim that the difference in the toxicity of Ag NPs synthesized by biological and chemical methods may be attributed to variations in particle size and chemical composition.

13.3.2.2 Disposal and Recycling of NPs

As the use and applications of NMs in industry and everyday products continue to expand, the production of waste containing these materials is also on the rise (Part et al., 2018). Mass flow analyses indicate that only a small proportion of NMs are directly released into the environment (Keller et al., 2013). Instead, the majority pass through managed waste systems such as wastewater infrastructure (including sewers and treatment plants), incineration facilities, and landfills (Keller et al., 2013; Bundschuh et al., 2018) (see Fig. 13.4).

13.3.2.3 Incineration

Incineration is a treatment procedure that involves the combustion of waste materials to reduce their volume. This technique could modify and control the NMs entering the waste streams generated by consumers' disposal as municipal solid waste as the residues generated from nanotechnology research and medical applications (Holder et al., 2013). There is little research existing on the control of pollutant formation through NMs incineration. Tobias Walser et al. (Walser et al., 2012) introduced CeO_2 GNPs into a full-scale waste incineration plant and observed that these particles loosely adhered to solid residues produced during combustion. They found that current filtration technologies effectively remove GNPs from flue gas. Notably, the GNPs attached to the surface of the solid residues without becoming an

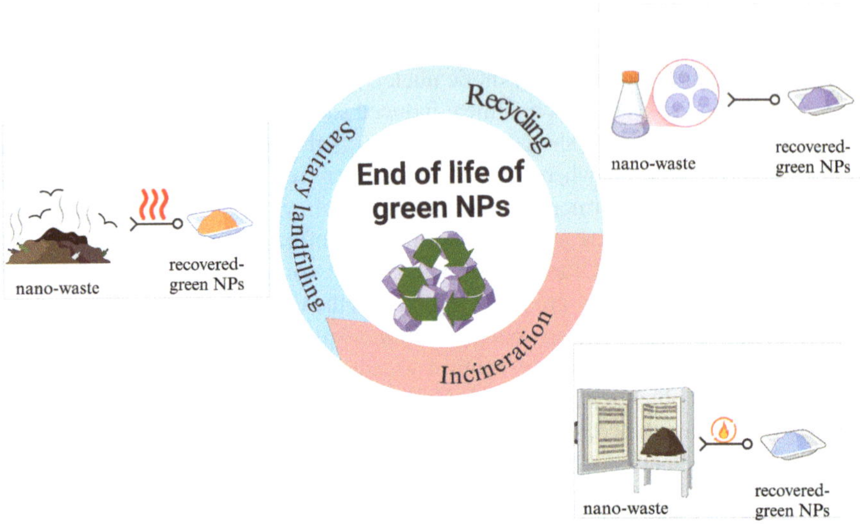

Fig. 13.4 End of life of the GNPs. (Constructed by A. Hernández-Rangel and A. Rosales-Pérez)

integral part of them or undergoing any physical or chemical transformations. However, the authors highlight that while incineration can prevent NPs from being released into the atmosphere, the residues to which they adhere ultimately end up in landfills or are processed as recovered raw materials. This underscores the environmental urgency of developing degradable NPs. In another study, Andrea Massari and colleagues (Massari et al., 2014) examined the behavior of TiO_2 NPs during the incineration of solid paint waste containing these particles. The solid paint debris was processed in a laboratory-scale incineration setup under oxidizing conditions at a combustion temperature of 950 °C. The researchers observed no release of TiO_2 NPs into the atmosphere. Instead, titanium was found to adhere to the surface of the resulting solid residues (ashes). Further analysis revealed that the TiO_2 NPs underwent a reaction during incineration, forming calcium titanate.

13.3.2.4 Sanitary Landfilling

A sanitary landfill is a controlled land area designated for the disposal of municipal solid waste, aiming to minimize pollution and protect the environment (Ceyda Senem Uyguner-Demirel). However, the operation of landfills raises significant health and environmental concerns, particularly due to the emissions of landfill gas and leachate (Ceyda Senem Uyguner-Demirel). Ralf Kaegi et al. (2017) conducted a year-long study collecting leachate samples from a landfill containing TiO_2 particles to evaluate their potential release from construction and demolition waste sites. Based on the 8000 m^3 of leachate produced in 2014, the total annual discharge of titanium into surface water was estimated at approximately 0.3 kg, equivalent to about 0.5 kg of TiO_2 NPs. Analysis revealed predominantly nanoscale TiO_2 particles with a spherical shape, characteristic of those used in white pigments. Florian Part and colleagues (2020) further explored the fate of NPs in landfills, emphasizing the role of NPs coatings in determining their environmental clearance pathways, such as sedimentation or degradation. They examined the behavior of quantum dots (QDs) exposed to leachates from Austrian landfills containing municipal solid and bulky wastes. Brij-58-coated CdSe QDs primarily sedimented, degrading more slowly over 6 months. Contrary to N-acetyl-L-cysteine-coated CdTe QDs underwent faster degradation mechanisms influenced by temperature, with 71–95% remaining dispersed in leachates at low temperatures after 6 months. The leachate temperature, dissolved organic carbon content, and NPs coatings were critical factors affecting whether sedimentation or degradation was the dominant clearance mechanism.

13.3.2.5 Recycling of NPs

Recycling is a way to recover nanoparticulated materials from waste and further use them in other applications. At laboratory scale, Paramjeet Pati et al. (2016) developed a method to recover nanowaste gold. The recovery process involved using α-cyclodextrin to form an inclusion complex with gold. The recovered gold was subsequently utilized to produce new NPs via the citrate-reduction method. However, the newly synthesized Au NPs exhibited colloidal instability and coalesced shortly after synthesis, likely due to unidentified impurities in the recovered gold. On the other hand, Oestreicher et al. (2019) proposed a simple and effective methodology for recovering gold from aqueous laboratory nanowaste and converting it into an aqueous $HAuCl_4$ solution. Their methodology utilized simple procedures and readily available chemical reagents, including NaCl, HCl, and H_2O_2, achieving a recovery efficiency of over 99% of the original gold. Zhang et al. (2016) investigated the fate of NMs during the paper recycling process. Their study focused on polylactic acid (PLA)-coated paper containing zinc oxide NPs in the coating layer. Using a lab-scale paper recycling system, they successfully recovered more than 99% of the solid material. The mass balance analysis revealed that 86–91% of the zinc oxide NPs were retained in the rejected material stream, primarily embedded in the polymer coating, while 7–16% of the NPs were found in the accepted material stream.

13.4 Conclusions

GNPs present significant potential for economic and sustainable advancements in agriculture by enhancing crop yields, reducing input costs, and minimizing waste through controlled-release mechanisms for fertilizers and pesticides. Produced using eco-friendly biological methods, such as plant extracts, fungi, and microorganisms,

GNPs align with GCPs, providing cost-effective and environmentally sustainable alternatives to conventional chemical synthesis. However, their widespread adoption faces significant challenges, including scalability of production, life cycle management, and addressing potential toxicity concerns. Variability in raw materials, inconsistent yields, and the lack of standardized processes hinder industrial-scale production. Furthermore, end-of-life issues, including aggregation, sedimentation, and bioavailability, require extensive research to mitigate potential environmental risks.

While the idea of repurposing waste materials to develop advanced environmental technologies is compelling, substantial knowledge gaps concerning engineered NMs must be addressed before these ideas can be practically realized. Ensuring that the ingredients and processing methods used in the synthesis of GNPs are economically viable is crucial for scaling up production to industrial levels. Advances in manufacturing techniques, including enzymatic biosynthesis and low-energy methods such as sonochemistry, show potential for improving scalability and cost-efficiency. Recycling nano-waste also presents an opportunity for sustainability, though economic and technical barriers remain significant.

Beyond fertilizers and pesticides, GNPs play a broader role in agriculture, contributing to stress tolerance in crops, enhancing soil health, and reducing the environmental impact of agricultural practices. Leveraging their biocompatibility and controlled-release properties, GNPs offer transformative potential for modern farming systems. To ensure their long-term viability and impact, future efforts must focus on optimizing synthesis processes, standardizing production protocols, and integrating sustainability goals. Green chemistry will continue to be a key driver in this journey, advancing cleaner production methods, fostering innovation, and supporting the global transition to a more environmentally responsible economy.

Acknowledgments The authors express their gratitude to the Mexican Council for the Humanities, Science and Technology (CONAHCyT-SECIHTI).

Competing Interest The authors have no conflicts of interest to declare that are relevant to the content of this chapter.

References

Abbas, Q., Yousaf, B., Amina, Ali, M. U., Munir, M. A. M., El-Naggar, A., Rinklebe, J., & Naushad, M. (2020). Transformation pathways and fate of engineered nanoparticles (ENPs) in distinct interactive environmental compartments: A review. *Environment International, 138*, 105646. https://doi.org/10.1016/j.envint.2020.105646

Abinaya, S., Kavitha, H. P., Prakash, M., & Muthukrishnaraj, A. (2021). Green synthesis of magnesium oxide nanoparticles and its applications: A review. *Sustainable Chemistry and Pharmacy, 19*, 100368. https://doi.org/10.1016/j.scp.2020.100368

Acharya, P., Jayaprakasha, G. K., Crosby, K. M., Jifon, J. L., & Patil, B. S. (2019). Green-synthesized nanoparticles enhanced seedling growth, yield, and quality of onion (Allium cepa

L.). *ACS Sustainable Chemistry & Engineering, 7*, 14580–14590. https://doi.org/10.1021/acssuschemeng.9b02180

Agarwal, H., Venkat Kumar, S., & Rajeshkumar, S. (2017). A review on green synthesis of zinc oxide nanoparticles – An eco-friendly approach. *Resource-Efficient Technologies, 3*, 406–413. https://doi.org/10.1016/j.reffit.2017.03.002

Ahmed, T., Noman, M., Manzoor, N., Ali, S., Rizwan, M., Ijaz, M., Allemailem, K. S., BinShaya, A. S., Alhumaydhi, F. A., & Li, B. (2021). Recent advances in nanoparticles associated ecological harms and their biodegradation: Global environmental safety from nano-invaders. *Journal of Environmental Chemical Engineering, 9*, 106093. https://doi.org/10.1016/j.jece.2021.106093

Ahmed, S. F., Mofijur, M., Rafa, N., Chowdhury, A. T., Chowdhury, S., Nahrin, M., Islam, A. B. M. S., & Ong, H. C. (2022). Green approaches in synthesising nanomaterials for environmental nanobioremediation: Technological advancements, applications, benefits and challenges. *Environmental Research, 204*, 111967. https://doi.org/10.1016/j.envres.2021.111967

Alharbi, N. S., & Alsubhi, N. S. (2022). Green synthesis and anticancer activity of silver nanoparticles prepared using fruit extract of Azadirachta indica. *Journal of Radiation Research and Applied Sciences, 15*, 335–345. https://doi.org/10.1016/j.jrras.2022.08.009

Ali, H., Dixit, S., Almutairi, B. O., & Yadav, N. (2023). Synthesis and characterization of eco-friendly TiO2 nanoparticle from combine extract of onion and garlic peel. *Journal of King Saud University – Science, 35*, 102918. https://doi.org/10.1016/j.jksus.2023.102918

Alshammari, S. O., Mahmoud, S. Y., & Farrag, E. S. (2023). Synthesis of green copper nanoparticles using medicinal plant Krameria sp. root extract and its applications. *Molecules, 28*, 4629. https://doi.org/10.3390/molecules28124629

Altammar, K. A. (2023). A review on nanoparticles: Characteristics, synthesis, applications, and challenges. *Frontiers in Microbiology, 14*, 1–20. https://doi.org/10.3389/fmicb.2023.1155622

Anju, T. R., Parvathy, S., Valiya, M., Rosemary, J., Ansalna, T. H., Shahzabanu, M. M., & Devika, S. (2021). Green synthesis of silver nanoparticles from Aloe vera leaf extract and its antimicrobial activity. *Materials Today: Proceedings, 43*, 3956–3960. https://doi.org/10.1016/j.matpr.2021.02.665

Araújo, R. G., Chavez-Santoscoy, R. A., Parra-Saldívar, R., Melchor-Martínez, E. M., & Iqbal, H. M. N. (2023). Agro-food systems and environment: Sustaining the unsustainable. *Current Opinion in Environmental Science & Health, 31*, 100413. https://doi.org/10.1016/j.coesh.2022.100413

Bahrulolum, H., Nooraei, S., Javanshir, N., Tarrahimofrad, H., Mirbagheri, V. S., Easton, A. J., & Ahmadian, G. (2021). Green synthesis of metal nanoparticles using microorganisms and their application in the agrifood sector. *Journal of Nanobiotechnology, 19*, 1–26. https://doi.org/10.1186/s12951-021-00834-3

Baig, N., Kammakakam, I., Falath, W., & Kammakakam, I. (2021). Nanomaterials: A review of synthesis methods, properties, recent progress, and challenges. *Materials Advances, 2*, 1821–1871. https://doi.org/10.1039/d0ma00807a

Banjara, R. A., Kumar, A., Aneshwari, R. K., Satnami, M. L., & Sinha, S. K. (2024). A comparative analysis of chemical vs green synthesis of nanoparticles and their various applications. *Environmental Nanotechnology, Monitoring and Management, 22*, 100988. https://doi.org/10.1016/j.enmm.2024.100988

Bapat, M. S., Singh, H., Shukla, S. K., Singh, P. P., Vo, D.-V. N., Yadav, A., Goyal, A., Sharma, A., & Kumar, D. (2022). Evaluating green silver nanoparticles as prospective biopesticides: An environmental standpoint. *Chemosphere, 286*, 131761. https://doi.org/10.1016/j.chemosphere.2021.131761

Baraketi, S., & Khwaldia, K. (2024). Nanoparticles from agri-food by-products: Green technology synthesis and application in food packaging. *Current Opinion in Green and Sustainable Chemistry, 49*, 100953. https://doi.org/10.1016/j.cogsc.2024.100953

Bartolucci, C., Antonacci, A., Arduini, F., Moscone, D., Fraceto, L., Campos, E., Attaallah, R., Amine, A., Zanardi, C., Cubillana-Aguilera, L. M., Palacios Santander, J. M., & Scognamiglio, V. (2020). Green nanomaterials fostering agrifood sustainability. *TrAC Trends in Analytical Chemistry, 125*, 115840. https://doi.org/10.1016/j.trac.2020.115840

Begum, T., Agarwal, S., Bhuyan, P., Das, J., Verma, A. K., Guha, A., & Ganguly, M. (2025). Aloe vera gel mediated green synthesis of ruthenium nanoparticles and their potential anticancer activity. *Next Nanotechnology, 7*, 100095. https://doi.org/10.1016/j.nxnano.2024.100095

Besenhard, M. O., Pal, S., Gkogkos, G., & Gavriilidis, A. (2023). Non-fouling flow reactors for nanomaterial synthesis. *Reaction Chemistry and Engineering, 8*, 955–977. https://doi.org/10.1039/d2re00412g

Boruah, J. S., Devi, C., Hazarika, U., Bhaskar Reddy, P. V., Chowdhury, D., Barthakur, M., & Kalita, P. (2021). Green synthesis of gold nanoparticles using an antiepileptic plant extract: In vitrobiological and photo-catalytic activities. *RSC Advances, 11*, 28029–28041. https://doi.org/10.1039/d1ra02669k

Budlayan, M. L. M., Patricio, J. N., Arco, S. D., Capangpangan, R. Y., & Alguno, A. C. (2020). Effects of precursor concentration on the properties of magnetic iron oxide nanoparticles synthesized using brown seaweed (Sargassum crassifolium) extract. *Materials Today: Proceedings, 46*, 1608–1612. https://doi.org/10.1016/j.matpr.2020.07.251

Bundschuh, M., Filser, J., Lüderwald, S., McKee, M. S., Metreveli, G., Schaumann, G. E., Schulz, R., & Wagner, S. (2018). Nanoparticles in the environment: Where do we come from, where do we go to? *Environmental Sciences Europe, 30*, 6. https://doi.org/10.1186/s12302-018-0132-6

Carbajal Arízaga, G. G., Quiñones Galván, J. G., Bail, A., Pérez González, A. L., Pereyra Nuñez, C., & López Álvarez, M. Á. (2022). Green approach to synthesize functional carbon nanoparticles at low temperature. *Sustainable Chemistry for Climate Action, 1*, 100002. https://doi.org/10.1016/j.scca.2022.100002

Carvalho, F. P. (2017). Pesticides, environment, and food safety. *Food and Energy Security, 6*, 48–60. https://doi.org/10.1002/fes3.108

Castillo-Henríquez, L., Alfaro-Aguilar, K., Ugalde-Álvarez, J., Vega-Fernández, L., Montes de Oca-Vásquez, G., & Vega-Baudrit, J. R. (2020). Green synthesis of gold and silver nanoparticles from plant extracts and their possible applications as antimicrobial agents in the agricultural area. *Nanomaterials, 10*, 1763. https://doi.org/10.3390/nano10091763

Choudhary, R. C., Kumaraswamy, R. V., Kumari, S., Sharma, S. S., Pal, A., Raliya, R., Biswas, P., & Saharan, V. (2019). Zinc encapsulated chitosan nanoparticle to promote maize crop yield. *International Journal of Biological Macromolecules, 127*, 126–135. https://doi.org/10.1016/j.ijbiomac.2018.12.274

Costa Silva, L. P., Pinto Oliveira, J., Keijok, W. J., da Silva, A. R., Aguiar, A. R., Guimarães, M. C. C., Ferraz, C. M., Araújo, J. V., Tobias, F. L., & Braga, F. R. (2017). Extracellular biosynthesis of silver nanoparticles using the cell-free filtrate of nematophagous fungus Duddingtonia flagrans. *International Journal of Nanomedicine, 12*, 6373–6381. https://doi.org/10.2147/IJN.S137703

Das, T., Mishra, S., Nag, S., & Saha, K. D. (2022). Green-synthesized gold nanoparticles from black tea extract enhance the chemosensitivity of doxorubicin in HCT116 cells via a ROS-dependent pathway. *RSC Advances, 12*, 8996–9007. https://doi.org/10.1039/d1ra08374k

de França Bettencourt, G. M., Degenhardt, J., Zevallos Torres, L. A., de Andrade Tanobe, V. O., & Soccol, C. R. (2020). Green biosynthesis of single and bimetallic nanoparticles of iron and manganese using bacterial auxin complex to act as plant bio-fertilizer. *Biocatalysis and Agricultural Biotechnology, 30*, 101822. https://doi.org/10.1016/j.bcab.2020.101822

Dey, S., Nath, S., Alam Ansari, T., Biswas, A., Barman, F., Mukherjee, S., Gopal, G., Bhattacharyya, A., Mukherjee, A., Kundu, R., & Paul, S. (2023). Application of green synthesized bimetallic nZVI-Cu nanoparticle as a sustainable alternative to chemical fertilizers to enhance growth and photosynthetic efficiency of rice seedlings. *Plant Physiology and Biochemistry, 201*, 107837. https://doi.org/10.1016/j.plaphy.2023.107837

Dowlath, M. J. H., Musthafa, S. A., Mohamed Khalith, S. B., Varjani, S., Karuppannan, S. K., Ramanujam, G. M., Arunachalam, A. M., Arunachalam, K. D., Chandrasekaran, M., Chang, S. W., Chung, W. J., & Ravindran, B. (2021). Comparison of characteristics and biocompatibility of green synthesized iron oxide nanoparticles with chemical synthesized nanoparticles. *Environmental Research, 201*, 111585. https://doi.org/10.1016/j.envres.2021.111585

Ehsan, M., Raja, N. I., Mashwani, Z.-R., Ikram, M., Zohra, E., Zehra, S. S., Abasi, F., Hussain, M., Iqbal, M., Mustafa, N., & Ali, A. (2021). Responses of bimetallic Ag/ZnO alloy nanoparticles and urea on morphological and physiological attributes of wheat. *IET Nanobiotechnology, 15*, 602–610. https://doi.org/10.1049/nbt2.12048

Elazab, N. T., Baka, Z. A. M., Saleh, H. H., & El-Zahed, M. M. (2024). Green synthesis of silver nanoparticles using Cakile maritima seed extract: Molecular, antifungal and physiological studies. *Physiological and Molecular Plant Pathology, 129*, 102183. https://doi.org/10.1016/j.pmpp.2023.102183

Elbasuney, S., El-Sayyad, G. S., Attia, M. S., & Abdelaziz, A. M. (2022). Ferric oxide colloid: Towards green nano-fertilizer for tomato plant with enhanced vegetative growth and immune response against Fusarium wilt disease. *Journal of Inorganic and Organometallic Polymers, 32*, 4270–4283. https://doi.org/10.1007/s10904-022-02442-6

Gallagher, M. J., Allen, C., Buchman, J. T., Qiu, T. A., Clement, P. L., Krause, M. O. P., & Gilbertson, L. M. (2017). Research highlights: Applications of life-cycle assessment as a tool for characterizing environmental impacts of engineered nanomaterials. *Environmental Science: Nano, 4*, 276–281. https://doi.org/10.1039/c7en90005h

Ganguly, P., Breen, A., & Pillai, S. C. (2018). Toxicity of nanomaterials: Exposure, pathways, assessment, and recent advances. *ACS Biomaterials Science and Engineering, 4*, 2237–2275. https://doi.org/10.1021/acsbiomaterials.8b00068

Gao, Y., Pinho, B., & Torrente-Murciano, L. (2020). Recent progress on the manufacturing of nanoparticles in multi-phase and single-phase flow reactors. *Current Opinion in Chemical Engineering, 29*, 26–33. https://doi.org/10.1016/j.coche.2020.03.008

Gianessi, L. P. (2013). The increasing importance of herbicides in worldwide crop production. *Pest Management Science, 69*, 1099–1105. https://doi.org/10.1002/ps.3598

Guilger-Casagrande, M., & de Lima, R. (2019). Synthesis of silver nanoparticles mediated by fungi: A review. *Frontiers in Bioengineering and Biotechnology, 7*, 1–16. https://doi.org/10.3389/fbioe.2019.00287

Gull, A., Lone, A. A., & Wani, N. U. I. (2019). Biotic and abiotic stresses in plants. In A. B. de Oliveira (Ed.), *Abiotic and biotic stress in plants*. IntechOpen.

Hano, C., & Abbasi, B. H. (2022). Plant-based green synthesis of nanoparticles: Production, characterization and applications. *Biomolecules, 12*, 31. https://doi.org/10.3390/biom12010031

Hashem, A. H., Al-Askar, A. A., Saeb, M. R., Abd-Elsalam, K. A., El-Hawary, A. S., & Hasanin, M. S. (2023). Sustainable biosynthesized bimetallic ZnO@SeO nanoparticles from pomegranate peel extracts: Antibacterial, antifungal and anticancer activities. *RSC Advances, 13*, 22918–22927. https://doi.org/10.1039/d3ra03260d

Haydar, M. S., Ghosh, S., & Mandal, P. (2022). Application of iron oxide nanoparticles as micronutrient fertilizer in mulberry propagation. *Journal of Plant Growth Regulation, 41*, 1726–1746. https://doi.org/10.1007/s00344-021-10413-3

Holder, A. L., Vejerano, E. P., Zhou, X., & Marr, L. C. (2013). Nanomaterial disposal by incineration. *Environmental Science: Processes & Impacts, 15*, 1652–1664. https://doi.org/10.1039/c3em00224a

Hossain, A., Abdallah, Y., Ali, M. A., Masum, M. M. I., Li, B., Sun, G., Meng, Y., Wang, Y., & An, Q. (2019). Lemon-fruit-based green synthesis of zinc oxide nanoparticles and titanium dioxide nanoparticles against soft rot bacterial pathogen Dickeya dadantii. *Biomolecules, 9*, 863. https://doi.org/10.3390/biom9120863

Hussain, M. S., Fareed, S., Ansari, S., Rahman, M. A., Ahmad, I. Z., & Saeed, M. (2012). Current approaches toward production of secondary plant metabolites. *Journal of Pharmacy and Bioallied Sciences, 4*, 10–20. https://doi.org/10.4103/0975-7406.92725

Hussain, I., Singh, N. B., Singh, A., Singh, H., & Singh, S. C. (2016). Green synthesis of nanoparticles and its potential application. *Biotechnology Letters, 38*, 545–560. https://doi.org/10.1007/s10529-015-2026-7

Hussain, Z., Jahangeer, M., Sarwar, A., Ullah, N., Aziz, T., Alharbi, M., Alshammari, A., & Alasmari, A. F. (2023). Synthesis and characterization of silver nanoparticles mediated by the mentha piperita leaves extract and exploration of its antimicrobial activities. *Journal of the Chilean Chemical Society, 68*, 5865–5870. https://doi.org/10.4067/s0717-97072023000205865

Iravani, S. (2011). Green synthesis of metal nanoparticles using plants. *Green Chemistry, 13*, 2638–2650. https://doi.org/10.1039/c1gc15386b

Islam, F., Shohag, S., Uddin, M. J., Islam, M. R., Nafady, M. H., Akter, A., Mitra, S., Roy, A., Bin Emran, T., & Cavalu, S. (2022). Exploring the journey of zinc oxide nanoparticles (ZnO-NPs) toward biomedical applications. *Materials, 15*, 1–31. https://doi.org/10.3390/ma15062160

Jamkhande, P. G., Ghule, N. W., Bamer, A. H., & Kalaskar, M. G. (2019). Metal nanoparticles synthesis: An overview on methods of preparation, advantages and disadvantages, and applications. *Journal of Drug Delivery Science and Technology, 53*, 101174. https://doi.org/10.1016/j.jddst.2019.101174

Jassim, A. Y., Wang, J., Chung, K. W., Loosli, F., Chanda, A., Scott, G. I., & Baalousha, M. (2022). Comparative assessment of the fate and toxicity of chemically and biologically synthesized silver nanoparticles to juvenile clams. *Colloids and Surfaces B: Biointerfaces, 209*, 18–21. https://doi.org/10.1016/j.colsurfb.2021.112173

Jeevanandam, J., Kiew, S. F., Boakye-Ansah, S., Lau, S. Y., Barhoum, A., Danquah, M. K., & Rodrigues, J. (2022). Green approaches for the synthesis of metal and metal oxide nanoparticles using microbial and plant extracts. *Nanoscale, 14*, 2534–2571. https://doi.org/10.1039/D1NR08144F

Jiang, Y., Zhou, P., Zhang, P., Adeel, M., Shakoor, N., Li, Y., Li, M., Guo, M., Zhao, W., Lou, B., Wang, L., Lynch, I., & Rui, Y. (2022). Green synthesis of metal-based nanoparticles for sustainable agriculture. *Environmental Pollution, 309*, 119755. https://doi.org/10.1016/j.envpol.2022.119755

Kaegi, R., Englert, A., Gondikas, A., Sinnet, B., von der Kammer, F., & Burkhardt, M. (2017). Release of TiO2 – (Nano) particles from construction and demolition landfills. *NanoImpact, 8*, 73–79. https://doi.org/10.1016/j.impact.2017.07.004

Kah, M., Kookana, R. S., Gogos, A., & Bucheli, T. D. (2018). A critical evaluation of nanopesticides and nanofertilizers against their conventional analogues. *Nature Nanotechnology, 13*, 677–684. https://doi.org/10.1038/s41565-018-0131-1

Kale, S. S., Chauhan, R., Nigam, B., Gosavi, S., & Chaudhary, I. J. (2024). Effectiveness of nanoparticles in improving soil fertility and eco-friendly crop resistance: A comprehensive review. *Biocatalysis and Agricultural Biotechnology, 56*, 103066. https://doi.org/10.1016/j.bcab.2024.103066

Keller, A. A., McFerran, S., Lazareva, A., & Suh, S. (2013). Global life cycle releases of engineered nanomaterials. *Journal of Nanoparticle Research, 15*, 1692. https://doi.org/10.1007/s11051-013-1692-4

Khan, S., Almarhoon, Z. M., Bakht, J., Mabkhot, Y. N., Rauf, A., & Shad, A. A. (2022). Single-step acer pentapomicum-mediated green synthesis of silver nanoparticles and their potential antimicrobial and antioxidant activities. *Journal of Nanomaterials, 2022*, 3783420. https://doi.org/10.1155/2022/3783420

Khorrami, S., Zarepour, A., & Zarrabi, A. (2019). Green synthesis of silver nanoparticles at low temperature in a fast pace with unique DPPH radical scavenging and selective cytotoxicity against MCF-7 and BT-20 tumor cell lines. *Biotechnology Reports, 24*, e00393. https://doi.org/10.1016/j.btre.2019.e00393

Kumar, S. A., Inbanathan, S. S. R., Manikandan, A., Topare, N. S., Srinivasan, S., Kumar, M. A., Mahmoud, M. H., Fouad, H., Ansari, A., & Priya, L. S. (2024). Green synthesis structural, morphological and sonocatalytic properties of CuO nanoparticles assisted by Camellia sinensis

(Tea) extract. *Industrial Crops and Products, 219*, 118915. https://doi.org/10.1016/j.indcrop.2024.118915

Lan, K., Wang, H. S. H., Lee, T., de Assis, C. A., Venditti, R. A., Zhu, Y., & Yao, Y. (2024). A modeling framework to identify environmentally greener and lower-cost pathways of nanomaterials. *Green Chemistry, 26*, 3466–3478. https://doi.org/10.1039/d3gc04036d

Loza, K., Epple, M., & Maskos, M. (2019). Stability of nanoparticle dispersions and particle agglomeration. In P. Gehr & R. Zellner (Eds.), *Biological responses to nanoscale particles: Molecular and cellular aspects and methodological approaches* (pp. 85–100). Springer International Publishing.

Manojkumar, M. S., Sivaprakash, B., Arumugam, N., & Almansour, A. I. (2024). Banana peel extract for CeO2 nanoflower synthesis: Enhancing photocatalytic activity for methyl orange dye removal and bactericidal effects. *Environmental Research, 252*, 118785. https://doi.org/10.1016/j.envres.2024.118785

Martínez-Barbosa, M. E., & Figueroa-Pizano, M. D. (2024). Chapter 2 – Green synthesis and methodologies of nanomaterials: State of the art. In B. Sharma, S. Thomas, P. Kumar Bajpai, K. Ghosal, & S. Shekhar (Eds.), *Advances in bionanocomposites* (pp. 17–54). Elsevier.

Mary Isabella Sonali, J., Kavitha, R., Kumar, P. S., Rajagopal, R., Gayathri, K. V., Ghfar, A. A., & Govindaraju, S. (2022). Application of a novel nanocomposite containing micro-nutrient solubilizing bacterial strains and CeO2 nanocomposite as bio-fertilizer. *Chemosphere, 286*, 131800. https://doi.org/10.1016/j.chemosphere.2021.131800

Massari, A., Beggio, M., Hreglich, S., Marin, R., & Zuin, S. (2014). Behavior of TiO2 nanoparticles during incineration of solid paint waste: A lab-scale test. *Waste Management, 34*, 1897–1907. https://doi.org/10.1016/j.wasman.2014.05.015

Mitrano, D. M., Motellier, S., Clavaguera, S., & Nowack, B. (2015). Review of nanomaterial aging and transformations through the life cycle of nano-enhanced products. *Environment International, 77*, 132–147. https://doi.org/10.1016/j.envint.2015.01.013

Molnár, Z., Bódai, V., Szakacs, G., Erdélyi, B., Fogarassy, Z., Sáfrán, G., Varga, T., Kónya, Z., Tóth-Szeles, E., Szucs, R., & Lagzi, I. (2018). Green synthesis of gold nanoparticles by thermophilic filamentous fungi. *Scientific Reports, 8*, 1–12. https://doi.org/10.1038/s41598-018-22112-3

Mousavi-Khattat, M., Keyhanfar, M., & Razmjou, A. (2018). A comparative study of stability, antioxidant, DNA cleavage and antibacterial activities of green and chemically synthesized silver nanoparticles. *Artificial Cells, Nanomedicine, and Biotechnology, 46*, S1022–S1031. https://doi.org/10.1080/21691401.2018.1527346

Ndaba, B., Roopnarain, A., Rama, H., & Maaza, M. (2022). Biosynthesized metallic nanoparticles as fertilizers: An emerging precision agriculture strategy. *Journal of Integrative Agriculture, 21*, 1225–1242. https://doi.org/10.1016/S2095-3119(21)63751-6

Noman, M., Shahid, M., Ahmed, T., Tahir, M., Naqqash, T., Muhammad, S., Song, F., Abid, H. M. A., & Aslam, Z. (2020). Green copper nanoparticles from a native *Klebsiella pneumoniae* strain alleviated oxidative stress impairment of wheat plants by reducing the chromium bioavailability and increasing the growth. *Ecotoxicology and Environmental Safety, 192*, 110303. https://doi.org/10.1016/j.ecoenv.2020.110303

Oestreicher, V., García, C. S., Soler-Illia, G. J. A. A., & Angelomé, P. C. (2019). Gold recycling at laboratory scale: From nanowaste to nanospheres. *ChemSusChem, 12*, 4882–4888. https://doi.org/10.1002/cssc.201901488

Ohiduzzaman, M., Khan, M. N. I., Khan, K. A., & Paul, B. (2024). Biosynthesis and characterizations of silver nanoparticles by using green banana peel extract: Evaluation of their antibacterial and electrical performances. *Heliyon, 10*, e31140. https://doi.org/10.1016/j.heliyon.2024.e31140

Osonga, F. J., Yazgan, I., Kariuki, V., Luther, D., Jimenez, A., Le, P., & Sadik, O. A. (2016). Greener synthesis and characterization, antimicrobial and cytotoxicity studies of gold nanoparticles of novel shapes and sizes. *RSC Advances, 6*, 2302–2313. https://doi.org/10.1039/c5ra22906e

Part, F., Berge, N., Baran, P., Stringfellow, A., Sun, W., Bartelt-Hunt, S., Mitrano, D., Li, L., Hennebert, P., Quicker, P., Bolyard, S. C., & Huber-Humer, M. (2018). A review of the fate of engineered nanomaterials in municipal solid waste streams. *Waste Management, 75*, 427–449. https://doi.org/10.1016/j.wasman.2018.02.012

Part, F., Zaba, C., Bixner, O., Zafiu, C., Lenz, S., Martetschläger, L., Hann, S., Huber-Humer, M., & Ehmoser, E. K. (2020). Mobility and fate of ligand stabilized semiconductor nanoparticles in landfill leachates. *Journal of Hazardous Materials, 394*, 122477. https://doi.org/10.1016/j.jhazmat.2020.122477

Pati, P., McGinnis, S., & Vikesland, P. J. (2016). Waste not want not: Life cycle implications of gold recovery and recycling from nanowaste. *Environmental Science: Nano, 3*, 1133–1143. https://doi.org/10.1039/c6en00181e

Perveen, R., Shujaat, S., Naz, M., Qureshi, M. Z., Nawaz, S., Shahzad, K., & Ikram, M. (2021). Green synthesis of antimicrobial silver nanoparticles with Brassicaceae seeds. *Materials Research Express, 8*, 055007. https://doi.org/10.1088/2053-1591/ac006b

Phan, T. T. V., Nguyen, Q. V., & Huynh, T. C. (2020). Simple, green, and low-temperature method for preparation of palladium nanoparticles with controllable sizes and their characterizations. *Journal of Nanoparticle Research, 22*, 1–11. https://doi.org/10.1007/s11051-020-04801-1

Rafique, M., Sadaf, I., Rafique, M. S., & Tahir, M. B. (2017). A review on green synthesis of silver nanoparticles and their applications. *Artificial Cells, Nanomedicine, and Biotechnology, 45*, 1272–1291. https://doi.org/10.1080/21691401.2016.1241792

Rahman, A., Kang, S., McGinnis, S., & Vikesland, P. J. (2022). Life cycle impact assessment of iron oxide ($Fe3O4/\gamma$-Fe2O3) nanoparticle synthesis routes. *ACS Sustainable Chemistry & Engineering, 10*, 3155–3165. https://doi.org/10.1021/acssuschemeng.1c05763

Rajput, V. D., Minkina, T., Suskova, S., Mandzhieva, S., Tsitsuashvili, V., Chapligin, V., & Fedorenko, A. (2018). Effects of copper nanoparticles (CuO NPs) on crop plants: A mini review. *BioNanoSci, 8*, 36–42. https://doi.org/10.1007/s12668-017-0466-3

Ramachandran, R., Krishnaraj, C., Kumar, V. K. A., Harper, S. L., Kalaichelvan, T. P., & Yun, S. I. (2018). In vivo toxicity evaluation of biologically synthesized silver nanoparticles and gold nanoparticles on adult zebrafish: A comparative study. *3 Biotech, 8*, 1–12. https://doi.org/10.1007/s13205-018-1457-y

Rana, A., Yadav, K., & Jagadevan, S. (2020). A comprehensive review on green synthesis of nature-inspired metal nanoparticles: Mechanism, application and toxicity. *Journal of Cleaner Production, 272*, 122880. https://doi.org/10.1016/j.jclepro.2020.122880

Rauwel, P., Küünal, S., Ferdov, S., & Rauwel, E. (2015). A review on the green synthesis of silver nanoparticles and their morphologies studied via TEM. *Advances in Materials Science and Engineering, 2015*, 682749. https://doi.org/10.1155/2015/682749

Razavi, R., Molaei, R., Moradi, M., Tajik, H., Ezati, P., & Shafipour Yordshahi, A. (2020). Biosynthesis of metallic nanoparticles using mulberry fruit (Morus alba L.) extract for the preparation of antimicrobial nanocellulose film. *Applied Nanoscience (Switzerland), 10*, 465–476. https://doi.org/10.1007/s13204-019-01137-8

Saad, A. M., Alabdali, A. Y. M., Ebaid, M., Salama, E., El-Saadony, M. T., Selim, S., Safhi, F. A., ALshamrani, S. M., Abdalla, H., Mahdi, A. H. A., & El-Saadony, F. M. A. (2022). Impact of green chitosan nanoparticles fabricated from shrimp processing waste as a source of nano nitrogen fertilizers on the yield quantity and quality of wheat (Triticum aestivum L.) cultivars. *Molecules, 27*, 5640. https://doi.org/10.3390/molecules27175640

Saif, S., Tahir, A., & Chen, Y. (2016). Green synthesis of iron nanoparticles and their environmental applications and implications. *Nanomaterials, 6*, 1–26. https://doi.org/10.3390/nano6110209

Saldanha, P. L., Lesnyak, V., & Manna, L. (2017). Large scale syntheses of colloidal nanomaterials. *Nano Today, 12*, 46–63. https://doi.org/10.1016/j.nantod.2016.12.001

Salem, S. S., & Fouda, A. (2021). Green synthesis of metallic nanoparticles and their prospective biotechnological applications: An overview. *Biological Trace Element Research, 199*, 344–370. https://doi.org/10.1007/s12011-020-02138-3

Salieri, B., Turner, D. A., Nowack, B., & Hischier, R. (2018). Life cycle assessment of manufactured nanomaterials: Where are we? *NanoImpact, 10*, 108–120. https://doi.org/10.1016/j.impact.2017.12.003

Saravanan, A., Kumar, P. S., Hemavathy, R. V., Jeevanantham, S., Jawahar, M. J., Neshaanthini, J. P., & Saravanan, R. (2022). A review on synthesis methods and recent applications of nanomaterial in wastewater treatment: Challenges and future perspectives. *Chemosphere, 307*, 135713. https://doi.org/10.1016/j.chemosphere.2022.135713

Semenova, N. A., Burmistrov, D. E., Shumeyko, S. A., & Gudkov, S. V. (2024). Fertilizers based on nanoparticles as sources of macro- and microelements for plant crop growth: A review. *Agronomy, 14*, 1646. https://doi.org/10.3390/agronomy14081646

Shah, M., Fawcett, D., Sharma, S., Tripathy, S. K., & Poinern, G. E. J. (2015). Green synthesis of metallic nanoparticles via biological entities. *Materials, 8*, 7278–7308. https://doi.org/10.3390/ma8115377

Shankar, S. S., Rai, A., Ankamwar, B., Singh, A., Ahmad, A., & Sastry, M. (2004). Biological synthesis of triangular gold nanoprisms. *Nature Materials, 3*, 482–488. https://doi.org/10.1038/nmat1152

Shankar, S. S., Rai, A., Ahmad, A., & Sastry, M. (2005). Controlling the optical properties of lemongrass extract synthesized gold nanotriangles and potential application in infrared-absorbing optical coatings. *Chemistry of Materials, 17*, 566–572. https://doi.org/10.1021/cm048292g

Sharma, P., Urfan, M., Anand, R., Sangral, M., Hakla, H. R., Sharma, S., Das, R., Pal, S., & Bhagat, M. (2022). Green synthesis of zinc oxide nanoparticles using Eucalyptus lanceolata leaf litter: Characterization, antimicrobial and agricultural efficacy in maize. *Physiology and Molecular Biology of Plants, 28*, 363–381. https://doi.org/10.1007/s12298-022-01136-0

Shinde, S., Paralikar, P., Ingle, A. P., & Rai, M. (2020). Promotion of seed germination and seedling growth of *Zea mays* by magnesium hydroxide nanoparticles synthesized by the filtrate from *Aspergillus niger*. *Arabian Journal of Chemistry, 13*, 3172–3182. https://doi.org/10.1016/j.arabjc.2018.10.001

Shukla, R., Bansal, V., Chaudhary, M., Basu, A., Bhonde, R. R., & Sastry, M. (2005). Biocompatibility of gold nanoparticles and their endocytotic fate inside the cellular compartment: A microscopic overview. *Langmuir, 21*, 10644–10654. https://doi.org/10.1021/la0513712

Singh, J., Dutta, T., Kim, K.-H., Rawat, M., Samddar, P., & Kumar, P. (2018). "Green" synthesis of metals and their oxide nanoparticles: Applications for environmental remediation. *Journal of Nanbiotechnology, 16*, 84. https://doi.org/10.1186/s12951-018-0408-4

Slimani, C., & Rais, C. (2024). Advances in green chemistry: Plant and microbial processes for sustainable nanoparticle synthesis, and their antimicrobial and anticancer applications. *Chemical Review and Letters, 7*, 573–583. https://doi.org/10.22034/crl.2024.457614.1337

Soliemanzadeh, A., & Fekri, M. (2021). Effects of green iron nanoparticles on iron changes and phytoavailability in a calcareous soil. *Pedosphere, 31*, 761–770. https://doi.org/10.1016/S1002-0160(21)60035-8

Su, Y., Zhou, X., Meng, H., Xia, T., Liu, H., Rolshausen, P., Roper, C., McLean, J. E., Zhang, Y., Keller, A. A., & Jassby, D. (2022). Cost–benefit analysis of nanofertilizers and nanopesticides emphasizes the need to improve the efficiency of nanoformulations for widescale adoption. *Nature Food, 3*, 1020–1030. https://doi.org/10.1038/s43016-022-00647-z

Suhail, M., & Prathibha, N. (2022). Green synthesis and characterisation of silver nanoparticles using Cassia tora seed extract and investigation of antibacterial potential. *Applied Biochemistry and Biotechnology, 194*, 464–478. https://doi.org/10.1007/s12010-021-03651-4

Tarakanov, R., Shagdarova, B., Lyalina, T., Zhuikova, Y., Il'ina, A., Dzhalilov, F., & Varlamov, V. (2023). Protective properties of copper-loaded chitosan nanoparticles against soybean pathogens Pseudomonas savastanoi pv. glycinea and Curtobacterium flaccumfaciens pv. flaccumfaciens. *Polymers, 15*, 1100. https://doi.org/10.3390/polym15051100

Tauseef, A., & Uddin, I. (2024). Novel insights on sustainable nanoparticles in crop protection: Current status and future prospectives. In I. Uddin (Ed.), *Sustainable nanomaterials: Synthesis and environmental applications* (pp. 249–270). Springer Nature.

Terenteva, E. A., Apyari, V. V., Dmitrienko, S. G., & Zolotov, Y. A. (2015). Formation of plasmonic silver nanoparticles by flavonoid reduction: A comparative study and application for determination of these substances. *Spectrochimica Acta – Part A: Molecular and Biomolecular Spectroscopy, 151*, 89–95. https://doi.org/10.1016/j.saa.2015.06.049

Thangavelu, R. M., da Silva, W. L., Zuverza-Mena, N., Dimkpa, C. O., & White, J. C. (2024). Nano-sized metal oxide fertilizers for sustainable agriculture: Balancing benefits, risks, and risk management strategies. *Nanoscale, 16*, 19998–20026. https://doi.org/10.1039/d4nr01354a

Tirumala, M. G., Anchi, P., Raja, S., Rachamalla, M., & Godugu, C. (2021). Novel methods and approaches for safety evaluation of nanoparticle formulations: A focus towards in vitro models and adverse outcome pathways. *Frontiers in Pharmacology, 12*, 612659. https://doi.org/10.3389/fphar.2021.612659

Vijayaram, S., Razafindralambo, H., Sun, Y.-Z., Vasantharaj, S., Ghafarifarsani, H., Hoseinifar, S. H., & Raeeszadeh, M. (2024). Applications of green synthesized metal nanoparticles—A review. *Biological Trace Element Research, 202*, 360–386. https://doi.org/10.1007/s12011-023-03645-9

Wahid, I., Kumari, S., Ahmad, R., Hussain, S. J., Alamri, S., Siddiqui, M. H., & Khan, M. I. R. (2020). Silver nanoparticle regulates salt tolerance in wheat through changes in ABA concentration, ion homeostasis, and defense systems. *Biomolecules, 10*, 1506. https://doi.org/10.3390/biom10111506

Walser, T., Limbach, L. K., Brogioli, R., Erismann, E., Flamigni, L., Hattendorf, B., Juchli, M., Krumeich, F., Ludwig, C., Prikopsky, K., Rossier, M., Saner, D., Sigg, A., Hellweg, S., Günther, D., & Stark, W. J. (2012). Persistence of engineered nanoparticles in a municipal solid-waste incineration plant. *Nature Nanotechnology, 7*, 520–524. https://doi.org/10.1038/nnano.2012.64

Wu, F., Zhou, Z., & Hicks, A. L. (2019). Life cycle impact of titanium dioxide nanoparticle synthesis through physical, chemical, and biological routes. *Environmental Science and Technology, 53*, 4078–4087. https://doi.org/10.1021/acs.est.8b06800

Xu, L., Zhu, Z., & Sun, D.-W. (2021). Bioinspired nanomodification strategies: Moving from chemical-based agrosystems to sustainable agriculture. *ACS Nano, 15*, 12655–12686. https://doi.org/10.1021/acsnano.1c03948

Yahyaei, B., Nouri, M., Bakherad, S., Hassani, M., & Pourali, P. (2019). Effects of biologically produced gold nanoparticles: Toxicity assessment in different rat organs after intraperitoneal injection. *AMB Express, 9*, 38. https://doi.org/10.1186/s13568-019-0762-0

Yang, Z., Xiao, Y., Jiao, T., Zhang, Y., Chen, J., & Gao, Y. (2020). Effects of copper oxide nanoparticles on the growth of rice (Oryza sativa L.) seedlings and the relevant physiological responses. *International Journal of Environmental Research and Public Health, 17*, 1260. https://doi.org/10.3390/ijerph17041260

Ying, S., Guan, Z., Ofoegbu, P. C., Clubb, P., Rico, C., He, F., & Hong, J. (2022). Green synthesis of nanoparticles: Current developments and limitations. *Environmental Technology and Innovation, 26*, 102336. https://doi.org/10.1016/j.eti.2022.102336

Younas, W., Khan, F. U., Zaman, M., Lin, D., Zuberi, A., & Wang, Y. (2022). Toxicity of synthesized silver nanoparticles in a widespread fish: A comparison between green and chemical. *Science of the Total Environment, 845*, 157366. https://doi.org/10.1016/j.scitotenv.2022.157366

Zhang, H., Bussini, D., Hortal, M., Elegir, G., Mendes, J., & Jordá Beneyto, M. (2016). PLA coated paper containing active inorganic nanoparticles: Material characterization and fate of nanoparticles in the paper recycling process. *Waste Management, 52*, 339–345. https://doi.org/10.1016/j.wasman.2016.03.045

Zhang, J., Guo, W., Li, Q., Wang, Z., & Liu, S. (2018). The effects and the potential mechanism of environmental transformation of metal nanoparticles on their toxicity in organisms. *Environmental Science: Nano, 5*, 2482–2499. https://doi.org/10.1039/c8en00688a

Zhang, H., Chen, S., Jia, X., Huang, Y., Ji, R., & Zhao, L. (2021). Comparation of the phytotoxicity between chemically and green synthesized silver nanoparticles. *Science of the Total Environment, 752*, 142264. https://doi.org/10.1016/j.scitotenv.2020.142264

Zhang, X., Sathiyaseelan, A., Naveen, K. V., Lu, Y., & Wang, M.-H. (2023). Research progress in green synthesis of manganese and manganese oxide nanoparticles in biomedical and environmental applications – A review. *Chemosphere, 337*, 139312. https://doi.org/10.1016/j.chemosphere.2023.139312

Zhao, P., Lu, Y., Li, L., Wang, J., El-kott, A. F., AlShehri, M. A., & Morsy, K. (2024). Biopreparation of titanium dioxide nanoparticles mediated by Calendula officinalis flower extract for catalytic reduction of 4-nitrophenol and treatment of allergic conjunctivitis. *Journal of Organometallic Chemistry, 1016*, 123260. https://doi.org/10.1016/j.jorganchem.2024.123260

Zuanazzi, N. R., Ghisi, N. D. C., & Oliveira, E. C. (2020). Analysis of global trends and gaps for studies about 2,4-D herbicide toxicity: A scientometric review. *Chemosphere, 241*, 125016. https://doi.org/10.1016/j.chemosphere.2019.125016

Chapter 14
Circular Economy Principles

Deepshikha Gupta, Pratibha Sharma, and Tejendra Kumar Gupta

14.1 Introduction

The world is moving forward with amazing new inventions and technologies, the climate is nevertheless becoming worse due to growing carbon footprint. The use of technology and the creation of possibilities that reduce waste would result in a 30% reduction in emissions worldwide. The United Nations Climate Change Conference (COP21) held in Parin in 2015 adopted a landmark international treaty, "The Paris Agreement," that aims to combat climate change by limiting global warming to well below 2 °C, preferably to 1.5 °C, compared to pre-industrial levels (United Nations Climate Change, 2025). The key objectives of this agreement are to reduce greenhouse gas (GHG) emissions by the member Countries, achieve climate neutrality (Net-Zero Emissions) by 2050, by ensuring greenhouse gas emissions are balanced by removal (e.g., through carbon capture, reforestation). Developed nations are committed to mobilizing $100 billion per year to provide financial and technological support to assist developing countries in climate mitigation and adaptation. They also commit to strengthening climate resilience by helping vulnerable nations adapt to climate impacts such as rising sea levels, extreme weather events, and food security challenges. The agreement entered into force in 2016, and 196 countries (including the USA, China, and India) have signed it. The USA temporarily withdrew under the Trump administration from the agreement in January 2025, as part of his efforts to increase US production of fossil fuels (The Hindu, Jan 22, 2025).

The circular economy is the need of the hour as the world grapples with global warming, resource depletion, and excessive waste generation. The traditional linear economy ("take-make-dispose") model is unsustainable, driving high carbon emissions, deforestation, and pollution, which accelerate climate change (Silva-Alvarado

D. Gupta (✉) · P. Sharma · T. K. Gupta
Department of Chemistry, Amity Institute of Applied Sciences, Amity University Uttar Pradesh, Noida, India
e-mail: dgupta2@amity.edu; psharma25@amity.edu; tkgupta@amity.edu

K. Esquivel Escalante et al. (eds.), *Green Nanomaterials*, Nanotechnology in the Life Sciences, https://doi.org/10.1007/978-3-032-04492-1_14

et al., 2023). By shifting to a circular approach, we can significantly reduce greenhouse gas emissions, minimize resource extraction, and curb methane emissions from landfills, all of which are major contributors to global warming. Industries like cement, steel, and plastics, which account for 70% of industrial CO_2 emissions, can be transformed through recycling, renewable energy, and sustainable manufacturing (Holappa, 2020). Additionally, regenerative agriculture and biodegradable materials can help sequester carbon and restore ecosystems. Studies show that a well-implemented circular economy can cut global CO_2 emissions by 39% and keep the temperature rise below 2 °C, aligning with the Paris Agreement goals. As we move toward a climate-resilient future, adopting circular strategies is no longer an option but an urgent necessity to decarbonize industries, conserve natural resources, and create a truly sustainable planet.

14.2 Why Circular Economy?

The linear economy follows a "take-make-dispose" model, where raw materials are extracted, processed into products, used, and then discarded as waste. This approach leads to resource depletion, excessive waste generation, and high carbon emissions, making it unsustainable in the long run (Marino & Pariso, 2016). As environmental issues worsen, these ideas have become more and more important. The necessity for a sustainable strategy is evident, given the yearly generation of over 2 billion tons of solid garbage and the estimated 8 million metric tons of plastic that reach the oceans. Recycling is only one aspect of this paradigm; it also rethinks how products are made, utilized, and discarded to ensure that their environmental impact is kept to a minimum.

In contrast, the circular economy focuses on minimizing waste and maximizing resource efficiency by promoting reuse, recycling, and regeneration. Instead of discarding products, materials are kept in circulation through repair, refurbishment, and sustainable design, significantly reducing environmental impact and greenhouse gas emissions. While the linear model accelerates climate change and pollution, the circular model supports sustainability, lowers carbon footprints, and conserves natural resources, making it essential for a greener and more resilient future (Kirchherr & Piscicelli, 2019; Kirchherr et al., 2023). By 2029 supply chains will not be allowed to produce any waste, and circular economy will be the only economy as per surveys (Henrik Hvid Jensen; Chief Technology Strategist NEE Jan 27, 2022).

There are numerous advantages to switching to a circular economy in terms of the environment, the economy, and society. It promises a future where economic growth is attained in balance with social well-being and environmental stewardship, offering a sustainable substitute for the conventional linear paradigm.

- *Environmental Benefits*: The circular economy's favorable effects on the environment rank among its most important advantages. Trash minimization lowers

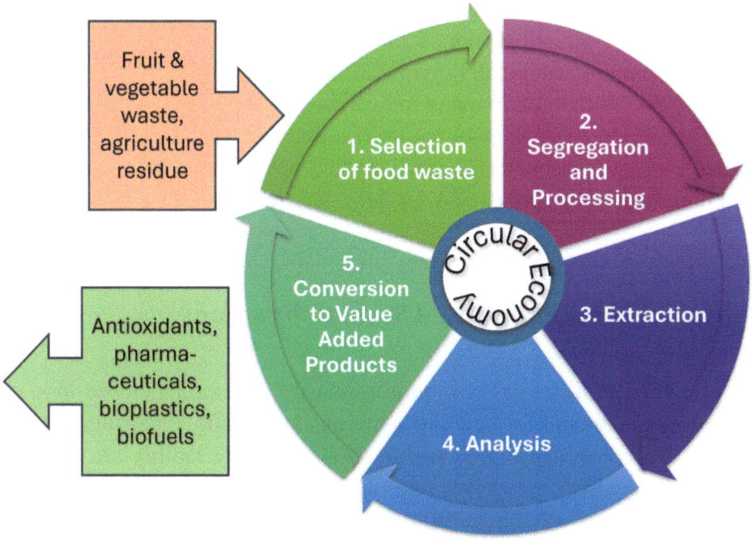

Fig. 14.1 A generic representation of circular economic framework starting from agriculture waste to value added products

pollutants and greenhouse gas emissions related to trash disposal in addition to relieving pressure on landfills and waste management systems. A circular economic approach for the utilization of fruit, vegetable, and agricultural waste and converting them into more useful products like antioxidants, biofuels, and pharmaceuticals is represented in Fig. 14.1. Additionally, the circular economy encourages resource conservation by extending the lifespan of materials and products, hence lowering the demand for raw material extraction and processing. The preservation of biodiversity and the fight against resource depletion depend heavily on this conservation.

By promoting the creation of more robust and easily repairable products, the circular economy also helps to reduce waste. As a result of this change, the environmental pollution linked to the production of new goods is significantly decreased, including carbon emissions and contamination of the air and water. The circular economy also aids in reducing the effects of climate change by emphasizing renewable energy and resources.

- *Economic Benefits:* The efficient use of materials and resources can reduce production costs and opens new business opportunities in the fields of recycling, refurbishing, and sustainable product design. The circular economy model requires a range of new skills and roles, leading to the creation of a substantial number of jobs.
- *Social Benefits:* The circular economy has several positive societal effects. Community relationships can be improved, and a sense of shared environmental

responsibility can be fostered through local recycling and renovation projects. These programs, which support social cohesiveness, local economic development, and creativity, frequently involve neighborhood businesses and community organizations. Innovation in many industries is fueled by the need for sustainable product design, economic resource utilization, and efficient waste management. New technology and business strategies that help the environment and provide consumers with better goods and services may result from this innovation.

14.3 Core Principles of the Circular Economy

The key principles of a circular economy emphasize developing closed-loop systems that minimize the demand for new raw materials by reusing, repairing, refurbishing, or recycling resources. Durability, repairability, and recyclability are designed into products to prolong their life cycles, and sustainable materials and renewable energy sources take the place of finite resources. In order to achieve the objective of a sustainable, circular economy, each of these principles contributes in a different but complementary way. Implementing CE principles is becoming more and more encouraged as a practical way to accomplish the Sustainable Development Goals (SDGs), a group of 17 global objectives that the UN approved in 2015 as part of the 2030 Agenda for Sustainable Development THE 17 GOALS (2015). The 17 UN Sustainable Development Goals are represented and listed in Fig. 14.2. They aim to address global challenges, including poverty, inequality, climate change, environmental degradation, peace, and justice. These goals impact societies by promoting economic growth, social inclusion, and environmental sustainability. For instance,

Fig. 14.2 17 UN Sustainable Development Goals that can be enabled by implementing circular economy either partially or strongly. (Reproduced from (Velenturf & Purnell, 2021, Henrik Hvid Jensen; Chief Technology Strategist NEE Jan 27, 2022))

SDG 1 (No Poverty) and SDG 2 (Zero Hunger) work to eradicate poverty and food insecurity, improving livelihoods. SDG 3 (Good Health and Well-being) enhances healthcare systems, while SDG 4 (Quality Education) ensures equitable learning opportunities. Gender equality (SDG 5) fosters empowerment, and clean water and sanitation (SDG 6) improve public health. SDGs 7–9 focus on affordable clean energy, economic development, and innovation, SDGs 10–12 promote reduced inequalities, sustainable cities, and responsible consumption. Climate action (SDG 13) and life-focused goals (SDGs 14 & 15) protect ecosystems, and SDGs 16 & 17 emphasize peace, justice, and global partnerships. These SDGs aim to create a sustainable and equitable world, addressing social, economic, and environmental. These goals are interconnected, aiming to end poverty, protect the planet, and ensure prosperity for all by 2030. They guide governments, businesses, and individuals toward a more sustainable and equitable future.

Nowadays, 42% of people on the planet make their living from agriculture, which also powers the economies of most emerging nations. We will now focus on each principle of the circular economy in detail, emphasizing the agriculture sector.

14.3.1 Principle 1: Design Out Waste and Pollution

Products and processes are designed to prevent waste and pollution from being created in the first place. This includes using non-toxic materials, designing for durability, repairability, and recyclability, and reducing resource consumption during manufacturing.

The core of circular agriculture is the reduction of resource input. It is about utilizing cutting-edge techniques to preserve resources like water, energy, and nutrients so that you can accomplish more with less. This lowers expenses for farmers and agribusinesses in addition to protecting the environment. In addition to increasing crop yields and decreasing waste, this also shields neighboring ecosystems from runoff. For example, the most effective irrigation method to minimize wastewater is drip irrigation which delivers water directly to the base of plants through a network of tubes and emitters, significantly reducing runoff and evaporation, and saving 80% wastage of water compared to overhead sprinkler systems (van der Kooij et al., 2013).

The system's effectiveness is fostered by identifying and eliminating negative externalities. For example, negative externalities in agriculture are responsible for soil contamination due to the inappropriate use of fertilizers, herbicides, and pesticides (Aznar-Sánchez et al., 2019). However, most developed countries have laws to limit or prohibit the use of these products, which has led to the substitution of chemical fertilizers for organic fertilizers or the development of biological pest-control systems (Cobo et al., 2019). The combined production of crops and livestock fisheries has proven effective in minimizing the use of harmful products. Another important issue is the conservation of bodies of water, which are currently overexploited and subject to severe degradation as a result of agricultural activity (Aznar-Sánchez et al., 2019).

14.3.2 Principle 2: Close Nutrient Loops

In traditional agriculture, declining soil fertility is due to the depletion of nutrients from the soil without being replenished. Returning valuable nutrients to the soil and ecosystems to improve biodiversity and resilience is known as "closing nutrient loops." Additionally, organic waste from food processing and farms is left unutilized, contributing to contamination in the environment. By recycling nutrients back into the system, circular agriculture provides a strategy that emulates natural ecosystems in which nothing is wasted. For example, for a vegetable farm using compost made from leftover produce and crop residues reduces chemical fertilizer dependency, enhances soil health, and minimizes the amount of organic waste sent to landfills. Practices like crop rotation, agroforestry, and intercropping help maintain soil fertility and reduce the need for external inputs, promoting a more sustainable system.

Key Practices

- Composting: You may make nutrient-rich organic fertilizer by composting food wastes, crop residues, and animal dung. This enhances microbiological health and soil structure in addition to lowering waste.
- Manure Management: Animal manure can be processed by livestock farms to create digestate, a by-product that can be used as fertilizer, and biogas.
- Waste-to-Technologies for Fertilizer: Agricultural waste can be converted into useful soil amendments using sophisticated techniques like anaerobic digestion or the creation of biochar.

14.3.3 Principle 3: Reduce and Repurpose Waste

The life of products can be extended by promoting reuse, repair, refurbishment, remanufacturing, and recycling. Shifting towards product-as-a-service models (e.g., leasing or renting instead of selling) will encourage a system for sharing, reselling, and repurposing items.

Waste is sometimes thrown away in traditional farming systems. Circular agriculture, on the other hand, promotes a change in perspective by turning food waste and agricultural waste into useful resources. This approach lowers the quantity of garbage dumped in landfills while opening doors to new sources of income and increasing farm profitability. By converting residues and by-products into useful resources, efficient waste management in circular agriculture decreases the impact of landfills while promoting sustainability and economic expansion. Technological development has enabled a variety of materials to be used in many processes before their permanent disposal, such as in the production of bioenergy and for soil amendment and bio-fertilizers, or as livestock feed. For example, a part of straw from harvested crops in a wheat farm can be used to produce biogas, which powers the farm's machinery, while other parts may be used as animal feed for nearby livestock farms, reducing waste and enhancing the farm's sustainability.

Key Practices

- Biofuel from Crop Residues: Straw and maize stalks are examples of leftover crop materials that can be converted into biogas or burned to produce bioenergy. This gives farms or nearby populations an alternate energy source while lowering dependency on fossil fuels.
- Animal Feed: Damaged or infected fruits and vegetables as well as other crop leftovers after harvesting, can be turned into nutritious animal feed, minimizing food waste and lowering the need to buy outside feed.
- Upcycling Food Waste: On a larger scale, food waste from retailers or processing facilities can be turned into high-value goods like fertilizer, compost, or even brand-new foods like snacks, jams, or juices.

14.3.4 Principle 4: Regenerate Natural Ecosystems

Using regenerative farming techniques to actively work toward the regeneration or replenishment of natural ecosystems is one of the basic principles of circular agriculture. Circular agriculture aims to build more resilient and sustainable farming systems by boosting biodiversity, enhancing water quality, and restoring soil health rather than destroying the environment. The regenerative concept aids in the restoration of the planet for coming generations.

Key Practices

- Agroforestry: This is a technique that incorporates trees and plants in farms. In addition to aiding in carbon sequestration, trees also enhance soil structure, lessen erosion, and offer wildlife habitat. Farmers can increase biodiversity and replicate natural ecosystems by growing crops alongside trees.
- Cover Cropping: Cover crops like clover, rye, or vetch can be planted between main crops, which helps to restore soil health. It prevents soil erosion, improves water infiltration, and adds organic matter to the soil. These crops also help to fix nitrogen in the soil, thereby reducing the need for chemical fertilizers.
- Integrated Pest Management (IPM): IPM employs a mix of mechanical, cultural, and biological methods to manage pests instead of depending on artificial pesticides. This strategy safeguards the environment and public health by reducing of use of chemical fertilizers.

Example
By growing shade trees next to coffee plants, farmers can practice agroforestry on coffee farms, which enhances biodiversity and soil quality. They might lessen their reliance on artificial fertilizers by using cover crops, such as legumes, to organically increase soil fertility. The farm might also implement IPM techniques, which would introduce beneficial insects that regulate pest populations and maintain the equilibrium of the ecosystem.

14.3.5 Principle 5: Foster Collaboration Across the Supply Chain and Innovation

A fundamental idea of circular agriculture is to promote collaboration across the whole supply chain. Establish circular value chains and encourage collaboration between businesses, governments, and communities in order to develop innovative, sustainable materials, technologies, and business models. Together, farmers, food processors, retailers, and even consumers may form strong alliances to put circular solutions into practice. Instead of operating independently, different stakeholders can collaborate to increase resource utilization, cut waste, and establish a sustainable food system by exchanging ideas, resources, and expertise. Understanding that each link in the supply chain contributes to sustainability and that the advantages increase when they cooperate is key.

Key Practices
- Collaborating with Food Processors: Direct collaboration between farmers and food processors may guarantee that by-products such as fruit peels, shells, or stems are recycled into useful goods like compost, animal feed, or biofuel. This enhances the supply chain's overall resource efficiency and reduces waste.
- Retailer Partnerships: Farmers and retailers may work together to better plan output to meet consumer demand, which will cut down on food waste and overproduction. Additionally, retailers can encourage consumers to make more sustainable purchases by promoting sustainable items.
- Consumer Engagement: In the context of circular agriculture, consumers are vital. Consumer education regarding trash reduction, packaging reuse, and sustainable choices increases demand for circular products and aids in closing the loop. Landfill trash can be decreased, for example, by encouraging the use of compostable packaging.

Example
One instance of effective cooperation is the farm-to-table program, which aims to decrease waste by bringing together farmers, nearby food processors, and retailers. To use the leftover components (such as skin and pulp) for other food products, farmers may sell their excess straight to processors. In order to create a more sustainable consumption cycle, retailers then assist in promoting these products to customers through messages centered around waste reduction.

14.3.6 Principle 6: Prioritize Renewable Energy and Materials

Adopting renewable energy sources, such as solar, wind, or bioenergy, reduces the carbon footprint of farms. Transition to renewable energy sources and materials that can regenerate over time is the demand of this time. By shifting away from fossil fuels to renewable energy sources like solar, wind, and bioenergy, industries can significantly lower carbon emissions and decrease reliance on finite resources. The renewable energy sources utilization to run different agricultural machines is depicted in Fig. 14.3. At the same time, using biodegradable, recyclable, and sustainably sourced materials minimizes waste and pollution while promoting resource regeneration. Innovations such as bio-based plastics, recycled metals, and sustainable textiles help close the material loop, reducing extraction pressure on the planet. A circular economy powered by renewable energy and eco-friendly materials not only mitigates climate change but also fosters economic resilience, energy security, and long-term sustainability. Using sustainable energy sources on farms, such as solar panels or biogas facilities, is another example. These alternatives lessen dependency on fossil fuels, cut greenhouse gas emissions, and even produce additional revenue when surplus energy is sold back to the grid. Circular economy in farming offers a sustainable and resilient approach to agriculture, promoting environmental stewardship, economic viability, and social equity. There is significant potential for lowering greenhouse gas emissions, increasing energy efficiency, and advancing sustainability in food production by switching to alternative energy sources for agricultural energy management (Majeed et al., 2023).

14.3.7 Principe 7: Shift to a Systems Perspective

It is important to recognize the interconnectedness of ecological, social, and economic systems. Intelligent assets such as artificial intelligence, machine learning, the internet of things, big data, data visualization, blockchain, and similar can greatly optimize the efficiency of processes. Resources are applied in exact proportions to specific regions rather than covering a whole field by precision agriculture practices such as using technologies like global positioning systems, sensors, and thermal imaging cameras.

Digitalization brings wonders in the circular economic implementation by addressing the entire lifecycle of products, including sourcing, production, distribution, and end-of-life management. Tracking products through their use cycle, efficiency of delivery routes where real-time data facilitates faster routing, reduced fuel, and minimal damages and asset loss, developing an efficient reverse logistics system where data allows asset owners to quantify the cost and benefits. Using predictive analysis like machine learning and other statistical techniques, the recycled materials

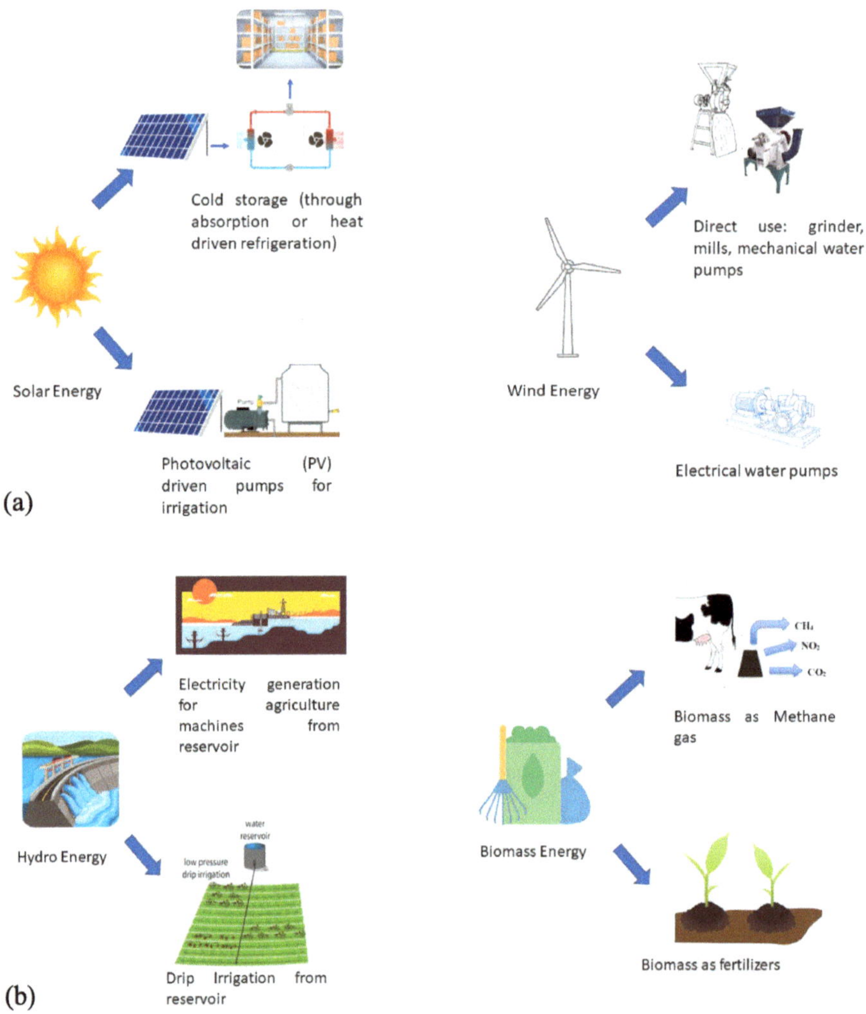

Fig. 14.3 (**a**) Usage of wind and solar as renewable energy sources to run different agricultural machines. (**b**) Efficient use of hydro and biomass as renewable energy and recycling of agricultural waste/produce. (Reproduced from (Majeed et al., 2023) under Creative Commons license)

can be sorted out for reuse and regeneration. A simple and straightforward recipe for circular economy transformation goes as follows:

- Building an understanding of resources, wastage, materials flows, and inefficiencies.
- Focusing on hotspots of the circular economy to align them with business strategy.
- Rethinking operations by breaking down silos, enabling connectivity, and optimizing the use of resources.

- Connecting data, devices, and partners using digital platforms.
- Creating and selling an experience using the circular economy by selling "value" instead of materials.
- Communicating with the customers by storytelling or creating a circular product system and community around the core product, thereby empowering the customers.
- Learning continuously and ensuring a smart, feedback loop for better eco-design.

Blockchain traceability systems and other technological solutions that offer accountability, efficiency, and transparency throughout the supply chain can help to solve many of the issues facing circular agriculture. Blockchain makes it possible to track agricultural resources in real time, including fertilizer, energy, and water. The difficulty of lowering resource input is addressed by making sure that these inputs are tracked and optimized. Agribusiness can reduce waste and minimize resource consumption by gathering information on where and how resources are used. Farmers can use blockchain-powered precision agriculture technologies to determine how much water or energy they are consuming and modify their methods accordingly. Blockchain technology allows farmers to monitor the exact amount of water used for each crop, which facilitates the adoption of water-saving techniques like rainwater collection and drip irrigation.

The transition might be tough for some, but we are about to face the generation of "*born-circulars*" as the new industry leaders during the 2030s (News, 2022). Born circulars refer to products, businesses, or systems that are designed from the outset to align with circular economic principles. Companies adopting a born-circular approach integrate sustainability into their core business model, fostering long-term economic and environmental benefits. To achieve its goal of a fully circular economy by 2050, the Netherlands hopes to transition to "sustainable and strong agriculture" as early as 2030 (Government.nl, 2020). Dutch laws and regulations will be changed in a number of ways, including by ending the fodder-manure cycle, reusing as many waste streams as possible, using natural manure instead of artificial fertilizers, allowing farms in experimentation areas to deviate from the law and regulations, establishing uniform methods for measuring soil quality, giving agricultural entrepreneurs opportunities, and introducing initiatives to boost farmers' incomes.

14.4 Examples of Circular Economy in Farming

The circular economy principles are applied in farming to promote sustainability and efficiency.

(a) *Composting and Organic Fertilizers*—Agricultural waste like crop residues, manure, and food scraps are composted to create nutrient-rich organic fertilizers, reducing reliance on synthetic fertilizers. Composting is a key circular economy practice in agriculture that transforms biowaste—such as food scraps, crop residues, and manure—into nutrient-rich organic fertilizers (Zhou & Yao, 2020). This process reduces landfill waste, lowers greenhouse gas emissions, and enhances soil health. Composting enhances soil fertility by providing essential nutrients like nitrogen, phosphorus, and potassium back to the soil. It increases the soil structure by water retention and aeration, reducing soil erosion. These methods indirectly reduced greenhouse gas emissions (methane) by diverting waste from landfills and encouraging growth of the microbiome that promotes plant growth. Composting methods may be traditional methods like layering and microbial composing (bacteria, fungi, and actinomycetes) (Mandal et al., 2015) or vermicomposting (using earthworms) (Ali et al., 2015), biogas fermenting (for liquid waste), or Bokashi composting (using anaerobic microbial fermentation) (Footer, 2013).

(b) *Precision Agriculture*—Precision agriculture is essential for advancing a circular economy by optimizing resource use, reducing waste, and promoting sustainable farming. By leveraging technologies like GPS, IoT sensors, AI, and drones, farmers can apply water, fertilizers, and pesticides with precision, minimizing excess use and environmental impact (Pierce & Nowak, 1999). Efficient irrigation systems conserve water, while targeted nutrient application prevents soil degradation and chemical runoff. Precision monitoring enhances soil health, supports biodiversity, and reduces greenhouse gas emissions by optimizing machinery use. Additionally, data-driven decision-making improves crop yields, prevents losses, and integrates well with other circular practices like composting and regenerative farming. By minimizing waste and maximizing efficiency, precision agriculture ensures a more sustainable and resilient food production system, aligning with circular economy principles (Yaqot et al., 2023).

(c) *Livestock Integration*—Animal manure is used to fertilize crops, and crop residues can be fed to animals. Livestock integration in agriculture plays a crucial role in the circular economy by creating closed-loop systems that enhance sustainability, reduce waste, and improve resource efficiency. Manure from livestock serves as a natural fertilizer, enriching soil nutrients and reducing the need for synthetic fertilizers. Crop residues and food waste can be repurposed as animal feed, minimizing waste and lowering feed costs. Integrated systems, such as agroforestry and silvopasture (Klopfenstein et al., 1997), combine livestock with crops and trees to improve biodiversity, enhance soil health, and sequester carbon. Additionally, biogas production from animal waste generates renewable energy, further reducing the environmental footprint of farming. By efficiently cycling nutrients and resources, livestock integration promotes a sustainable agricultural model that aligns with circular economy principles, ensuring long-term productivity while preserving ecosystems.

(d) *Agroforestry*—Planting trees alongside crops provides benefits like carbon sequestration, improved soil health, and habitat for wildlife. The trees also produce fruits, nuts, or timber, adding an extra source of income for farmers. It is a management approach that incorporates grassland or crops with trees. It blends forestry and agricultural technology. Fruits, nuts, other edible plant products, medicinal plants, ornamental plants, timber and wood products, animals and animal products, and other products from both wild and domesticated species can all be produced in an agroforestry system, which is a polyculture system (Dagar et al., 2014). Sustainable agriculture can include agroforestry, which has social, environmental, and economic advantages. Increased farm productivity, healthier environments, lower farmer risk, beauty and aesthetics, higher farm profits, less pollution, managing animal waste, increased biodiversity, better soil structure, carbon sequestration, and reduced soil erosion are all advantages of agroforestry in addition to production. Agroforestry techniques are particularly common in tropical regions, particularly in places with subsistence smallholdings, and they are especially significant in sub-Saharan Africa (Wato & Amare, 2020). It has been embraced in the USA and Europe because of its many advantages, including as its benefits for the nitrogen cycle and its capacity to lessen droughts.

(e) *Bioenergy Production*—Agricultural residues and waste, such as stalk, leaves, and roots of crops after harvest or animal waste such as manure, bedding, are converted into biofuels or biogas, reducing fossil fuel dependency (Kim & Dale, 2004). Farms can convert organic waste (e.g., crop residues, animal manure) into biogas, or biohydrogen, which can be used as a source of energy (Fig. 14.4). The by-product, digestate, can be used as a nutrient-rich fertilizer. Biodiesel, which is obtained by transesterification, and bioethanol produced from various biological sources like wood, straw, or even household wastes, are blended with petrol/diesel, leading to 80% less CO_2 emission. Yeast species *S. cerevisiae* converts starch to glucose. *Aspergillus niger* and *S. cerevisiae* mixed culture produced higher ethanol upon hydrolyzing and fermentation of coffee pulp. *S. cerevisiae* fermented fruit and vegetable peels that were discarded as kitchen waste in a shaking environment, producing 7.3% (v/v) ethanol after 48 hours. The naturally occurring cellulose enzyme derived from cellulolytic microbes like *Cellulomonas, Clostridium, Bacillus, Bacteroides, Erwinia, Thermospora,* and *Streptomyces* (Nair et al., 2022). Biohydrogen is hydrogen gas (H_2) produced biologically through microbial processes, offering a renewable and clean energy source. There are several main methods for biohydrogen production. Dark fermentation uses anaerobic bacteria like *Clostridium* and *Enterobacter* to break down organic matter in the absence of light, producing hydrogen along with organic acids and CO_2, though yields are relatively low. Photofermentation relies on purple non-sulfur bacteria such as *Rhodobacter* to convert organic acids into hydrogen using light, but scaling up remains challenging. Photobiological water splitting utilizes photosynthetic microorganisms like cyanobacteria and green algae (*Chlamydomonas reinhardtii*) to split water into hydrogen and oxygen via sunlight and hydrogenase enzymes, though efficiency is limited due to oxygen sensitivity (Levin et al., 2004).

(f) *Aquaponics and Hydroponics*—Combines aquaponics/aquaculture (raising fish) with hydroponics (growing plants without soil). Closed-loop farming methods that use fish excrement to fertilize plants and the plants' ability to purify the water result in an effective and sustainable agricultural system. With their roots submerged in the nutrient-rich effluent water, plants are cultivated in hydroponic

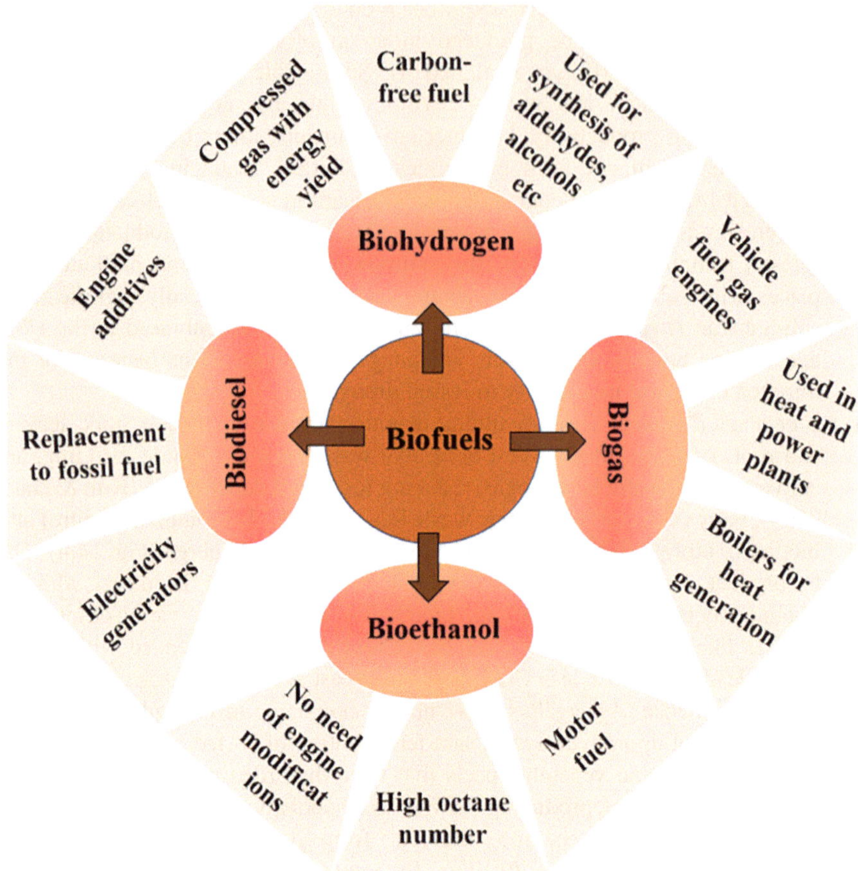

Fig. 14.4 Schematic representation of different biofuels and their major applications. (Reproduced from Nair et al. (2022) under Creative Commons license)

systems. Because of this, they can filter out ammonia or its metabolites that are harmful to aquatic life. The water is cleansed, oxygenated, and ready to be used again in the aquaculture vessels after passing through the hydroponic subsystem. Aquaponics is mostly used to cultivate tilapia, koi, goldfish, carp, catfish, barramundi, and various ornamental fish (Hussain & Brown, 2025). Squash, broccoli, cauliflower, cabbage, tomatoes, peppers, cucumbers, beans, peas, pak choi, kale, lettuce, watercress, basil, and mint are among the primary crops grown (Yavuzcan Yildiz et al., 2017). Aquaponics consists of three primary components: fish, plants, and microbes. Microbes serve as a bridge by transforming fish waste into plant nutrients.

(g) *Water Recycling and Rainwater Harvesting*—Efficient irrigation systems, such as drip irrigation and rainwater collection, minimize freshwater consumption. Water recycling and rainwater harvesting are essential circular economy practices in agriculture that optimize water use, reduce waste, and enhance sustainability (Zhu et al., 2015). Rainwater harvesting involves collecting and storing rainwater from rooftops, fields, or other surfaces for irrigation, livestock, and other farm needs, reducing dependence on groundwater and municipal water supplies (Velasco-Muñoz et al., 2019). Water recycling repurposes used water from agricultural processes, such as irrigation runoff or livestock operations, through filtration and treatment for reuse. These practices help conserve freshwater resources, reduce water pollution, and improve resilience against droughts and climate change. By efficiently managing water within a closed-loop system, farmers can minimize waste, lower costs, and promote sustainable agricultural practices that support long-term environmental health (Nahar et al., 2019).

These practices contribute to sustainability by reducing waste, improving resource efficiency, and enhancing resilience in agricultural systems.

14.5 Challenges in Implementing CE Practices

Implementing circular economy practices faces several challenges, including high initial costs, lack of infrastructure, and resistance to change. Transitioning from a linear to a circular model requires significant investment in recycling facilities, reverse logistics, and sustainable product design, which many businesses find financially demanding (Chhimwal et al., 2022). Additionally, inadequate waste management systems and limited access to advanced recycling technologies make it difficult to close material loops effectively. Consumer awareness and behavioural shifts also play a crucial role, as many still prefer disposable products due to convenience and cost factors. Regulatory and policy gaps further slow progress, as clear incentives and frameworks are needed to drive circular initiatives at macro, meso, and micro levels (Nadeem et al., 2018). Overcoming these barriers requires collaborative efforts between governments, industries, and consumers to create a supportive ecosystem for circular practices. It can be expensive, particularly at the beginning of the transition, to set up a circular supply chain model since it frequently requires integrating new logistics structures, renewable energy sources, and expensive waste processing technologies.

- Facilities that may not be available everywhere are necessary for proper composting or the use of crop leftovers for bioenergy. Similar to this, waste-to-fertilizer technology requires certain machinery and processing facilities that not all farmers may be able to afford.
- There is frequently little chance to reuse food waste or crop by-products if various supply chain linkages are not in sync or do not efficiently exchange data. Circular systems fail when there is inadequate coordination and communication.
- It may be difficult for a farmer to transition to circular practices without proper training or technical assistance if they are not familiar with composting methods or bio-based fertilizers.

- Although the European Union has policies such as the Circular Economy Action Plan to encourage circular practices, companies that operate across borders may have difficulties if local laws do not encourage or even acknowledge circular agriculture.
- Sometimes consumers are not aware of or appreciate the circular processes that go into the things they purchase. Businesses could have a hard time convincing the final customer of the additional expenses of circular activities without effective marketing and consumer education.

One of the biggest challenges is the need for systemic change. Changes in consumer behavior, company models, production methods, and even regulations are necessary to make the shift to a circular economy. Collaboration between many sectors and governmental levels is necessary for this transformation, which can be complicated and resource intensive. The designing of products and systems poses another challenge. Products need to be designed for durability, repairability, and recyclability for an effective CE. It is also necessary to develop and enhance the infrastructure for recycling and renovation. Large-scale implementation of circular practices is difficult in many areas because they lack the infrastructure and technology needed to handle recycled materials or refurbish used goods effectively. Lastly, consumer behavior and cultural attitudes play a significant role (Koistila, 2020). Although it is crucial, changing consumer preferences toward more environmentally friendly goods and methods can be difficult because of ingrained customs and beliefs about the relative merits of new vs repurposed goods.

14.6 Role of Block Chain Technology in CE

Facilitating real-time connectivity among machines, digital devices, transactions, and multiple stakeholders, such as suppliers, purchasers, and customers, is a primary goal and characteristic of Industry 4.0. This networked communication has been partially made feasible by the "Internet of Things (IoT)" (Wang et al., 2020; Upadhyay et al., 2021). Real-time tracking and data-driven insights throughout the agricultural value chain are made possible by a blockchain-powered system, which strengthens circular agriculture. By ensuring accountability and transparency, it minimizes waste and inefficiencies while enabling companies to track products from farm to fork (F2F). The F2F strategy consists of the following specific actions: ensuring sustainable food production; encouraging sustainable food processing, promoting sustainable food consumption, facilitating the transition to healthy, sustainable diets; wholesale, retail, hospitality, and food services practices; and minimizing food loss and waste (Wesseler, 2022). By incorporating solutions for resource optimization, trash repurpose, and nutrient recycling, the platform assists agribusinesses in adhering to circular principles. Additionally, its sustainability features promote closed-loop systems and regenerative practices by tracking important indicators, including biodiversity, soil health, and carbon footprint. This

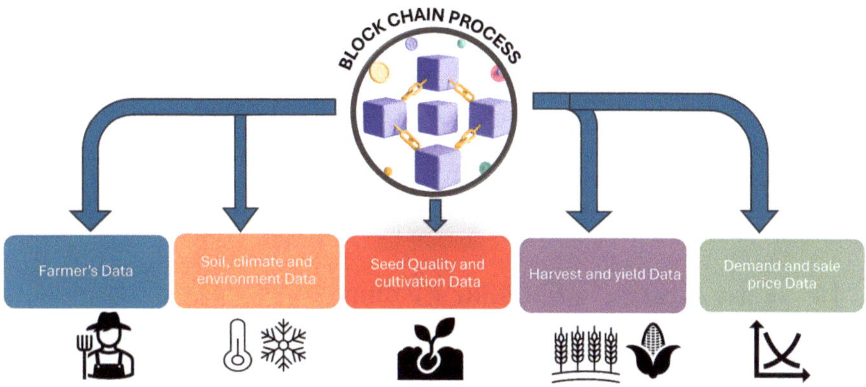

Fig. 14.5 A block chain process in agriculture

all-encompassing strategy fosters the shift to sustainable food systems while enhancing resilience and generating benefits for everybody.

Blockchain can trace the path of compost, manure, and waste-to-fertilizer products, which can assist in managing the movement of organic waste and by-products within the agricultural system. This ensures that nutrients are recycled effectively and do not end up in the waste stream, which contributes to the creation of a circular loop. Through transparent tracking, businesses can make sure that fertilizers come from sustainable practices like composting or turning agricultural waste into bio-based fertilizers. Because blockchain tracks agricultural by-products and makes sure they are sent to the right places, like bioenergy or animal feed, it makes waste repurposing more efficient. By guaranteeing that waste materials are being employed to create value rather than ending up in landfills, the system offers a clear and safe means of tracking their progress.

Blockchain technology in agriculture enhances transparency, traceability, and efficiency in the supply chain, improves trust between farmers, distributors, retailers, and consumers, while reducing food fraud and inefficiencies (Pal et al., 2024; Böckel et al., 2021). An example of a blockchain process in agriculture is farm-to-table traceability for organic produce (Fig. 14.5). Farmers register crop details (seed quality, fertilizers, pesticides used) on a blockchain ledger. IoT sensors can also log soil quality and weather conditions. When crops are harvested, a unique digital record (QR code or RFID tag) is assigned, containing batch details stored on the blockchain. As the produce moves through the supply chain (from farm to distributor to retailer), each transaction is recorded in the blockchain, ensuring real-time tracking of storage conditions, transport routes, and handling. At the retail stage, customers can scan a QR code on the packaging to view the entire history of the product, ensuring authenticity, safety, and adherence to organic standards. Smart contracts can automatically verify compliance with agricultural regulations, release payments upon successful delivery, and prevent fraud by ensuring that records cannot be altered.

Techniques like no-till farming, cover cropping, and agroforestry help regenerate soil, enhance water retention, and support biodiversity (Pieri et al., 2002). These practices lead to healthier ecosystems that can sustain agriculture for the long term, while also mitigating climate change. Circular farming focuses on maximizing the use of available resources. Efficient water use, nutrient recycling, and sustainable soil management are essential elements.

14.7 Case Studies in Circular Economy Practices

14.7.1 Naandi Foundation's Araku Coffee Project

Araku Coffee Project, an initiative by the Naandi Foundation, is in Araku Valley, Andhra Pradesh, India. It is a leading example of regenerative agriculture, transforming the Araku Valley's tribal farming communities in the eastern ghats through sustainable practices. The Araku Coffee Project is an inspiring model for how a circular economy can create positive environmental, economic, and social impacts while promoting global sustainability (Fig. 14.6). It links farming to education and community bonds using science-based knowledge focusing on land productivity, crop and pest management, and market connection. By implementing agroforestry, organic composting, and natural soil enhancers, the initiative restores soil health, enhances biodiversity, and sequesters carbon, making coffee cultivation both productive and environmentally responsible. This project is based on closed-loop system principles discussed below:

14.7.1.1 Regenerative Agriculture and Soil Health

- Organic Farming Practices: The Araku Coffee project trains tribal farmers in organic and biodynamic farming techniques, eliminating chemical fertilizers and pesticides.
- Agroforestry Model: Coffee plants are grown in harmony with native trees and other crops, creating a biodiverse ecosystem. This promotes carbon sequestration, improves soil fertility, and prevents soil erosion.
- Compost and Mulching: Farmers use organic compost and crop residues to enrich the soil, ensuring a closed-loop nutrient cycle.

14.7.1.2 Waste Utilization

- Coffee Pulp and Wastewater: Waste products from coffee processing, like pulp and wastewater, are repurposed as organic compost or recycled for irrigation. This minimizes pollution and supports nutrient cycling.

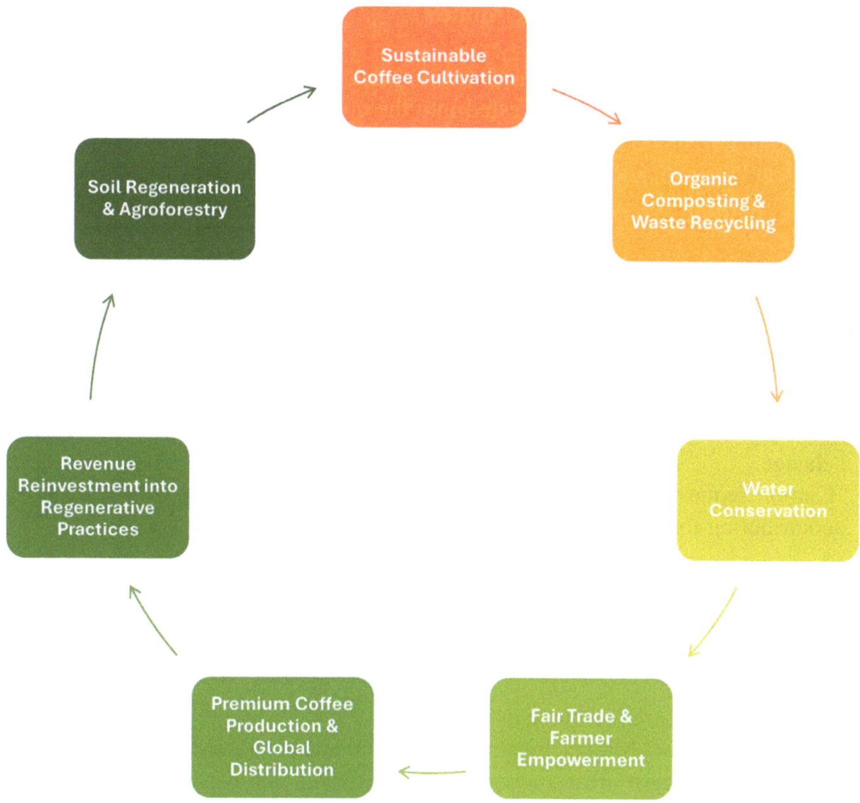

Fig. 14.6 CE loop in sustainable model of Araku Coffee Project

- By-Products: Coffee by-products, such as parchment and husks, can be used for bioenergy, packaging, or even as a growing medium for mushrooms, closing the waste loop.

14.7.1.3 Empowering Communities and Creating Value

- Fair Trade Model: The project ensures tribal farmers receive fair prices for their premium-quality coffee, empowering them economically while encouraging sustainable practices.
- Skill Development: Farmers are trained in organic farming, quality control, and processing techniques, creating a self-sufficient and skilled workforce.
- Value Addition at Source: Coffee is processed and packaged locally, retaining value within the community and reducing the carbon footprint associated with transportation.

14.7.1.4 Sustainable Economic Models

- Premium Coffee Brand: Araku Coffee is marketed as a high-end, organic brand globally. The profits are reinvested into the community, funding health, education, and livelihood initiatives.
- Product Diversification: Beyond coffee, the agroforestry model allows farmers to grow fruits, spices, and vegetables, diversifying income sources and increasing resilience.

14.7.1.5 Environmental Regeneration

- Reforestation Efforts: The project has planted millions of trees in the Araku Valley, restoring degraded land, conserving biodiversity, and mitigating climate change.
- Carbon Sequestration: Through agroforestry, the coffee farms act as carbon sinks, contributing to global climate goals.

14.7.1.6 Circular Business Practices

- Closed-Loop System: From cultivation to processing, the project minimizes waste and optimizes resource use.
- Local and Global Market Integration: By linking local production with global markets, the project creates a circular flow of value that benefits both producers and consumers.

14.7.1.7 Social Impact

- The project has transformed the livelihoods of over 100,000 tribal farmers, improving income, health, and education. Women are actively involved in farming and decision-making, promoting gender equality and social inclusion.
- Araku Coffee's premium organic produce has gained international recognition, demonstrating how regenerative agriculture can create a sustainable, climate-resilient, and socially empowering agricultural model.

14.7.2 CE Model in Banana Cropping

A "circular economy in the banana industry" refers to a system where banana production waste, like peels, stems, and leaves, are not discarded but are instead reused and repurposed as valuable inputs in different stages of the banana production process, minimizing waste and maximizing resource utilization, thus contributing to a more sustainable agricultural practice; this can involve using banana waste to create bioplastics for packaging, animal feed, compost for soil enrichment, or even fiber for textiles. In 2021, global banana fruit production was estimated at 114.08 million tons per year (Alzate Acevedo et al., 2021). Notably, bananas are cultivated in 130 countries; India accounts for 11% of the world's banana cultivation area, followed by Brazil, Tanzania, the Philippines, and China (Provin et al., 2024). After fruit harvest, the pseudo-stem ceases to be useful to the farmer and is typically discarded or incinerated. Banana peel has been used primarily to create bioplastics with high rates of disintegration and biofuels like ethanol and diesel. The banana plant's leaves are used to make organic fertilizers, biodegradable containers, and utensils. The pseudostems of bananas are used to make bioplastics that require less time for biodegradation because of their fibrous character (Faradilla, 2017). The waste-loss content of bananas' cellulose and hemicellulose is crucial for nanotechnological operations, particularly for creating green nanoparticles. It has been demonstrated that the banana peel, rachis, pseudostems, and leaves have a great potential for reuse in a variety of recovery procedures.

Banana fiber has significant potential, and it is an underutilized biological resource. Banana peel contains 12% of raw fiber by weight along with starch, protein, calcium, phosphorous, zinc, lipids, and ash. In comparison to other natural fibers such as jute, coconut, or palm, it exhibits mechanical strength ten times greater. Banana fibers are biodegradable, environmentally friendly, odorless, and chemically inert and regarded as a positive alternative to synthetic fibers. Furthermore, they are resistant to high temperatures, water, fire, tension, and grease, making them suitable for a variety of applications. When combined with other natural fibers, banana fibers demonstrate good performance. Banana peel-polystyrene composites represent an innovative solution for thermal insulation in buildings (Mohamed et al., 2023).

Banana fiber–polymer composites are eco-friendly materials made by reinforcing banana fibers into various polymer matrices (biodegradable or synthetic). These composites offer high strength, durability, and biodegradability, making them a sustainable alternative to synthetic fiber composites. Important properties of Banana Fiber-Polymer Composites are high tensile strength, lightweight, biodegradable, and thermal resistance. However, it has a high moisture retention capacity. Its composites with polymers may increase its water resistance. Banana fibers can be blended with natural polymers like polylactic acid, starch, or polyhydroxyalkanoates to obtain eco-friendly, biodegradable packaging materials, containers, spoons, forks, plates, trays etc., and compostable bags for waste management (Fig. 14.7) (Raghunathan et al., 2022). Banana fibers may also be incorporated into synthetic polymers like polyesters, polypropylene or epoxy resins to make highly durable composites for applications in sports goods, automotives, or building materials. In the automotive sector, they are used in dashboards, door panels, and seat backs to reduce vehicle weight and improve fuel efficiency. Construction applications include wall panels, roofing sheets, and eco-friendly furniture, while the packaging industry benefits from biodegradable food trays, containers, and compostable bags. In

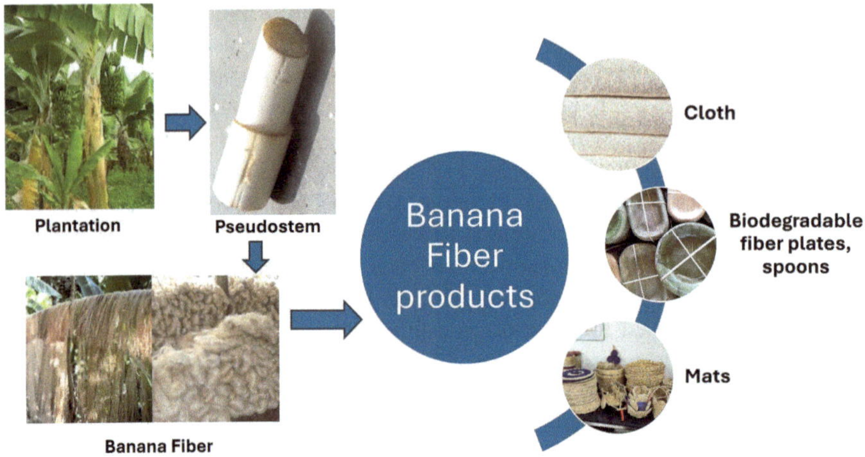

Fig. 14.7 Conversion of banana waste into useful products via extraction of fires from pseudostem

aerospace and sports, banana fiber composites are used in aircraft interiors, bicycle frames, helmets, and cricket bats due to their durability and lightweight properties. Additionally, medical applications involve biodegradable surgical implants and disposable medical trays. Compared to synthetic fiber composites, banana fiber-polymer materials offer advantages such as lower cost, reduced carbon footprint, and enhanced sustainability, making them an eco-friendly choice for various engineering and industrial applications (Begum et al., 2020). Ripe banana peels as a source of nutrients for livestock as animal feed. Minimizing waste may lead to reduced environmental impact by lowering greenhouse gas emissions from waste disposal.

Key aspects of a circular economy in the banana industry are waste valorization, closed-loop system formulation for energy recovery, and innovation in biodegradable packaging for sustainable development. Utilizing banana peels, pseudo stems, and leaves as raw materials, which are waste after fruit harvest, to produce various products like bioplastics, biofuel, compost, or fiber for textiles is the first step in CE goals. Optimal use of resourcese by reusing and recycling waste materials within the production chain increases resource efficiency. To establish economically viable markets for waste banana-based products, focus on diverse applications like high-fiber food products, biodegradable packaging, natural textiles, and even renewable energy advanced technologies are needed. Ensuring proper collection and transportation of banana waste to processing facilities by improving the logistic facilities can help to overcome challenges in implementing CE in the banana industry.

Energy can be generated from banana waste through anaerobic digestion to power farm operations. Developing sustainable packaging materials using banana-based bioplastics reduces reliance on conventional plastic. In an experiment, an insulating composite was prepared from banana peels and polystyrene polymer (Fig. 14.8). These results indicate the possibility of creating novel recyclable and eco-friendly materials for use as insulation in buildings. All the bio-composites displayed

Fig. 14.8 The production process of the Banana peels-polystyrene composites. (Reproduced from (Mohamed et al., 2023) under a common creative license)

comparable thermal conductivity values between 0.028 and 0.030 W/m.K. On the other hand, the pure peel powder provides great crystallinity (56.1%) and high thermal stability (66.4% total weight loss). On the other hand, the pure peel powder provides great crystallinity (56.1%) and high thermal stability (66.4% total weight loss) (Mohamed et al., 2023).

14.7.3 CE in the Sugarcane Industry

The sugarcane industry has significant potential to adopt circular economy principles, turning waste into valuable resources and promoting sustainability. About 20% of the world's sugar is obtained from sugar beet, which is primarily grown in the temperate regions of the Northern Hemisphere, while the remaining 80% is produced from sugar cane in tropical and subtropical regions. India is the world's second largest sugar producer. By utilizing every part of the sugarcane plant, reducing emissions, and creating closed-loop systems, the industry can achieve economic, environmental, and social benefits. India produced about 34 million metric tons of sugar in 2023/2024 (Keelery, 2024). Sugarcane plants are a perfect model of zero-waste utilization as every part of the sugarcane plant can be repurposed. The sugar industry produces four substances, like sugar & molasses are used for food, sweeteners, and fermentation. Bagasse (fibrous residue) is used for bioenergy, paper, biodegradable packaging, and biochar. Press Mud (filter cake) is used as organic fertilizer or animal feed. Vinasse (distillery waste) can be used for biogas production or irrigation in fields. Using bagasse and vinasse to generate bioelectricity and biogas for sugar mills, reduces the dependence on fossil fuels. Ethanol from sugarcane can replace gasoline, and sugarcane fiber can be used for biodegradable plastics. Sugarcane-based green chemicals like lactic acid, furfural, and bio-based polymers reduce reliance on petrochemicals. Sugar derivatives are widely used in medicinal syrups, skincare products, and bio-based surfactants due to their versatility and beneficial properties like mildness, biodegradability, and emulsifying capabilities. Sugar mills collaborate with biofuel, paper, and fertilizer industries to utilize by-products. Encouraging small-scale biorefineries to process sugarcane waste into high-value products. Despite being one of the huge agricultural sectors, the sugar industry is facing a major economic obstruction, where providing fair prices to the farmers have been considered as a challenge (Gopinath et al., 2018). In view of the same, initiatives should be taken by the policy makers to integrate a variety of sectors such as energy, materials, or other food sectors, with the sugarcane industry, to enhance the outcome and improve the economic condition. In this direction, Patil et al. published their opinion on employing the waste generated by the sugarcane industry to produce compressed biogas. It is to be noted that domestic production of biofuels, which can meet the standards, will play an important role in substituting the country's energy demand as well as will successfully reduce the import dependency in the energy sector. It has been mentioned that India is presently equipped with approximately 540 sugar mills, 450 distilleries and the integration of 50% of these industries with CBG production technology will produce 3.75 lakh tonne of CBG per annum, thereby contributing Rs. 1750 crores in the nation's revenue. The route to biomass generated from the sugarcane industry to CBG production is well known, which usually involves the process of anaerobic digestion, resulting in the production of biogas, which is further purified and compressed for its utilization by the customers. Further, the digestate released from the digester can be employed as an organic manure for agricultural production, as it enhances the soil texture and improves the water holding capacity, assisting the microbial activities. It is to be noted that this action will help in reducing the utilization of chemical-based fertilizers and contribute towards sustainable agriculture. Further, carbon dioxide is produced as a by-product and can be isolated during the purification of biogas. The carbon dioxide can be liquified and can be supplied to the fire-extinguisher industry (Patil et al., 2022).

Additionally, Gopinath et al. suggested that sugar industry waste can be upcycled for the production of energy from sugarcane bagasse, which is obtained as fibrous residue after the juice is extracted from the sugarcane (Gopinath et al., 2018). This fiber comprises cellulose, hemi-cellulose, lignin, and is usually disposed of in fertile lands. Instead, cogeneration, which involves the production of electricity and heat simultaneously from a single source of fuel, has been referred to as a combined heat and power cycle. In accordance with waste-to-energy and circular bioeconomy principles, Hassa et al. have suggested turning sugarcane bagasse into single cell oil, sulfonated carbon-based catalyst, and biodiesel (Fig. 14.9) (Hassa et al., 2024).

Modern sugar industries are equipped with cogeneration boilers, which generate heat and electricity by using the principle of the Rankine cycle. The generated heat is supplied to the process of producing sugar from cane juice, while the electricity is

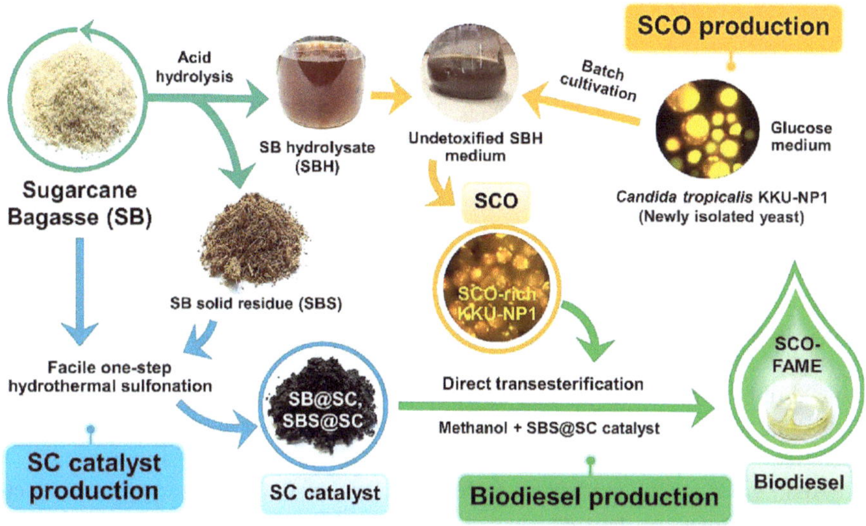

Fig. 14.9 Sugarcane bagasse valorization through an integrated process for single cell oil, sulfonated carbon-based catalyst, and biodiesel co-production. (Reproduced from (Hassa et al., 2024) under a common creative licence)

exported to the power grid station for external consumer use. In 2008, the cogeneration industry of Brazil reportedly produced 3000 MW of power during the sugarcane harvest season (Granville et al., 2009). In another report published by Alonso Pippo et al., it was mentioned that sugar factories in Cuba exhibit potential to generate electricity to satisfy their internal energy requirements by using cogeneration plants having low-efficiency steam turbines and can be further improved by employing a condensing-extraction steam turbine, which is estimated to generate 130 kWh per tonne of sugar cane (Alonso Pippo et al., 2007). In view of the same, Thailand promoted small power producers and independent power producers' schemes, and it was reported that electricity production in a cogeneration sugar plant in Thailand generated approximately 124 kWh energy (Silalertruksa et al., 2017).

The bio-waste generated from the sugarcane industry has emerged as an attractive source for the production of biofuels, non-condensable gases, and biochar to reduce the dependence of industries on the petroleum feedstock (Carpenter et al., 2014). The technologies employed for biomass conversion are either assisted by biological processes or thermochemical methods, where the former has been considered as time-consuming. Thermochemical methods usually comprise processes such as pyrolysis, gasification, or combustion, where pyrolysis has been reported to be more efficient for the production of gas, char, as well as oil. Biomass pyrolysis, which is performed at high temperature in the absence of oxygen, produces char, tar, and gases. Further, biomass tar comprises benzene and a variety of polycyclic hydrocarbon moieties, which can be employed to generate hydrogen in the presence of steam or oxygen by using natural, alkaline, and metal-based catalysts (Alauddin et al., 2010). Ahmed and Gupta reported that one kilogram of bagasse exhibits the potential to produce approximately 60–100 g of hydrogen (Ahmed & Gupta, 2009). Further, the production of hydrogen has been found to be dependent on raw materials, reaction conditions, catalyst as well as on the reactor design. In a study published by Peng and Wu, it was reported that pyrolysis of hemicellulose obtained from sugarcane at 550–850 °C, in inert conditions, yielded H_2, CO, CO_2, and CH_4, where the fraction of hydrogen and methane increased on increasing the temperature of the process (Peng & Wu, 2011). Moreover, the fast pyrolysis of sugarcane bagasse, which involves fast heating rates, higher,temperatures and less residence time, results in the formation of bio-oils. In 2017, Sohaib et al. reported that fast pyrolysis of bagasse in a bench-scale reactor resulted in a production of bio-oil with a yield of 60.4%, where the temperature was kept at 500 °C. The density of bio-oil thus obtained was reported to be 1.162 g/cc and was observed to 38% higher than recorded for the diesel (Sohaib et al., 2017).

Apart from these valorization techniques, one of the most important products which is produced by the sugarcane industry, apart from sugar, is ethanol. It is to be noted that ethanol can be obtained from the sugarcane juice, sugarcane bagasse, and molasses via fermentation of cellulose hydrolysate obtained from the hydrolysis of lignocellulosic bagasse or molasses. In addition, the process comprises pretreatments, microorganisms, and selective conditions to maximize the production of ethanol from bagasse (Cardona et al., 2010), and the alternative fuel, thus obtained, is referred to as second-generation bio-ethanol. Presently, ethanol-blended fuels are commercially available in countries such as Brazil and the USA. A report suggests that the Brazilian government has passed an ordinance to blend gasoline with anhydrous ethanol with a replacement of 25% (Marcoccia, 2007). In addition, the Government of India is also targeting to blend petrol and diesel with 22.5% and 15% ethanol, respectively (Sakthivel et al., 2018).

Utilization of secondary by-products from the sugar industry, such as sugarcane bagasse ash, sugarcane filter cake, in the production of cement, bricks, paver blocks, activated binder, and other construction products have been reported to cut down the carbon footprint of the construction industry, while improving the properties of the final products. An overview of the circular framework focusing on the sugar industry and its conjunction with the construction and energy sectors has been provided by Gopinath et al. and the arrangement eliminates challenges of the industry in terms of the disposal of waste and opportunities for the generation of additional revenue (Gopinath et al., 2018).

14.8 Conclusion

The implementation of a circular economy in agricultural practices enhances sustainability by minimizing waste, improving resource efficiency, and promoting regenerative farming methods. By recycling organic waste into compost, utilizing precision farming to optimize inputs, and integrating sustainable water management techniques, agriculture can reduce its environmental impact while maintaining productivity. Additionally, adopting renewable energy sources, such as biogas from agricultural residues, and utilizing blockchain for transparent supply chains further strengthen circular economy principles. These approaches not only enhance soil health and biodiversity but also create economic opportunities for farmers by reducing dependency on synthetic inputs and minimizing losses. Successfully integrating circular economy practices in agriculture requires collaboration among policymakers, farmers, and agribusinesses to develop supportive policies, invest in innovative technologies, and promote awareness. Ultimately, transitioning to a circular economy in agriculture can lead to a more resilient, efficient, and environmentally friendly food system.

CE implementation faces several challenges, including high initial investment costs, lack of infrastructure for waste recycling, resistance to change from traditional farming methods, and limited awareness among stakeholders. Additionally, integrating circular practices such as composting, biogas production, and precision agriculture requires technological advancements and skilled labor. Policy and regulatory barriers also hinder widespread adoption. However, the future direction involves leveraging digital technologies, improving supply chain traceability, and enhancing collaboration between farmers, industries, and research institutions. Governments play a crucial role by formulating supportive policies, providing financial incentives, investing in sustainable infrastructure, and promoting education on circular practices. Strengthening public-private partnerships and integrating sustainability standards into agricultural policies will accelerate the transition toward a more resilient and resource-efficient agricultural sector.

Acknowledgments The authors are thankful to Amity University Uttar Pradesh for providing the infrastructure and IT support for carrying out this work.

Conflicts of Interest The authors have no conflicts of interest to declare that are relevant to the content of this chapter.

References

Ahmed, I., & Gupta, A. K. (2009). Syngas yield during pyrolysis and steam gasification of paper. *Applied Energy, 86*(9), 1813–1821.

Alauddin, Z. A. B. Z., Lahijani, P., Mohammadi, M., & Mohamed, A. R. (2010). Gasification of lignocellulosic biomass in fluidized beds for renewable energy development: A review. *Renewable and Sustainable Energy Reviews, 14*(9), 2852–2862.

Ali, U., Sajid, N., Khalid, A., Riaz, L., Rabbani, M. M., Syed, J. H., & Malik, R. N. (2015). A review on vermicomposting of organic wastes. *Environmental Progress & Sustainable Energy, 34*(4), 1050–1062.

Alonso Pippo, W., Garzone, P., & Cornacchia, G. (2007). Agro-industry sugarcane residues disposal: The trends of their conversion into energy carriers in Cuba. *Waste Management, 27*(7), 869–885.

Alzate Acevedo, S., Díaz Carrillo, Á. J., Flórez-López, E., & Grande-Tovar, C. D. (2021). Recovery of banana waste-loss from production and processing: A contribution to a circular economy. *Molecules, 26*(17), 5282.

Aznar-Sánchez, J. A., Piquer-Rodríguez, M., Velasco-Muñoz, J. F., & Manzano-Agugliaro, F. (2019). Worldwide research trends on sustainable land use in agriculture. *Land Use Policy, 87*, 104069.

Begum, S., Fawzia, S., & Hashmi, M. (2020). Polymer matrix composite with natural and synthetic fibres. *Advances in Materials and Processing Technologies, 6*(3), 547–564.

Böckel, A., Nuzum, A.-K., & Weissbrod, I. (2021). Blockchain for the circular economy: Analysis of the research-practice gap. *Sustainable Production and Consumption, 25*, 525–539.

Cardona, C. A., Quintero, J. A., & Paz, I. C. (2010). Production of bioethanol from sugarcane bagasse: Status and perspectives. *Bioresource Technology, 101*(13), 4754–4766.

Carpenter, D., Westover, T. L., Czernik, S., & Jablonski, W. (2014). Biomass feedstocks for renewable fuel production: A review of the impacts of feedstock and pretreatment on the yield and product distribution of fast pyrolysis bio-oils and vapors. *Green Chemistry, 16*(2), 384–406.

Chhimwal, M., Agrawal, S., & Kumar, G. (2022). Challenges in the implementation of circular economy in manufacturing industry. *Journal of Modelling in Management, 17*(4), 1049–1077.

Cobo, S., Levis, J. W., Domínguez-Ramos, A., & Irabien, Á. (2019). Economics of enhancing nutrient circularity in an organic waste valorization system. *Environmental Science & Amp Technology, 53*(11), 6123–6132.

Dagar, J., Pandey, C., & Chaturvedi, C. (2014). Agroforestry: A way forward for sustaining fragile coastal and Island agro-ecosystems. *Agroforestry systems in India: livelihood security & ecosystem services*, 185–232.

Faradilla, R. (2017). *Biodegradable food packaging using agricultural wastes (Banana Pseudostem)*. UNSW Sydney.

Footer, A. (2013). *Bokashi composting: Scraps to soil in weeks*. New society publishers.

Gopinath, A., Bahurudeen, A., Appari, S., & Nanthagopalan, P. (2018). A circular framework for the valorisation of sugar industry wastes: Review on the industrial symbiosis between sugar, construction and energy industries. *Journal of Cleaner Production, 203*, 89–108.

Government.nl (2020). Circular Dutch economy by 2050.

Granville, S., Lino, P., Ralston, F., Barroso, L. A., & Pereira, M. (2009). *Recent advances of sugarcane biomass cogeneration in Brazil*. 2009 IEEE Power & Energy Society General Meeting.

Hassa, W., Fiala, K., Apiraksakorn, J., & Leesing, R. (2024). Sugarcane bagasse valorization through integrated process for single cell oil, sulfonated carbon-based catalyst and biodiesel co-production. *Carbon Resources Conversion, 100245*.

Henrik Hvid Jensen; Chief Technology Strategist NEE, D. T. (Jan 27, 2022). *5 circular economy business models that offer a competitive advantage*. from https://www.weforum.org/stories/2022/01/5-circular-economy-business-models-competitive-advantage/

Holappa, L. (2020). A general vision for reduction of energy consumption and CO2 emissions from the steel industry. *Meta, 10*(9), 1117.

Hussain, A. S., & Brown, P. B. (2025). Current status of the propagation of basil in Aquaponic systems: A literature review. *Aquaculture Research, 2025*(1), 1320019.

Keelery, S. (2024). Sugar production volume in India 2018/19-2023/24. from https://www.statista.com/statistics/761818/india-sugar-production-volume/.

Kim, S., & Dale, B. E. (2004). Global potential bioethanol production from wasted crops and crop residues. *Biomass and Bioenergy, 26*(4), 361–375.

Kirchherr, J., & Piscicelli, L. (2019). Towards an education for the circular economy (ECE): Five teaching principles and a case study. *Resources, Conservation and Recycling, 150*, 104406.

Kirchherr, J., Yang, N.-H. N., Schulze-Spüntrup, F., Heerink, M. J., & Hartley, K. (2023). Conceptualizing the circular economy (revisited): An analysis of 221 definitions. *Resources, Conservation and Recycling, 194*, 107001.

Klopfenstein, N. B., Rietveld, W., Carman, R. C., Clason, T. R., Sharrow, S. H., Garrett, G., & Anderson, B. (1997). Silvopasture: an agroforestry practice

Koistila, K. (2020). Cultural barriers to circular economy adoption: Consumer awareness and attitudes towards refurbished ICT devices.

Levin, D. B., Pitt, L., & Love, M. (2004). Biohydrogen production: Prospects and limitations to practical application. *International Journal of Hydrogen Energy, 29*(2), 173–185.

Majeed, Y., Khan, M. U., Waseem, M., Zahid, U., Mahmood, F., Majeed, F., Sultan, M., & Raza, A. (2023). Renewable energy as an alternative source for energy management in agriculture. *Energy Reports, 10*, 344–359.

Mandal, S., Van Treuren, W., White, R. A., Eggesbø, M., Knight, R., & Peddada, S. D. (2015). Analysis of composition of microbiomes: A novel method for studying microbial composition. *Microbial Ecology in Health and Disease, 26*(1), 27663.

Marcoccia, R. (2007). *A participação do etanol brasileiro em uma nova perspectiva na matriz energética mundial*. Universidade de São Paulo.

Marino, A., & Pariso, P. (2016). From linear economy to circular economy: Research agenda. *International Journal of Research in Economics and Social Sciences, 6225*(6), 2249–7382.

Mohamed, G. R., Mahmoud, R. K., Shaban, M., Fahim, I. S., Abd El-Salam, H. M., & Mahmoud, H. M. (2023). Towards a circular economy: Valorization of banana peels by developing bio-composites thermal insulators. *Scientific Reports, 13*(1), 12756.

Nadeem, S. P., Garza-Reyes, J. A., & Glanville, D. (2018). The challenges of the circular economy. In *Contemporary issues in accounting: The current developments in accounting beyond the numbers* (pp. 37–60). Springer.

Nahar, J., Liu, S., Mao, Y., Liu, J., & Shah, S. L. (2019). Closed-loop scheduling and control for precision irrigation. *Industrial & Engineering Chemistry Research, 58*(26), 11485–11497.

Nair, L. G., Agrawal, K., & Verma, P. (2022). An overview of sustainable approaches for bioenergy production from agro-industrial wastes. *Energy Nexus, 6*, 100086.

News, P. [LinkedIn]. (2022). *The shift from linear to circular economy is the road to sustainable digitalization*. Caroline Macdonald.

Pal, T., Bommisetti, V., Suvarnakanti, P., Dudala, S. S., Asha, S., & Jaiswal, V. (2024). The revolutionary role of blockchain technology in the agri-food sector focusing on the food supply chain. *Food Reviews International, 41*, 770–801.

Patil, S., Konde, K., & Behera, S. (2022). Bio-circular economy: An opportunity for diversification for sugar Industries in Compressed Biogas (CBG) and organic fertilizer production. *Sugar Tech, 24*(4), 1079–1092.

Peng, Y., & Wu, S. (2011). Fast pyrolysis characteristics of sugarcane bagasse hemicellulose. *Cellulose Chemistry and Technology, 45*(9), 605.

Pierce, F. J., & Nowak, P. (1999). Aspects of precision agriculture. *Advances in Agronomy, 67*, 1–85.

Pieri, C., Evers, G., Landers, J., O'Connell, P., & Terry, E. (2002). No-till farming for sustainable rural development.

Provin, A. P., Medeiros d'Alva, A., de Aguiar Dutra, A. R., Salgueirinho Osório de Andrade Guerra, J. B., & Leal Vieira Cubas, A. (2024). Closing the cycle: Circular economy strategies for the textile industry using banana farming waste. *Journal of Cleaner Production, 470*, 143352.

Raghunathan, R., Nelluri, P., Rajendran, D., Pandiselvam, R., Thulasiraman, V., Sahoo, S. K., Pillai, S., & Jerome, R. E. (2022). Biodegradable products from renewable sources: impact on replacing single use plastic for protecting the environment. *International Journal of Environmental Science and Technology, 22*, 6181–6208.

Sakthivel, P., Subramanian, K., & Mathai, R. (2018). Indian scenario of ethanol fuel and its utilization in automotive transportation sector. *Resources, Conservation and Recycling, 132*, 102–120.

Silalertruksa, T., Pongpat, P., & Gheewala, S. H. (2017). Life cycle assessment for enhancing environmental sustainability of sugarcane biorefinery in Thailand. *Journal of Cleaner Production, 140*, 906–913.

Silva-Alvarado, P. M., Orozco-Crespo, E., Verduga-Alcívar, D. A., Diéguez-Santana, K., Ruiz-Cedeño, S. d. M., & Sablón-Cossío, N. (2023). Prospective of the circular economy in a banana Agri-food chain. *Tec Empresarial, 17*(1), 34–52.

Sohaib, Q., Muhammad, A., & Younas, M. (2017). Fast pyrolysis of sugarcane bagasse: Effect of pyrolysis conditions on final product distribution and properties. *Energy Sources, Part A: Recovery, Utilization, and Environmental Effects, 39*(2), 184–190.

"THE 17 GOALS." (2015) from https://sdgs.un.org/goals

The Hindu. (2025, January 22, 01:32 am IST – Washington). Trump signs executive order directing U.S. withdrawal from the Paris climate agreement — again.

United Nations Climate Change. (2025). The Paris Agreement. What is the Paris Agreement?, from https://unfccc.int/process-and-meetings/the-paris-agreement

Upadhyay, A., Mukhuty, S., Kumar, V., & Kazancoglu, Y. (2021). Blockchain technology and the circular economy: Implications for sustainability and social responsibility. *Journal of Cleaner Production, 293*, 126130.

van der Kooij, S., Zwarteveen, M., Boesveld, H., & Kuper, M. (2013). The efficiency of drip irrigation unpacked. *Agricultural Water Management, 123*, 103–110.

Velasco-Muñoz, J. F., Aznar-Sánchez, J. A., Batlles-delaFuente, A., & Fidelibus, M. D. (2019). Rainwater harvesting for agricultural irrigation: An analysis of global research. *Water, 11*(7), 1320.

Velenturf, A. P., & Purnell, P. (2021). Principles for a sustainable circular economy. *Sustainable Production and Consumption, 27*, 1437–1457.

Wang, E., Attard, S., Linton, A., McGlinchey, M., Xiang, W., Philippa, B., & Everingham, Y. (2020). Development of a closed-loop irrigation system for sugarcane farms using the internet of things. *Computers and Electronics in Agriculture, 172*, 105376.

Wato, T., & Amare, M. (2020). Opportunities and challenges of scaling up agroforestry practices in Sub-Saharan Africa: A review. *Agricultural Reviews, 41*(3), 216–226.

Wesseler, J. (2022). The EU'S farm-to-fork strategy: An assessment from the perspective of agricultural economics. *Applied Economic Perspectives and Policy, 44*(4), 1826–1843.

Yaqot, M., Menezes, B. C., & Al-Ansari, T. (2023). Roadmap to precision agriculture under circular economy constraints. *Journal of Information & Knowledge Management, 22*(05), 2250092.

Yavuzcan Yildiz, H., Robaina, L., Pirhonen, J., Mente, E., Domínguez, D., & Parisi, G. (2017). Fish welfare in aquaponic systems: Its relation to water quality with an emphasis on feed and faeces—A review. *Water, 9*(1), 13.

Zhou, Z., & Yao, H. (2020). Effects of composting different types of organic fertilizer on the microbial community structure and antibiotic resistance genes. *Microorganisms, 8*(2), 268.

Zhu, Q., Gould, J., Li, Y., & Ma, C. (2015). *Rainwater harvesting for agriculture and water supply*. Springer.

Chapter 15
Circular Economy and Sustainability

José L. Navarro, Gustavo Acosta-Santoyo, Irma Robles, Josué D. García Espinoza, and Luis A. Godínez

15.1 Sustainability

Humanity's dependence on ecosystem services has accompanied us since the development of the first communities in the Paleolithic. These common goods of natural origin have provided humanity with a suitable environment for its survival, as well as raw materials that are necessary for productive activities. Ecological stability and nature appreciation began to degrade during the development of larger societies and more complex networks of trade and interdependence catalyzed by the first industrial revolution. As a result of the fast dynamics of these unprecedented economic changes, environmental anthropogenic problems began to be perceptible. Several artists, scientists, and policymakers have, for instance, analyzed from a technical perspective and disseminated in an expressionist way, possible scenarios resulting from a catastrophic ecological modification. Impressionist painters such as Pizarro, Monet, and Van Gogh for example, illustrated industrialized gray machinery or scenery that contrasted with the green fields (Rubin, 2008); writers such as Ralph Waldo Emerson and Henry David Thoreau, with their works "Nature" (1836) and "Walden" (1854), respectively, have also invited us to a reflection on the connection between human nature and self-sufficiency as a measure to reduce negative impacts on the environment. These artistic and literary contributions constitute important

J. L. Navarro · G. Acosta-Santoyo · J. D. García Espinoza · L. A. Godínez (✉)
Centro de Investigación en Química para la Economía Circular, CIQEC, Facultad de Química, Universidad Autónoma de Querétaro, Centro Universitario, Santiago de Querétaro, Querétaro, Mexico
e-mail: jnavarro17@alumnos.uaq.mx; gustavo.acosta@uaq.mx; josue.daniel.garcia@uaq.mx; luis.godinez@uaq.mx

I. Robles
Centro de Investigación y Desarrollo Tecnológico en Electroquímica, S.C., Parque Tecnológico Querétaro-Sanfandila, Pedro Escobedo, Querétaro, Mexico
e-mail: irobles@cideteq.mx

K. Esquivel Escalante et al. (eds.), *Green Nanomaterials*, Nanotechnology in the Life Sciences, https://doi.org/10.1007/978-3-032-04492-1_15

stepping-stones for the development of a sense of appreciation of nature as a common good on which any and every society depends.

Meanwhile, in academia, various contributions by scientists were important during the nineteenth century to affirm our interrelationship with nature. Among them, Alexander von Humboldt, in his work, Cosmos (1828), described how climate and ecosystems were interrelated, anticipating the concept of ecology. He warned about the human impact of the overuse of natural resources, such as deforestation and its relationship with soil erosion. His interpretation of the cosmos through the two aspects of "order" and "adornment," promoted human feelings about nature to broaden and deepen. Years later, Ernst Haeckel, who took up Darwinian and von Humboldt concepts, in 1869, defined ecology as "the study of the interdependence and interaction between living organisms—animals and plants—and their environment—inorganic beings" (Reyes, 2007). Subsequently, in 1896, Svante Arrhenius proposed one of the first theories on the greenhouse effect by calculating how carbon dioxide levels in the atmosphere could influence the global climate. This work has become a precursor of our modern climate change understanding (Arrhenius, 1896).

The artistic and epistemological contributions to the understanding of nature, the use of common goods, together with the increasingly evident dependence on ecosystem services, led to the creation of local regulatory tools whose objective was to regulate the use of natural resources. In the case of Mexico, one of the first decision-makers to speak about the importance of the correct management of the commons, using the scientific contributions of the aforementioned Ernst Haeckel, was the constituent congressman Andrés Molina Enríquez, in his work "The Great Problems of Nationals" (1909) (Andrés Molina Enríquez, 1909). The argumentation and discussion carried out by the congress promoted the creation of Article 27 of the Political Constitution of the United Mexican States in 1917, an article in which the Nation is granted powers to regulate the use of natural elements susceptible to appropriation, to make an equitable distribution of public wealth, and to take care of their conservation.

During the second half of the twentieth century, after the catastrophic post-war scenarios that demonstrated the self-destructive capacity of human beings towards themselves and nature, concern for the alteration of common goods became a constant. An example of this was the dissemination work done by Rachel Carson with her book Silent Spring (1962) (Carson, 2000), which served as a key catalyst in global environmental awareness. In this work, she denounced the impact of pesticides (such as DDT) on human health and ecosystems, warning about the dangers of chemical pollution derived from anthropogenic activities. In this regard, it is worth pointing out recent ecocides resulting from the mismanagement of products and wastes such as oil spills. An important example in this category is the Torrey Canyon spill in 1967, which devastated coastlines and marine ecosystems by spilling 94–164 million liters of oil (Burrows et al., 1974). Similarly, the eutrophication of water bodies due to changes in unregulated land use, as well as wastewater discharges, showed the ability of urban areas to modify the stability of complex ecosystems, such as Erie Lake in North America, which, due to nutrient pollution and poor land management, allowed invasive aquatic species to enter and thrive. As a consequence of this problem, and due to the collaborative work between scientists and policymakers, the Great Lakes Water Quality Agreement emerged, the formation of USEPA, NOAA, and Environment and Climate Change Canada, the first Earth Day, and the approval of the Clean Water Act (Reutter, 2019).

In historical and contextual analogy with the events of the last century, scientific discoveries, as well as their interaction and utilization with culture-arts, and economy-politics, have allowed the detection of environmental changes as never before. Satellite photographs of the planet, for example, have shown the expansion of deserted areas previously destined for agriculture, and the uncontrolled and unplanned growth of cities. The use of radioisotopes in oceanography and meteorology on the other hand, has unraveled the mechanisms of ocean currents and winds. It has also been discovered that the stratospheric ozone (O3) layer has been reducing its thickness due to anthropogenic effects (McNeill, 2010). For this reason, during the 1960s and 1970s, multiple environmental movements in the United States and Europe, backed by groups of scientists, pressured governments to include environmental issues in their political agendas (Lear, 1997).

As a result of this scenario, the United Nations began to recognize that developing countries faced environmental problems related to poverty and lack of access to basic technologies. This led to the proposal that economic development should be integrated with environmental protection. Subsequently, in 1968, the First International Conference on the Biosphere took place. The main theme focused on the scientific basis for the rational use and conservation of biosphere resources. As a result of these dialogues, in 1970, UNESCO launched its "Man and the Biosphere Program" (MAB) to protect areas representing the planet's core ecosystems as "biosphere reserves" (Caldwell, 1996). Years later, as lead author Donella Meadows and in collaboration with 17 professionals, the report "The Limits to Growth" was generated. This report reflected the growing interest in studying the interactions between the economy, environment, and society. In addition, the computer simulation of the World3 program warned that exponential economic and population growth could lead to resource depletion and environmental collapse.

The publication of the United Nations Brundtland report, entitled "Our Common Future", also raised common concerns, tasks, and efforts at the global level. In addition, the concept of sustainability acquired the now widely recognized definition that highlights the interconnectedness of economic growth, social equity, and environmental protection. This concept refers to the ability to meet the needs of the present without compromising the ability of future generations to meet their own needs. This premise helps to maintain a balance between human development and the earth's ecosystems, which focuses on ensuring the long-term viability and health of natural and man-made systems.

This systemic interconnection complies with the three widely known pillars that include the environment, which focuses on the preservation of natural resources, biodiversity, and ecosystems, aspiring to achieve the reduction of pollution, generating alternatives and solutions to combat climate change, and the promotion of renewable energy sources. This aspect is frequently related to practices such as reforestation, water conservation, and above all, waste minimization. The second pillar, which relates to social issues, emphasizes equity, justice, and the well-being of individuals and communities, as it involves access to basic needs such as education, health care, and clean water while promoting cultural diversity and the protection of human rights by addressing the problems of poverty, gender equality, and social inclusion. The third pillar of sustainability refers to economic issues that seek to create viable and equitable systems in the long term, involving practices that support economic growth while minimizing environmental degradation and ensuring a fair distribution of wealth, ethical investments, and sustainable business practices.

The time following the publication of this agreement was marked by the collaboration of UN member countries, as well as the creation of the Intergovernmental Panel on Climate Change (IPCC) (Houghton et al., 1990), which produced its first Assessment Report in 1990 and catalyzed the creation of the Conference of the Parties (COP) to the Climate Convention, held in Berlin in 1995. In 1992, in the city of Rio de Janeiro, the Earth Summit was held, where Agenda 21 was approved, a global action plan to promote sustainable development, in addition to the creation of the UN Commission on Sustainable Development. In 1997, the Kyoto Protocol emerged, which sought to reduce greenhouse gas emissions, in addition to committing developed countries to limit or reduce their emissions.

15.1.1 Transition to Circular Economy Models That Support Sustainability

The new millennium laid the foundations for the development of an action plan marked by the eight Millennium Development Goals (MDGs), a global initiative of the United Nations from 2000 to 2015 whose purpose was to address humanity's most urgent challenges around poverty, hunger, health, education, gender equality, and environmental sustainability. Although this global tool was not fully implemented, at the end of the period, the importance of clear goals, comprehensive approaches, and local participation was highlighted. Likewise, the results showed persistent inequalities and the need for reliable data. In addition, they underscored the importance of strengthening partnerships and diversifying sources of financing to ensure sustainable results (Naciones Unidad, 2007). At the end of the MDG period, the Paris Agreement (2015) and the result of the Conference of the Parties (COP 21), were implemented. Using this tool, countries agreed to limit global temperature rise to less than 2 °C (with efforts to keep it to 1.5 °C). To this end, Nationally Determined Contributions (NDCs) were established to reduce air emissions. It also recognized that technology transfer is crucial for developing countries to meet their climate goals and identified local challenges such as innovation capacities, knowledge transfer, and interaction between actors (Coninck, 2015).

In 2015, the 2030 Agenda emerged, a global action plan adopted by the 193 Member States of the UN. This is a conducting document that establishes 17 Sustainable

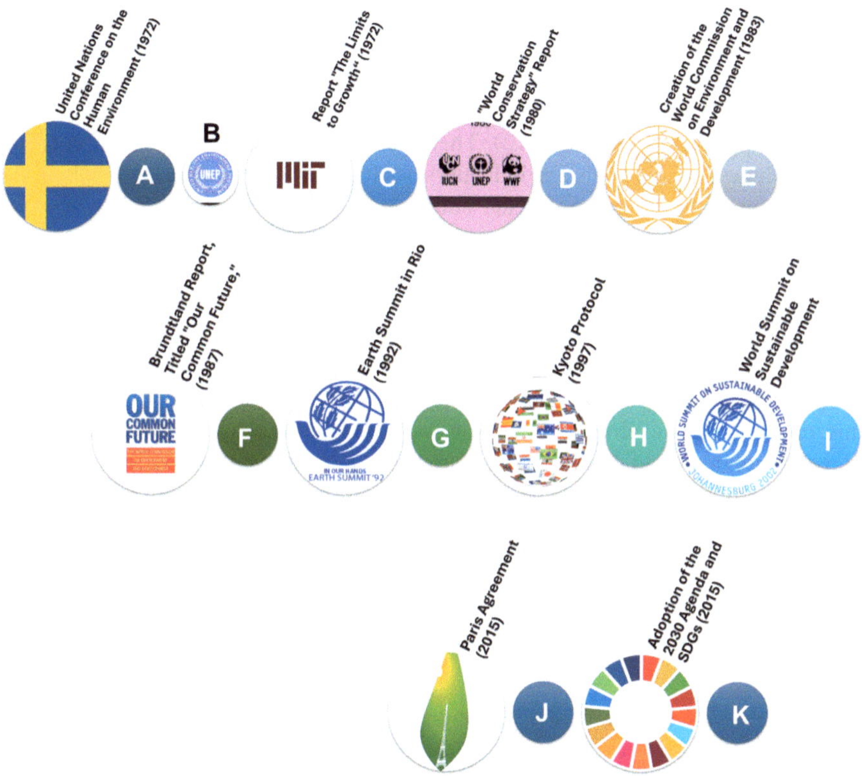

Fig. 15.1 Pathway to Sustainability: Global Events and Agreements. (**a**) Notion of the Interrelationship between environment and development. (**b**) Creation of UNEP. (**c**) Computational models to predict consequences of unlimited economic and population growth. (**d**) Introduced the concept of sustainable development, linking natural resource conservation with economic and social development. (**e**) UN General Assembly established this commission to study the interactions between economic development and the environment. (**f**) Introduced the concept of sustainable development. (**g**) Agreements like Agenda 21, the Climate Change Convention, and the Convention on Biological Diversity. (**h**) The first binding agreement requiring industrialized countries to reduce their greenhouse gas emissions. (**i**) Also known as Rio + 10, it reviewed progress since 1992 and emphasized integrating social, economic, and environmental goals. (**j**) Achieved during COP21, it marked a milestone in climate action, setting global goals to limit temperature increases below 2 °C. (**k**) A global framework to address social, economic, and environmental challenges through 17 interconnected goals. (Constructed by J.L. Navarro)

Development Goals (SDGs) and 169 specific targets to address the most notable challenges in environmental, social, and economic sustainability. These goals, unlike the 8 MDGs, range from poverty eradication to climate action, promoting development that guarantees the needs of the present without compromising those of future generations. A representation of global events and international agreements that gave rise to the term sustainability is shown in Fig. 15.1. This holistic perspective recognizes the interdependence of natural and human systems. Its implementation has inspired the formulation of local policies, projects, and multidisciplinary research aimed at promoting sustainability in various sectors, including the concept of circular economy, which invites us to rethink traditional (linear) production and consumption models toward regenerative and restorative (circular) approaches (Sachs, 2012). This opens a wide gap of innovation opportunities with measurable goals and indicators, such as SDG 9 (Industry, Innovation and Infrastructure), which promotes the adoption of clean technologies and resilient infrastructure; or SDG 13 (Climate Action), which highlights efforts to mitigate the environmental impact of human activities. One of the most important for this work is SDG 12 (Responsible Consumption and Production), fundamental for the transition to a circular economy. Among its goals is to develop sustainable consumption and production patterns, applying strategies that consider the life cycle of the product, such as the recovery of materials and the design of durable products. In this context, the development of new materials from agro-industrial waste and recycled polymers, among others, promotes an innovative scenario for R + D + I models focused on reducing the ecological footprint and exploring resilient local economic models.

15.2 Agricultural Circular Models

Agriculture is the backbone of civilization, and from its origins until today, agriculture has flourished and developed over millennia to the current form we know (Jararweh et al., 2023). Industrialized agriculture is heavily criticized as unsustainable in its practices based on the extensive use of synthetic inputs, considerable amounts of greenhouse gas emissions, and diverse implications for the degradation of natural ecosystems (Molinero-Gerbeau et al., 2021). A new concept—circular agriculture—has emerged in recent decades, which is based on the principles of circular economy and sustainability (Dagevos & de Lauwere, 2021; Velasco-Muñoz et al., 2022). This approach provides an alternative to commonly used industrial agriculture systems, since it is based on regenerative cycles of nature, decreasing and optimizing waste, reducing resource consumption, and promoting ecological balance (Velasco-Muñoz et al., 2021).

Circular agriculture works in three ways: Unlike linear models, where food is disposed of at the end of the process, this model treats waste as a resource by using organic waste as compost or using these materials to generate energy, closing the nutrient loops (Peng & Pivato, 2019). The second principle stresses systemic thinking based on specific links between the agricultural systems and the interconnections among crop residues, livestock manure, and other by-products in a formulation of reincorporated nutrient resources into specific geographies or systems that enhance soil health, improve nutrient cycling, and lower the dependence on external inputs (Khatri et al., 2023). Lastly, circular models in agriculture seek to regenerate degraded soils, restore biodiversity, and improve water cycles through practices like agroforestry, crop rotation, and cover cropping (Trigo et al., 2021; Khangura et al.,

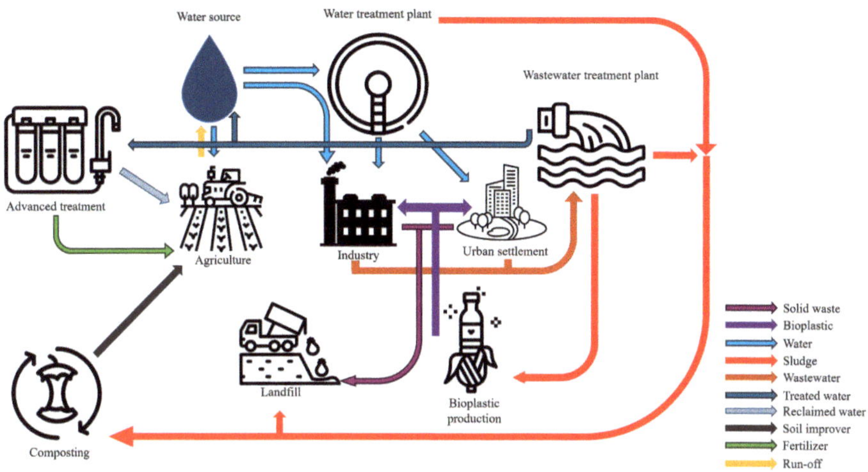

Fig. 15.2 Wastewater alternative applications toward an agricultural circular economy. (Adapted from (Mannina et al., 2022)

2023). Circular agriculture has a lot of benefits for sustainability. This approximation can, for instance, reduce pollution, enhance soil health, and preserve water and environmental conditions through the mitigation of climate change (Sgroi, 2022). Consequently, this strategy has the potential to create a more resilient economic system through dependable supply chains (Dushkova & Ivlieva, 2024).

In this context, the generation of value-added products from agro-industrial residues has given rise to the development of agro-industrial by-product-derived materials and nanomaterials that aim to not only reduce waste but also contribute to the advancement of several sectors, such as biomedicine, energy storage, and environmental remediation (Mohamed et al., 2021; Meral et al., 2022). The primary feedstocks are available from millions of tons of organic waste including crop residues, fruit peels, husks, and processing by-products, which are often discarded or barely used, contributing to methane emissions to the atmosphere and potentially usable as precursors of advanced materials (Ganesh et al., 2022; Karić et al., 2022). Many of such precursors are cellulose and hemicellulose which are widely available in plant-based waste and can be utilized to prepare bioplastics, hydrogels, and nanocellulose (Mujtaba et al., 2023; Mathura et al., 2024). For example, the complex polymer lignin from woody biomass has been used to produce carbon nanostructures, resins, and adhesives (Jędrzejczak et al., 2021; Mili et al., 2022; Zhao et al., 2023), and proteins and polysaccharides in by-products, including soy, wheat, and corn, can be processed into biocomposites, coatings, and films (Álvarez-Castillo et al., 2021; Mihalca et al., 2021; Mayilswamy & Kandasubramanian, 2022; Thivya et al., 2024). Important examples point to biomedicine and nanotechnology, which constitute two areas of application for phenolic compounds and oils that are obtained from seeds and fruit peels (Ahmed et al., 2022; Priya et al., 2023; Suhag et al., 2023). Figure 15.2 shows a schematic route for the applications and interconnections of wastewater alternatives in agriculture for a circular economy.

15.3 Development of Materials and Nanomaterials from Agro-industrial Waste Toward a Circular Economy Model

As the pace of agro-industrial activities accelerates, a substantial amount of waste is produced, and this means that sustainable solutions must be found (Singh et al., 2021b; Nair et al., 2022). Researchers and industries are now developing alternative methods to transform these residues into added-value materials and nanomaterials that can be used in a variety of fields such as water treatment, energy storage, and environmental remediation (Duan et al., 2022; Wang et al., 2024). In this section, some of the applications and methodologies for the development of materials and nanomaterials from agro-industrial waste are explored, emphasizing their potential to address pressing environmental and industrial needs.

15.3.1 Biochar and Activated Carbon from Agro-industrial Waste

Biochar is generated from carbon-rich products from the pyrolysis processes applied to agricultural waste (Khedulkar et al., 2023). This type of charcoal is stable with a porous structure and a large surface area that make it an excellent soil modifier that increases fertility by enabling the soil to retain water, make more nutrients bioavailable, and boost microbial activity (Khadem et al., 2021; Leng et al., 2021; Zhang et al., 2021). Also, biochar acts as a carbon sink that can lock away atmospheric CO2 for centuries, helping to fight and mitigate climate change (Yadav et al., 2023). This material falls in line with the principles of closed-loop farming and enhances systemic resilience by reducing inputs and waste accumulation, thus reducing the demand for chemical fertilizers and irrigation (Babalola et al., 2022; Wang et al., 2023). Biochar also reduces CO_2 emissions, making the carbon balance more sustainable. Dispersion and uptake of greenhouse gases may also work to counterbalance each other, while releasing into the soil any nutrients that could otherwise have leached away, thus making plant fertilization more efficient (Mona et al., 2021; Shoudho et al., 2024). In addition, during biochar production, energy emerges as a by-product that can be utilized to power agricultural operations and thus preserve resources (Nie et al., 2021; Vuppaladadiyam et al., 2023; Ravenni et al., 2023).

Another example is activated carbon, which is a form of processed carbon characterized by a high surface area and an enhanced porosity (Sabzehmeidani et al., 2021; Srivastava et al., 2021). This allows it to take in a wide range of organic and inorganic compounds. Such materials are made by thermal or chemical activation of biomass such as wood, coconut shells, and agricultural residues. They are useful in more industries than just the manufacture of watering machines (e.g., air purifiers or medical cupping utensils) (Lewoyehu, 2021; Gayathiri et al., 2022). Activated carbon can lift soil quality by removing pollutants such as heavy metals, pesticides, and excessive nutrients; this keeps these pollutants from getting into water, while at the same time promoting healthier crops (Sharma et al., 2022; Mohammed & Abdulhameed, 2024). When activated carbon is applied to soil or worked into compost systems, it can take up nutrients and slowly release them, reducing the demand for chemical fertilizers (Goldan et al., 2023). In addition, this material is used in the pretreatment of water for irrigation, allowing it to filter out harmful substances that would otherwise degrade soil quality when they enter the earth, both helping to recycle wastewater from agricultural back into use around farms and preserving water resources (Kucserka et al., 2023; Al-Hazmi et al., 2023). Finally, activated carbon can absorb and hold gases such as methane or ammonia, making it a useful tool for managing emissions at livestock operations and composting stations; it can also extend the shelf-life of harvested crops by taking up ethylene gas and water vapor so that decomposition is slowed down which leads to less wastefulness and better storage techniques (Ebrahimi et al., 2022; Sosa et al., 2023).

15.3.2 Nanomaterials from Agro-industrial Waste

Nanocellulose is a polymer consisting of nanometer-scaled cellulose particles (Dufresne, 2019). These have high strength, lightweight, and biodegradability properties, which lead to excellent mechanical properties, making them an ideal substance for a wide range of applications, including agriculture and environmental remediation (Zinge & Kandasubramanian, 2020). This material is bio-based and of renewable plant origin, such as wood, agricultural residues, and algae (Srivastava, 2024). Nanocellulose products can be broadly divided into two categories: cellulose nanocrystals (CNCs) and cellulose nanofibers (CNFs) (Xu et al., 2013). Nanocellulose can be produced from waste biomass, such as crop residues, and therefore, it is a part of circular agriculture, being used to reduce waste and create new sources of income and sustainability support (Mateo et al., 2021).

As it was previously noted, nanocellulose improves soil health by promoting soil porous structures and water retention, thus increasing water infiltration and root penetration, and assisting in the treatment of arid or degraded soil (Malekzadeh et al., 2024). A similar field where nanocellulose could make alternative use, replacing plastic in the weed control practice, and retaining soil moisture, is with nanocellulose-based biodegradable films that can be used as plastic mulching (Cárcamo et al., 2021). Nanocellulose polymers also play an important role in crop protection: they can be used as a carrier for slow-release pesticides or fertilizers so that the chemicals are only administered in small amounts and do not need to be frequently sprayed (Sales et al., 2024). Additionally, nanocellulose itself has been explored for its potential antimicrobial properties, providing an alternative to synthetic pesticides that are both natural and biodegradable (Ahankari et al., 2021).

Incorporating such materials into agricultural practices can yield substantial eco-friendly outcomes, minimizing reliance on detrimental plastics, reducing the carbon footprint of farm life, and encouraging the reestablishment of soil health (Skrzypczak et al., 2023). From an economic perspective, this material provides opportunities for farmers to earn additional income by producing value-added products that can be used in areas like packing, biocomposites, and medical devices (Norizan et al., 2022). However, there are some issues to be solved for the implementation of this material. One of them is the required scaling of cellulose production that involves important investments in technology and infrastructure, which can be perceived as a burden for the smallholders (Norizan et al., 2022). Secondary disadvantages comprise the high cost of nanocellulose production, which may not be affordable to farmers in developing regions, thus requiring production optimization techniques and improvements in the economy scale to reduce costs (Farooq et al., 2024).

15.3.3 Biodegradable Composites from Agro-industrial Waste

Biodegradable composites are defined as composite materials made from natural fibers (e.g., jute, hemp, and flax) or biopolymers (e.g., polylactic acid, starch-based plastics) that can be 100% renewable or that are obtained from renewable and biodegradable synthetic components that decompose via biological processes without leaving toxic residues (Elfaleh et al., 2023). Its applications range from the widely used plastic mulch films employed in the agricultural sector to retain soil moisture, suppress weeds, and control soil temperature, to biodegradable mulch films (Khalid et al., 2021). These compounds are also used in the manufacturing of plant pots and seed trays. The traditional plastic ones can be replaced with compostable alternatives that can be planted directly into the soil and help with waste removal, reduce labor costs, and increase soil fertility (Paudel et al., 2022). In addition, they can be used as packing material for agricultural products, especially for storing and transporting fresh products. As these products decompose naturally, the packaging lifecycle—created from the soil—is returned to the soil to cycle nutrients (Samir et al., 2022).

Another application of circular agriculture is related to the use of hydrogels, which are three-dimensional polymeric networks that can absorb several hundred times their weight while keeping their shape (Singh et al., 2021a). The hydrogel material is not only biodegradable but also environmentally friendly, which is one of the main characteristic advantages for circular agriculture (Gul et al., 2022). Hydrogel materials can also be mixed with soil to absorb surplus water during rain or irrigation by acting as a reserve and slowly releasing water source as it is required by the plants, thereby minimizing water wastage and improving plant growth and resistance during periods of drought (Wu et al., 2023). In addition, hydrogels can reduce water runoff, which is a leading cause of soil erosion and nutrient loss, thereby promoting soil conservation—a key aspect of circular agriculture (Patra et al., 2022). Table 15.1 shows some circular economy case studies reported in the literature for processes that involve biomass waste. It is important to point out that the information in this table is not an exhaustive but rather a representative collection of works since there are many other laboratory-level studies that explore some type of valorization of biomass waste with a long-term circular economy approach.

15.3.4 Applications of Agro-industrial Based Materials and Nanomaterials for Water Treatment

Fresh water is without doubt one of our most basic resources, and its demand is in continuous increase due to population growth, industrialization, and climate change (e.g., an abundance of water, but uncertain quality). In this context, innovative solutions are needed to address water quality and availability challenges, which will further stabilize sustainable agricultural practices (Srivastav et al., 2021; Gavrilescu, 2021). Materials used in water treatment range from conventional adsorbents to advanced nanomaterials like graphene oxide and metal-organic frameworks (MOFs) (Saleh, 2021). As an example, these materials are used for adsorption and filtration, and nanomaterials such as graphene oxide may outperform traditional materials by showing superior adsorption capacities and selectivity for specific contaminants (Chai et al., 2021). Their use in membrane-based systems also benefits from these reasons (Ibrahim et al., 2023). Electrochemical water treatment processes, including electrocoagulation, electrooxidation, and electro-catalysis, may benefit from the high conductivity and stability properties of new materials, which may improve the efficiency of these processes (Shokri et al., 2023). For instance, several kinds of catalytic nanomaterials, including zero-valent ion nanoparticles and metal-organic frameworks (MOFs), are effective in AOPs. Along with their chemical composition, the high specific surface area of these materials has proven their ability to generate reactive oxygen species (ROS) to remove organic pollutants (Liu et al., 2023). Thus, the coupling of electrochemical systems with catalytic nanomaterials may enhance performance for water treatment, as is the case for hybrid electro-Fenton systems (Nidheesh et al., 2023).

Optimizing efficiency in terms of ion selectivity and power consumption can enable electrochemical systems coupled with nanomaterials to be used for recovering nutrients such as nitrogen and phosphorus from agricultural runoff or wastewater (Devda et al., 2021).

Table 15.1 Various impact of nanotechnology on the quantity indicators of crops

Waste biomass	Circular economy approach	Findings of cases studies	Reference
Pineapple processing	A case study of pineapple processing industrialization in a circular economy	By-products are generated (106% peels, cores and crowns and 58.2% pomace), and with this implementation, two hypotheses can be suggested. From the liquid fraction, no waste is generated, everything is valorized; from the solid fraction (press cake), it is possible to produce biofuels, leading to the generation of energy and 20% (dry mass) of biomass (waste). On the other hand, the press cake can be totally dried, powdered and be used to produce a new product—pineapple flour. Pineapple flour is full of dietary fibre and complex soluble sugars.	Campos et al. (2020)
Agro-industrial biomass biorefinery products are obtained by employing different conversion technologies.	Biorefinery process and circular bioeconomy for the generation of renewable energy products for the systemic approach to economic development	The production of biofuels and energy from lignocellulosic biomass depends on two primary processes: (i) biochemical (ii) thermochemical methodologies for obtaining products from agro-industrial wastes 1-sources of waste and energy production 2-sources of agro-industrial waste and value-added products 3-balorization strategies 4-management of agro-industrial waste 5- products from (4) 6-application of LCA in agro-industrial CE Applications of LCA in agro-industrial circular bioeconomy.	Yaashikaa et al. (2022)
Waste biomass	Mapping of research lines on circular economy	The economic (costs, investments), technical,	Barros et al. (2020)

(continued)

Table 15.1 (continued)

Waste biomass	Circular economy approach	Findings of cases studies	Reference
	practices in agriculture: From waste to energy. Particularly the electricity generation and biofuel produced from biogas	and social aspects (social development, job creation, and income) appear as demands for future research. New circular business models are also expected to be designed to facilitate business management. In terms of tools and indicators to measure the circular economy in rural properties, these need further elaboration to facilitate the management of material and energy that circulates inside and outside the organization.	
Fodder production Livestock Cheese-making	Integration of life cycle Assessment and energy accounting methods (LEAF) in the agro-industry	Two different scenarios were built based on the identified hotspots (cleaning products and electricity): (i) a technological improvement (dealing with cleaning processes methods), (ii) an eco-efficiency perspective (fossil energy replaced with renewable alternatives). Additionally, viewpoint shifting scenarios based on (iii) different allocation procedures were proposed to discuss crucial methodological issues. Results showed that technological improvements provide the best environmental performance, with lower emissions and better energy indicators, and better work conditions.	Oliveira et al. (2021)
Hyacinth	Water hyacinth biochar: a circular economy case study	The environmental impacts represented by life-cycle GHG emission of three WH gasification biochar applications. The GHG mitigation	He et al. (2022)

(continued)

Table 15.1 (continued)

Waste biomass	Circular economy approach	Findings of cases studies	Reference
		performances in soil and concrete applications are mainly due to the stabilization of biogenic carbon contents, and the AC generation was based on the coal-derived product substitution. Although there was a significant GHG emission reduction when employing WH biochar for AC generation when compared to coal-derived AC, the economic performance of WH-derived AC (represented by specific surface area) was not ideal.	
Waste biomass	Waste to biochar as a by-product via pyrolysis/gasification The study analyzed four key uses of biochar: (1) in agriculture and horticulture, (2) as construction material, (3) as activated carbon, and (4) in anaerobic digestion.	A consolidated standard is needed for the healthy and sustainable development of the biochar industry. In this paper, the specification for production and application of biochar was proposed as the Singapore biochar standard which regulates the specific requirements of the production and subsequent application in agriculture and horticulture, and in anaerobic digestion, as construction material and activated carbon in Singapore. The proposed standard can act as the guideline for biochar market and ensure a safe and healthy environment. The code of practice for biochar application in Singapore is proposed	Hu et al. (2021)
Tomato crop	Urban agriculture solid waste in the frame of circular economy: case study of tomato crop in	Use of tomato stems from an integrated rooftop greenhouse as a substrate for lettuce crops.	Manríquez-Altamirano et al. (2020)

(continued)

Table 15.1 (continued)

Waste biomass	Circular economy approach	Findings of cases studies	Reference
	integrated rooftop greenhouse	Reduction of the urban agriculture solid waste within the cities and closure of the UA life cycle Is possible to increase the environmental benefits of i-RTG by taking advantage of its biomass locally, helping to foresee a possible future problem regarding the management of the residual biomass of integrated rooftop greenhouse within cities.	
Pruning residues of olive groves	Circular economy model: Insights from a case study in South Italy, from nine municipalities in southern Italy	The whole agro-energetic chain, namely the power plant and the collection company, has both a good return on owner's equity ratio (15.89% and 31.48%, respectively) and ROI ratio (4.34% and 6.14% respectively). Moreover, the power plant's ecoefficiency ratio (2.64€ per 1 Mg of GHG) is slightly lower than harvest firm one (2.91€ per 1 Mg of GHG). The findings could be useful to develop new business models based on the circular economy concept.	Palmieri et al. (2020)

15.4 Water Use in Agricultural Systems

Circular agriculture is a vision for a regenerative process of water use on the spectrum between treatment and recycling (i.e., used water), following ecological and economic goals. Because conventional linear systems are based on continual extraction and disposal, they are not sustainable, considering the increasing stress on freshwater systems around the world. Circular systems need a reconsideration of water utilization trends: prioritizing, recycling, and implementing solutions (Lee et al., 2023). For this to be possible, modern agriculture must develop in terms of precision irrigation via modern technology, for example, soil moisture sensors, remote sensing, and data analytics, which need to be used so that water can be delivered to crops most efficiently, reducing wastage while improving yield

(Lu et al., 2022a). Nanomaterials-based sensors can give real-time information about the soil and environmental conditions for making accurate irrigation schedules (Chaudhary et al., 2024). Drip irrigation systems paired with nutrient delivery mechanisms to simplify the application process fall as another alternative, reducing water consumption by delivering to specific root zones, and avoiding evaporation and runoff (Sidhu et al., 2021). Water recycling and reuse, which are aspects of circular agriculture, could be addressed by employing membrane filtration, reverse osmosis, and advanced oxidation processes that allow converting agricultural runoff and municipal wastewater into irrigation water (Mainardis et al., 2022).

In this context, Chen et al., reported the parameters that are important in determining the quality of water used in agriculture. Among these, pH, SS (suspended solids), DO (dissolved oxygen), BOD/COD (biological oxygen demand/chemical oxygen demand, ratio), oil and grease, bacteria, EC (electrical conductivity), SAR (sodium adsorption ratio), TN (total nitrogen), Cl-, F-, SO4-2, RSC (residual sodium carbonate index), Co, Cu, Pb, Li, Mn, Hg, Mo, Ni, Se, V, Zn, Al, As, Be, B, Cd, Cr, and Fe stand out (Chen et al., 2021). A self-sustaining water-use-and-treatment cycle can also reduce waste and provide continuity (Manikandan et al., 2022). For example, runoff from irrigated fields can be collected, purified, and reused in subsequent irrigation cycles, ensuring water availability while protecting ecosystems downstream (Kalboussi et al., 2022). Coupling water management with renewable energy usage through solar-powered water treatment systems utilizing photocatalytic materials (e.g., titanium dioxide) could also be employed to provide decentralized solutions for rural farming communities (Lu et al., 2022b). Moreover, wind and biogas energy can also provide water treatment systems, connecting the dots between renewable energy and sustainable agriculture (Rahman et al., 2022). However, transitioning to circular water management will need the support of innovation, investment, education, and public policy (Salminen et al., 2022). Circular agriculture is a key approach to sustainable agriculture for producers and end-consumers with water security and preservation of the ecosystem of our planet for generations to come (Kumar et al., 2022).

15.5 Contribution of Advanced Wastewater Treatments Toward a Circular Economy Model

Aging infrastructure, increasing environmental regulations, and environmental issues related to water use dynamics make it necessary the need efficient and widely used advanced wastewater treatment processes.

Advanced wastewater treatment processes are useful not only for the removal of organic carbon but also for the removal of nutrients. The recovery of valuable products is possible, as well as the removal of organic matter, emerging pollutants, and pathogens from wastewater using advanced oxidation processes (AOPs). AOPs are efficient alternatives due to the highly reactive oxidant species that are produced to ensure the destruction of non-biodegradable organic pollutants. The appropriate AOP is frequently selected according to the emerging pollutant to be eliminated. The most common types of advanced oxidation processes are Fenton, solar Fenton, photo-Fenton, and electro-Fenton, ultraviolet mediated processes, photocatalysis, nanofiltration, ozonation, membrane filtration, and chlorination (Collivignarelli et al., 2020; Mannina et al., 2022; Tarpani & Azapagic, 2023; Rodríguez-Peña et al., 2024). On the other hand, the main products obtained from EAOPs are energy, nutrients, and organic sludges that can be treated to obtain value-added products. The main alternatives for waste recycling from EAOPs are: fertilizer and electricity by anaerobically digested or composted sludge. Sludge recovery alternatives include the agricultural application of anaerobically digested sludge, as well as the use of composted sludge, incineration, pyrolysis, and wet air oxidation (Tarpani & Azapagic, 2023).

Life cycle assessment (LCA) is a common tool for quantifying sustainability processes, and consists of a systematic analysis of the environmental impact of a product, material, or activity over its entire life cycle. The LCA considers (Navarro et al., 2020; Feijoo et al., 2023; Tarpani & Azapagic, 2023):

- Inventory data.
- Functional units.
- Unit process.
- System boundaries.
- Estimation of environmental impacts.
- Midpoint and endpoint impact categories.
  - Representative of energy, toxicity, and water effects: global warming potential (GWP), terrestrial acidification (TA), freshwater eutrophication (FE), marine eutrophication (ME), terrestrial ecotoxicity (TET), and fossil resource scarcity (FRS).
- Indicators.
  - Environmental indicators: Climate change, Fossil fuel depletion, Metal depletion, Water depletion, Ozone depletion, Freshwater eutrophication, Marine eutrophication, Terrestrial acidification, Ionizing radiation, Freshwater ecotoxicity, Marine ecotoxicity, Terrestrial ecotoxicity, Human toxicity, Natural land transformation, Urban land occupation, Agricultural land occupation, Particulate matter formation, Photochemical oxidant formation.
  - Economic costs indicators: construction, replacements, operation, waste management, and transport, and revenue.
  - Social indicators: Net water use, Freshwater pollution, Net energy use, Electricity consumption, Land use, Synthetic fertilizer production, Adoption risk, Public opposition, Dissimilarity from traditional products, Damage to human health, Presence of harmful substances. Social indicators are classified under the issues of water security, Energy security, Food security, Technology adoption, Acceptance, and Human health.

- Impact results.
- Certification schemes and compliance standards.

The most important aspects to consider here are the degradation effectiveness, treatment time, energy consumption and generation, chemical dosage use, footprint, sludge generation, vapor production, operation costs, and environmental impact (Nath, 2024).

Electrochemical Advanced Oxidation treatments, on the other hand, use electrochemical reactions to facilitate the production of reactive species. These processes have some advantages when compared to conventional methods that may help to achieve circular processes. For example, the effectiveness in removing a wide range of pollutants has been validated for these processes, and although electricity is required, the use of solar systems is an alternative to fulfill those requirements. In addition, the usage of chemicals is usually lower when compared to conventional methods. The EAOPs can also be easily adapted to traditional methods in the disinfection stage, frequently requiring reduced spaces due to compact designs.

In spite of their great potential, it is clear that some disadvantages need to be overcome so that these processes can be considered as part of the traditional wastewater treatments. For example, sludge generation and operation costs, which are higher when compared to traditional methods, are some evident constraints that must be dealt with in view of the growing production of more and new emerging compounds (Nath, 2024). Electrochemical wastewater treatment technologies therefore offer significant promises to environmental and economic challenges associated to water quality requirements. There is a need to integrate economic and environmental sustainability processes (Nath, 2024). Recent reports from the literature state as future research directions, the development of novel materials for electrodes that may improve durability and efficiency, and the use of renewable energy sources, which may enhance sustainability and reduce operational costs.

Finally, the active collaboration among government, academia, industry, and civil society is important to overcome emerging barriers to merge EAOP with traditional wastewater technologies and facilitate the transition to circular economy processes.

15.6 Relevant Wastewater Treatment Applications Toward a Circular Economy Transition

Table 15.2 compares those reports that have implemented the LCA methodology to study the performance of electrochemical wastewater treatment processes. This comparison gives an overview of the type of water used. The trend is the use of real wastewater, mainly from the textile industry, although wastewater from domestic, hospital, and petroleum distillery origins has also been employed. This table points out the most used LCA methodologies as well as the focus of each study.

As pointed out by Tarpani et al., the following LCA elements must be considered: definition of the scope, selection and definition of treatment methods for water and

Table 15.2 Comparison of LCA approaches applied to advanced wastewater treatment

LCSA approach and scope	Wastewater treatment	Water type	Scale	LCA tools Indicators and methods	Important aspects Opportunity area	Reference
Sewage sludge treatment methods for resource recovery considering ecotoxicity of heavy metals and pharmaceutical and personal care products	Anaerobic digestion form: (i) Agricultural application of anaerobically digested sludge; (ii) agricultural application of composted sludge; (iii) incineration; iv) pyrolysis; and v) wet air oxidation	Sludge handling from urban wastewater treatment plant	Laboratory	Multicriteria decision analysis	PPCPs have a negligible contribution to freshwater ecotoxicity when compared to heavy metals in the anaerobically digested sludge.	Tarpani et al. (2020)
Urban wastewater reuse and sewage sludge resource recovery	Nanofiltration (after a comparison) anaerobic digestion for sludge recovery	Urban wastewater	Laboratory	Multicriteria decision analysis	Social aspects were considered	Tarpani and Azapagic (2023)
Laboratory scale to industrial scale using prospective life cycle assessment (LCA) on a quantitative basis	Electrochemical oxidation comparing three representative anode materials (SnO_2, PbO_2, and boron-doped diamond) and other two typical processes (adsorption and Fenton method)	Synthetic wastewater	Laboratory	OpenLCA 1.10.3 was employed as LCA software to calculate the environmental impacts of each individual elements using the ReCiPe midpoint (H) method. All the relevant materials and energy inputs were taken from the Ecoinvent database.	LCA of scale-up EO was performed to offer guidance for practical application, and the life cycle inventory was compiled upon thermodynamic and kinetic simulations, empirical calculation rules, and similar technical information	Sun et al. (2023)

(continued)

Table 15.2 (continued)

LCSA approach and scope	Wastewater treatment	Water type	Scale	LCA tools Indicators and methods	Important aspects Opportunity area	Reference
Scale-up modelling and life cycle assessment of electrochemical oxidation in wastewater treatment	Electrochemical oxidation pilot-scale eAOP	Synthetic wastewater with carbamazepine	Laboratory to prototype	The LCA was performed using the ReCiPe MidPoint (H) V1.06/World (2010) and EndPoint (H/H) V1.06/World (2010) methods in SimaPro 9.3.0.2. software.	Mitigating the impacts of energy-intensive pieces of equipment	Feijoo et al. (2023)
Comparative life cycle assessment of sequential chemical and electrochemical processes	Coagulation-flocculation-Fenton-neutralization (CF-F-N), and coagulation-flocculation-electro-Fenton neutralization (CF-EF-N)	Industrial textile wastewater	Optimized at laboratory scale	Comparison through LCA, using the IPCC-2013 and ReCiPe-2016 midpoint and endpoint methods	The CF-EF-N can be considered more environmentally sustainable technology.	Salazar-Sogamoso et al. (2024)
Prospective life cycle Assessment and cost analysis of novel Electrochemical Struvite recovery	Electro-oxidation Electrochemical struvite Reactor	Domestic wastewater	Existing treatment train	A life cycle impact assessment (LCIA) was generated in SimaPro using the IMPACT world+ MidPoint 1.22 characterization method. Software GPS-X and CapdetWorks assuming its integration in a full-scale existing wastewater treatment	Struvite recovery reduced Aquatic eutrophication impacts as well as terrestrial acidification impacts.	Morrissey et al. (2022)

				plant with and without anaerobic digestion.		
Techno-economic assessment of microbial fuel cells, microbial electrolysis cells, microbial electrosynthesis cells, and microbial desalination cells	Microbial fuel cells, microbial electrolysis cells, microbial electrosynthesis cells, and microbial desalination cells	Synthetic	Laboratory	The ISO 14040 LCA standard categorizes LCA in four stages: (1) goal and scope definition; (2) inventory analysis; (3) impact assessment; and (4) interpretation.	Authors conclude further research is required in order to scale up and reducing the overall costs of these MES for efficient incorporating for practical usage.	Savla et al. (2021)
Advances about EAOPs and LCA	Electrooxidation, ozone, electrocoagulation, electroFenton, electro-photocatalysis, and ultrasound-assisted electrochemical treatment	Pharmaceutical/hospital wastewater, wastewater of petroleum industry, distillery wastewater, textile wastewater, real high salinity wastewater	Prototype (laboratory scale	IPCC 2007 ReCiPe version Ecoinvent v3.1 and CLCD database Tool for reduction and assessment of chemicals and other environmental Impacts (TRACI) CML 2000 baseline		Rodríguez-Peña et al. (2024)
Development and life cycle assessment of an auto circulating bio-electrochemical reactor for energy positive continuous wastewater treatment	Bio-electrochemical reactor Microbial fuel cell	Domestic wastewater (DW) from primary effluent collection tank	Laboratory	SimaPro 8.3 software. Ecoinvent 3.0 database was used for background processes	One of the most important challenges is to find a balance between treatment efficiency, electricity production, and capital costs, which requires multidimensional study, including cost-benefit analysis,	Singh Mathuriya et al. (2020)

(continued)

Table 15.2 (continued)

LCSA approach and scope	Wastewater treatment	Water type	Scale	LCA tools Indicators and methods	Important aspects Opportunity area	Reference
					change in liquid volume due to absorption on anodes in anode tank and evaporation in cathode chamber, fluid dynamic analysis, etc.	
Performance evaluation and life cycle assessment of electrocoagulation process for manganese removal from wastewater using titanium electrodes	Electrocoagulation	Synthetic and real wastewater	Laboratory	ISO 14044 guidelines LCA software (OpenLCA 1.11) European reference life Cycle (ELCD) databases. ReCiPe 2016 midpoint (H) method	The system's viability in large-scale applications can be improved by the availability of these data	Singh Mathuriya et al. (2020)
Life cycle assessment (LCA) of treating PFASs with ion exchange and electrochemical Oxidation technology	Ion exchange (IX) followed by electrochemical oxidation (EO)	Per- and polyfluoroalkyl substance (PFASs)	Laboratory	LCA methodology in ISO 14040–14,044 Ecoinvent unit processes ELCD USLCI Ecoinvent unit processes	Empirical equations were developed to calculate the global warming potential for the three scenarios. They can be used in different cases after modifying the parameters and constants according to site-specific conditions.	Li et al. (2022)
		Textile wastewater	Laboratory			

Life cycle assessment of sequential and simultaneous combination of electrocoagulation and ozonation for textile wastewater treatment	Combination of EC and O_3 were analyzed and compared to EC and O_3 alone processes			ReCiPe midpoint and endpoint methods.	Sequential EC/O_3 treatment can be sustainably applied to textile wastewater.	Ahangarnokolaei et al. (2021)

LCSA life cycle sustainability assessment, *PPCPs* pharmaceutical and personal care products, *eAOP* electrochemical advanced oxidation processes, *MES* microbial electrochemical systems, *ELCD database* European reference Life Cycle Database, *USLCI database* U.S. Life Cycle Inventory (USLCI) Database sludge generation, and calculation a life cycle assessment, life cycle costing, and social life cycle assessment (Tarpani & Azapagic, 2023). Zhai et al. also reported a review that cites the most relevant aspects for Microbial electrochemical technologies for assisted nitrogen recovery from different wastewater sources. Their results show that the microbial electrolysis cell (MEC) has a better potential engineering application when compared to a microbial fuel cell (MFC), which could reduce the overall environmental burden and infrastructure costs (Zhai et al., 2023). Rodríguez-Peña, on the other hand, reported the use of LCA on electrochemical advanced oxidation processes. In this study, the authors compared the advantages and limitations of the following wastewater treatments: electrooxidation, ozone, electrocoagulation, electro-Fenton, electro-photocatalysis, and Ultrasound-assisted electrochemical treatments, for which real wastewater (pharmaceutical/hospital wastewater, wastewater of petroleum industry, distillery wastewater, textile wastewater, real high salinity wastewater) was employed (Rodríguez-Peña et al., 2024). Acevedo-García et al. also reported the development of biomaterials from agricultural waste that can be used for bioadsorption and subsequently regenerated by means of Fenton reactions (Acevedo-García et al., 2020). In this regard, it is important to note that the regeneration of carbonaceous materials is an approach that recently has become increasingly relevant (Fernández et al., 2018; Robles et al., 2020; Salazar-López et al., 2023; Banda-Alemán et al., 2023).

Two main routes for the development of more sustainable and circular economy-based wastewater treatment systems would necessarily involve (a) innovating and integrating anaerobic wastewater treatment systems that efficiently use energy and resources and (b) improving carbon capture with an energy recovery approach (Ghimire et al., 2021).

15.7 Challenges and Opportunities to Achieve Process Sustainability

The circular economy framework aims to eliminate waste, retain assets, and renew the natural world (Velenturf & Purnell, 2021). In the agricultural sector, this implies the implementation of some practices like nutrient recycling, precision farming, and bioenergy production. When organic waste is mixed with water, farmers can develop resilience by reconnecting flow streams through sustainable practices such as composting, anaerobic digestion and biochar application, among other strategies that increase soil fertility, reduce greenhouse emissions and generate sustainable food systems (Toplicean & Datcu, 2024). However, several challenges still need to be overcome before circularity can be fully realized. Among the problems to be dealt with are the following: (i) the recovery and processing of resources from agricultural by-products that are often hinging upon advanced technologies; (ii) agricultural waste streams from various sources that are seasonallyvariable and difficult to process; and (iii) the organic waste that is often contaminated with plastic or heavy metals, a situation that jeopardizes soil health and food safety (Iacovidou et al., 2021; Haque et al., 2023).

Moreover, the shift to agricultural circularity frequently requires significant initial expenditures in infrastructure, research, and capacity-building ad therefore, small and medium-sized farms may not have the means to adopt advanced recycling technologies or scale sustainable practices. Although technologies close to food systems such as bioenergy systems or precision agriculture are being developed, they also require specific knowledge and training for their application at a more appropriate level (Medhekar, 2024; Sreekumar et al., 2024). An additional challenge that needs to be addressed is related to policies and regulatory barriers. Since linear agricultural systems are incentivized by policies (e.g., subsidies on chemical fertilizers), these regulatory frameworks can impede or promote the widespread adoption of circular and sustainable practices (Vermunt et al., 2022). Another important issue is associated with cultural behavior, which implies overcoming the agricultural status quo which is usually combined with little understanding of the advantages of circular agriculture (Ajwani-Ramchandani et al., 2021). If the perceived benefits are properly transmitted, however, it is obviously more likely to have circular economy approaches to be implemented in the near future (Morseletto, 2023). The transition requires not only new policies and global commitments that seek to limit the impact on land and water but also rethinking consumption in an enabling environment for circular agriculture, where incentives, subsidies, and public/private partnerships drive the transition (Yang et al., 2023).

15.7.1 Resource Management and Cross-Sector Collaboration

The success of circular agriculture is coupled to a good coordination process between sectors where stakeholders from agriculture, water management, waste recycling, and technology collaborate to design and implement integrated solutions (Adami & Schiavon, 2021). Some initiatives can improve specialist collaboration; innovative hubs, for example, integrate various expertise individuals, enabling stakeholders to co-develop and pilot solutions, ensuring they are technically sound and socially relevant (Cicculo et al., 2021). So do digital platforms, which are playing an ever-greater role in bringing new actors together and sharing relevant data and tools (such as resource mapping and stakeholder networks) to improve connection and collaboration (Voulgaridis et al., 2022). Moreover, policy frameworks can stimulate partnerships via government funds through the offer of grants, subsidies, and tax breaks for projects that incorporate sustainable resource management practices, as well as establishing guidelines that promote transparency and accountability among partners to ensure that the circular initiatives can be aligned with wider environmental and social goals (Cong & Thomsen, 2021). Finally, in terms of how cross-sector collaboration can be promoted, some of the non-governmental organizations and educational institutions can create interdisciplinary institutes focused on research in agricultural production, chemistry, engineering, and social sciences (Guerrero & Hansen, 2021; Pedersen et al., 2021; Sadabadi & Rahimi Rad, 2022).

15.8 Final Remarks

- The production of advanced materials derived from agro-industrial waste aligns with circular economy principles by promoting waste valorization and reducing environmental burdens.
- Many challenges need to be addressed, such as the scalability of laboratory studies and the derived cost-effectiveness. There is a need to monitor consistent quality in the production of materials and nanomaterials.
- The role of wastewater treatment applications in facilitating the transition to a circular economy is fundamental to achieving sustainability goals, particularly from the evolution of the scope of wastewater treatment from a disposal-driven process to a resource-oriented one.
- Wastewater treatment applications can significantly reduce greenhouse gas emissions and pollution while helping to achieve climate change mitigation and the preservation of ecosystems.
- The circular economy concept has been applied to a wide variety of studies of the recovery or valorization of biomass waste, not always considering the whole perspective of a circular economy model.
- The use of Advanced oxidation processes is a relevant strategy for the circular economy of water, since it is possible to obtain water for reuse, it is also possible the recover nutrients and materials, and reduce waste generation.
- There is a great opportunity area for: a) integrating those materials obtained from the recovery of waste generated in various stages of agriculture. b) The water requirements in agriculture must be redirected from industrial processes along the global circular economy. c) In agriculture, it is important to take into account water quality.
- It is necessary to develop wastewater treatment plants that incorporate advanced treatment systems, where it is possible to recover energy as well as valuable substances and nutrients to be used in other sectors of the agricultural production chains.
- The use of tools that facilitate the transition to the circular economy such as life cycle sustainability assessment.

Acknowledgments The authors express their gratitude to the Mexican National Council of Humanities, Sciences, and Technologies (CF-2023-I- 939), to the CONAHCyT National Laboratory of Water, Science, and Technology, and to the Autonomous University of Queretaro (FQU-202415,; FQ-202533 and FQU-202308) for financial support of this work.

Conflict of Interest The authors have no conflicts of interest to declare relevant to this chapter's content.

References

Acevedo-García, V., Rosales, E., Puga, A., et al. (2020). Synthesis and use of efficient adsorbents under the principles of circular economy: Waste valorisation and electroadvanced oxidation process regeneration. *Separation and Purification Technology, 242*, 116796. https://doi.org/10.1016/j.seppur.2020.116796

Adami, L., & Schiavon, M. (2021). From circular economy to circular ecology: A review on the solution of environmental problems through circular waste management approaches. *Sustainability (Switzerland), 13*, 1–20.

Ahangarnokolaei, M. A., Attarian, P., Ayati, B., et al. (2021). Life cycle assessment of sequential and simultaneous combination of electrocoagulation and ozonation for textile wastewater treatment. *Journal of Environmental Chemical Engineering, 9*, 106251. https://doi.org/10.1016/j.jece.2021.106251

Ahankari, S. S., Subhedar, A. R., Bhadauria, S. S., & Dufresne, A. (2021). Nanocellulose in food packaging: A review. *Carbohydrate Polymers, 255*, 117479. https://doi.org/10.1016/j.carbpol.2020.117479

Ahmed, M. M., Badawy, M. T., Ahmed, F. K., et al. (2022). Fruit peel waste-to-wealth: Bionanomaterials production and their applications in agroecosystems. In *Agri-waste and microbes for production of sustainable nanomaterials* (pp. 231–257). Elsevier.

Ajwani-Ramchandani, R., Figueira, S., Torres de Oliveira, R., & Jha, S. (2021). Enhancing the circular and modified linear economy: The importance of blockchain for developing economies. *Resources, Conservation and Recycling, 168*, 105468. https://doi.org/10.1016/j.resconrec.2021.105468

Al-Hazmi, H. E., Mohammadi, A., Hejna, A., et al. (2023). Wastewater reuse in agriculture: Prospects and challenges. *Environmental Research, 236*, 116711. https://doi.org/10.1016/j.envres.2023.116711

Álvarez-Castillo, E., Felix, M., Bengoechea, C., & Guerrero, A. (2021). Proteins from agri-food industrial biowastes or co-products and their applications as green materials. *Food, 10*.

Andrés Molina Enríquez. (1909). *Los grandes problemas nacionales*. Secretaría de Cultura.

Arrhenius, S. (1896). On the influence of carbonic acid in the air upon the temperature of the ground. *Philosophical Magazine and Journal of Science, 41*, 237–276.

Babalola, T. E., Adabembe, B. A., & Faloye, O. T. (2022). Water use – yield relationship of maize as influenced by biochar and inorganic fertilizer applications in a tropical sandy clay loam soil. *Agricultural Water Management, 271*, 107801. https://doi.org/10.1016/j.agwat.2022.107801

Banda-Alemán, J. A., Camacho-Callejas, M., Salazar-López, M. L., et al. (2023). Alternating polarization of Fe modified carbon fiber materials. Towards the development of single-electrode electro-Fenton batch reactors. *Electrochimica Acta, 450*, 142282. https://doi.org/10.1016/j.electacta.2023.142282

Barros, M. V., Salvador, R., de Francisco, A. C., & Piekarski, C. M. (2020). Mapping of research lines on circular economy practices in agriculture: From waste to energy. *Renewable and Sustainable Energy Reviews, 131*, 109958. https://doi.org/10.1016/j.rser.2020.109958

Burrows, P., Rowley, C., & Owen, D. (1974). Torrey canyon: A case study in accidental pollution*. *Scottish Journal of Political Economy, 21*, 237–258. https://doi.org/10.1111/j.1467-9485.1974.tb00195.x

Caldwell, L. K. (1996). *Nternational environmental policy: From the twentieth to the twenty-first century* (3rd ed.). Duke University Press.

Campos, D. A., Gómez-García, R., Vilas-Boas, A. A., et al. (2020). Management of Fruit Industrial by-products—A case study on circular economy approach. *Molecules, 25*, 320. https://doi.org/10.3390/molecules25020320

Cárcamo, L., Sierra, S., Osorio, M., et al. (2021). Bacterial Nanocellulose mulch as a potential greener alternative for urban gardening in the small-scale food production of onion plants. *Agricultural Research, 10*, 66–71. https://doi.org/10.1007/s40003-020-00479-y

Carson, R. (2000). *Silent Spring*. Penguin Classics.

Chai, W. S., Cheun, J. Y., Kumar, P. S., et al. (2021). A review on conventional and novel materials towards heavy metal adsorption in wastewater treatment application. *Journal of Cleaner Production, 296*, 126589. https://doi.org/10.1016/j.jclepro.2021.126589

Chaudhary, V., Gaur, P., & Rustagi, S. (2024). Sensors, society, and sustainability. *Sustainable Materials and Technologies, 40*, e00952. https://doi.org/10.1016/j.susmat.2024.e00952

Chen, C.-Y., Wang, S.-W., Kim, H., et al. (2021). Non-conventional water reuse in agriculture: A circular water economy. *Water Research, 199*, 117193. https://doi.org/10.1016/j.watres.2021.117193

Cicullo, F., Cagliano, R., Bartezzaghi, G., & Perego, A. (2021). Implementing the circular economy paradigm in the agri-food supply chain: The role of food waste prevention technologies. *Resources, Conservation and Recycling, 164*, 105114. https://doi.org/10.1016/j.resconrec.2020.105114

Collivignarelli, M. C., Abbà, A., Miino, M. C., et al. (2020). Disinfection of wastewater by UV-based treatment for reuse in a circular economy perspective. Where are we at? *International Journal of Environmental Research and Public Health, 18*, 77. https://doi.org/10.3390/ijerph18010077

Cong, R.-G., & Thomsen, M. (2021). Review of ecosystem services in a bio-based circular economy and governance mechanisms. *Ecosystem Services, 50*, 101298. https://doi.org/10.1016/j.ecoser.2021.101298

Coninck, H. (2015). *Technology in the 2015 Paris climate agreement and beyond*. International Centre for Trade and Sustainable Development (ICTSD).

Dagevos, H., & de Lauwere, C. (2021). Circular business models and circular agriculture: Perceptions and practices of dutch farmers. *Sustainability (Switzerland), 13*, 1–15. https://doi.org/10.3390/su13031282

Devda, V., Chaudhary, K., Varjani, S., et al. (2021). Recovery of resources from industrial wastewater employing electrochemical technologies: Status, advancements and perspectives. *Bioengineered, 12*, 4697–4718. https://doi.org/10.1080/21655979.2021.1946631

Duan, Y., Tarafdar, A., Kumar, V., et al. (2022). Sustainable biorefinery approaches towards circular economy for conversion of biowaste to value added materials and future perspectives. *Fuel, 325*, 124846. https://doi.org/10.1016/j.fuel.2022.124846

Dufresne, A. (2019). Nanocellulose processing properties and potential applications. *Current Forestry Reports, 5*, 76–89. https://doi.org/10.1007/s40725-019-00088-1

Dushkova, D., & Ivlieva, O. (2024). Empowering communities to act for a change: A review of the community empowerment programs towards sustainability and resilience. *Sustainability (Switzerland), 16*.

Ebrahimi, A., Zabihzadeh Khajavi, M., Ahmadi, S., et al. (2022). Novel strategies to control ethylene in fruit and vegetables for extending their shelf life: A review. *International journal of Environmental Science and Technology, 19*, 4599–4610. https://doi.org/10.1007/s13762-021-03485-x

Elfaleh, I., Abbassi, F., Habibi, M., et al. (2023). A comprehensive review of natural fibers and their composites: An eco-friendly alternative to conventional materials. *Results in Engineering, 19*.

Farooq, A., Islam, S. R., Al-Amin, M., et al. (2024). From farm to function: Exploring new possibilities with jute nanocellulose applications. *Carbohydrate Polymers, 342*, 122423. https://doi.org/10.1016/j.carbpol.2024.122423

Feijoo, S., Estévez, S., Kamali, M., et al. (2023). Scale-up modelling and life cycle assessment of electrochemical oxidation in wastewater treatment. *Chemical Engineering Journal, 455*, 140627. https://doi.org/10.1016/j.cej.2022.140627
Fernández, D., Robles, I., Rodríguez-Valadez, F. J., & Godínez, L. A. (2018). Novel arrangement for an electro-Fenton reactor that does not require addition of iron, acid and a final neutralization stage. Towards the development of a cost-effective technology for the treatment of wastewater. *Chemosphere, 199*, 251–255. https://doi.org/10.1016/j.chemosphere.2018.02.036
Ganesh, K. S., Sridhar, A., & Vishali, S. (2022). Utilization of fruit and vegetable waste to produce value-added products: Conventional utilization and emerging opportunities–A review. *Chemosphere, 287*, 132221. https://doi.org/10.1016/j.chemosphere.2021.132221
Gavrilescu, M. (2021). Water, soil, and plants interactions in a threatened environment. *Water (Switzerland), 13.*
Gayathiri, M., Pulingam, T., Lee, K. T., & Sudesh, K. (2022). Activated carbon from biomass waste precursors: Factors affecting production and adsorption mechanism. *Chemosphere, 294*, 133764. https://doi.org/10.1016/j.chemosphere.2022.133764
Ghimire, U., Sarpong, G., & Gude, V. G. (2021). Transitioning wastewater treatment plants toward circular economy and energy sustainability. *ACS Omega, 6*, 11794–11803. https://doi.org/10.1021/acsomega.0c05827
Goldan, E., Nedeff, V., Barsan, N., et al. (2023). Assessment of manure compost used as soil amendment—A review. *PRO, 11.*
Guerrero, J. E., & Hansen, E. (2021). Company-level cross-sector collaborations in transition to the bioeconomy: A multi-case study. *Forest Policy and Economics, 123*, 102355. https://doi.org/10.1016/j.forpol.2020.102355
Gul, K., Gan, R.-Y., Sun, C.-X., et al. (2022). Recent advances in the structure, synthesis, and applications of natural polymeric hydrogels. *Critical Reviews in Food Science and Nutrition, 62*, 3817–3832. https://doi.org/10.1080/10408398.2020.1870034
Haque, F., Fan, C., & Lee, Y.-Y. (2023). From waste to value: Addressing the relevance of waste recovery to agricultural sector in line with circular economy. *Journal of Cleaner Production, 415*, 137873. https://doi.org/10.1016/j.jclepro.2023.137873
He, X., Wang, Y., Tai, M. H., et al. (2022). Integrated applications of water hyacinth biochar: A circular economy case study. *Journal of Cleaner Production, 378*, 134621. https://doi.org/10.1016/j.jclepro.2022.134621
Houghton, J. T., Jenkins, G. J., & Ephraums, J. J. (1990). *Climate change: The IPCC scientific assessment.* Cambridge University Press.
Hu, Q., Jung, J., Chen, D., et al. (2021). Biochar industry to circular economy. *Science of the Total Environment, 757*, 143820. https://doi.org/10.1016/j.scitotenv.2020.143820
Iacovidou, E., Hahladakis, J. N., & Purnell, P. (2021). A systems thinking approach to understanding the challenges of achieving the circular economy. *Environmental Science and Pollution Research, 28*, 24785–24806. https://doi.org/10.1007/s11356-020-11725-9
Ibrahim, M., Nawaz, M. H., Rout, P. R., et al. (2023). Advances in produced water treatment technologies: An in-depth exploration with an emphasis on membrane-based systems and future perspectives. *Water (Switzerland), 15.*
Jararweh, Y., Fatima, S., Jarrah, M., & AlZu'bi, S. (2023). Smart and sustainable agriculture: Fundamentals, enabling technologies, and future directions. *Computers and Electrical Engineering, 110*, 108799. https://doi.org/10.1016/j.compeleceng.2023.108799
Jędrzejczak, P., Collins, M. N., Jesionowski, T., & Klapiszewski, Ł. (2021). The role of lignin and lignin-based materials in sustainable construction—A comprehensive review. *International Journal of Biological Macromolecules, 187*, 624–650. https://doi.org/10.1016/j.ijbiomac.2021.07.125
Kalboussi, N., Biard, Y., Pradeleix, L., et al. (2022). Life cycle assessment as decision support tool for water reuse in agriculture irrigation. *Science of the Total Environment, 836*, 155486. https://doi.org/10.1016/j.scitotenv.2022.155486

Karić, N., Maia, A. S., Teodorović, A., et al. (2022). Bio-waste valorisation: Agricultural wastes as biosorbents for removal of (in)organic pollutants in wastewater treatment. *Chemical Engineering Journal Advances, 9*. https://doi.org/10.1016/j.ceja.2021.100239

Khadem, A., Raiesi, F., Besharati, H., & Khalaj, M. A. (2021). The effects of biochar on soil nutrients status, microbial activity and carbon sequestration potential in two calcareous soils. *Biochar, 3*, 105–116. https://doi.org/10.1007/s42773-020-00076-w

Khalid, M. Y., Imran, R., Arif, Z. U., et al. (2021). Developments in chemical treatments, manufacturing techniques and potential applications of natural-fibers-based biodegradable composites. *Coatings, 11*, 1–18.

Khangura, R., Ferris, D., Wagg, C., & Bowyer, J. (2023). Regenerative agriculture—A literature review on the practices and mechanisms used to improve soil health. *Sustainability (Switzerland), 15*.

Khatri, P., Kumar, P., Shakya, K. S., et al. (2023). Understanding the intertwined nature of rising multiple risks in modern agriculture and food system. *Environment, Development and Sustainability, 26*, 24107–24150. https://doi.org/10.1007/s10668-023-03638-7

Khedulkar, A. P., Pandit, B., Dang, V. D., & Doong, R. (2023). Agricultural waste to real worth biochar as a sustainable material for supercapacitor. *Science of the Total Environment, 869*, 161441. https://doi.org/10.1016/j.scitotenv.2023.161441

Kucserka, T., Németh, G. I., Pálfi, I., et al. (2023). Adsorption-based pretreatment of irrigation water to prevent water quality issues. *Separations, 10*.

Kumar, M., Sharma, M., Raut, R. D., et al. (2022). Performance assessment of circular driven sustainable agri-food supply chain towards achieving sustainable consumption and production. *Journal of Cleaner Production, 372*, 133698. https://doi.org/10.1016/j.jclepro.2022.133698

Lear, L. (1997). *Rachel Carson: Witness for nature*. Henry Holt.

Lee, Y.-C., Leite, F., & Lieberknecht, K. (2023). Prioritizing selection criteria of distributed circular water systems: A fuzzy based multi-criteria decision-making approach. *Journal of Cleaner Production, 417*, 138073. https://doi.org/10.1016/j.jclepro.2023.138073

Leng, L., Xiong, Q., Yang, L., et al. (2021). An overview on engineering the surface area and porosity of biochar. *Science of the Total Environment, 763*, 144204. https://doi.org/10.1016/j.scitotenv.2020.144204

Lewoyehu, M. (2021). Comprehensive review on synthesis and application of activated carbon from agricultural residues for the remediation of venomous pollutants in wastewater. *Journal of Analytical and Applied Pyrolysis, 159*, 105279. https://doi.org/10.1016/j.jaap.2021.105279

Li, G., Dunlap, J., Wang, Y., et al. (2022). Environmental life cycle assessment (LCA) of treating PFASs with ion exchange and electrochemical oxidation technology. *ACS ES&T Water, 2*, 1555–1564. https://doi.org/10.1021/acsestwater.2c00196

Liu, M., Ye, Y., Xu, L., et al. (2023). Recent advances in nanoscale zero-Valent iron (nZVI)-based advanced oxidation processes (AOPs): Applications, mechanisms, and future prospects. *Nanomaterials, 13*.

Lu, Y., Liu, M., Li, C., et al. (2022a). Precision fertilization and irrigation: Progress and applications. *AgriEngineering, 4*, 626–655.

Lu, Y., Zhang, H., Fan, D., et al. (2022b). Coupling solar-driven photothermal effect into photocatalysis for sustainable water treatment. *Journal of Hazardous Materials, 423*, 127128. https://doi.org/10.1016/j.jhazmat.2021.127128

Mainardis, M., Cecconet, D., Moretti, A., et al. (2022). Wastewater fertigation in agriculture: Issues and opportunities for improved water management and circular economy. *Environmental Pollution, 296*, 118755. https://doi.org/10.1016/j.envpol.2021.118755

Malekzadeh, E., Tatari, A., & Dehghani Firouzabadi, M. (2024). Effects of biodegradation of starch-nanocellulose films incorporated with black tea extract on soil quality. *Scientific Reports, 14*. https://doi.org/10.1038/s41598-024-69841-2

Manikandan, S., Subbaiya, R., Saravanan, M., et al. (2022). A critical review of advanced nanotechnology and hybrid membrane based water recycling, reuse, and wastewater treatment processes. *Chemosphere, 289*, 132867. https://doi.org/10.1016/j.chemosphere.2021.132867

Mannina, G., Gulhan, H., & Ni, B.-J. (2022). Water reuse from wastewater treatment: The transition towards circular economy in the water sector. *Bioresource Technology, 363*, 127951. https://doi.org/10.1016/j.biortech.2022.127951

Manríquez-Altamirano, A., Sierra-Pérez, J., Muñoz, P., & Gabarrell, X. (2020). Analysis of urban agriculture solid waste in the frame of circular economy: Case study of tomato crop in integrated rooftop greenhouse. *Science of the Total Environment, 734*, 139375. https://doi.org/10.1016/j.scitotenv.2020.139375

Mateo, S., Peinado, S., Morillas-Gutiérrez, F., et al. (2021). Nanocellulose from agricultural wastes: Products and applications—A review. *PRO, 9*.

Mathura, S. R., Landázuri, A. C., Mathura, F., et al. (2024). *Hemicelluloses from bioresidues and their applications in the food industry – Towards an advanced bioeconomy and a sustainable global value chain of chemicals and materials*. Sustainable Food Technology.

Mayilswamy, N., & Kandasubramanian, B. (2022). Green composites prepared from soy protein, polylactic acid (PLA), starch, cellulose, chitin: A review. *Emergent Materials, 5*, 727–753. https://doi.org/10.1007/s42247-022-00354-2

McNeill, J. R. (2010). The environment, environmentalism, and International Society in the Long 1970s. In *The shock of the global* (pp. 263–278). Harvard University Press.

Medhekar, A. (2024). Circular economy in agriculture and sustainable development. In *Circular economy and manufacturing* (pp. 15–31). Elsevier.

Meral, R., Kose, Y. E., Ceylan, Z., & Cavidoglu, İ. (2022). The potential use of agro-industrial by-products as sources of bioactive compounds: A nanotechnological approach. *Studies in Natural Products Chemistry, 73*, 435–466.

Mihalca, V., Kerezsi, A. D., Weber, A., et al. (2021). Protein-based films and coatings for food industry applications. *Polymers (Basel), 13*, 1–24.

Mili, M., Hashmi, S. A. R., Ather, M., et al. (2022). Novel lignin as natural-biodegradable binder for various sectors—A review. *Journal of Applied Polymer Science, 139*.

Mohamed, H., Shah, A. M., & Song, Y. (2021). *Conversion of agro-industrial wastes into value-added products* (pp. 197–217). Springer.

Mohammed, O. N., & Abdulhameed, H. M. (2024). Groundwater remediation for irrigation at Al-Raaed Station using Banana Peel-derived activated carbon. *International Journal of Design and Nature and Ecodynamics, 19*, 1887–1899. https://doi.org/10.18280/ijdne.190605

Molinero-Gerbeau, Y., López-Sala, A., & Şerban, M. (2021). On the social sustainability of industrial agriculture dependent on migrant workers. Romanian workers in Spain's seasonal agriculture. *Sustainability (Switzerland), 13*, 1–17. https://doi.org/10.3390/su13031062

Mona, S., Malyan, S. K., Saini, N., et al. (2021). Towards sustainable agriculture with carbon sequestration, and greenhouse gas mitigation using algal biochar. *Chemosphere, 275*, 129856. https://doi.org/10.1016/j.chemosphere.2021.129856

Morrissey, K. G., English, L., Thoma, G., & Popp, J. (2022). Prospective life cycle assessment and cost analysis of novel electrochemical Struvite recovery in a U.S. Wastewater Treatment Plant. *Sustainability, 14*, 13657. https://doi.org/10.3390/su142013657

Morseletto, P. (2023). Sometimes linear, sometimes circular: States of the economy and transitions to the future. *Journal of Cleaner Production, 390*.

Mujtaba, M., Fernandes Fraceto, L., Fazeli, M., et al. (2023). Lignocellulosic biomass from agricultural waste to the circular economy: A review with focus on biofuels, biocomposites and bioplastics. *Journal of Cleaner Production, 402*.

Naciones Unidad. (2007). *Los objetivos de desarrollo del Milenio y los desafíos para América Latina y el Caribe para avanzar en mayor bienestar, mejor capital humano y más igualdad de oportunidades*. New York.

Nair, L. G., Agrawal, K., & Verma, P. (2022). An overview of sustainable approaches for bioenergy production from agro-industrial wastes. *Energy Nexus, 6*. https://doi.org/10.1016/j.nexus.2022.100086

Nath, S. (2024). Electrochemical wastewater treatment technologies through life cycle assessment: A review. *ChemBioEng Reviews, 11*. https://doi.org/10.1002/cben.202400016

Navarro, D., Cantero, R., Valls, E., & Puig, R. (2020). Circular economy: The case of a shared wastewater treatment plant and its adaptation to changes of the industrial zone over time. *Journal of Cleaner Production, 261*, 121242. https://doi.org/10.1016/j.jclepro.2020.121242
Nidheesh, P. V., Ganiyu, S. O., Martínez-Huitle, C. A., et al. (2023). Recent advances in electro-Fenton process and its emerging applications. *Critical Reviews in Environmental Science and Technology, 53*, 887–913. https://doi.org/10.1080/10643389.2022.2093074
Nie, A., Kung, S. S., Li, H., et al. (2021). An environmental and economic assessment from bioenergy production and biochar application. *Journal of Saudi Chemical Society, 25*. https://doi.org/10.1016/j.jscs.2020.11.006
Norizan, M. N., Shazleen, S. S., Alias, A. H., et al. (2022). Nanocellulose-based nanocomposites for sustainable applications: A review. *Nanomaterials, 12*.
Oliveira, M., Cocozza, A., Zucaro, A., et al. (2021). Circular economy in the agro-industry: Integrated environmental assessment of dairy products. *Renewable and Sustainable Energy Reviews, 148*, 111314. https://doi.org/10.1016/j.rser.2021.111314
Palmieri, N., Suardi, A., Alfano, V., & Pari, L. (2020). Circular economy model: Insights from a case study in South Italy. *Sustainability, 12*, 3466. https://doi.org/10.3390/su12083466
Patra, S. K., Poddar, R., Brestic, M., et al. (2022). Prospects of hydrogels in agriculture for enhancing crop and water productivity under water deficit condition. *International Journal of Polymeric Science, 2022*.
Paudel, B., Basak, J. K., Kaushalya Madhavi, B. G., et al. (2022). Properties of paper-based biodegradable pots for growing seedlings. *Horticulture, Environment and Biotechnology, 63*, 793–807. https://doi.org/10.1007/s13580-022-00457-z
Pedersen, E. R. G., Lüdeke-Freund, F., Henriques, I., & Seitanidi, M. M. (2021). Toward collaborative cross-sector business models for sustainability. *Business & Society, 60*, 1039–1058. https://doi.org/10.1177/0007650320959027
Peng, W., & Pivato, A. (2019). Sustainable Management of Digestate from the organic fraction of municipal solid waste and food waste under the concepts of Back to earth alternatives and circular economy. *Waste and Biomass Valorization, 10*, 465–481. https://doi.org/10.1007/s12649-017-0071-2
Priya, Ashique, S., Afzal, O., et al. (2023). Biogenic nanoparticles from waste fruit peels: Synthesis, applications, challenges and future perspectives. *International Journal of Pharmaceutics, 643*, 123223. https://doi.org/10.1016/j.ijpharm.2023.123223
Rahman, M. M., Khan, I., Field, D. L., et al. (2022). Powering agriculture: Present status, future potential, and challenges of renewable energy applications. *Renewable Energy, 188*, 731–749. https://doi.org/10.1016/j.renene.2022.02.065
Ravenni, G., Thomsen, T. P., Smith, A. M., et al. (2023). Integration of a drying and pyrolysis system in a green biorefinery: Biochar product quality and impacts on the overall energy balance and climate footprint. *Biomass Conversion and Biorefinery*. https://doi.org/10.1007/s13399-023-04877-4
Reutter, J. M. (2019). Lake Erie: Past, present, and future. In *Encyclopedia of water* (pp. 1–15). Wiley.
Reyes, L. M. (2007). *Historia de la ecología*. Universidad de San Carlos de Guatemala Facultad de Humanidades.
Robles, I., Moreno-Rubio, G., Garciá-Espinoza, J. D., et al. (2020). Study of polarized activated carbon filters as simultaneous adsorbent and 3D-type electrode materials for electro-Fenton reactors. *Journal of Environmental Chemical Engineering, 8*. https://doi.org/10.1016/j.jece.2020.104414
Rodríguez-Peña, M., Natividad, R., Barrera-Díaz, C. E., et al. (2024). Current perspective of advanced electrochemical oxidation processes in wastewater treatment and life cycle analysis. *International Journal of Electrochemical Science, 19*, 100589. https://doi.org/10.1016/j.ijoes.2024.100589
Rubin, J. H. (2008). *Impressionism and the modern landscape: Productivity, technology, and urbanization from Manet to Van Gogh*. Illustrated. Cal.

Sabzehmeidani, M. M., Mahnaee, S., Ghaedi, M., et al. (2021). Carbon based materials: A review of adsorbents for inorganic and organic compounds. *Materials Advances, 2*, 598–627.
Sachs, J. D. (2012). From millennium development goals to sustainable development goals. *The Lancet, 379*, 2206–2211. https://doi.org/10.1016/S0140-6736(12)60685-0
Sadabadi, A. A., & Rahimi Rad, Z. (2022). How can cross-sector partnership promote social innovation? *Systemic Practice and Action Research, 35*, 471–490. https://doi.org/10.1007/s11213-021-09581-w
Salazar-López, M. L., Robles, I., Martínez, R. J., et al. (2023). Surface oxidation pre-treatment on activated carbon: Effect on its cathode performance in electro-Fenton processes. *Materials Today Communications, 34*, 105290. https://doi.org/10.1016/j.mtcomm.2022.105290
Salazar-Sogamoso, L. M., Gómez-García, M.-Á., & Dobrosz-Gómez, I. (2024). Comparative life cycle assessment of sequential chemical and electrochemical processes for the treatment of industrial textile wastewater. *Journal of Solid State Electrochemistry*. https://doi.org/10.1007/s10008-024-05976-6
Saleh, T. A. (2021). Protocols for synthesis of nanomaterials, polymers, and green materials as adsorbents for water treatment technologies. *Environmental Technology and Innovation, 24*, 101821. https://doi.org/10.1016/j.eti.2021.101821
Sales, H. B. E., Carolino, A. d. S., Nunes, R. Z. d. A., et al. (2024). Advances in agricultural technology: A review of slow-release Nanofertilizers and innovative carriers. *Communications in Soil Science and Plant Analysis, 55*, 1849–1882. https://doi.org/10.1080/00103624.2024.2326145
Salminen J, Määttä K, Haimi H, et al (2022) Water-smart circular economy—Conceptualisation, transitional policy instruments and stakeholder perception. Journal of Cleaner Production 334:. https://doi.org/10.1016/j.jclepro.2021.130065
Samir, A., Ashour, F. H., Hakim, A. A. A., & Bassyouni, M. (2022). Recent advances in biodegradable polymers for sustainable applications. *npj Materials Degradation*, 6.
Savla, N., Suman, Pandit, S., et al. (2021). Techno-economical evaluation and life cycle assessment of microbial electrochemical systems: A review. *Current Research in Green and Sustainable Chemistry, 4*, 100111. https://doi.org/10.1016/j.crgsc.2021.100111
Sgroi, F. (2022). The circular economy for resilience of the agricultural landscape and promotion of the sustainable agriculture and food systems. *Journal of Agriculture and Food Research, 8*. https://doi.org/10.1016/j.jafr.2022.100307
Sharma, G., Sharma, S., Kumar, A., et al. (2022). Activated carbon as Superadsorbent and sustainable material for diverse applications. *Adsorption Science and Technology, 2022*.
Shokri, A., Nasernejad, B., & Sanavi Fard, M. (2023). Challenges and future roadmaps in heterogeneous electro-Fenton process for wastewater treatment. *Water, Air, and Soil Pollution, 234*.
Shoudho, K. N., Khan, T. H., Ara, U. R., et al. (2024). Biochar in global carbon cycle: Towards sustainable development goals. *Current Research in Green and Sustainable Chemistry, 8*.
Sidhu, R. K., Kumar, R., Rana, P. S., & Jat, M. L. (2021). Automation in drip irrigation for enhancing water use efficiency in cereal systems of South Asia: Status and prospects. *Advances in Agronomy, 167*, 247–300.
Singh Mathuriya, A., Hiloidhari, M., Gware, P., et al. (2020). Development and life cycle assessment of an auto circulating bio-electrochemical reactor for energy positive continuous wastewater treatment. *Bioresource Technology, 304*, 122959. https://doi.org/10.1016/j.biortech.2020.122959
Singh, N., Agarwal, S., Jain, A., & Khan, S. (2021a). 3-dimensional cross linked hydrophilic polymeric network "hydrogels": An agriculture boom. *Agricultural Water Management, 253*, 106939. https://doi.org/10.1016/j.agwat.2021.106939
Singh, R., Das, R., Sangwan, S., et al. (2021b). Utilisation of agro-industrial waste for sustainable green production: A review. *Environmental Sustainability, 4*, 619–636. https://doi.org/10.1007/s42398-021-00200-x

Skrzypczak, D., Izydorczyk, G., Taf, R., et al. (2023). Cellulose-based fertilizers for sustainable agriculture: Effective methods for increasing crop yield and soil health. *Industrial Crops and Products, 205*, 117500. https://doi.org/10.1016/j.indcrop.2023.117500

Sosa, J. A., Laines, J. R., García, D. S., et al. (2023). Activated carbon: A review of residual precursors, synthesis processes, characterization techniques, and applications in the improvement of biogas. *Environmental Engineering Research, 28*. https://doi.org/10.4491/eer.2022.100

Sreekumar, N. M., Sudheep, N. M., & Radhakrishnan, E. K. (2024). Framework for implementing circular economy in agriculture. In *The potential of microbes for a circular economy* (pp. 25–52). Elsevier.

Srivastav, A. L., Dhyani, R., Ranjan, M., et al. (2021). Climate-resilient strategies for sustainable management of water resources and agriculture. *Environmental Science and Pollution Research, 28*, 41576–41595. https://doi.org/10.1007/s11356-021-14332-4

Srivastava, S. (2024). Advancements in Nanocellulose derived from plant waste: Strategies for sustainable innovations and applications. *The Journal of Solid Waste Technology and Management, 50*, 543–563. https://doi.org/10.5276/jswtm/iswmaw/503/2024.543

Srivastava, A., Gupta, B., Majumder, A., et al. (2021). A comprehensive review on the synthesis, performance, modifications, and regeneration of activated carbon for the adsorptive removal of various water pollutants. *Journal of Environmental Chemical Engineering, 9*, 106177. https://doi.org/10.1016/j.jece.2021.106177

Suhag, R., Kumar, R., Dhiman, A., et al. (2023). Fruit peel bioactives, valorisation into nanoparticles and potential applications: A review. *Critical Reviews in Food Science and Nutrition, 63*, 6757–6776. https://doi.org/10.1080/10408398.2022.2043237

Sun, Y., Bai, S., Wang, X., et al. (2023). Prospective life cycle assessment for the electrochemical oxidation wastewater treatment process: From laboratory to industrial scale. *Environmental Science & Technology, 57*, 1456–1466. https://doi.org/10.1021/acs.est.2c04185

Tarpani, R. R. Z., & Azapagic, A. (2023). Life cycle sustainability assessment of advanced treatment techniques for urban wastewater reuse and sewage sludge resource recovery. *Science of the Total Environment, 869*, 161771. https://doi.org/10.1016/j.scitotenv.2023.161771

Tarpani, R. R. Z., Alfonsín, C., Hospido, A., & Azapagic, A. (2020). Life cycle environmental impacts of sewage sludge treatment methods for resource recovery considering ecotoxicity of heavy metals and pharmaceutical and personal care products. *Journal of Environmental Management, 260*, 109643. https://doi.org/10.1016/j.jenvman.2019.109643

Thivya, P., Gururaj, P. N., Reddy, N. B. P., & Rajam, R. (2024). Recent advances in protein-polysaccharide based biocomposites and their potential applications in food packaging: A review. *International Journal of Biological Macromolecules, 268*, 131757. https://doi.org/10.1016/j.ijbiomac.2024.131757

Toplicean, I. M., & Datcu, A. D. (2024). An overview on bioeconomy in agricultural sector, biomass production, recycling methods, and circular economy considerations. *Agriculture (Switzerland), 14*.

Trigo, A., Marta-Costa, A., & Fragoso, R. (2021). Principles of sustainable agriculture: Defining standardized reference points. *Sustainability (Switzerland), 13*. https://doi.org/10.3390/su13084086

Velasco-Muñoz, J. F., Mendoza, J. M. F., Aznar-Sánchez, J. A., & Gallego-Schmid, A. (2021). Circular economy implementation in the agricultural sector: Definition, strategies and indicators. *Resources, Conservation and Recycling, 170*, 105618. https://doi.org/10.1016/j.resconrec.2021.105618

Velasco-Muñoz, J. F., Aznar-Sánchez, J. A., López-Felices, B., & Román-Sánchez, I. M. (2022). Circular economy in agriculture. An analysis of the state of research based on the life cycle. *Sustainable Production and Consumption, 34*, 257–270. https://doi.org/10.1016/j.spc.2022.09.017

Velenturf, A. P. M., & Purnell, P. (2021). Principles for a sustainable circular economy. *Sustainable Production and Consumption, 27*, 1437–1457.

Vermunt, D. A., Wojtynia, N., Hekkert, M. P., et al. (2022). Five mechanisms blocking the transition towards 'nature-inclusive' agriculture: A systemic analysis of Dutch dairy farming. *Agricultural Systems, 195*. https://doi.org/10.1016/j.agsy.2021.103280

Voulgaridis, K., Lagkas, T., Angelopoulos, C. M., & Nikoletseas, S. E. (2022). IoT and digital circular economy: Principles, applications, and challenges. *Computer Networks, 219*, 109456. https://doi.org/10.1016/j.comnet.2022.109456

Vuppaladadiyam, A. K., Vuppaladadiyam, S. S. V., Sahoo, A., et al. (2023). Bio-oil and biochar from the pyrolytic conversion of biomass: A current and future perspective on the trade-off between economic, environmental, and technical indicators. *Science of the Total Environment, 857*, 159155. https://doi.org/10.1016/j.scitotenv.2022.159155

Wang, L., Leghari, S. J., Wu, J., et al. (2023). Interactive effects of biochar and chemical fertilizer on water and nitrogen dynamics, soil properties and maize yield under different irrigation methods. *Frontiers in Plant Science, 14*. https://doi.org/10.3389/fpls.2023.1230023

Wang, Z., Ahmad, W., Zhu, A., et al. (2024). Recent advances review in tea waste: High-value applications, processing technology, and value-added products. *Science of the Total Environment, 946*, 174225. https://doi.org/10.1016/j.scitotenv.2024.174225

Wu, Y., Li, S., & Chen, G. (2023). Hydrogels as water and nutrient reservoirs in agricultural soil: A comprehensive review of classification, performance, and economic advantages. *Environment, Development and Sustainability, 26*, 24653–24685. https://doi.org/10.1007/s10668-023-03706-y

Xu, X., Liu, F., Jiang, L., et al. (2013). Cellulose nanocrystals vs. cellulose Nanofibrils: A comparative study on their microstructures and effects as polymer reinforcing agents. *ACS Applied Materials & Interfaces, 5*, 2999–3009. https://doi.org/10.1021/am302624t

Yaashikaa, P. R., Senthil Kumar, P., & Varjani, S. (2022). Valorization of agro-industrial wastes for biorefinery process and circular bioeconomy: A critical review. *Bioresource Technology, 343*, 126126. https://doi.org/10.1016/j.biortech.2021.126126

Yadav, S., Sonkar, V., & Malyan, S. K. (2023). Soil carbon sequestration strategies: Application of biochar an option to combat global warming. In *Bio-inspired land remediation* (pp. 353–374). Springer.

Yang, M., Chen, L., Wang, J., et al. (2023). Circular economy strategies for combating climate change and other environmental issues. *Environmental Chemistry Letters, 21*, 55–80.

Zhai, S., Zhang, D., Liu, W., et al. (2023). Microbial electrochemical technologies assisted nitrogen recovery from different wastewater sources: Performance, life cycle assessment, and challenges. *Resources, Conservation and Recycling, 194*, 107000. https://doi.org/10.1016/j.resconrec.2023.107000

Zhang, Y., Wang, J., & Feng, Y. (2021). The effects of biochar addition on soil physicochemical properties: A review. *Catena (Amst), 202*, 105284. https://doi.org/10.1016/j.catena.2021.105284

Zhao, X., Zhang, Z., Pang, J., & Su, L. (2023). Preparation of carbon fibre-reinforced composite panels from epoxy resin matrix of nano lignin polyol particles. *Journal of Cleaner Production, 428*, 139170. https://doi.org/10.1016/j.jclepro.2023.139170

Zinge, C., & Kandasubramanian, B. (2020). Nanocellulose based biodegradable polymers. *European Polymer Journal, 133*, 109758. https://doi.org/10.1016/j.eurpolymj.2020.109758

Chapter 16
Sustainability and Economics

Jean Vinícius Moreira

16.1 Introduction

The shift toward adopting a circular economy (CE) in agriculture presents a range of solutions to address the increasing issues of resource scarcity, environmental harm, and the need for sustainable economic practices. This approach moves away from the traditional linear economic model, instead focusing on a regenerative system where waste is reduced and resources are continuously cycled (Velasco-Muñoz et al., 2022).

When incorporating CE principles into conventional agriculture, several factors must be considered, but three core elements stand out. First, optimizing resource efficiency is crucial, requiring agricultural systems to minimize resource input while maximizing output through more efficient processes. Second, achieving sustainability across environmental, economic, and social dimensions is essential for long-term resilience. Finally, regenerative agricultural methods are necessary to close nutrient loops, avoid resource wastage, and support biodiversity (Morseletto, 2020).

Velasco-Muñoz et al. (2021) described CE in agriculture as a set of activities that ensure resource efficiency and regeneration within the agroecosystem. This aligns with the three foundational principles of CE: (i) design out waste and pollution, (ii) keep products and materials in use, and (iii) regenerate natural systems (Velasco-Muñoz et al., 2021).

Mitigating externalities such as soil degradation, water pollution, and biodiversity loss must be pursued to achieve the initial stage of CE in agriculture. Furthermore, promoting regenerative practices using renewable resources will help to maintain and improve natural systems (Martinho, 2020).

J. V. Moreira (✉)
Coordination of Biotechnology, FIRJAN SENAI SESI, Rio de Janeiro, Brazil
e-mail: jean@jvmoreiraconsultoria.com.br

K. Esquivel Escalante et al. (eds.), *Green Nanomaterials*, Nanotechnology in the Life Sciences, https://doi.org/10.1007/978-3-032-04492-1_16

Implementing circular agriculture requires careful planning and organization of production cycles, from estimating food demand and planning land use to optimizing nutrient cycles and minimizing the use of harmful inputs. Every phase of production must prioritize sustainability (Velasco-Muñoz et al., 2021). However, the full implementation of CE in agriculture still needs to be improved, not only in regulatory affairs but also in reverse logistics, consumer awareness, and the geographical dispersion of agricultural enterprises (Ada et al., 2021).

Despite these challenges, economic sustainability in agriculture can be achieved by adopting circular models. Some key economic indicators, such as profitability, liquidity, and productivity, may be used to evaluate sustainability. The CE still offers opportunities to improve these metrics by reducing input costs and creating new markets for sustainable products (Martinho, 2020; Velasco-Muñoz et al., 2022).

Legislative frameworks have begun to support the adoption of CE in agriculture. For example, Germany, Japan, and China have implemented national laws promoting a circular economy. At the same time, the European Union's 2020 Circular Economy Action Plan aims to integrate these principles across all sectors, including agriculture. These policy efforts are crucial for creating an enabling environment that encourages innovation and investment in sustainable agricultural practices, especially by research centers, integrating academic knowledge into farms (Ogunmakinde, 2019; Farmer, 2020; Bleischwitz et al., 2022; Steenmans & Lesniewska, 2023).

Applying CE principles in agriculture offers a transformative path toward sustainability, rethinking resource use, production processes, and economic viability, enabling agriculture to contribute to broader goals of environmental conservation, economic resilience, and social well-being, ensuring that agriculture will remain a pillar of economic development while addressing pressing environmental and societal challenges (Batlles-delaFuente et al., 2022).

16.2 Economic Impact of Environmental and Sustainability Goals

As sustainability becomes essential across industries, agriculture is evolving to embrace both environmental stewardship and economic vitality. New approaches, like green nanotechnology and circular economy practices, aim to tackle big challenges—soil depletion, water pollution, and resource inefficiency—while supporting a thriving agricultural economy (Li et al., 2022; Sgroi, 2023). Though these innovations, such as nano-fertilizers, often involve higher initial investments, research shows they can lead to valuable long-term benefits like higher yields, lower costs, and less environmental strain (Su et al., 2022; Kumar et al., 2023).

Green nanotechnology, for instance, is transforming how nutrients are delivered to crops, cutting down on waste and pollution by improving nutrient uptake. Meanwhile, circular economy strategies are creating fresh economic opportunities by turning agricultural by-products into useful resources, which builds resilience and profitability within the sector (Ghorbanpour et al., 2020; Cahyadi et al., 2024).

16.2.1 Cost-Benefit Analysis

The analysis of costs and benefits of utilizing nanoparticle-based fertilizers (NPs) is still in its early stages, but evolving with many promising findings regarding the lower dosage and higher efficiency. The question of whether they offer a cost-effective alternative or complement to traditional fertilizers remains unresolved (Su et al., 2022).

A key factor in evaluating the cost-benefit ratio of nano-fertilizers is their unit cost, which is currently higher than conventional fertilizers. However, the higher upfront costs of nano-fertilizers could be offset by their enhanced efficiency, meaning that less material is required to achieve the same or better productivity outcomes (Kumar et al., 2023; Yadav et al., 2023a). This reduction in the quantity of fertilizer not only lowers the direct cost of input over time but also reduces transportation and application expenses. Furthermore, it reduces the greenhouse gas (GHG) emissions associated with fertilizer production. Additionally, they lower the risk of nutrients leaching into groundwater or causing runoffs that lead to eutrophication—major concerns linked to conventional fertilizers (Saraiva et al., 2023; Sun et al., 2024).

The co-application involves strategies combining the use of traditional fertilizers with nano-fertilizers to enhance agricultural productivity and maintain the economic and environmental balance. This approach could reduce the use of both conventional fertilizers and nanomaterials, optimizing nutrient delivery. Such a combination could prove particularly efficient in transitioning to more sustainable agriculture, keeping the existing fertilizer supply chains (Upadhyay et al., 2023; Channab et al., 2024).

However, it is important to note that despite the significant potential of nano-fertilizers, widespread adoption depends on more than just efficacy. The agricultural sector is highly sensitive to cost, and any substantial increase in input prices must be justified by a commensurate increase in productivity and profitability. As it stands, the long-term economic benefits of nano-fertilizers are still being investigated. Current research suggests that while the immediate return on investment (ROI) might be lower due to higher initial costs, the cumulative effects of enhanced crop yields, reduced fertilizer consumption, and environmental savings could lead to a more favorable ROI over time (Kumar et al., 2023; Yadav et al., 2023a).

While nano-fertilizers hold significant promises for improving agricultural efficiency and sustainability, their high costs, varying efficacy, and the need for further research make it difficult to predict their large-scale economic viability. Nonetheless, as more data becomes available, nano-fertilizers could become an integral part of sustainable farming practices, contributing to the balance between environmental preservation and agricultural productivity (Zain et al., 2024).

16.2.2 Short-Term Versus Long-Term Benefits

The preliminary analysis of the use of green nano-fertilizers (NFs) in agriculture, compared to conventional fertilizers, highlights significant short- and long-term benefits from both economic and environmental perspectives, aligning with broader sustainability goals. NFs aim to prevent environmental risks while promoting long-term well-being. Within the framework of the circular economy (CE), which emphasizes resource efficiency and waste minimization, green nanotechnologies offer a clear path to more sustainable agricultural practices (Gade et al., 2023).

Historically, chemical fertilizers have been essential to modern agriculture, driving significant increases in productivity but also causing negative externalities, such as water pollution, soil degradation, and greenhouse gas (GHG) emissions, contributing to climate change (Saha & Bauddh, 2020).

When it comes to environmental externalities, green nano-fertilizers present a better outlook than conventional fertilizers. Furthermore, there are also short- and long-term economic gains. In the short term, the initial cost of adoption might be higher due to the advanced technology involved in their synthesis and application. However, the reduced quantity of fertilizer needed and the enhanced nutrient use efficiency can offset these costs, leading to substantial savings for producers (Gottardo et al., 2021; Sah et al., 2024).

In the short term, the adoption of green nano-fertilizers can offer improvements in the efficiency of the input and productivity of the crop. In the long term, these benefits expand as the need for frequent soil amendments decreases, soil health improves, and more sustainable production cycles are established. This results in a reduction of GHG emissions associated with fertilizer application, mitigating externalities related to climate change and lowering future costs linked to climate impacts (Akhtar et al., 2022; Saraiva et al., 2023).

Another important aspect of green nanotechnologies is the potential to improve product quality and food security. It can be used to develop controlled-release nutrient systems, maximizing plant growth while minimizing environmental losses. Thus, while there is still much to explore regarding their economic feasibility and environmental safety, green nano-fertilizers emerge as a promising tool in the transition toward more sustainable and efficient agricultural practices (Haris et al., 2023; Munir et al., 2023).

16.3 The Impact of Green Nanotechnology on Agricultural Productivity

Nanoparticles (NPs) play a significant role in enhancing plant growth and development by promoting seed germination and vigor due to their different properties and characteristics, such as antibacterial, antiviral, and anti-inflammatory, which help strengthen plant resilience against pathogens and environmental stressors (Guleria et al., 2023). Titanium dioxide (TiO_2) nanoparticles have demonstrated notable effects on spinach seed germination, boosting dry weight and photosynthetic rates through photoactivation, which generates reactive oxygen species that stimulate cellular growth and improve stress tolerance. Similarly, silicon dioxide (SiO_2) NPs have shown promising results in enhancing seed germination, vigor index, and biomass in plants such as *Solanum lycopersicum* by improving nutrient uptake and photosynthetic efficiency (Awasthi et al., 2020; Zhao et al., 2020).

In agricultural applications, green nanotechnology—prioritizing eco-friendly materials and processes—presents a transformative approach for sustainably increasing crop yields while minimizing environmental impact. This heightened productivity is particularly valuable in dense crop systems where rapid and robust germination reduces competition from weeds, thus naturally lowering the need for herbicides (Pirzadah et al., 2020).

16.4 Economic and Environmental Benefits

Green nanotechnology is making farming both more cost-effective and eco-friendly by boosting how well plants use nutrients. This means farmers can rely less on traditional fertilizers and pesticides, which tend to be expensive and often lead to environmental problems over time. By cutting these input costs, green nanotech helps keep soils healthier and reduces pollution in groundwater. It also strengthens crop resilience against tough conditions like drought and soil salinity, offering farmers a more reliable income and the chance to get good yields even when growing conditions are not ideal (Saritha et al., 2022; Yadav et al., 2023c).

What sets green nanotechnology apart is its ability to release nutrients gradually and in a way that plants can better absorb, unlike conventional chemical methods. This targeted release minimizes waste, reduces the amount of chemicals ending up in the environment, and keeps soil and water quality intact over the long haul. Not only does this model help conserve resources but it also reduces dependence on non-renewable inputs, making agriculture more sustainable, especially in nutrient-poor soils where it can restore productivity season after season (Chand Mali et al., 2020; Yadav et al., 2023b).

But the impact goes beyond just better crop yields. Green nanotechnology plays a direct role in advancing a circular economy for agriculture. By recycling nutrients and creating slow-release fertilizers, it minimizes waste and creates a self-sustaining loop where resources are continually reused. This kind of approach is essential for building an agricultural future that is resilient to climate challenges and resource limitations (Govil et al., 2024).

16.5 Sustainability and Precision Farming as a Competitive Advantage

With controlled release mechanisms, nanomaterials offer a gradual and targeted delivery of nutrients and active ingredients directly to plants. This approach significantly reduces waste and minimizes environmental damage. A major advantage of green nanotechnology is its sustainable production process (Zain et al., 2023). Through biological synthesis methods—utilizing plants or microorganisms to generate nanoparticles—these materials are created without the need for harmful chemicals or energy-intensive procedures. The green nanotechnology, if applied, provides a sustainable alternative that meets the increasing demand for eco-friendly agricultural practices. This model not only enhances input efficiency but also bolsters crop resilience while maintaining the ecological balance (Usman et al., 2020; Moulick et al., 2020).

The accurate information provided by precision farming empowers farmers not to waste resources and have a real-time analysis of the soil farm's health, water use, and fertilizer application. By doing so, they can cut down on costs and lessen the environmental footprint. Precision farming allows for targeted application, which helps avoid the usual pitfalls of overuse and waste (Chugh et al., 2021; Younis et al., 2021).

By reducing input waste, preserving natural resources, and improving crop yields, the combination of green nanotechnology and precision farming transforms sustainability into a viable and profitable endeavor. This (re)evolution helps farmers not only comply with regulations and respond to market demands for sustainable practices but also underscores that sustainability is more than a responsible choice—it is the most effective strategy for ensuring the future resilience of agriculture (Acharya & Pal, 2020).

16.6 Policy Frameworks from an Economic Perspective

In today's world, sustainable agriculture is becoming essential to meet the challenges posed by climate change and environmental degradation. Economic policies and regulations are crucial to this transition, with global and local initiatives like the European Green Deal providing the resources and framework needed to help farmers adopt sustainable practices (Dönmez et al., 2024). Incentives such as tax breaks, grants, and carbon credits are now encouraging more farmers to invest in green technologies. Meanwhile, policies like the EU's Common Agricultural Policy (CAP) bring both opportunities and challenges, as compliance costs and regulations can impact farmers' economic stability (Ikram et al., 2021).

Market-based approaches add further support. Through carbon markets, for example, farmers using eco-friendly nanotechnologies can participate in carbon offset programs, creating new revenue streams. Eco-labels and certifications add value to environmentally responsible products, often allowing these items to sell at higher prices. Additionally, green finance tools, including green bonds, offer vital funding to support sustainable practices in agriculture (Zhang et al., 2021; Sharma et al., 2023).

16.7 Global and Local Policy Influences

Sustainable agriculture is no longer just a lofty ideal; it is a necessity in the face of climate change, population growth, and dwindling natural resources. To make this vision a reality, global and local policies are crucial in driving innovation, regulating safety, and creating economic opportunities. From the European Green Deal to Brazil's bioeconomy strategy, policies are shaping how nanotechnology and precision agriculture are transforming farming worldwide (Verma et al., 2022; de Queiroz-Stein et al., 2024; Brandão & Santos, 2024).

16.7.1 The Role of Global and Local Policies

Global frameworks are setting the stage for innovation by turning sustainability into action. For example, the European Green Deal (EGD) combines funding with strict regulations, driving the development of environmentally and economically sound technologies. Programs like Horizon Europe are helping accelerate the creation of nanomaterials, including controlled-release fertilizers and nano pesticides. These technologies not only enhance resource efficiency but also reduce harmful runoff, showing how policy can move from theory to impactful practice (Zhang et al., 2021, 2023; Kumari et al., 2023).

Organizations like the Food and Agriculture Organization (FAO) and the United Nations are also promoting nanotechnology as a tool to tackle global challenges, such as food insecurity and soil degradation. These efforts align with the Sustainable Development Goals (SDGs), emphasizing innovative, practical solutions to agricultural problems (Zhang et al., 2021).

Local Policies Leading the Way

Global policies provide the framework, but it is local action that truly drives change. Each region adapts these principles to meet its own needs and challenges:

Brazil has emerged as a leader in blending nanotechnology with its bioeconomy. The country's policies focus on converting agricultural waste into valuable nanomaterials, such as those that enhance soil fertility. Institutions like the Brazilian Agricultural Research Corporation (EMBRAPA) collaborate closely with local farmers to ensure these technologies are both practical and widely accessible. This innovative approach not only improves crop productivity but also strengthens

Brazil's position as a global leader in the circular economy (Zhang et al., 2021; Carbonell et al., 2021; Tiwari et al., 2024).

India is faced with rising food demands; India is leveraging nanotechnology to reshape its agricultural landscape. Through collaborations between the government and the private sector, the country is advancing the development of nano-based fertilizers and pesticides. These innovations are designed to reduce costs, lower environmental impact, and boost crop yields. Crucially, they also prioritize the needs of smallholder farmers, ensuring that progress in technology translates into equitable benefits across all levels of farming (Pandey, 2018; Singh & Gurjar, 2022).

Across Africa, initiatives such as the Comprehensive Africa Agriculture Development Programme (CAADP) are reshaping how farmers adopt and apply advanced technologies. These programs focus on equipping and empowering farmers to thrive in a rapidly changing agricultural landscape with the knowledge and tools they need to integrate innovations into their practices. By prioritizing equity, CAADP ensures that even smallholder farmers with limited resources can participate in and benefit from the global transition to sustainable agriculture (Babu et al., 2016; Makombe et al., 2020).

Indonesia is embracing a progressive approach to agriculture by combining technology forecasting with strategic planning. Through the integration of nanotechnology and artificial intelligence, the country is tackling pressing challenges such as soil degradation and inefficient water use. This forward-looking model highlights the potential of emerging economies to lead in agricultural modernization. By blending cutting-edge science with practical solutions, Indonesia is setting an example of how innovation can drive sustainable and impactful progress in farming (Thangamani et al., 2024; Soni et al., 2024).

16.7.2 Where Policy Meets Technology: Nanotechnology and AI

The combination of nanotechnology and artificial intelligence is transforming farming, making it smarter, more efficient, and less wasteful by enabling:

- Real-Time Monitoring—nano sensors embedded in fields measure soil health, and moisture levels, and even detect crop diseases. This gives farmers the ability to make real-time decisions and to know how micro-regions of the farm are performing (Yadav et al., 2023b; El-Chaghaby & Rashad, 2024).
- Precision at Scale—AI-driven systems analyze data from nano sensors and satellite imagery to optimize resource allocation. The result? Higher yields with fewer inputs—saving money and protecting the planet (Altabbaa et al., 2023).
- Circular Solutions—it enables to conversion of agricultural residues into valuable materials, like nanoparticles that enhance soil nutrient retention, biopesticides, and biofertilizers. This closed-loop system reduces waste, costs, and carbon footprint while improving efficiency (Flores-Contreras et al., 2024).

Challenges to Overcome

While these technologies hold immense potential, several hurdles need to be addressed to ensure their benefits reach everyone:

- Cost—nano-enabled tools and precision farming technologies are often too expensive for smallholder farmers, leaving them at risk of falling behind in the race for productivity. Governments need to step up with targeted solutions, such as subsidies, tax incentives, and public-private partnerships, to make these innovations accessible and affordable for all (Gumbi et al., 2023; Dhillon & Moncur, 2023).
- Regulations—the lack of harmonized global standards for nanotechnology in agriculture creates trade barriers and slows adoption. To address this, international regulatory bodies must collaborate to establish consistent guidelines that ensure safety and streamline the global exchange of products and ideas (Mishra et al., 2022).
- Awareness: for these technologies to succeed, farmers need proper training to understand and use them effectively. Hands-on workshops and demonstration projects are essential for building trust and encouraging adoption, helping farmers see how these innovations can improve their livelihoods and practices (Hatzenbuehler & Peña-Lévano, 2022).

16.7.3 Success Stories: Policies in Action

Several regions around the world are already showing how strong policies can turn innovative ideas into real-world impacts. Europe, supported by the European Green Deal, has seen precision farming projects reduce fertilizer use by 20% while increasing yields. This approach demonstrates how sustainability can go together with economic benefits—helping farmers and the planet alike (Malik et al., 2023; Boix-Fayos & de Vente, 2023).

Brazil's collaboration with local farmers, EMBRAPA has proven that nanotechnology can enhance soil health and resilience in tropical regions. This success sets a powerful example for other countries looking to integrate similar approaches (Souza et al., 2024).

Smallholder farmers in India have embraced nano-based fertilizers and pesticides, cutting costs while improving productivity. These efforts highlight how targeted policies can drive widespread impact, especially in regions with high agricultural demands (Pramanik & Pramanik, 2016; Alhassan et al., 2024).

16.8 Market-Based Approaches

Green nanomaterials are paving the way for significant progress in sustainable agriculture. They provide innovative solutions to reduce greenhouse gas (GHG) emissions, improve carbon sequestration, and enable farmers to actively participate in carbon offset programs. These programs incentivize practices that combat climate change, combining environmental responsibility with economic resilience. By advancing soil management, optimizing fertilizer use, and cutting emissions, green nanomaterials align agricultural operations with global climate objectives (Narayanan & Bhaskar, 2024).

A notable application lies in soil carbon sequestration. Biochar, a carbon-dense product from biomass, becomes even more effective when enhanced with nanotechnology. This nano-biochar retains carbon in the soil for extended periods, boosts fertility, minimizes fertilizer requirements, and reduces emissions linked to soil management (Velasco-Muñoz et al., 2022; Soni et al., 2024).

These attributes also meet carbon offset standards, as measurable carbon storage is a key criterion for earning carbon credits. Nanotechnology also targets major GHG sources in agriculture. Slow-release nano-fertilizers, for instance, regulate nutrient delivery, significantly lowering losses through leaching and volatilization, which are primary contributors to nitrous oxide emissions. Research suggests that nitrogen volatilization drops by more than 30% with nano-fertilizers compared to traditional alternatives, effectively reducing farming's carbon footprint (Ganachari et al., 2019; Abhiram, 2023; Sun et al., 2024).

Similarly, nanomaterials are proving effective in curbing methane emissions. In livestock systems, nanotechnology applied to manure management disrupts microbial pathways that generate methane, cutting emissions by up to 25%. Such innovations support carbon market requirements, reducing GHG emissions while preserving productivity (Sales et al., 2024; Sun et al., 2024).

Nano-sensors in precision agriculture further enhance soil health monitoring and emission control. These sensors provide real-time data, helping farmers measure carbon sequestration and track GHG reductions, ensuring alignment with carbon offset certification standards. Moreover, they strengthen transparency and accountability in reporting, which is critical for carbon market participation (Kumar et al., 2024).

From an economic standpoint, adopting green nanomaterials offers substantial long-term benefits, despite higher upfront costs. Lower fertilizer use reduces environmental risks, and potential revenue from carbon credits can outweigh initial investments. For example, combining nano-fertilizers with conventional inputs optimizes nutrient efficiency, cutting costs while improving yields. This dual approach enhances both sustainability metrics and profitability (Zhao et al., 2020).

Nanotechnology also integrates seamlessly with circular economy principles. Agricultural waste, such as by-products, can be transformed into valuable resources like biochar or nano-fertilizers. This minimizes waste, maximizes efficiency, and contributes to carbon reduction goals while aligning with market incentives for sustainability (Singla & Kapoor, 2024).

Despite its vast potential, the widespread adoption of green nanotechnology faces challenges, including regulatory hurdles, cost barriers, and the need for farmer education. However, supportive measures like tax breaks, subsidies, and public-private collaborations can help overcome these obstacles (Ganachari et al., 2019; Soni et al., 2024). Examples from Brazil and India highlight how partnerships between researchers and farmers can scale these technologies effectively (Bhandari et al., 2023).

16.9 Green Finances

Green finance has become a cornerstone in advancing sustainable practices across industries, particularly in agriculture. By leveraging financial tools such as green bonds and loans, stakeholders can channel investments toward environmentally friendly technologies that address pressing challenges like resource depletion and carbon emissions. This integration aligns economic growth with sustainability objectives, creating a pathway for a more resilient agricultural sector (van Veelen, 2021; Guo et al., 2022; Agrawal et al., 2024).

16.9.1 Green Bonds: Transforming Agricultural Financing

Green bonds are pivotal in mobilizing resources for sustainable agricultural practices. These bonds provide funding for projects with clear environmental benefits, such as implementing precision agriculture systems, adopting renewable energy in irrigation, and transitioning to organic fertilizers. In China, green bonds have effectively driven reductions in agricultural carbon emissions, highlighting their potential to support low-carbon innovations (Huang et al., 2022; Wan et al., 2022). Despite their promise, challenges persist, particularly regarding the need for greater additionality in bond-financed projects. Regulatory refinements, such as enhanced "green" standards, could further optimize their impact (Sethi et al., 2024; Cao & Gao, 2024).

16.9.2 Green Loans and Linked Incentives

Complementing green bonds, green loans offer targeted financing for sustainable projects. These loans have funded technologies that improve water efficiency and soil health while reducing dependence on chemical inputs. Sustainability-linked loans incentivize borrowers to meet predefined environmental benchmarks and are particularly promising in driving change. For instance, some programs reward reduced water usage or carbon footprints, directly supporting the adoption of green innovations (Bhatnagar & Sharma, 2022; Du & Wang, 2023; Agrawal et al., 2024).

16.9.3 Challenges in Scaling Green Finance

Despite significant advancements, green finance faces barriers to broader adoption in agriculture. Developing economies, which stand to benefit the most, often grapple with insufficient financial infrastructure and inconsistent regulatory frameworks. Moreover, ensuring that financial instruments genuinely support transformative practices rather than business-as-usual projects remains a concern (van Veelen, 2021; Huang et al., 2022). Collaborative efforts between policymakers and financial institutions could bridge these gaps and expand access to green finance tools (Desalegn & Tangl, 2022; Cao & Gao, 2024).

16.9.4 Global Impacts and Lessons

The global application of green finance demonstrates its potential. In Brazil, green loans have driven the widespread adoption of sustainable farming practices, while similar initiatives in India have improved rural livelihoods and reduced emissions. These examples underscore the importance of tailored financial solutions that address regional challenges and opportunities (Bhatnagar & Sharma, 2022; Du & Wang, 2023).

Green bonds and loans are critical to aligning agriculture with sustainability goals. By funding projects that reduce emissions and enhance efficiency, these tools drive innovation while addressing environmental concerns. To maximize their impact, ongoing efforts should focus on improving regulatory clarity, expanding access in underserved regions, and fostering collaborations that prioritize sustainability (Guo et al., 2022; Wan et al., 2022).

16.9.5 Challenges and Future Directions

16.9.5.1 Barriers to the Adoption of Green Nanomaterials in Agriculture

The promise of green nanomaterials in agriculture is immense, offering novel solutions to improve crop yields, optimize resource use, and reduce environmental footprints. However, the path to their widespread adoption is far from straightforward. Farmers, policymakers, and researchers face intertwined challenges that stem from economic limitations, technical complexities, and an uneven market landscape. Addressing these barriers is crucial for integrating these materials into sustainable agricultural practices (Saritha et al., 2022; Wahab et al., 2024).

Economic Challenges

The high costs of developing and producing green nanomaterials often place them out of reach for many agricultural stakeholders. Manufacturing processes rely on advanced technologies, which require significant capital investment. This expense translates into products that are often priced higher than traditional alternatives. For small-scale farmers, who already operate on tight margins, the financial burden is simply too great (Usman et al., 2020; Hofmann et al., 2020).

Moreover, the long-term economic advantages of these materials, such as reduced input costs and higher efficiency, are not immediately evident. This time lag between investment and return discourages adoption, particularly in regions where short-term survival often outweighs longer-term sustainability goals. The lack of government incentives or subsidies exacerbates the problem, leaving farmers with limited options to experiment with innovative technologies (Usman et al., 2020).

Technical Complexities

The production and deployment of green nanomaterials involves technical hurdles that are not easily overcome. Variations in particle size, shape, and chemical properties affect their consistency and effectiveness. Farmers may experience different outcomes depending on soil type, climate, and crop species. These inconsistencies erode trust and make it harder to build confidence in technology (Hofmann et al., 2020; Singh et al., 2021).

Scaling these materials from laboratory research to field-ready products presents additional challenges. While promising results are often reported under controlled conditions, real-world applications can be far less predictable. The technical expertise required to handle and apply nanomaterials effectively is another barrier, particularly in regions where access to training and extension services is limited (Preethi et al., 2024).

Market Barriers and Awareness

Even when technical and economic barriers are addressed, the market for green nanomaterials is still in its infancy. Regulatory hurdles slow the approval process, creating uncertainty for manufacturers and potential users. These delays are compounded by inconsistent definitions and standards for nanomaterials across countries, which complicates international trade (Younis et al., 2021; Yadav et al., 2023a; Peng & Qin, 2024).

Public perception also plays a crucial role. Many farmers and consumers remain skeptical of nanotechnology, associating it with unknown risks rather than tangible benefits. This skepticism is often fueled by a lack of accessible information and limited success stories from the field. Without targeted outreach and education, these misconceptions will continue to hinder adoption (Hofmann et al., 2020).

Moving Forward
For green nanomaterials to fulfill their potential in agriculture, concerted efforts must address these barriers. Governments and industry stakeholders need to collaborate to lower production costs and provide financial incentives. Standardizing production methods and offering targeted training programs can resolve technical issues. Meanwhile, transparent communication strategies can bridge the gap between scientific advances and public understanding. With these steps, the integration of green nanomaterials into agriculture can move from an ideal to a reality (Bartolucci et al., 2020; Parra-López et al., 2024).

16.10 Scaling Up: How to Economically Scale Green Nanotechnology in Agriculture for Global Impact

Scaling green nanotechnology in agriculture to achieve a global impact requires a balanced approach that combines technical feasibility with economic practicality. While nano-fertilizers and nano-biofertilizers have shown immense potential in enhancing crop yields and sustainability, their widespread adoption faces challenges related to cost, infrastructure, and farmer acceptance (Saratale et al., 2018; Indira et al., 2021).

One of the critical hurdles is the expense associated with nanoparticle production. Traditional chemical methods, often energy-intensive and reliant on toxic reagents, are not economically viable for large-scale use. Green synthesis methods, using biological agents like microbes and plants, offer a promising alternative by reducing costs and environmental impact. For instance, microbial processes have demonstrated the ability to produce metal nanoparticles such as zinc oxide and silver at lower costs, which could be pivotal for agricultural applications (Raj et al., 2021; Soni et al., 2024).

Integrating green nanotechnology with precision agriculture further enhances its scalability. Nano-enabled delivery systems, like slow-release fertilizers, not only improve nutrient-use efficiency but also reduce input costs for farmers. Additionally, nano-sensors for real-time soil and crop monitoring can optimize resource application, minimizing waste and maximizing productivity (Acharya & Pal, 2020; Soni et al., 2024).

Expanding these technologies also requires a supportive framework of public-private partnerships and policy incentives. Subsidies for sustainable farming practices and collaborations between research institutions and industries can help make nano-based products more accessible to small-scale farmers. Success stories in emerging markets, where partnerships have led to the affordable production of nano-biofertilizers, highlight the potential of such collaborative efforts (Bairwa et al., 2023; Soni et al., 2024).

However, scaling up must be guided by sustainability principles. While the environmental benefits of green nanotechnology are evident, life-cycle assessments are necessary to address potential risks and ensure that large-scale applications do not harm ecosystems. Prioritizing biodegradable and biocompatible nanoparticles, derived from renewable sources, is essential for maintaining this balance (Gilbertson et al., 2020).

The economic scaling of green nanotechnology in agriculture demands an interdisciplinary approach. By reducing production costs, aligning with precision farming, fostering strategic partnerships, and addressing environmental concerns, this innovation can transform global agriculture. It offers a pathway not only to increased productivity but also to a more sustainable and resilient food system (Preethi et al., 2024; Dönmez et al., 2024).

16.11 The Role of Sustained R&D Investment in Enhancing Sustainable Technologies

The advancement of sustainable technologies is intrinsically tied to continuous research and development (R&D). The high costs of R&D can only be borne by the big agricultural industries. To address this challenge for developing economies, both public and private sector investments are critical to evaluate scalability and efficiency. Nevertheless, government policies of developed countries must enable technological innovation and foster global collaboration for sustained R&D to achieve sustainability goals and reduce climate change (Rosegrant et al., 2022; Bathaei & Štreimikienė, 2023).

The huge investments in R&D to reduce the costs of renewable energy technologies took years to come out of laboratories and developed countries; however, these investments paid off and now solar and wind power are a reality, with competitive prices. Similarly, in agriculture, developments in nanotechnology have introduced fertilizers that enhance nutrient delivery, reduce waste, and minimize environmental impacts. However, this technology is not available for small farmers. This is why strong government lines of finance must be established; without them, the green revolution will never be fully achieved (Pandey et al., 2022; Bathaei & Štreimikienė, 2023).

Collaborative Efforts in R&D

Public-private partnerships (PPPs) are particularly effective in accelerating the development and dissemination of sustainable technologies. Governments often provide the foundational funding for high-risk research, while private enterprises contribute expertise in scaling and commercialization. For instance, PPPs have enabled the creation of biofertilizers that simultaneously improve soil health and reduce dependence on chemical inputs, demonstrating both economic and environmental benefits (Pandey et al., 2022; Lee & Kim, 2022; Alonso-Martínez et al., 2024).

These collaborations also align technological innovations with global policy frameworks, such as the Sustainable Development Goals (SDGs), ensuring that R&D investments address pressing societal challenges (Pandey et al., 2022).

Bridging Disparities in R&D

Despite global progress, disparities in R&D investment remain a significant barrier, particularly for low- and middle-income countries. These regions often lack the financial and technical resources to fully engage in sustainable innovation, leading to slower adoption of critical technologies. Addressing these gaps requires targeted international funding mechanisms and strategic collaboration that aligns public and private priorities (War et al., 2020; Bathaei & Štreimikienė, 2023).

Frameworks that focus on capacity building and equitable resource distribution can enhance regional capabilities, ensuring that sustainable technologies are accessible globally (Cristóbal et al., 2021).

Sustained investment in R&D is essential for advancing sustainable technologies, reducing their costs, and improving their efficiency. It will only be achieved by collaborative efforts between public and private sectors, which amplify these impacts, driving innovations that support environmental and economic objectives. By addressing disparities in R&D capacity, particularly in underfunded regions, we can accelerate the global transition to sustainability. Furthermore, a global initiative must be implemented not only on paper and through agreements but also in practice. This is the way that small farmers will gain access to technologies that will help them overcome the duality of productivity and sustainability (Wani et al., 2020; Bollington et al., 2021; Rosegrant et al., 2022; Hurduzeu et al., 2022).

16.12 Conclusion

The transition to sustainable agriculture is not just an ideal but a necessary shift in response to pressing global challenges such as resource depletion, climate change, and food insecurity. This chapter has explored the pivotal role of circular economy principles, green nanotechnology, and precision farming in redefining agricultural practices for a more resilient future.

Central to this transformation is continuous investment in research and development (R&D), but this transformation will only occur if public-private partnerships emerge from the offices to the farms. This pivotal new course in the road of sustainability, translating innovative ideas into practical solutions, bridging gaps in funding and expertise, will merge technologies such as nano-fertilizers and advanced monitoring systems, promising remarkable efficiency gains. However, their accessibility remains uneven, particularly in regions with limited resources. Addressing these disparities calls for global action, not just through policy but through targeted financial mechanisms and local engagement.

The economic potential of sustainable practices is equally compelling. Though initial costs may deter adoption, the long-term savings from reduced input use, higher yields, and environmental benefits underscore their value. Initiatives like the European Green Deal and Brazil's integration of nanotechnology into agricultural policies highlight how strategic frameworks can catalyze progress, offering lessons for other regions.

Education and training stand out as critical components in ensuring that farmers can adopt and adapt these innovations effectively. Bridging the gap between lab-based discoveries and real-world applications requires a concerted effort to build trust and understanding within farming communities.

Ultimately, sustainable agriculture is about more than technology; it is about fostering a system that values efficiency, equity, and environmental health. By embracing innovative practices and committing to collaborative efforts, the agricultural sector can play a transformative role in addressing global sustainability challenges while securing its place as a cornerstone of economic growth and resilience.

Acknowledgments The author would like to express heartfelt gratitude to Dr. Ricardo Valle Ladewig Zappala and Dr. Carlos Alberto Xavier Gonçalves for their strategic collaboration and invaluable contributions to the development of this chapter.

Special thanks are also extended to the FIRJAN SENAI SESI for providing an exceptional environment and opportunities to conduct innovative research in biotechnology. The support and resources made available through this institution have significantly enriched the research outcomes and have been pivotal in advancing this work.

Conflict of Interest The author has no conflicts of interest to declare that are relevant to the content of this chapter.

References

Abhiram, G. (2023). Contributions of nano-nitrogen fertilizers to sustainable development goals: A comprehensive review. *Nitrogen, 4*, 397–415. https://doi.org/10.3390/nitrogen4040028

Acharya, A., & Pal, P. K. (2020). Agriculture nanotechnology: Translating research outcome to field applications by influencing environmental sustainability. *NanoImpact, 19*, 100232. https://doi.org/10.1016/j.impact.2020.100232

Ada, N., Kazancoglu, Y., Sezer, M. D., et al. (2021). Analyzing barriers of circular food supply chains and proposing industry 4.0 solutions. *Sustainability, 13*, 6812. https://doi.org/10.3390/su13126812

Agrawal, R., Agrawal, S., Samadhiya, A., et al. (2024). Adoption of green finance and green innovation for achieving circularity: An exploratory review and future directions. *Geoscience Frontiers, 15*, 101669. https://doi.org/10.1016/j.gsf.2023.101669

Akhtar, N., Ilyas, N., Meraj, T. A., et al. (2022). Improvement of plant responses by nanobiofertilizer: A step towards sustainable agriculture. *Nanomaterials, 12*, 965. https://doi.org/10.3390/nano12060965

Alhassan, A. M., Bodunrinde, R. E., Oyewole, O. A., et al. (2024). The role of policy maker, relevant stakeholders and government agency in translating nanobiofertilizer research into policy. In *Handbook of agricultural biotechnology* (pp. 99–116). Wiley.

Alonso-Martínez, D., Jiménez-Parra, B., & Cabeza-García, L. (2024). Theoretical framework to foster and assess sustainable agriculture practices: Drivers and key performance indicators.

Environmental and Sustainability Indicators, 23, 100434. https://doi.org/10.1016/j.indic.2024.100434

Altabbaa, S., Mann, N. A., Chauhan, N., et al. (2023). Era connecting nanotechnology with agricultural sustainability: Issues and challenges. *Nanotechnology For Environmental Engineering, 8*, 481–498. https://doi.org/10.1007/s41204-022-00289-3

Awasthi, G., Singh, T., Tiwari, Y., et al. (2020). A review on nanotechnological interventions for plant growth and production. *Materials Today Proceedings, 31*, 685–693. https://doi.org/10.1016/j.matpr.2020.07.255

Babu, S. C., Manvatkar, R., & Kolavalli, S. (2016). Strengthening capacity for agribusiness development and management in Sub-Saharan Africa. *Africa Journal of Management, 2*, 1–30. https://doi.org/10.1080/23322373.2015.1112714

Bairwa, P., Kumar, N., Devra, V., & Abd-Elsalam, K. (2023). Nano-biofertilizers synthesis and applications in agroecosystems. *Agrochemicals, 2*, 118–134. https://doi.org/10.3390/agrochemicals2010009

Bartolucci, C., Antonacci, A., Arduini, F., et al. (2020). Green nanomaterials fostering agrifood sustainability. *TrAC Trends in Analytical Chemistry, 125*, 115840. https://doi.org/10.1016/j.trac.2020.115840

Bathaei, A., & Štreimikienė, D. (2023). Renewable energy and sustainable agriculture: Review of indicators. *Sustainability, 15*, 14307. https://doi.org/10.3390/su151914307

Batlles-delaFuente, A., Abad-Segura, E., González-Zamar, M.-D., & Cortés-García, F. J. (2022). An evolutionary approach on the framework of circular economy applied to agriculture. *Agronomy, 12*, 620. https://doi.org/10.3390/agronomy12030620

Bhandari, G., Dhasmana, A., Chaudhary, P., et al. (2023). A perspective review on green nanotechnology in agro-ecosystems: Opportunities for sustainable agricultural practices & environmental remediation. *Agriculture, 13*, 668. https://doi.org/10.3390/agriculture13030668

Bhatnagar, S., & Sharma, D. (2022). Evolution of green finance and its enablers: A bibliometric analysis. *Renewable and Sustainable Energy Reviews, 162*, 112405. https://doi.org/10.1016/j.rser.2022.112405

Bleischwitz, R., Yang, M., Huang, B., et al. (2022). The circular economy in China: Achievements, challenges and potential implications for decarbonisation. *Resources, Conservation and Recycling, 183*, 106350. https://doi.org/10.1016/j.resconrec.2022.106350

Boix-Fayos, C., & de Vente, J. (2023). Challenges and potential pathways towards sustainable agriculture within the European Green Deal. *Agricultural Systems, 207*, 103634. https://doi.org/10.1016/j.agsy.2023.103634

Bollington, A., DeLonge, M., Mungra, D., et al. (2021). Closing research investment gaps for a global food transformation. *Frontiers in Sustainable Food Systems, 5*. https://doi.org/10.3389/fsufs.2021.794594

Brandão, A. S., & Santos, J. M. R. C. A. (2024). Sustainability from policy to practice: Assessing the impact of European research and innovation frameworks on circular bioeconomy. *Sustainability, 16*, 2355. https://doi.org/10.3390/su16062355

Cahyadi, E. R., Hidayati, N., Zahra, N., & Arif, C. (2024). Integrating circular economy principles into agri-food supply chain management: A systematic literature review. *Sustainability, 16*, 7165. https://doi.org/10.3390/su16167165

Cao, L., & Gao, J. (2024). The impact of green finance on agricultural pollution and carbon reduction: The case of China. *Sustainability, 16*, 5832. https://doi.org/10.3390/su16145832

Carbonell, S. A. M., Cortez, L. A. B., Madi, L. F. C., et al. (2021). Bioeconomy in Brazil: Opportunities and guidelines for research and public policy for regional development. *Biofuels, Bioproducts and Biorefining, 15*, 1675–1695. https://doi.org/10.1002/bbb.2263

Chand Mali, S., Raj, S., & Trivedi, R. (2020). Nanotechnology a novel approach to enhance crop productivity. *Biochemistry and Biophysics Reports, 24*, 100821. https://doi.org/10.1016/j.bbrep.2020.100821

Channab, B.-E., El Idrissi, A., Ammar, A., et al. (2024). Recent advances in nano-fertilizers: Synthesis, crop yield impact, and economic analysis. *Nanoscale, 16*, 4484–4513. https://doi.org/10.1039/D3NR05012B

Chugh, G., Siddique, K. H. M., & Solaiman, Z. M. (2021). Nanobiotechnology for agriculture: Smart technology for combating nutrient deficiencies with nanotoxicity challenges. *Sustainability, 13*, 1781. https://doi.org/10.3390/su13041781

Cristóbal, J., Ehrenstein, M., Domínguez-Ramos, A., et al. (2021). Unraveling the links between public spending and Sustainable Development Goals: Insights from data envelopment analysis. *Science of the Total Environment, 786*, 147459. https://doi.org/10.1016/j.scitotenv.2021.147459

de Queiroz-Stein, G., Martinelli, F. S., Dietz, T., & Siegel, K. M. (2024). Disputing the bioeconomy-biodiversity nexus in Brazil: Coalitions, discourses and policies. *Forest Policy and Economics, 158*, 103101. https://doi.org/10.1016/j.forpol.2023.103101

Desalegn, G., & Tangl, A. (2022). Enhancing green finance for inclusive green growth: A systematic approach. *Sustainability, 14*, 7416. https://doi.org/10.3390/su14127416

Dhillon, R., & Moncur, Q. (2023). Small-scale farming: A review of challenges and potential opportunities offered by technological advancements. *Sustainability, 15*, 15478. https://doi.org/10.3390/su152115478

Dönmez, D., Isak, M. A., İzgü, T., & Şimşek, Ö. (2024). Green horizons: Navigating the future of agriculture through sustainable practices. *Sustainability, 16*, 3505. https://doi.org/10.3390/su16083505

Du, Y., & Wang, W. (2023). The role of green financing, agriculture development, geopolitical risk, and natural resource on environmental pollution in China. *Resources Policy, 82*, 103440. https://doi.org/10.1016/j.resourpol.2023.103440

El-Chaghaby, G. A., & Rashad, S. (2024). Nanosensors in agriculture: Applications, prospects, and challenges. In *Handbook of nanosensors* (pp. 1–29). Springer Nature Switzerland.

Farmer, A. (2020). Developing the circular economy in the European Union. In *Circular economy: Global perspective* (pp. 389–412). Springer Singapore.

Flores-Contreras, E. A., González-González, R. B., Pablo Pizaña-Aranda, J. J., et al. (2024). Agricultural waste as a sustainable source for nanoparticle synthesis and their antimicrobial properties for food preservation. *Frontiers in Nanotechnology, 6*. https://doi.org/10.3389/fnano.2024.1346069

Gade, A., Ingle, P., Nimbalkar, U., et al. (2023). Nanofertilizers: The next generation of agrochemicals for long-term impact on sustainability in farming systems. *Agrochemicals, 2*, 257–278. https://doi.org/10.3390/agrochemicals2020017

Ganachari, S. V., Yaradoddi, J. S., Somappa, S. B., et al. (2019). Green nanotechnology for biomedical, food, and agricultural applications. In *Handbook of ecomaterials* (pp. 2681–2698). Springer International Publishing.

Ghorbanpour, M., Bhargava, P., Varma, A., & Choudhary, D. K. (Eds.). (2020). *Biogenic nanoparticles and their use in agro-ecosystems*. Springer Singapore.

Gilbertson, L. M., Pourzahedi, L., Laughton, S., et al. (2020). Guiding the design space for nanotechnology to advance sustainable crop production. *Nature Nanotechnology, 15*, 801–810. https://doi.org/10.1038/s41565-020-0706-5

Gottardo, S., Mech, A., Drbohlavová, J., et al. (2021). Towards safe and sustainable innovation in nanotechnology: State-of-play for smart nanomaterials. *NanoImpact, 21*, 100297. https://doi.org/10.1016/j.impact.2021.100297

Govil, S., Van Duc, L. N., Escribà-Gelonch, M., & Hessel, V. (2024). Controlled-release fertiliser: Recent developments and perspectives. *Industrial Crops and Products, 219*, 119160. https://doi.org/10.1016/j.indcrop.2024.119160

Guleria, G., Thakur, S., Shandilya, M., et al. (2023). Nanotechnology for sustainable agro-food systems: The need and role of nanoparticles in protecting plants and improving crop productivity. *Plant Physiology and Biochemistry, 194*, 533–549. https://doi.org/10.1016/j.plaphy.2022.12.004

Gumbi, N., Gumbi, L., & Twinomurinzi, H. (2023). Towards sustainable digital agriculture for smallholder farmers: A systematic literature review. *Sustainability, 15*, 12530. https://doi.org/10.3390/su151612530

Guo, L., Zhao, S., Song, Y., et al. (2022). Green finance, chemical fertilizer use and carbon emissions from agricultural production. *Agriculture, 12*, 313. https://doi.org/10.3390/agriculture12030313

Haris, M., Hussain, T., Mohamed, H. I., et al. (2023). Nanotechnology – A new frontier of nano-farming in agricultural and food production and its development. *Science of the Total Environment, 857*, 159639. https://doi.org/10.1016/j.scitotenv.2022.159639

Hatzenbuehler, P., & Peña-Lévano, L. (2022). Adoption potential of sustainability-Related agriculture technologies for smallholder farmers in the global south. *Sustainability, 14*, 13176. https://doi.org/10.3390/su142013176

Hofmann, T., Lowry, G. V., Ghoshal, S., et al. (2020). Technology readiness and overcoming barriers to sustainably implement nanotechnology-enabled plant agriculture. *Nature Food, 1*, 416–425. https://doi.org/10.1038/s43016-020-0110-1

Huang, Y., Chen, C., Lei, L., & Zhang, Y. (2022). Impacts of green finance on green innovation: A spatial and nonlinear perspective. *Journal of Cleaner Production, 365*, 132548. https://doi.org/10.1016/j.jclepro.2022.132548

Hurduzeu, G., Pânzaru, R. L., Medelete, D. M., et al. (2022). The development of sustainable agriculture in EU countries and the potential achievement of sustainable development goals specific targets (SDG 2). *Sustainability, 14*, 15798. https://doi.org/10.3390/su142315798

Ikram, M., Sroufe, R., Awan, U., & Abid, N. (2021). Enabling progress in developing economies: A novel hybrid decision-making model for green technology planning. *Sustainability, 14*, 258. https://doi.org/10.3390/su14010258

Indira, M., Krupanidhi, S., Venkateswarulu, T. C., et al. (2021). *Current aspects of nanotechnology: Applications in agriculture* (pp. 73–99). Springer.

Kumar, N., Samota, S. R., Venkatesh, K., & Tripathi, S. C. (2023). Global trends in use of nano-fertilizers for crop production: Advantages and constraints – A review. *Soil and Tillage Research, 228*, 105645. https://doi.org/10.1016/j.still.2023.105645

Kumar, A., Tyagi, P. K., Tyagi, S., & Ghorbanpour, M. (2024). Integrating green nanotechnology with sustainable development goals: A pathway to sustainable innovation. *Discover Sustainability, 5*, 364. https://doi.org/10.1007/s43621-024-00610-x

Kumari, R., Suman, K., Karmakar, S., et al. (2023). Regulation and safety measures for nanotechnology-based agri-products. *Frontiers in Genome Editing, 5*. https://doi.org/10.3389/fgeed.2023.1200987

Lee, D., & Kim, K. (2022). National investment framework for revitalizing the R&D collaborative ecosystem of sustainable smart agriculture. *Sustainability, 14*, 6452. https://doi.org/10.3390/su14116452

Li, J., Song, G., Cai, M., et al. (2022). Green environment and circular economy: A state-of-the-art analysis. *Sustainable Energy Technologies and Assessments, 52*, 102106. https://doi.org/10.1016/j.seta.2022.102106

Makombe, T., Tefera, W., & Ulimwengu, J. M. (2020). *Tracking key CAADP indicators and implementation processes*. International Food Policy Research Institute.

Malik, A., Lafortune, G., Dahir, S., et al. (2023). Global environmental and social spillover effects of EU's food trade. *Global Sustainability, 6*, e6. https://doi.org/10.1017/sus.2023.4

Martinho, V. J. P. D. (2020). Exploring the topics of soil pollution and agricultural economics: Highlighting good practices. *Agriculture, 10*, 24. https://doi.org/10.3390/agriculture10010024

Mishra, A. K., Das, R., Sahoo, S., & Biswal, B. (2022). Global regulations and legislations on nanoparticles usage and application in diverse horizons. In *Comprehensive analytical chemistry* (pp. 261–290). Elsevier.

Morseletto, P. (2020). Targets for a circular economy. *Resources, Conservation and Recycling, 153*, 104553. https://doi.org/10.1016/j.resconrec.2019.104553

Moulick, R. G., Das, S., Debnath, N., & Bandyopadhyay, K. (2020). Potential use of nanotechnology in sustainable and 'smart' agriculture: Advancements made in the last decade. *Plant Biotechnology Reports, 14*, 505–513. https://doi.org/10.1007/s11816-020-00636-3

Munir, N., Gulzar, W., Abideen, Z., et al. (2023). Nanotechnology improves disease resistance in plants for food security: Applications and challenges. *Biocatalysis and Agricultural Biotechnology, 51*, 102781. https://doi.org/10.1016/j.bcab.2023.102781

Narayanan, K. B., & Bhaskar, R. (2024). Green nanotechnology: Paving the way for environmental sustainability. *Sustainability, 16*, 6262. https://doi.org/10.3390/su16146262

Ogunmakinde, O. E. (2019). A review of circular economy development models in China, Germany and Japan. *Recycling, 4*, 27. https://doi.org/10.3390/recycling4030027

Pandey, G. (2018). Challenges and future prospects of agri-nanotechnology for sustainable agriculture in India. *Environmental Technology and Innovation, 11*, 299–307. https://doi.org/10.1016/j.eti.2018.06.012

Pandey, N., de Coninck, H., & Sagar, A. D. (2022). Beyond technology transfer: Innovation cooperation to advance sustainable development in developing countries. *WIREs Energy and Environment, 11*. https://doi.org/10.1002/wene.422

Parra-López, C., Ben Abdallah, S., Garcia-Garcia, G., et al. (2024). Integrating digital technologies in agriculture for climate change adaptation and mitigation: State of the art and future perspectives. *Computers and Electronics in Agriculture, 226*, 109412. https://doi.org/10.1016/j.compag.2024.109412

Peng, S., & Qin, X. (2024). Application of phytonanotechnology for modern sustainable green agriculture: Promising opportunities and scientific challenges. *Industrial Crops and Products, 218*, 119001. https://doi.org/10.1016/j.indcrop.2024.119001

Pirzadah, B., Pirzadah, T. B., Jan, A., & Hakeem, K. R. (2020). *Nanofertilizers: A way forward for green economy* (pp. 99–112). Springer.

Pramanik, S., & Pramanik, G. (2016). *Nanotechnology for sustainable agriculture in India* (pp. 243–280). Springer.

Preethi, B., Karmegam, N., Manikandan, S., et al. (2024). Nanotechnology-powered innovations for agricultural and food waste valorization: A critical appraisal in the context of circular economy implementation in developing nations. *Process Safety and Environmental Protection, 184*, 477–491. https://doi.org/10.1016/j.psep.2024.01.100

Raj, N. B., Swamy, M. K., Purushotham, B., & Sukrutha, S. K. (2021). *Applications of microbe-based nanoparticles in agriculture: Present state and future challenges* (pp. 343–382). Springer.

Rosegrant, M. W., Sulser, T. B., & Wiebe, K. (2022). Global investment gap in agricultural research and innovation to meet Sustainable Development Goals for hunger and Paris Agreement climate change mitigation. *Frontiers in Sustainable Food Systems, 6*. https://doi.org/10.3389/fsufs.2022.965767

Sah, M. K., Thakuri, B. S., Pant, J., et al. (2024). The multifaceted perspective on the role of green synthesis of nanoparticles in promoting a sustainable green economy. *Sustainable Chemistry, 5*, 40–59. https://doi.org/10.3390/suschem5020004

Saha, L., & Bauddh, K. (2020). Sustainable agricultural approaches for enhanced crop productivity, better soil health, and improved ecosystem services. In *Ecological and practical applications for sustainable agriculture* (pp. 1–23). Springer Singapore.

Sales, H. B. E., Carolino, A. d. S., Nunes, R. Z. d. A., et al. (2024). Advances in agricultural technology: A review of slow-release nanofertilizers and innovative carriers. *Communications in Soil Science and Plant Analysis, 55*, 1849–1882. https://doi.org/10.1080/00103624.2024.2326145

Saraiva, R., Ferreira, Q., Rodrigues, G. C., & Oliveira, M. (2023). Nanofertilizer use for adaptation and mitigation of the agriculture/climate change dichotomy effects. *Climate, 11*, 129. https://doi.org/10.3390/cli11060129

Saratale, R. G., Karuppusamy, I., Saratale, G. D., et al. (2018). A comprehensive review on green nanomaterials using biological systems: Recent perception and their future applications.

Colloids and Surfaces. B, Biointerfaces, 170, 20–35. https://doi.org/10.1016/j.colsurfb.2018.05.045

Saritha, G. N. G., Anju, T., & Kumar, A. (2022). Nanotechnology – Big impact: How nanotechnology is changing the future of agriculture? *Journal of Agriculture and Food Research, 10*, 100457. https://doi.org/10.1016/j.jafr.2022.100457

Sethi, L., Behera, B., & Sethi, N. (2024). Do green finance, green technology innovation, and institutional quality help achieve environmental sustainability? Evidence from the developing economies. *Sustainable Development, 32*, 2709–2723. https://doi.org/10.1002/sd.2811

Sgroi, F. (2023). Circular economy and sustainable agri-food systems. *Journal of Agriculture and Food Research, 14*, 100815. https://doi.org/10.1016/j.jafr.2023.100815

Sharma, B., Tiwari, S., Kumawat, K. C., & Cardinale, M. (2023). Nano-biofertilizers as bio-emerging strategies for sustainable agriculture development: Potentiality and their limitations. *Science of the Total Environment, 860*, 160476. https://doi.org/10.1016/j.scitotenv.2022.160476

Singh, D., & Gurjar, B. R. (2022). Nanotechnology for agricultural applications: Facts, issues, knowledge gaps, and challenges in environmental risk assessment. *Journal of Environmental Management, 322*, 116033. https://doi.org/10.1016/j.jenvman.2022.116033

Singh, R. P., Handa, R., & Manchanda, G. (2021). Nanoparticles in sustainable agriculture: An emerging opportunity. *Journal of Controlled Release, 329*, 1234–1248. https://doi.org/10.1016/j.jconrel.2020.10.051

Singla, D., & Kapoor, K. (2024). *Agricultural waste availability for nanomaterial synthesis: Recent advances* (pp. 129–144). Springer.

Soni, S. K., Dogra, S., Sharma, A., et al. (2024). Nanotechnology in agriculture: Enhancing crop productivity with sustainable nano-fertilizers and nano-biofertilizers. *Journal of Soil Science and Plant Nutrition*. https://doi.org/10.1007/s42729-024-01988-3

Souza, V. S., da Silva Vanolli, B., Schiebelbein, B. E., et al. (2024). Cover crops and soil health in Brazilian agricultural systems. In *Soil health series: Volume 3 soil health and sustainable agriculture in Brazil* (pp. 103–144). https://doi.org/10.1002/9780891187448.ch4

Steenmans, K., & Lesniewska, F. (2023). Limitations of the circular economy concept in law and policy. *Frontiers in Sustainability, 4*. https://doi.org/10.3389/frsus.2023.1154059

Su, Y., Zhou, X., Meng, H., et al. (2022). Cost–benefit analysis of nanofertilizers and nanopesticides emphasizes the need to improve the efficiency of nanoformulations for widescale adoption. *Nature Food, 3*, 1020–1030. https://doi.org/10.1038/s43016-022-00647-z

Sun, Y., Jiang, Y., Wei, H., et al. (2024). Nano-enabled strategies for greenhouse gases emission mitigation: A comprehensive review. *Nano Today, 57*, 102378. https://doi.org/10.1016/j.nantod.2024.102378

Thangamani, R., Sathya, D., Kamalam, G. K., & Lyer, G. N. (2024). AI green revolution: Reshaping agriculture's future. In *Intelligent robots and drones for precision agriculture* (pp. 421–461). Springer.

Tiwari, T., Kaur, G. A., Singh, P. K., et al. (2024). Emerging bio-capture strategies for greenhouse gas reduction: Navigating challenges towards carbon neutrality. *Science of the Total Environment, 929*, 172433. https://doi.org/10.1016/j.scitotenv.2024.172433

Upadhyay, P. K., Singh, V. K., Rajanna, G. A., et al. (2023). Unveiling the combined effect of nano fertilizers and conventional fertilizers on crop productivity, profitability, and soil well-being. *Frontiers in Sustainable Food Systems, 7*. https://doi.org/10.3389/fsufs.2023.1260178

Usman, M., Farooq, M., Wakeel, A., et al. (2020). Nanotechnology in agriculture: Current status, challenges and future opportunities. *Science of the Total Environment, 721*, 137778. https://doi.org/10.1016/j.scitotenv.2020.137778

van Veelen, B. (2021). Cash cows? Assembling low-carbon agriculture through green finance. *Geoforum, 118*, 130–139. https://doi.org/10.1016/j.geoforum.2020.12.008

Velasco-Muñoz, J. F., Mendoza, J. M. F., Aznar-Sánchez, J. A., & Gallego-Schmid, A. (2021). Circular economy implementation in the agricultural sector: Definition, strategies and indicators. *Resources, Conservation and Recycling, 170*, 105618. https://doi.org/10.1016/j.resconrec.2021.105618

Velasco-Muñoz, J. F., Aznar-Sánchez, J. A., López-Felices, B., & Román-Sánchez, I. M. (2022). Circular economy in agriculture. An analysis of the state of research based on the life cycle. *Sustainable Production and Consumption, 34*, 257–270. https://doi.org/10.1016/j.spc.2022.09.017

Verma, K. K., Song, X.-P., Joshi, A., et al. (2022). Recent trends in nano-fertilizers for sustainable agriculture under climate change for global food security. *Nanomaterials, 12*, 173. https://doi.org/10.3390/nano12010173

Wahab, A., Muhammad, M., Ullah, S., et al. (2024). Agriculture and environmental management through nanotechnology: Eco-friendly nanomaterial synthesis for soil-plant systems, food safety, and sustainability. *Science of the Total Environment, 926*, 171862. https://doi.org/10.1016/j.scitotenv.2024.171862

Wan, Q., Qian, J., Baghirli, A., & Aghayev, A. (2022). Green finance and carbon reduction: Implications for green recovery. *Economic Analysis and Policy, 76*, 901–913. https://doi.org/10.1016/j.eap.2022.09.022

Wani, T. A., Rather, G. A., Ahmad, M., & Kaloo, Z. A. (2020). *Embodiment of nanobiotechnology in agriculture: An overview* (pp. 113–128). Springer.

War, J. M., Fazili, M. A., Mushtaq, W., et al. (2020). Role of nanotechnology in crop improvement. In *Nanobiotechnology in agriculture* (pp. 63–97). Springer.

Yadav, A., Yadav, K., & Abd-Elsalam, K. A. (2023a). Exploring the potential of nanofertilizers for a sustainable agriculture. *Plant Nano Biology, 5*, 100044. https://doi.org/10.1016/j.plana.2023.100044

Yadav, A., Yadav, K., Ahmad, R., & Abd-Elsalam, K. (2023b). Emerging frontiers in nanotechnology for precision agriculture: Advancements, hurdles and prospects. *Agrochemicals, 2*, 220–256. https://doi.org/10.3390/agrochemicals2020016

Yadav, N., Garg, V. K., Chhillar, A. K., & Rana, J. S. (2023c). Recent advances in nanotechnology for the improvement of conventional agricultural systems: A review. *Plant Nano Biology, 4*, 100032. https://doi.org/10.1016/j.plana.2023.100032

Younis, S. A., Kim, K.-H., Shaheen, S. M., et al. (2021). Advancements of nanotechnologies in crop promotion and soil fertility: Benefits, life cycle assessment, and legislation policies. *Renewable and Sustainable Energy Reviews, 152*, 111686. https://doi.org/10.1016/j.rser.2021.111686

Zain, M., Ma, H., Nuruzzaman, M., et al. (2023). Nanotechnology based precision agriculture for alleviating biotic and abiotic stress in plants. *Plant Stress, 10*, 100239. https://doi.org/10.1016/j.stress.2023.100239

Zain, M., Ma, H., Ur Rahman, S., et al. (2024). Nanotechnology in precision agriculture: Advancing towards sustainable crop production. *Plant Physiology and Biochemistry, 206*, 108244. https://doi.org/10.1016/j.plaphy.2023.108244

Zhang, P., Guo, Z., Ullah, S., et al. (2021). Nanotechnology and artificial intelligence to enable sustainable and precision agriculture. *Nature Plants, 7*, 864–876. https://doi.org/10.1038/s41477-021-00946-6

Zhang, N., Sun, J., Yin, L., et al. (2023). Silver nanoparticles: From in vitro green synthesis to in vivo biological effects in plants. *Advanced Agrochem, 2*, 313–323. https://doi.org/10.1016/j.aac.2023.08.004

Zhao, L., Lu, L., Wang, A., et al. (2020). Nano-biotechnology in agriculture: Use of nanomaterials to promote plant growth and stress tolerance. *Journal of Agricultural and Food Chemistry, 68*, 1935–1947. https://doi.org/10.1021/acs.jafc.9b06615

Chapter 17
Innovative Applications in Sustainable Agriculture

Aldo Cordoba Guerrero, Diego F. Monroy-Macias, Sergio Ontiveros Seoane, Arturo M. Castillo-Rico, and Karen Esquivel Escalante

17.1 Introduction

Today, agriculture faces unprecedented challenges due to increasing food demand, diminishing natural resources, and the effects of climate change (Oluwole et al., 2023). In this context, nanotechnology has emerged as a novel tool with promising applications for sustainable agriculture. Specifically, nanomaterials (NMs) green synthesis has captured the attention due to its potential to improve the efficiency of agricultural inputs, reduce environmental impact, and promote a better approach to crop management (Huston et al., 2021; Maity et al., 2022).

Green synthesis of nanomaterials is based on the use of biological precursors, such as plant extracts, microorganisms, and agricultural by-products, to generate NMs with specific properties (Jadoun et al., 2021). This approach eliminates the use of toxic chemical agents and reduces energy consumption, contributing to environmental and human safety (Gour & Jain, 2019). In the agricultural field, these NMs have proven to be effective as nanofertilizers, nanopesticides, and nanosensors, improving nutrient use efficiency and protecting crops from pests and diseases (Dimkpa & Bindraban, 2018; Kumari et al., 2023).

A. Cordoba Guerrero
División de Investigación y Posgrado, Facultad de Ingeniería, Universidad Autónoma de Querétaro, Cerro de las Campanas S/N, Santiago de Querétaro, Querétaro, Mexico

Departamento de Métodos Cuantitativos, Universidad Anáhuac Campus Querétaro, Circuito Universidades l, Kilómetro 7, Fracción 2, Querétaro, Mexico
e-mail: acordoba07@alumnos.uaq.mx

D. F. Monroy-Macias · S. Ontiveros Seoane · A. M. Castillo-Rico · K. Esquivel Escalante (✉)
División de Investigación y Posgrado, Facultad de Ingeniería, Universidad Autónoma de Querétaro, Cerro de las Campanas S/N, Santiago de Querétaro, Querétaro, Mexico
e-mail: dmonroy30@alumnos.uaq.mx; sontiveros20@alumnos.uaq.mx; acastillo39@alumnos.uaq.mx; karen.esquivel@uaq.mx

K. Esquivel Escalante et al. (eds.), *Green Nanomaterials*, Nanotechnology in the Life Sciences, https://doi.org/10.1007/978-3-032-04492-1_17

This chapter explores the role of environmentally friendly nanotechnology in sustainable agricultural practices. Its applications in improving agricultural productivity, reducing resource waste, optimizing water use, and developing new crop protection strategies are discussed. Specific studies and examples are presented to illustrate the impact of these NMs in modern agriculture. Finally, the challenges and prospects of this technology are addressed, considering regulatory, economic, and scalability aspects for its large-scale implementation.

17.2 Green-Synthesized Nanotechnology for Sustainable Farming Practices

To discuss how nanomaterials can be applied and assist in agricultural practices, it is first necessary to explain what green synthesis is. Green synthesis is a sustainable approach, making it an innovative method within the field of nanomaterials (NMs) synthesis. It enables the synthesis of nanoparticles (NPs) derived from natural extract precursors such as plants, fungi, bacteria, and algae (Huston et al., 2021). This method aims to replace traditional methods, such as those using sodium borohydride (Esquivel et al., 2023), and other substances that are typically toxic or require high energy inputs (Gour & Jain, 2019). The process relies on the use of reducing and stabilizing agents found in biological extracts. These precursors facilitate the transformation of metal ions to its zero valent state (Nadaroglu et al., 2017) and a subsequent stabilization phase. The precursors control both the particle size and morphology, which can be mainly spherical. (Hano & Abbasi, 2022). This is exemplified in more detail in Fig. 17.1 (Jadoun et al., 2021).

Agricultural practices encompass all activities performed by producers aimed at cultivating and managing soil resources, water, and crop systems to obtain food and other agricultural products (Feregrino-Pérez et al., 2024). These activities include land preparation, planting, pest management, disease control, weed elimination, and storage of the final product.

Such practices are regulated depending on the country and its standards and regulations. Some of those regulations address soil contamination, production quality, and the handling and transportation of food products (Xu et al., 2021). With an understanding of biosynthesis and agricultural practices, it can be examined how they interconnect, and the advances achieved so far (Maity et al., 2022).

17.2.1 Relationship Between Green Synthesis and Agricultural Activities

Nanotechnology has been proven to have a significant impact in the agricultural sector with applications such as fertilizers, pesticides, growth regulators, and

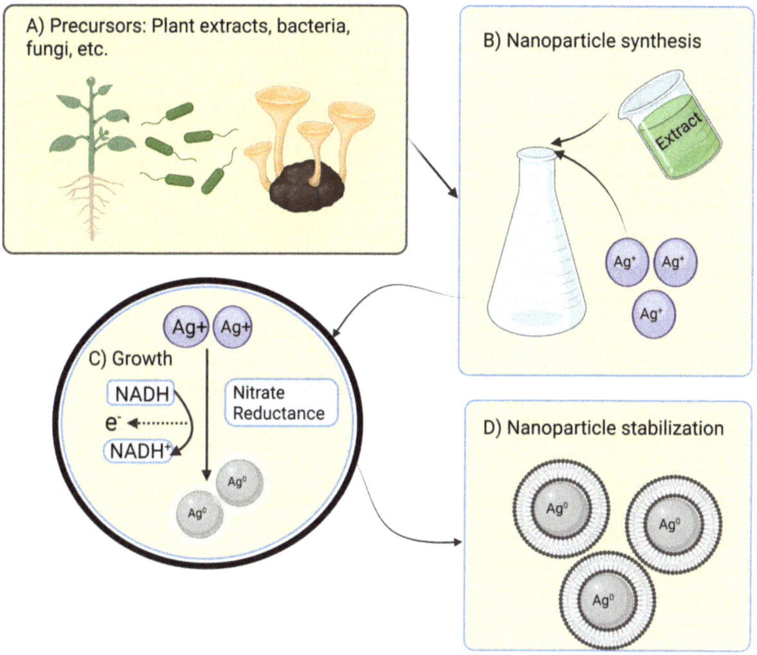

Fig. 17.1 (**a**) Different biomass extract precursors, (**b**) bio-synthesis method, (**c**) metal ions are reduced to metal NPs by the active molecules through an intracellular enzymatic redox process. (**d**) Nanoparticle stabilization. (Constructed by Diego F. Monroy-Macias)

transgenic plants that enhance yield and nutritional value (He et al., 2019b; Oliveira et al., 2015). Conventional fertilizers suffer from significant nutrient loss and low nutrient absorption. Nanofertilizers ranging a size <100 nm can reduce these losses and improve nutrient absorption efficiency (Babu et al., 2022; Dimkpa & Bindraban, 2018). Regarding plant micronutrients, such as zinc, iron, and copper (Dimkpa & Bindraban, 2018; Soni et al., 2023). Copper NPs improved tomato fruit quality at relatively low concentrations of 250 mg/L. Nevertheless, increasing the concentration to 500 mg/L proved to be toxic (López-Vargas et al., 2018). This highlights the need for further research before commercialization. Conventional pesticides are more likely to harm the environment, whereas nanopesticides, being soluble and controllably released, mitigate these disadvantages (Kumari et al., 2023). For instance, silicon nanoparticles demonstrated high efficacy against stored grain pests (El-Naggar et al., 2020). Their use is increasing, but studies on cytotoxicity are still needed. In agriculture, plant growth indicates overall health. NPs are now employed to achieve controlled and protected release, enhancing efficiency (Pereira et al., 2017). For example, chitosan NPs increased the bioactivity of gibberellic acid in beans.

Another agricultural activity where nanotechnology is related is the use of nanosensors for real-time monitoring of environmental conditions, diseases, pests, and nutrients (He et al., 2019b). Sensors based on gold nanoparticles and carbon nanotubes have detected volatile compounds from insect-attacked plants, soil nutrients, and low-concentration pesticides (Pan et al., 2016).

On the other hand, there has been an increase in studies on genetic modification in plants using nanomaterials (NMs), offering advantages over traditional methods such as biolistic or agrobacterium use (Fiaz et al., 2021). Nanoparticles have shown higher transformation activity and better nucleic acid protection (Altpeter et al., 2016). Furthermore, combining this with CRISPR-Cas can lead to significant advances in crop engineering (Ahmar et al., 2021; Watson et al., 2018).

17.2.2 Examples of Biosynthesized Nanomaterials in Agricultural Activities

One example of a biosynthesized nanomaterial is a nanofertilizer synthesized via *Streptomyces griseus*, producing copper nanoparticles. These not only improved tea leaf yield but also increased plant resistance to fungal agents, causing root rot (Ponmurugan et al., 2016).

In biosensors, gold nanoparticles are revolutionizing detection and monitoring in agriculture. Specifically, these nanoparticles have been used to detect Potato Leaf Roll Virus (Zhan et al., 2018) and to diagnose *Phytophthora infestans*, the cause of late blight, through induced genomic changes (Panferov et al., 2018).

Another application of biosynthesized nanomaterials is in crop protection and disease management. For example, silver nanoparticles AgNPs obtained by *halophilic* bacteria extract demonstrated antibacterial activity against *Dickeya dadantii*, which affects sweet potato roots (Hossain et al., 2023). In the same way, AgNPs produced by *Streptomyces* sp. extract effectively inhibited spore germination and destroyed the fungal membranes of *Fusarium verticillioides*, a maize crop pathogen (Marathe et al., 2021). There are more studies on bacterial and fungal extracts to obtain metallic NPs and semiconductor NPs, as presented in Table 17.1.

One of the main applications for the biosynthesized NMs in agriculture involves antimicrobial activity. For example, silver nanoparticles obtained from *Azadirachta indica* leaf extract showed activity against *Staphylococcus aureus* and *Escherichia coli* (Castillo-Henríquez et al., 2020). AuNPs from *Eclipta alba* leaf extract were effective against *Pseudomonas aeruginosa* and *Bacillus subtilis* (Vijayakumar et al., 2020). Nanofertilizers have also been developed using Ag and Au NPs for controlled nutrient release, reducing losses and environmental impact while improving soil quality (Panpatte et al., 2016). A suspension of 5 mg/L of AuNPs fertilizer was applied to red ginseng sprouts for 14 days, resulting in increased ginsenoside content (Kang et al., 2016). There are applications in disease-specific cases. For example, AgNPs synthesized from *Rhizophora mucronata* leaf extract showed larvicidal activity against *Aedes aegypti* and *Culex quinquefasciatus*, which cause dengue and filariasis, diseases that affect agricultural workers (Castillo-Henríquez et al., 2020). More studies are shown in Table 17.2.

Table 17.1 Green-synthesized nanoparticles by microorganism extracts

Microorganism	NPs	Synthesis type	Size (nm)	Application	References
Escherichia coli k12	Au	Extracellular	50	Antimicrobial activity	Srivastava et al. (2013)
L. casei (strain JCM1134)	Au	Intracellular	ca. 29.6	Antimicrobial activity	Kikuchi et al. (2016)
Lactobacillus kimchicus DCY51T (from kimchi)	Au	Intracellular	5–30	Antimicrobial activity	Markus et al. (2016)
Lactobacillus acidophilus DSMZ 20079T	CdS	Extracellular	2.5–5.5	Nanobiosensors	Riaz et al. (2020)
Fusarium solani	Ag	Extracellular	5–35	Textile fabrics, antifungal	Ingle et al. (2009)
Escherichia coli ATCC 8739	Cu	Extracellular	30–75	Plant micronutrient	Mehri et al. (2020)
Bacillus subtilis ATCC 6633	Cu	Extracellular	30–75	Plant micronutrient	
Lactobacillus casei	Cu	Extracellular	30–75	Plant micronutrient	

Table 17.2 Antimicrobial activity of biosynthesized Ag-Au NPs

Plant extract	Synthesized NPs	Target pathogen	References
Citrus limetta peel	Ag	*Micrococcus luteus*, *Streptococcus mutans*, *Staphylococcus epidermidis*, *S. aureus*, *E. coli*, *Candida* spp.	Kah et al. (2019)
Luffa acutangula leaf	Ag	*E. coli*, *Saccharomyces cerevisiae*	Costa et al. (2020)
Parkia speciosa leaf	Ag	*E. coli*, *S. aureus*, *Pseudomonas aeruginosa*, *Bacillus subtilis*	Singh et al. (2016)
Gomphrena globosa leaf	Ag	*S. aureus*, *B. subtilis*, *M. luteus*, *E. coli*, *P. aeruginosa*, *Klebsiella pneumoniae*	Fouda et al. (2020)
Pedalium murex leaf	Ag	*E. coli*, *K. pneumoniae*, *Micrococcus flavus*, *P. aeruginosa*, *B. subtilis*, *Bacillus pumilus*, *S. aureus*	Hossain et al. (2019)
Musa acuminata peel	Ag	*B. subtilis*, *S. aureus*, *P. aeruginosa*, *E. coli*	Marathe et al. (2021)
Caulerpa racemosa	Au	*Aeromonas veronii*, *Streptococcus agalactiae*	Hassan et al. (2018)
Eclipta alba	Au	*E. coli*, *P. aeruginosa*, *B. subtilis*, *S. aureus*	López-Vargas et al. (2018)
Nepenthes khasiana leaf	Au	*E. coli*, *Bacillus* spp., *Aspergillus niger*, *Candida albicans*	Youssef et al. (2017)
Abelmoschus esculentus pulp	Au	*B. subtilis*, *Bacillus cereus*, *E. coli*, *M. luteus*, *P. aeruginosa*	Gade et al. (2010)

17.3 Enhancing Resource Efficiency

Agriculture is one of the principal economic activities of the world, and food production is fundamental for the population's survival. In this way, it is very important to look for innovative solutions to improve the efficiency and quality of the product (Ragazou et al., 2022).

The global demand for water and other resources, on the whole shortage of fresh water and natural resources, highlights the emergency to find a form to enhance resource efficiency and sustainable practice across different industries, especially in food production (Eslami et al., 2023). It is estimated that by the year 2050, it will be necessary to see an improvement of 70% in food production to maintain food safety for the 9.8 billion people that are expected to be living in that year (Oluwole et al., 2023). One of the improvements that is being implemented is super absorbent polymers (SAPs) in use for water retention because SAPs can absorb more than 1000 times their weight and retain liquids, no matter that they are under pressure (Hou et al., 2018). In this case, there exist two types of SAPs: natural-based and synthetic-based.

1. Based on natural polymers like cellulose and chitosan, these types of polymers have the advantage of the fact that they are biodegradable, but they suffer from low water absorption, so they must be used in larger amounts (Chang et al., 2021).
2. Based on synthetic polymers, such as polyacrylic acid and polyacrylamide, are cheaper than those based on natural polymers, and have a long life of service and high water absorption, but their non-degradation characteristic may express adverse effects on the environment and plant growth (Chang et al., 2021).

Also, SAPs are being used to control the release of fertilizers. About 40%–70% nitrogen and 80%–90% phosphorus cannot be absorbed by crops due to their high solubility and diffusivity (Tubert et al., 2018). This is the reason why SAPs not only improve the fertilizer's efficacy but also avoid unnecessary environmental pollution. Another option for enhancing resource efficiency is the use of the Internet of Things (IoT) (Fig. 17.2). This tool is key to optimizing the use of water through the implementation of sensors and automated systems, specifically on three points:

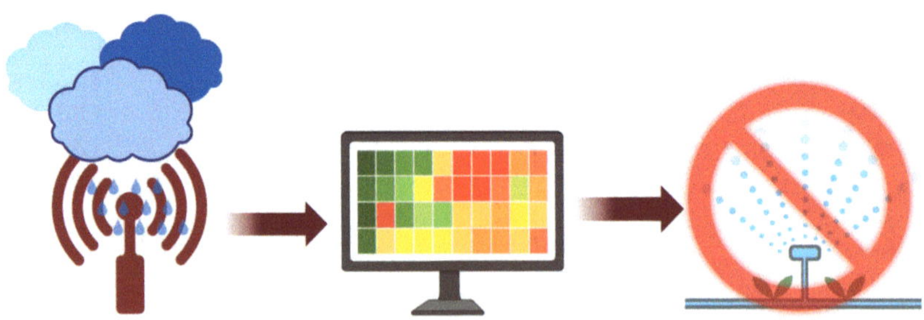

Fig. 17.2 Operation of IoT technology. (Constructed by S. O. Seoane)

1. Precise measurement of the humidity on the ground, the IoT sensors will provide real-time data about the humidity levels, and it will make prevention and planning against the risk easier by basing it on specific needs, thus reducing water wastage (Xu et al., 2022).
2. Automation of irrigation systems, along with the recompilation of environmental data to ensure efficient application of water resources and adapt to climatic conditions (Nawandar & Satpute, 2019).

The IoT systems can detect leaks and,malfunctions which makes its timely detection easier to avoid losses of water (Jan et al., 2022). As mentioned, in addition to water, fertilizers are also essential for agriculture. Plants need 14 different types of nutrients but nitrogen is one of the most used nutrients in fertilizers, so most of the improvements made to fertilizers have been nitrogen-based like the urea which transforms into ammonia and CO_2 very quickly in moist soils resulting in a loss of the majority of the fertilizer causing environmental implications like nitrate concentration on drinking water exceeding the limit of 50 mgL^{-1} which caused between 2300 and 12,594 cancer cases (Dimkpa et al., 2020b). One of the options to reduce this problem is the use of nanotechnology, in this case, a slow-release nano fertilizer comprised of hydroxyapatite (HAP)-urea-wood powder composite (Dimkpa et al., 2020a, b). This nano-urea-HAP was tested on *Gliricidia sepium* and showed a reduction in the N-release of 44% after 60 days on a pH 7 soil compared to traditional urea, this suggests that nano-urea-HAP can reduce the rate of application without compromising the crops, which implies a reduction of the pollution due to the excess of fertilizer and a better efficiency of the fertilizer (Dimkpa et al., 2020b).

One of the most abundant yet poorly utilized resources is the waste generated by crops. It's known that sometimes they're used like fertilizers, but this resource has a lot of different usages, like being treated to create biochar, which can be used for the water splitting process to create hydrogen as a form of green energy (Nikitin et al., 2024).

The way this happens is through biomass of agricultural waste with high silicon content, such as oat straw, which undergoes oxidative carbonization with a fluidized bed of a deep oxidation catalyst. The resulting biochar is subjected to high-energy treatment using a pulsed plasma arc, which produces a nano-dispersed material that will contain silicon carbide in its hexagonal form. This compound was modified with 5% Pt and used as a catalyst and co-catalyst in the hydrogen evolution reaction (HER). The data showed similar performance to the Pt/C commercial samples, which indicates that this material can be used as a more ecological and cost-effective alternative (Nikitin et al., 2024; Sun et al., 2022).

17.4 Reducing Agricultural Waste

The agricultural production system produces a large quantity of solid waste; worldwide agricultural production has placed great pressure on the environment, causing negative impacts on soil, air, and water resources, and thus jeopardizing the sustainability of ecosystems and human health (Karić et al., 2022). The residues of raw agricultural products constitute agricultural waste. Such residues include manure and animal waste; corn stalks, sugar cane bagasse, drops and culls from fruits and vegetables, crop waste; pesticides, insecticides, and herbicides (hazardous and toxic agricultural waste); and food processing waste obtained during growing and processing in liquid, slurry, or solid forms (Capanoglu et al., 2022). They often create environmental and economic burdens in the farming and primary processing sectors, which can be amplified by regional specialization of either crop or animal production (Lal, 2021).

Improper disposition of agricultural wastes not only results in environmental pollution but also wastes a lot of valuable biomass resources (Wang et al., 2016). The rapid growth in the volume and variety of agricultural waste biomass, driven by population growth and intensive farming practices, has made these wastes a significant source of pollution, with an average annual increase of 5%–10% (Karić et al., 2022). The term biomass refers to organic matter of plant or animal origin, whether naturally occurring or cultivated by humans, on land, or in water, and produced directly or indirectly through photosynthesis involving chlorophyll. In general, biomass can be defined as anything having an organic matrix. Thus, the term biomass identifies various heterogeneous materials and matrices (Bonechi et al., 2017).

17.4.1 Biomass Uses

A wide range of sectors benefit from biomass, either as food/feed to generate electricity, heat, and fuels or as material and resources in the industrial fields, mainly in the wood, paper, and chemical industries (Antar et al., 2021). The composition of biomass is largely diverse. For example, residues of plant origin are mainly composed of cellulose, hemicellulose, and lignin with varying percentages (Tursi, 2019).

Biomass energy is by far the largest renewable energy source, representing 10.4% of the world's total primary energy supply or 77.4% of the global renewable energy supply. However, biomass energy represents only 1% of the total fuel used for electricity production (Monteiro & Ferreira, 2022). Through efficient biochemical and thermochemical processes, these biomass resources can be converted into valuable products. Biomass is considered highly suitable for energy production due to its significant fuel-related properties (Kamran, 2018). The quantity of bioenergy generated from bioresources depends entirely on the type of feedstock. Woody biomass, characterized by a higher lignin content (23.9%–32.0%), is more solid and denser (350–490 kg/m^3) compared to agricultural biomass (108–130 kg/m^3). This inherent density and structural composition make woody biomass more resistant to microbial and enzymatic degradation than agricultural biomass (Zhu et al., 2010). For bioenergy production, particularly biofuels, a high

Table 17.3 Different types of agricultural solid waste, their by-products, and their possible uses (Ginni et al., 2021)

Source	Agricultural waste or by-products	Possible uses
Rice	Straw Husk Bran	Ethanol production
Wheat	Straw Bran	Absorbents
Maize	Chaff Cobs Straw	Biofuel production Adsorbents
Cassava	Peelings Stems	Carbon precursors Adsorbents Filter medium
Groundnut	Shell	Carrier material for the inoculation of fungus Feedstock and filler in fertilizer
Sugarcane	Sugarcane bagasse	Electricity generation Production of ethylene glycol, 1,2-propylene glycol
Soybean	Straw Pod husk	Inexpensive and renewable source of energy and chemicals

cellulose content in biomass is essential to generate more volatile compounds. But, lower levels of hemicellulose and lignin are desirable in feedstock, as lignin functions as a chemical binder, complicating the separation of cellulose from hemicellulose (Antar et al., 2021).

The conversion of biomass into methane offers high energy efficiency and can significantly reduce greenhouse gas emissions. Methane does not permeate through the metal lattice of most materials used in gas transportation pipelines. This characteristic allows methane to be transported using the existing natural gas infrastructure, making it a viable option. Methane production from biomass can be achieved through anaerobic digestion (Wu et al., 2023).

Other than the production of bioenergy and biofuels, biomass has been employed to produce specific biomass-derived chemicals (bioproducts) (Antar et al., 2021). Table 17.3 presents different types of agricultural waste that must be revalued to minimize the environmental risks they may cause.

Different from the side effects caused by the improper disposal of agricultural waste, the recycling and reuse of agricultural waste are conducive to environmental protection (He et al., 2019a). When agricultural waste is managed and used optimally, it has the potential to become a sustainable resource for value addition and a major contributor to energy security and ecological sustainability (Awogbemi & Kallon, 2022).

17.5 Economical Evaluation of Green-Synthesized Nanotechnology for Sustainable Farming Practices

The growing acceptance of green NPs systems for sustainable agriculture applications presents a promising tool to revolutionize farming practices. These technologies have demonstrated the potential to enhance crop quality, combat plant pathogens, and manage diseases and stress (Sundararajan et al., 2024). While the environmental and technological advantages of these NPs systems have been examined, understanding their economic viability is crucial for a smart implementation (Sah et al., 2024). Current research emphasizes the importance of developing cost-effective and scalable green synthesis methods to facilitate the transition from laboratory to large-scale applications.

Economic evaluation of these emerging processes should be done. Such analyses must account for critical factors, including production costs, scalability, return on investment (ROI), and long-term economic impacts (Su et al., 2022). This section explores the green synthesis cost-effectiveness (Fatmarin & Dani Nandiyanto, 2020), offers a comparative analysis with traditional agricultural methods (Gupta et al., 2023), and identifies potential barriers to widespread adoption (Ying et al., 2022). Figure 17.3 schematically illustrates the primary processes to consider in the green synthesis of NPs when conducting an economic evaluation, encompassing stages from raw material acquisition to large-scale production and agricultural application.

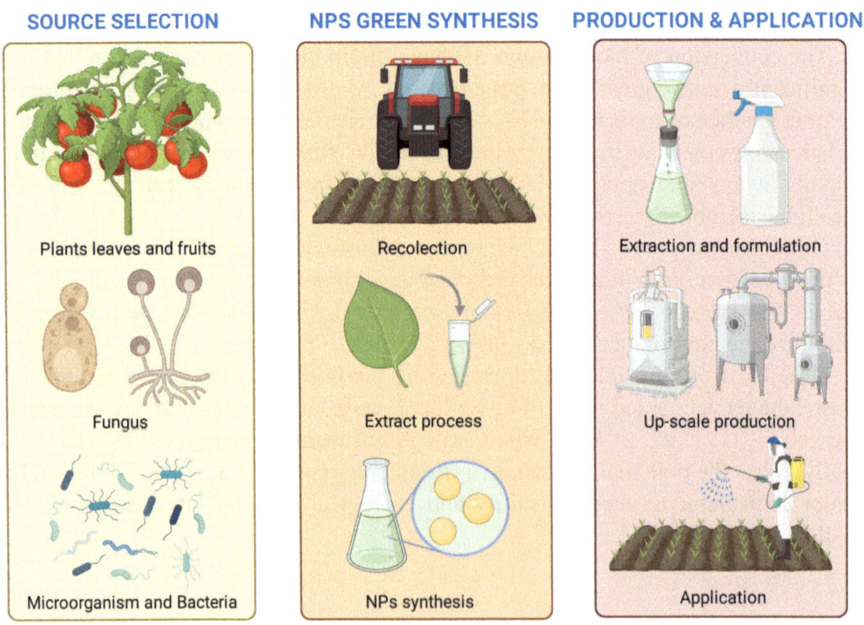

Fig. 17.3 Schematic representation of the green NPs production and large-scale application in agriculture. (Constructed by A. Cordoba)

17.5.1 Analysis of Green Nanoparticle Synthesis Processes

A critical aspect of evaluating the economic feasibility of green-synthesized nanomaterials for agricultural applications lies in analyzing production costs and principal processes. This analysis covers the entire production chain, from sourcing raw precursors to the fabrication (Gacem & Abd-Elsalam, 2022). Green synthesis methods rely on biological precursors as mentioned, providing not only an eco-friendly alternative to conventional chemical and physical synthesis techniques but also a potential source of low-cost or easily accessible materials derived from agricultural and industrial waste (Sundararajan et al., 2024). However, the economic viability of these methods depends on the availability of raw materials, synthesis efficiency, and scalability.

Plant-mediated NPs synthesis, including AgNPs (Zulfiqar et al., 2024), AuNPs (Castillo-Henríquez et al., 2020), and metal oxide nanoparticles (MONPs) (Alabdallah et al., 2021), has been extensively explored. Studies have demonstrated that several nanosystems, such as green-synthesized AgNPs can be obtained through various plant extracts (Liaqat et al., 2022), microorganisms (Singh et al., 2023), and biological waste (Thangadurai et al., 2021), serving as nontoxic antimicrobial agents with potential agricultural applications (Antunes Filho et al., 2023).

Nevertheless, the cost-effectiveness of these methods depends on the local availability and abundance of the chosen biological resources and the processes to extract the key agents from these raw materials. Extraction methods include organic solvent extraction, acid or base activation, microwave-assisted extraction, and Soxhlet extraction (Câmara et al., 2022). The selection of appropriate extraction techniques is crucial, as it influences the efficiency and scalability of the green synthesis method. The synthesis efficiency and yield are essential factors during the economical evaluation of the use of these NMs. The efficiency of the synthesis process, including reaction time, energy consumption, and yield, directly influences production costs. Green synthesis methods often operate under ambient conditions, reducing energy consumption compared to traditional methods (Sah et al., 2024). Achieving high yields with consistent NPs characteristics remains a challenge. For example, the preparation of 1 kg of AgNPs has been estimated to cost USD 4 million, while 1 kg of raw silver costs around USD 14,000 (Castillo-Henríquez et al., 2020).

The green synthesis of nanosystems involves variables such as the plant species used, extraction techniques, and reaction conditions, which can lead to variations in reaction efficiency compared to conventional nanoparticle synthesis methods that offer greater control. For instance, green synthesis of iron nanoparticles (FeNPs) has reported efficiency values ranging from 20% to 88% due to different plant extract sources, whereas conventional methods can achieve efficiencies exceeding 90% (Ying et al., 2022). This consideration is also a fundamental aspect in the analysis cost of these nanomaterials, requiring greater use of available raw materials for their production. Despite the potential for slightly lower yields, green synthesis methods provide significant advantages in terms of sustainability, reduced toxicity, and environmental compatibility, making them promising alternatives to traditional techniques (Liaqat et al., 2022).

17.5.2 Scalability and Industrial Production

In addition to the viability of precursors and the efficiency of green synthesis processes for NPs production, a comprehensive economic evaluation must consider the demand and production scale required for these materials, particularly when aiming to address the high demands of sectors like agriculture (Yadav et al., 2023). Current research has extensively explored processes for scaling up green NPs synthesis to an industrial level, with significant progress reported in fields such as energy (Khan et al., 2023), biomedicine (Abuzeid et al., 2023), and pharmaceuticals (Mukherjee et al., 2024).

Transitioning from laboratory-scale synthesis to industrial-scale production is a complex process that can significantly influence the cost of material synthesis. This transition requires careful consideration of factors such as process optimization, quality control, and regulatory compliance (Shegokar & Nakach, 2020). Numerous studies emphasize that, although green-synthesized NPs hold great promise for agricultural applications, the scalability of these methods presents economic challenges (Gacem & Abd-Elsalam, 2022). Key issues include the purification of nanoparticles and the regulation of their geometry, which are complicated by the limited understanding of biosynthetic mechanisms (Bhandari et al., 2023).

Economic analyses of green-synthesized silver nanoparticles (AgNPs) suggest that while their initial production costs may be higher than those of bulk silver, their superior efficacy in agricultural applications, particularly their potent antimicrobial properties, can justify the investment (Fahim et al., 2024). Multiple traditional methods, including solvothermal approaches (Wang et al., 2018) and electron beam evaporation techniques (Betancourt et al., 2021), have been explored to adapt these nanosystems for large-scale production.

The scalability potential of nanosystems highlights a significant economic advantage of green synthesis methods over conventional techniques (Santhosh et al., 2022). By utilizing synthesis processes that typically operate under ambient conditions, numerous studies have focused on developing large-scale production strategies for green-synthesized NPs (Table 17.4).

The green synthesis of gold nanoparticles (AuNPs) using plant extracts has garnered significant attention for various applications. Although gold is inherently expensive, green synthesis approaches have been shown to reduce the environmental remediation costs associated with conventional synthesis methods, thereby partially offsetting the high material costs (Santhosh et al., 2022). Moreover, Su et al. conducted a comprehensive cost-benefit analysis of nanopesticides, underscoring the necessity of improving the efficiency of formulations to enable widescale adoption. Their findings highlighted the potential of nanotechnology while emphasizing the critical need for cost reductions and meticulous evaluation of formulation efficacy (Su et al., 2022).

Table 17.4 Provides various proposals for large-scale green synthesis of the most extensively studied NPs for agricultural applications

NPs	Source	Application	References
Ag	*Plantago major* L. leaf	AgNP demonstrates broad-spectrum antimicrobial activity through bacterial membrane disruption and oxidative stress	Fahim et al. (2024)
Au	Palm oil mill	Analysis of reaction parameters (pH, gold precursor concentration, and reaction time) on the morphology and size of AuNPs. Address the feasibility of using agro-waste as a raw material for a scalable and cost-effective synthesis method	Gan et al. (2012)
ZnO	*Streptomyces albus* cell-free extract	Study of optimized precursors and fermentation methods to enhance bioactive metabolite production and maximize zinc oxide nanoparticle (ZnO NP) yields	El-Moslamy et al. (2023)
Fe	Green tea and leaf extracts	This review highlights green synthesis pathways for nanoscale zerovalent iron particles, emphasizing their potential for environmental remediation and catalysis. Sustainable methods are compared with conventional techniques	Monga et al. (2020)
CuO	Forest industrial residue	Green-synthesized CuO NPs showed strong antibacterial activity and offer a sustainable, economical method for revaluing forest residues	Varshney and Gupta (2024)

17.5.3 Potential Economic Benefits of Green Nanoparticles

In addition to the previously discussed properties of these materials, the primary economic potential of nanoparticulate systems lies in their capacity to replace traditional pesticides and fertilizers in various agricultural applications. This approach not only enhances efficiency but also substantially reduces environmental impact (Sah et al., 2024). The reduction in environmental harm is achieved through the lower concentrations of materials required and the minimized toxicity, attributes enabled by the intrinsic biocompatibility of green synthesis methods (Tonelli & Tonelli, 2023). Collectively, these features aim to decrease the financial costs associated with food and crop production, contributing to more sustainable agricultural practices.

Another significant economic advantage of green-synthesized NMs lies in their reliance on readily available, low-cost biological resources, such as plant extracts, microbial cultures, and agricultural residues. These resources not only reduce dependency on expensive chemical precursors but also offer opportunities for waste stream valorization, further enhancing sustainability and economic efficiency (Jacob et al., 2021; Varshney & Gupta, 2024).

For instance, previous sections of this chapter discuss the synthesis of AuNPs using plant extracts. While gold itself is a costly material, green synthesis methods have been shown to significantly lower the environmental remediation costs compared to conventional approaches, thereby partially offsetting the high material expenses (Santhosh et al., 2022). Additionally, Su et al. conducted a comprehensive cost-benefit analysis of nanopesticides, highlighting the necessity of improving formulation efficiency to facilitate widespread adoption (Su et al., 2022). Their findings emphasize the potential of nanotechnology while emphasizing the need for cost reductions and rigorous evaluation of formulation effectiveness to ensure economic and practical feasibility (Abuzeid et al., 2023).

Another critical economic factor is the potential reduction in crop losses, as the FAO reports that up to 40% of global agricultural yields are lost annually due to pests and diseases (Bigini et al., 2021). Enhancing antimicrobial and antifungal activity significantly beyond the capabilities of conventional pesticides and fungicides is a topic of high ecological importance today. This advancement can substantially minimize the release of fertilizers and pesticides toxic compounds into soils (Hu et al., 2024), aquifers (Liu et al., 2021), and fauna (Dhankhar & Kumar, 2023).

Green-synthesized NPs, such as copper oxide (CuO) and silver (Ag) nanoparticles, have demonstrated superior antimicrobial and pesticidal efficacy in crops like rice (Shireen Akhter Jahan et al., 2024) and tomato (Yan et al., 2025). These properties can significantly mitigate crop losses while reducing dependence on conventional agrochemicals, thereby contributing to both economic efficiency and environmental sustainability.

Finally, the costs associated with the application and irrigation of fertilizers and pesticides can be significantly reduced with NPs. The ability of nanomaterials to achieve controlled and targeted release of active ingredients onto the leaves and stems of crops minimizes environmental losses during application and reduces their absorption and accumulation in soil and roots. This capability lowers expenses on agricultural inputs by decreasing both the frequency of application and the amount of material required (Liu et al., 2024).

For example, lignin NPs synthesized via green methods have demonstrated controlled-release properties for pyraclostrobin, effectively managing diseases caused by *Botrytis cinerea* in strawberries. This approach has the potential to reduce application frequency and improve material efficiency, thereby lowering pest control and application costs throughout the crop growth cycle (Hou et al., 2021).

The economic benefits of green nanoparticles in agriculture extend beyond direct cost savings to include enhanced productivity and access to high-value markets. These advantages potentially not only contribute to individual farm profitability but also support broader goals of sustainable development and environmental preservation, making green nanotechnology a pivotal tool in addressing global agricultural challenges.

17.6 Conclusion and Perspectives

Green nanotechnology offers a sustainable and innovative approach to tackling modern agricultural challenges. By improving the efficiency of agricultural inputs, reducing environmental impact, and boosting crop productivity, it has the potential to shape the future of sustainable farming. Through green synthesis, NMs can be produced using renewable natural resources while minimizing the need for harmful chemicals. These NMs have valuable applications, such as controlled-release nanofertilizers, biodegradable nanopesticides, and nanosensors that provide real-time soil and crop monitoring. Despite these benefits, large-scale adoption still faces challenges. Optimizing synthesis methods, ensuring nanoproduct safety, and establishing clear regulations are essential for their successful integration into agriculture. Additionally, assessing economic feasibility and gaining farmers' trust are key factors for widespread acceptance. Future research should focus on improving production efficiency and understanding the long-term effects of nanomaterials on the environment and human health. Combining green nanotechnology with circular economy strategies and precision agriculture could strengthen food security and environmental sustainability, creating a more resilient agricultural system for the future.

Acknowledgments Aldo Cordoba thanks to Secretaría de Ciencia, Humanidades, Tecnología e Innovación (SECIHTI) for the scholarship granted. The authors thank FONFIVE-2024 (FIN202413) for the financial support.

Conflict of Interest The authors declare no conflict of interest.

References

Abuzeid, H. M., Julien, C. M., Zhu, L., & Hashem, A. M. (2023). Green synthesis of nanoparticles and their energy storage, environmental, and biomedical applications. *Crystals, 13*, 1576.

Ahmar, S., Mahmood, T., Fiaz, S., Mora-Poblete, F., Shafique, M. S., Chattha, M. S., & Jung, K.-H. (2021). Advantage of nanotechnology-based genome editing system and its application in crop improvement. *Frontiers in Plant Science, 12*, 663849.

Alabdallah, N. M., Hasan, M. M., Hammami, I., Alghamdi, A. I., Alshehri, D., & Alatawi, H. A. (2021). Green synthesized metal oxide nanoparticles mediate growth regulation and physiology of crop plants under drought stress. *Plants, 10*, 1730.

Altpeter, F., Springer, N. M., Bartley, L. E., Blechl, A. E., Brutnell, T. P., Citovsky, V., Conrad, L. J., Gelvin, S. B., Jackson, D. P., Kausch, A. P., Lemaux, P. G., Medford, J. I., Orozco-Cárdenas, M. L., Tricoli, D. M., Van Eck, J., Voytas, D. F., Walbot, V., Wang, K., Zhang, Z. J., & Stewart, C. N., Jr. (2016). Advancing crop transformation in the era of genome editing. *Plant Cell, 28*, 1510–1520.

Antar, M., Lyu, D., Nazari, M., Shah, A., Zhou, X., & Smith, D. L. (2021). Biomass for a sustainable bioeconomy: An overview of world biomass production and utilization. *Renewable and Sustainable Energy Reviews, 139*, 110691.

Antunes Filho, S., Dos Santos, M. S., Dos Santos, O. A. L., Backx, B. P., Soran, M.-L., Opriş, O., Lung, I., Stegarescu, A., & Bououdina, M. (2023). Biosynthesis of nanoparticles using plant extracts and essential oils. *Molecules, 28*, 3060.

Awogbemi, O., & Kallon, D. V. V. (2022). Pretreatment techniques for agricultural waste. *Case Studies in Chemical and Environmental Engineering, 6*, 100229.

Babu, S., Singh, R., Yadav, D., Rathore, S. S., Raj, R., Avasthe, R., Yadav, S. K., Das, A., Yadav, V., Yadav, B., Shekhawat, K., Upadhyay, P. K., Yadav, D. K., & Singh, V. K. (2022). Nanofertilizers for agricultural and environmental sustainability. *Chemosphere, 292*, 133451.

Betancourt, A. P., Goswami, D. Y., Bhethanabotla, V. R., & Kuhn, J. N. (2021). Scalable and stable silica-coated silver nanoparticles, produced by electron beam evaporation and rapid thermal annealing, for plasmon-enhanced photocatalysis. *Catalysis Communications, 149*, 106213.

Bhandari, G., Dhasmana, A., Chaudhary, P., Gupta, S., Gangola, S., Gupta, A., Rustagi, S., Shende, S., Rajput, V., Minkina, T., Malik, S., & Slama, P. (2023). A perspective review on green nanotechnology in agro-ecosystems: Opportunities for sustainable agricultural practices & environmental remediation. *Agriculture, 13*, 668.

Bigini, V., Camerlengo, F., Botticella, E., Sestili, F., & Savatin, D. (2021). Biotechnological resources to increase disease-resistance by improving plant immunity: A sustainable approach to save cereal crop production. *Plants, 10*, 1146.

Bonechi, C., Consumi, M., Donati, A., Leone, G., Magnani, A., Tamasi, G., & Rossi, C. (2017). 1 – Biomass: An overview. In F. Dalena, A. Basile, & C. Rossi (Eds.), *Bioenergy systems for the future*. Woodhead Publishing.

Câmara, J. S., Perestrelo, R., Berenguer, C. V., Andrade, C. F. P., Gomes, T. M., Olayanju, B., Kabir, A., Rocha, C. M. R., Teixeira, J. A., & Pereira, J. A. M. (2022). Green extraction techniques as advanced sample preparation approaches in biological, food, and environmental matrices: A review. *Molecules, 27*, 2953.

Capanoglu, E., Nemli, E., & Tomas-Barberan, F. (2022). Novel approaches in the valorization of agricultural wastes and their applications. *Journal of Agricultural and Food Chemistry, 70*, 6787–6804.

Castillo-Henríquez, L., Alfaro-Aguilar, K., Ugalde-Álvarez, J., Vega-Fernández, L., Montes De Oca-Vásquez, G., & Vega-Baudrit, J. R. (2020). Green synthesis of gold and silver nanoparticles from plant extracts and their possible applications as antimicrobial agents in the agricultural area. *Nanomaterials, 10*, 1763.

Chang, L., Xu, L., Liu, Y., & Qiu, D. (2021). Superabsorbent polymers used for agricultural water retention. *Polymer Testing, 94*, 107021.

Costa, M. I., Álvarez-Cerimedo, M. S., Urquiza, D., Ayude, M. A., Hoppe, C. E., Fasce, D. P., De Castro, R. E., & Giménez, M. I. (2020). Synthesis, characterization and kinetic study of silver and gold nanoparticles produced by the archaeon Haloferax volcanii. *Journal of Applied Microbiology, 129*, 1297–1308.

Dhankhar, N., & Kumar, J. (2023). Impact of increasing pesticides and fertilizers on human health: A review. *Materials Today: Proceedings*. https://doi.org/10.1016/j.matpr.2023.03.766

Dimkpa, C. O., & Bindraban, P. S. (2018). Nanofertilizers: New products for the industry? *Journal of Agricultural and Food Chemistry, 66*, 6462–6473.

Dimkpa, C. O., Andrews, J., Fugice, J., Singh, U., Bindraban, P. S., Elmer, W. H., Gardea-Torresdey, J. L., & White, J. C. (2020a). Facile coating of urea with low-dose ZnO nanoparticles promotes wheat performance and enhances Zn uptake under drought stress. *Frontiers in Plant Science, 11*, 168.

Dimkpa, C. O., Fugice, J., Singh, U., & Lewis, T. D. (2020b). Development of fertilizers for enhanced nitrogen use efficiency – Trends and perspectives. *Science of the Total Environment, 731*, 139113.

El-Moslamy, S. H., Elnouby, M. S., Rezk, A. H., & El-Fakharany, E. M. (2023). Scaling-up strategies for controllable biosynthetic ZnO NPs using cell free-extract of endophytic Streptomyces albus: Characterization, statistical optimization, and biomedical activities evaluation. *Scientific Reports, 13*, 3200.

El-Naggar, M. E., Abdelsalam, N. R., Fouda, M. M. G., Mackled, M. I., Al-Jaddadi, M. A. M., Ali, H. M., Siddiqui, M. H., & Kandil, E. E. (2020). Soil application of nano silica on maize yield and its insecticidal activity against some stored insects after the post-harvest. *Nanomaterials, 10*, 739.

Eslami, E., Abdurrahman, E., Ferrari, G., & Pataro, G. (2023). Enhancing resource efficiency and sustainability in tomato processing: A comprehensive review. *Journal of Cleaner Production, 425*, 138996.

Esquivel, K., Martínez-Chávez, L. A., Rosales-Pérez, A., & Hernández-Rangel, R. (2023). Chapter 1. Nanomaterials: Classification, synthesis methods, and physicochemical

characterization. In F. Vasileios & G. Gholamreza (Eds.), *Engineered nanoparticles in agriculture*. De Gruyter.
Fahim, M., Shahzaib, A., Nishat, N., Jahan, A., Bhat, T. A., & Inam, A. (2024). Green synthesis of silver nanoparticles: A comprehensive review of methods, influencing factors, and applications. *JCIS Open, 16*, 100125.
Fatmarin, Z., & Dani Nandiyanto, A. B. (2020). Economic evaluation of zinc oxide nanoparticle production through green synthesis method using Cassia fistula plant extract. *International Journal of Energetica, 5*. https://doi.org/10.47238/ijeca.v5i2.133
Feregrino-Pérez, A. A., Rosales-Pérez, A., Hernández-Rangel, R., Juárez-Maldonado, A., Gohari, G., & Esquivel, K. (2024). Chapter 21 – Agricultural benefits, concerns, fate, and risks of the nanofertilizer's application to plants and the environment. In K. A. Abd-Elsalam (Ed.), *Nanofertilizer delivery, effects and application methods*. Elsevier.
Fiaz, S., Khan, S. A., Anis, G. B., Gaballah, M. M., & Riaz, A. (2021). Chapter 11 – CRISPR/Cas techniques: A new method for RNA interference in cereals. In K. A. Abd-Elsalam & K.-T. Lim (Eds.), *CRISPR and RNAi systems*. Elsevier.
Fouda, A., Hassan, S. E.-D., Abdo, A. M., & El-Gamal, M. S. (2020). Antimicrobial, antioxidant and larvicidal activities of spherical silver nanoparticles synthesized by Endophytic Streptomyces spp. *Biological Trace Element Research, 195*, 707–724.
Gacem, M. A., & Abd-Elsalam, K. A. (2022). *Strategies for scaling up of green-synthesized nanomaterials: Challenges and future trends*. Elsevier.
Gade, A., Ingle, A., Whiteley, C., & Rai, M. (2010). Mycogenic metal nanoparticles: Progress and applications. *Biotechnology Letters, 32*, 593–600.
Gan, P. P., Ng, S. H., Huang, Y., & Li, S. F. Y. (2012). Green synthesis of gold nanoparticles using palm oil mill effluent (POME): A low-cost and eco-friendly viable approach. *Bioresource Technology, 113*, 132–135.
Ginni, G., Kavitha, S., Ravi, Y. K., Bhatia, S. K., Adish Kumar, S., Rajkumar, M., Kumar, G., Pugazhendhi, A., Chi, N. T. L., & Banu, R. J. (2021). Valorization of agricultural residues: Different biorefinery routes. *Journal of Environmental Chemical Engineering, 9*, 105435.
Gour, A., & Jain, N. K. (2019). Advances in green synthesis of nanoparticles. *Artificial Cells, Nanomedicine, and Biotechnology, 47*, 844–851.
Gupta, D., Boora, A., Thakur, A., & Gupta, T. K. (2023). Green and sustainable synthesis of nanomaterials: Recent advancements and limitations. *Environmental Research, 231*, 116316.
Hano, C., & Abbasi, B. H. (2022). Plant-based green synthesis of nanoparticles: Production, characterization and applications. *Biomolecules, 12*, 31.
Hassan, S. E. L. D., Salem, S. S., Fouda, A., Awad, M. A., El-Gamal, M. S., & Abdo, A. M. (2018). New approach for antimicrobial activity and bio-control of various pathogens by biosynthesized copper nanoparticles using endophytic actinomycetes. *Journal of Radiation Research and Applied Sciences, 11*, 262–270.
He, K., Zhang, J., & Zeng, Y. (2019a). Knowledge domain and emerging trends of agricultural waste management in the field of social science: A scientometric review. *Science of the Total Environment, 670*, 236–244.
He, X., Deng, H., & Hwang, H. M. (2019b). The current application of nanotechnology in food and agriculture. *Journal of Food and Drug Analysis, 27*, 1–21.
Hossain, A., Hong, X., Ibrahim, E., Li, B., Sun, G., Meng, Y., Wang, Y., & An, Q. (2019). Green synthesis of silver nanoparticles with culture supernatant of a bacterium Pseudomonas rhodesiae and their antibacterial activity against soft rot pathogen Dickeya dadantii. *Molecules, 24*, 2303.
Hossain, A., Luo, J., Ali, M. A., Chai, R.-Y., Shahid, M., Ahmed, T., Hassan, M., Kadi, R., An, Q., Li, B., & Wang, Y. (2023). Synergistic action of biosynthesized silver nanoparticles and culture supernatant of Bacillus amyloliquefacience against the soft rot pathogen Dickeya dadantii. *Plants, 12*, 1817.
Hou, X., Li, R., He, W., Dai, X., Ma, K., & Liang, Y. (2018). Superabsorbent polymers influence soil physical properties and increase potato tuber yield in a dry-farming region. *Journal of Soils and Sediments, 18*, 816–826.

Hou, Q., Zhang, H., Bao, L., Song, Z., Liu, C., Jiang, Z., & Zheng, Y. (2021). NCs-delivered pesticides: A promising candidate in smart agriculture. *International Journal of Molecular Sciences, 22*, 13043.

Hu, J., Wang, Z., Williams, G. D. Z., Dwyer, G. S., Gatiboni, L., Duckworth, O. W., & Vengosh, A. (2024). Evidence for the accumulation of toxic metal(loid)s in agricultural soils impacted from long-term application of phosphate fertilizer. *Science of the Total Environment, 907*, 167863.

Huston, M., Debella, M., Dibella, M., & Gupta, A. (2021). Green synthesis of nanomaterials. *Nanomaterials, 11*, 2130.

Ingle, A., Rai, M., Gade, A., & Bawaskar, M. (2009). Fusarium solani: A novel biological agent for the extracellular synthesis of silver nanoparticles. *Journal of Nanoparticle Research, 11*, 2079–2085.

Jacob, J. M., Ravindran, R., Narayanan, M., Samuel, S. M., Pugazhendhi, A., & Kumar, G. (2021). Microalgae: A prospective low cost green alternative for nanoparticle synthesis. *Current Opinion in Environmental Science & Health, 20*, 100163.

Jadoun, S., Arif, R., Jangid, N. K., & Meena, R. K. (2021). Green synthesis of nanoparticles using plant extracts: A review. *Environmental Chemistry Letters, 19*, 355–374.

Jan, F., Min-Allah, N., Saeed, S., Iqbal, S., & Ahmed, R. (2022). IoT-based solutions to monitor water level, leakage, and motor control for smart water tanks. *Water*. https://doi.org/10.3390/w14030309

Kah, M., Tufenkji, N., & White, J. C. (2019). Nano-enabled strategies to enhance crop nutrition and protection. *Nature Nanotechnology, 14*, 532–540.

Kamran, M. (2018). Current status and future success of renewable energy in Pakistan. *Renewable and Sustainable Energy Reviews, 82*, 609–617.

Kang, H., Hwang, Y. G., Lee, T. G., Jin, C. R., Cho, C. H., Jeong, H. Y., & Kim, D. O. (2016). Use of gold nanoparticle fertilizer enhances the ginsenoside contents and anti-inflammatory effects of red ginseng. *Journal of Microbiology and Biotechnology, 26*, 1668–1674.

Karić, N., Maia, A. S., Teodorović, A., Atanasova, N., Langergraber, G., Crini, G., Ribeiro, A. R. L., & Đolić, M. (2022). Bio-waste valorisation: Agricultural wastes as biosorbents for removal of (in)organic pollutants in wastewater treatment. *Chemical Engineering Journal Advances, 9*, 100239.

Khan, F., Karimi, M. N., & Khan, O. (2023). Exploring the scalability and commercial viability of biosynthesized nanoparticles for cooling panels with the help of Artificial Intelligence and solar energy systems. *Green Technologies and Sustainability, 1*, 100036.

Kikuchi, F., Kato, Y., Furihata, K., Kogure, T., Imura, Y., Yoshimura, E., & Suzuki, M. (2016). Formation of gold nanoparticles by glycolipids of Lactobacillus casei. *Scientific Reports, 6*, 34626.

Kumari, R., Suman, K., Karmakar, S., Mishra, V., Lakra, S. G., Saurav, G. K., & Mahto, B. K. (2023). Regulation and safety measures for nanotechnology-based agri-products. *Frontiers in Genome Editing, 5*, 1200987.

Lal, R. (2021). Climate change and agriculture. *Climate Change*. https://doi.org/10.1007/springerreference_60242

Liaqat, N., Jahan, N., Khalil Ur, R., Anwar, T., & Qureshi, H. (2022). Green synthesized silver nanoparticles: Optimization, characterization, antimicrobial activity, and cytotoxicity study by hemolysis assay. *Frontiers in Chemistry, 10*, 952006.

Liu, L., Zheng, X., Wei, X., Kai, Z., & Xu, Y. (2021). Excessive application of chemical fertilizer and organophosphorus pesticides induced total phosphorus loss from planting causing surface water eutrophication. *Scientific Reports, 11*, 23015.

Liu, J., Wang, X., Chang, J., Du, P., Wu, J., Hou, R., Zhu, S., Liu, P., Miao, X., Zhang, P., & Zhang, Z. (2024). Green synthesized lignin nanoparticles for the sustainable delivery of pyraclostrobin to control strawberry diseases caused by Botrytis cinerea. *International Journal of Biological Macromolecules, 274*, 133488.

López-Vargas, E. R., Ortega-Ortíz, H., Cadenas-Pliego, G., De Alba Romenus, K., Cabrera De La Fuente, M., Benavides-Mendoza, A., & Juárez-Maldonado, A. (2018). Foliar application of copper nanoparticles increases the fruit quality and the content of bioactive compounds in tomatoes. *Applied Sciences, 8*, 1020.

Maity, D., Gupta, U., & Saha, S. (2022). Biosynthesized metal oxide nanoparticles for sustainable agriculture: Next-generation nanotechnology for crop production, protection and management. *Nanoscale*. https://doi.org/10.1039/d2nr03944c

Marathe, K., Naik, J., & Maheshwari, V. (2021). Biogenic synthesis of silver nanoparticles using Streptomyces spp. and their antifungal activity against Fusarium verticillioides. *Journal of Cluster Science, 32*, 1299–1309.

Markus, J., Mathiyalagan, R., Kim, Y.-J., Abbai, R., Singh, P., Ahn, S., Perez, Z. E. J., Hurh, J., & Yang, D. C. (2016). Intracellular synthesis of gold nanoparticles with antioxidant activity by probiotic Lactobacillus kimchicus DCY51T isolated from Korean kimchi. *Enzyme and Microbial Technology, 95*, 85–93.

Mehri, K., Parinaz, A., Leila Ashrafi, B., & Maryam, A.-D. (2020). Biosynthesis of copper oxide nanoparticles using Lactobacillus casei subsp. casei and its anticancer and antibacterial activities. *Current Nanoscience, 16*, 101–111.

Monga, Y., Kumar, P., Sharma, R. K., Filip, J., Varma, R. S., Zbořil, R., & Gawande, M. B. (2020). Sustainable synthesis of nanoscale zerovalent iron particles for environmental remediation. *ChemSusChem, 13*, 3288–3305.

Monteiro, E., & Ferreira, S. (2022). Biomass waste for energy production. *Energies*. https://doi.org/10.3390/en15165943

Mukherjee, S., Verma, A., Kong, L., Rengan, A. K., & Cahill, D. M. (2024). Advancements in green nanoparticle technology: Focusing on the treatment of clinical phytopathogens. *Biomolecules, 14*, 1082.

Nadaroglu, H., Alaylı Güngör, A., & İnc, S. (2017). Synthesis of nanoparticles by green synthesis method. *International Journal of Innovative Research and Reviews, 1*, 6–9.

Nawandar, N. K., & Satpute, V. R. (2019). IoT based low cost and intelligent module for smart irrigation system. *Computers and Electronics in Agriculture, 162*, 979–990.

Nikitin, D. S., Shanenkov, I. I., Yeletsky, P. M., Nassyrbayev, A., Tabakaev, R. B., Shanenkova, Y. L., Ryskulov, D. N., Tsimmerman, A. I., & Sivkov, A. A. (2024). Agricultural waste derived silicon carbide composite nanopowders as efficient coelectrocatalysts for water splitting. *Journal of Cleaner Production, 442*, 140890.

Oliveira, H. C., Stolf-Moreira, R., Martinez, C. B., Grillo, R., De Jesus, M. B., & Fraceto, L. F. (2015). Nanoencapsulation enhances the post-emergence herbicidal activity of atrazine against mustard plants. *PLoS One, 10*, e0132971.

Oluwole, O., Ibidapo, O., Arowosola, T., Raji, F., Zandonadi, R., Alasqah, I., Lho, L. H., Han, H., & Raposo, A. (2023). Sustainable transformation agenda for enhanced global food and nutrition security: A narrative review. *Frontiers in Nutrition, 10*, 1226538.

Pan, W. J., Tao, J. J., Cheng, T., Bian, X. H., Wei, W., Zhang, W. K., Ma, B., Chen, S. Y., & Zhang, J. S. (2016). Soybean miR172a improves salt tolerance and can function as a long-distance signal. *Molecular Plant, 9*, 1337–1340.

Panferov, V. G., Safenkova, I. V., Byzova, N. A., Varitsev, Y. A., Zherdev, A. V., & Dzantiev, B. B. (2018). Silver-enhanced lateral flow immunoassay for highly-sensitive detection of potato leafroll virus. *Food and Agricultural Immunology, 29*, 445–457.

Panpatte, D. G., Jhala, Y. K., Shelat, H. N., & Vyas, R. V. (2016). Nanoparticles: The next generation technology for sustainable agriculture. In D. P. Singh, H. B. Singh, & R. Prabha (Eds.), *Microbial inoculants in sustainable agricultural productivity: Vol. 2: Functional applications*. Springer India.

Pereira, A. E. S., Sandoval-Herrera, I. E., Zavala-Betancourt, S. A., Oliveira, H. C., Ledezma-Pérez, A. S., Romero, J., & Fraceto, L. F. (2017). γ-Polyglutamic acid/chitosan nanoparticles for the plant growth regulator gibberellic acid: Characterization and evaluation of biological activity. *Carbohydrate Polymers, 157*, 1862–1873.

Ponmurugan, P., Manjukarunambika, K., Elango, V., & Gnanamangai, B. M. (2016). Antifungal activity of biosynthesised copper nanoparticles evaluated against red root-rot disease in tea plants. *Journal of Experimental Nanoscience, 11*, 1019–1031.

Ragazou, K., Garefalakis, A., Zafeiriou, E., & Passas, I. (2022). Agriculture 5.0: A new strategic management mode for a cut cost and an energy efficient agriculture sector. *Energies*. https://doi.org/10.3390/en15093113

Riaz, S., Raza, Z. A., & Majeed, M. I. (2020). Preparation of cadmium sulfide nanoparticles and mediation thereof across poly(hydroxybutyrate) nanocomposite. *Polymer Bulletin, 77*, 775–791.

Sah, M. K., Thakuri, B. S., Pant, J., Gardas, R. L., & Bhattarai, A. (2024). The multifaceted perspective on the role of green synthesis of nanoparticles in promoting a sustainable green economy. *Sustainable Chemistry, 5*, 40–59.

Santhosh, P. B., Genova, J., & Chamati, H. (2022). Green synthesis of gold nanoparticles: An eco-friendly approach. *Chemistry, 4*, 345–369.

Shegokar, R., & Nakach, M. (2020). *Large-scale manufacturing of nanoparticles—An industrial outlook*. Elsevier.

Shireen Akhter Jahan, Q., Sultana, Z., Ud-Daula, M. A., Ashikuzzaman, M., Reja, M. S., Rahman, M. M., Khaton, A., Tang, M. A. K., Rahman, M. S., Faruquee, H. M., Lee, S. J., & Rahman, A. T. M. M. (2024). Optimization of green silver nanoparticles as nanofungicides for management of rice bakanae disease. *Heliyon, 10*, e27579.

Singh, P., Singh, H., Kim, Y. J., Mathiyalagan, R., Wang, C., & Yang, D. C. (2016). Extracellular synthesis of silver and gold nanoparticles by Sporosarcina koreensis DC4 and their biological applications. *Enzyme and Microbial Technology, 86*, 75–83.

Singh, N. A., Narang, J., Garg, D., Jain, V., Payasi, D., Suleman, S., & Swami, R. K. (2023). Nanoparticles synthesis via microorganisms and their prospective applications in agriculture. *Plant Nano Biology, 5*, 100047.

Soni, A. T., Rookes, J. E., & Arya, S. S. (2023). Chitosan nanoparticles as seed priming agents to alleviate salinity stress in rice (Oryza sativa L.) seedlings. *Polysaccharides, 4*, 129–141.

Srivastava, S. K., Yamada, R., Ogino, C., & Kondo, A. (2013). Biogenic synthesis and characterization of gold nanoparticles by Escherichia coli K12 and its heterogeneous catalysis in degradation of 4-nitrophenol. *Nanoscale Research Letters, 8*, 70.

Su, Y., Zhou, X., Meng, H., Xia, T., Liu, H., Rolshausen, P., Roper, C., Mclean, J. E., Zhang, Y., Keller, A. A., & Jassby, D. (2022). Cost–benefit analysis of nanofertilizers and nanopesticides emphasizes the need to improve the efficiency of nanoformulations for widescale adoption. *Nature Food, 3*, 1020–1030.

Sun, K., Wang, T., Gong, W., Lu, W., He, X., Eddings, E. G., & Fan, M. (2022). Synthesis and potential applications of silicon carbide nanomaterials/nanocomposites. *Ceramics International, 48*, 32571–32587.

Sundararajan, N., Habeebsheriff, H. S., Dhanabalan, K., Cong, V. H., Wong, L. S., Rajamani, R., & Dhar, B. K. (2024). Mitigating global challenges: Harnessing green synthesized nanomaterials for sustainable crop production systems. *Global Challenges, 8*, 2300187.

Thangadurai, D., Naik, J., Sangeetha, J., Al-Tawaha, A. R. M. S., Adetunji, C. O., Islam, S., David, M., Shettar, A. K., & Adetunji, J. B. (2021). *Nanomaterials from agrowastes: Past, present, and the future*. Springer International Publishing.

Tonelli, F. M. P., & Tonelli, F. C. P. (2023). *Biocompatibility of green synthesized nanomaterials*. Elsevier.

Tubert, E., Vitali, V. A., Alvarez, M. S., Tubert, F. A., Baroli, I., & Amodeo, G. (2018). Synthesis and evaluation of a superabsorbent-fertilizer composite for maximizing the nutrient and water use efficiency in forestry plantations. *Journal of Environmental Management, 210*, 239–254.

Tursi, A. (2019). A review on biomass: Importance, chemistry, classification, and conversion. *Biofuel Research Journal, 6*, 962–979.

Varshney, S., & Gupta, A. (2024). Forest industrial biomass residue-mediated green synthesized multifunctional copper oxide nanoparticles for efficient wastewater treatment and biomedical applications. *Journal of Cleaner Production, 434*, 140109.

Vijayakumar, S., Vinayagam, R., Anand, M. A. V., Venkatachalam, K., Saravanakumar, K., Wang, M.-H., Sangeetha Casimeer, C., Gothandam, K. M., & David, E. (2020). Green synthesis of gold nanoparticle using Eclipta alba and its antidiabetic activities through regulation of Bcl-2 expression in pancreatic cell line. *Journal of Drug Delivery Science and Technology, 58*, 101786.

Wang, B., Dong, F., Chen, M., Zhu, J., Tan, J., Fu, X., Wang, Y., & Chen, S. (2016). Advances in recycling and utilization of agricultural wastes in China: Based on environmental risk, crucial pathways, influencing factors, policy mechanism. *Procedia Environmental Sciences, 31*, 12–17.

Wang, C., Yang, K., Wei, X., Ding, S., Tian, F., & Li, F. (2018). One-pot solvothermal synthesis of carbon dots/Ag nanoparticles/TiO2 nanocomposites with enhanced photocatalytic performance. *Ceramics International, 44*, 22481–22488.

Watson, A., Ghosh, S., Williams, M. J., Cuddy, W. S., Simmonds, J., Rey, M. D., Asyraf Md Hatta, M., Hinchliffe, A., Steed, A., Reynolds, D., Adamski, N. M., Breakspear, A., Korolev, A., Rayner, T., Dixon, L. E., Riaz, A., Martin, W., Ryan, M., Edwards, D., Batley, J., Raman, H., Carter, J., Rogers, C., Domoney, C., Moore, G., Harwood, W., Nicholson, P., Dieters, M. J., Delacy, I. H., Zhou, J., Uauy, C., Boden, S. A., Park, R. F., Wulff, B. B. H., & Hickey, L. T. (2018). Speed breeding is a powerful tool to accelerate crop research and breeding. *Nature Plants, 4*, 23–29.

Wu, Y., Ye, X., Wang, Y., & Wang, L. (2023). Methane production from biomass by thermochemical conversion: A review. *Catalysts, 13*, 771.

Xu, L., Zhu, Z., & Sun, D. W. (2021). Bioinspired nanomodification strategies: Moving from chemical-based agrosystems to sustainable agriculture. *ACS Nano, 15*, 12655–12686.

Xu, J., Gu, B., & Tian, G. (2022). Review of agricultural IoT technology. *Artificial Intelligence in Agriculture, 6*, 10–22.

Yadav, A., Yadav, K., & Abd-Elsalam, K. A. (2023). Exploring the potential of nanofertilizers for a sustainable agriculture. *Plant Nano Biology, 5*, 100044–100044.

Yan, T., Ren, Y., Zhang, R., Li, K., Yang, B., Tong, M., & He, J. (2025). Biodegradable chitosan-based films decorated with biosynthetic copper oxide nanoparticle for post-harvest tomato preservation. *International Journal of Biological Macromolecules, 295*, 139595.

Ying, S., Guan, Z., Ofoegbu, P. C., Clubb, P., Rico, C., He, F., & Hong, J. (2022). Green synthesis of nanoparticles: Current developments and limitations. *Environmental Technology & Innovation, 26*, 102336.

Youssef, K., Hashim, A. F., Hussien, A., & Abd-Elsalam, K. A. (2017). Fungi as ecosynthesizers for nanoparticles and their application in agriculture. In R. Prasad (Ed.), *Fungal nanotechnology: Applications in agriculture, industry, and medicine*. Springer International Publishing.

Zhan, F., Wang, T., Iradukunda, L., & Zhan, J. (2018). A gold nanoparticle-based lateral flow biosensor for sensitive visual detection of the potato late blight pathogen, Phytophthora infestans. *Analytica Chimica Acta, 1036*, 153–161.

Zhu, J. Y., Pan, X., & Zalesny, R. S. (2010). Pretreatment of woody biomass for biofuel production: Energy efficiency, technologies, and recalcitrance. *Applied Microbiology and Biotechnology, 87*, 847–857.

Zulfiqar, Z., Khan, R. R. M., Summer, M., Saeed, Z., Pervaiz, M., Rasheed, S., Shehzad, B., Kabir, F., & Ishaq, S. (2024). Plant-mediated green synthesis of silver nanoparticles: Synthesis, characterization, biological applications, and toxicological considerations: A review. *Biocatalysis and Agricultural Biotechnology, 57*, 103121.

Index

A
Abiotic, 5, 19, 79, 90, 101–110, 137, 138, 162, 194, 286
Active packaging, 18, 214, 219, 220, 229
Aggregation, 35, 36, 43, 45, 79, 83, 85, 113, 120, 195, 208, 226, 228, 243–248, 281, 288, 292
Agricultural innovation, 13, 17, 130, 190, 370, 376, 379, 383
Agricultural nanotechnology, 55, 58, 67, 71
Agricultural productivity, 17, 103, 105, 114, 132, 140, 150, 157, 165, 166, 175, 199, 273, 274, 371–373, 394
Agricultural sustainability, 70, 103, 148, 199
Agricultural waste, 10, 18, 140, 202, 213, 273, 305, 308, 319, 340, 356, 375, 378, 399–401
Agriculture, 2, 71, 79, 106, 114, 129, 148, 190, 209, 242, 274, 304, 335, 369, 393
Agrochemical delivery, 4, 63
Agrochemicals, 4, 5, 16, 63, 69, 114, 116, 129, 136, 148–150, 157, 159, 163, 165, 273, 280, 406
Agrochemicals protection, 149
Agroforestry, 308–310, 314, 315, 320, 322, 338
Agro-industrial waste, 338, 340–344, 358
Airborne nanoparticles, 244, 247, 264
Algal, 130, 208, 246, 247
Antimicrobial nanomaterials, 166–167
Antioxidant system, 79, 80, 91, 116
Application, 1, 29, 55, 79, 106, 114, 129, 157, 191, 209, 242, 257, 274, 314, 339, 371, 393
Aquaponics, 315, 316
Atom economy, 30
Auxiliaries, safer alternatives, 30

B
Bacteria, 5, 11, 12, 14, 19, 28, 33, 34, 36–39, 41, 59, 84, 102, 116, 119, 120, 152, 153, 155, 167–170, 173, 174, 193, 196, 198, 208, 209, 212, 213, 217, 219–222, 258, 279, 282, 283, 286, 314, 315, 348, 394, 396
Bacteria-mediated synthesis, 38
Banana cropping, 322–325
Benefits, 4, 6, 10, 13, 16, 33, 36, 41, 56, 59, 61–65, 68, 70, 80, 115, 121, 122, 138–140, 148, 164, 165, 167, 172, 210, 224–225, 242, 250, 252, 261, 264, 273–275, 278–281, 304, 305, 311, 313–315, 319, 322, 323, 325, 339, 343, 347, 357, 370–373, 376–383, 385, 400, 407
Bioaccumulation, 41, 66, 114, 120, 149, 164, 212, 224–226, 228, 242, 246–249, 252, 257, 259
Bioactive compounds, 90, 108, 213, 216
Bioavailability, 37, 115, 116, 138, 149, 156, 157, 164, 191, 193, 198–200, 213, 223, 243, 246, 288, 292
Bio-based synthesis, 196
Biocatalysis, 32
Biocompatibility, 8, 30, 34, 37, 57, 60, 62, 66, 120, 157, 167, 197, 200, 278, 281, 282, 292, 405
Biocompatible carriers, 151

K. Esquivel Escalante et al. (eds.), *Green Nanomaterials*, Nanotechnology in the Life Sciences, https://doi.org/10.1007/978-3-032-04492-1

Biodegradable materials, 30, 32, 304
Bioeconomy, 326, 344, 375
Bioenergy production, 356, 400
Biofuels, 305, 309, 310, 315, 316, 323, 324, 326, 327, 344, 345, 400, 401
Bio-nanotechnology, 242
Biopolymers, 12, 120, 136, 213, 216–218, 221, 342
Bio-reduction, 279
Biosafety, 257–268
Biosorption, 34, 36
Biostimulants, 172
Biosynthesis, 28, 33, 34, 36, 37, 39, 41–45, 91, 92, 105, 107–109, 196, 197, 212, 283, 292, 394
Biosynthesis of nanoparticles, 34–35, 41
Biotic, 19, 79, 90, 101–109, 138, 162, 286
Blockchain, 311, 313, 319, 329
Bottom-up synthesis, 1, 80–83, 131, 282

C
Carbon-based nanomaterials (CBNs), 3, 60–62, 135, 167
Carbon nanotubes (CNTs), 3, 6, 20, 60, 61, 64, 68, 70, 91, 120, 121, 132, 134, 135, 163, 168, 170, 172, 246, 247, 258, 259, 262, 396
Cation exchange capacity (CEC), 17, 102, 115, 117
Cellular uptake, 89, 162, 170, 259
Challenges in nanofertilizer adoption, 203
Chitosan, 5, 62, 63, 91, 92, 120, 154–156, 158–162, 168, 172, 174, 194, 196, 213, 215–221, 280, 395, 398
Circular chemistry, 32
Circular economy, 32, 44, 55–71, 148, 193, 213, 242, 252, 303–329, 333–358, 369, 370, 372, 373, 376, 378, 384, 407
Climate change adaptation, 71
Climate resilience, 303
Coffee farming, 310, 322
Compliance, 67, 228, 251, 263–268, 319, 350, 374, 404
Composting, 308, 313, 314, 317, 319, 320, 329, 341, 356
Controlled-release, 16, 63, 64, 120, 129, 131, 136, 139, 140, 148, 149, 151, 156, 157, 162, 168, 172, 189, 192, 199, 200, 202, 223, 274, 280, 291, 292, 372, 374, 375, 406
Controlled release fertilizers, 375
Controlled release systems, 63
Copper nanoparticles (CuNPs), 16, 59, 107, 109, 119, 155, 159, 166, 167, 172, 217, 259, 396
Cost-benefit analysis, 68, 275–286, 353, 371, 404, 406
Cost-effectiveness of nanofertilizers, 140
Cost-efficiency, 175, 198, 292
Crop productivity, 17, 113, 139, 140, 149, 202, 243, 375, 407
Crop protection, 3, 5, 58, 60, 65, 69, 71, 166, 167, 341, 394, 396
Crop quality improvement, 19, 402
Crop yield improvements, 275
Crop yields, 6, 13, 56, 103, 106, 129, 139, 148, 168, 189, 191, 193, 198–201, 249, 273, 274, 280, 286, 291, 307, 314, 371, 373, 374, 376, 380, 382
Current challenges, 37, 45

D
Decision-making, 250, 314, 322
Disease control, 59–60, 394
Dissolution, 151, 243–246

E
Eco-friendly fertilizers, 131, 136, 195, 198
Eco-friendly inputs, 374
Eco-friendly nanomaterials, 57, 62
Economic assessments, 58
Economic benefits, 275, 281, 305, 371, 377, 405–406
Economics, 29, 31, 39, 45, 57, 67–69, 71, 147, 193, 224–225, 261, 266, 273–292, 304–307, 309, 311, 313, 315, 320, 322, 325, 326, 329, 333, 335–339, 342, 344, 346, 347, 349, 350, 369–385, 394, 398, 400, 402–407
Economic viability, 46, 147, 193, 228, 274, 280, 286, 311, 370, 371, 402, 403
Edible coatings, 215, 216, 218–219, 229
Enhanced nutrient uptake, 135
Environment, 28–31, 34, 36, 38, 39, 41, 57–59, 65, 67–71, 79, 80, 82, 84, 101, 102, 104, 110, 114, 117, 118, 122, 123, 130, 136, 138–140, 149, 163, 171, 193, 195, 208, 209, 219, 224, 226, 229, 242–247, 258, 261–265, 267, 273, 277, 283, 288–290, 304, 306–309, 315, 333–335, 337, 346, 357, 370, 373, 395, 398, 399, 407

Environmental, 1, 27, 57, 89, 101, 119, 130, 157, 191, 209, 242, 257, 273, 304, 333, 369, 393
Environmental economics, 320, 370–372
Environmental fate, 165, 191, 251
Environmental impacts, 8, 13, 16, 27–33, 44, 62, 93, 108, 109, 147–149, 157, 163–166, 189, 191–199, 209, 213, 228, 229, 242, 250, 257, 258, 260, 264, 278, 279, 285, 292, 304, 314, 324, 329, 338, 345, 349–351, 353, 373, 376, 382, 383, 393, 396, 405, 407
Environmental modeling, 245, 313, 320
Environmental monitoring, 242
Environmental persistence, 259
Environmental release, 242, 250, 264
Environmental remediation, 7, 17, 168, 198, 257, 282, 286, 339–341, 404–406
Environmental safety, 13, 156, 198, 264, 278, 372
Environmental sustainability, 46, 71, 201, 242, 251, 283, 285, 306, 336, 350, 406, 407
Environmental transformations, 242–244, 246
Enzyme-mediated synthesis, 45
Extracellular nanoparticle synthesis, 39

F
Fertilizer efficiency, 6, 16, 70, 140, 148, 192, 195, 198–200, 275, 371, 372, 374, 382, 393, 399
Fertilizer runoff, 16, 130, 191, 371
Field crops productivity, 139
Flavor enhancer, 222
Foliar applications, 88, 89, 92, 107, 163
Food analysis, 221, 229
Food security, 18, 21, 70, 71, 129, 140, 148, 149, 166, 175, 189, 190, 274, 303, 349, 372, 407
Fungi, 10, 14, 17, 18, 28, 33, 34, 36, 37, 39–41, 44, 59, 84, 102, 116, 120, 121, 137, 140, 149, 152–155, 158, 160, 161, 167–169, 193, 196, 208, 209, 211–213, 217, 219, 276, 277, 279, 282, 283, 286, 287, 291, 314, 394, 401
Fungi in nanoparticle synthesis, 39
Future of nanofertilizers, 140, 203
Future perspectives, 19–20

G
Gene expressions, 4, 80, 90–92, 101, 105, 113, 164, 166
Genetic improvement, 56
Gold nanoparticles (AuNPs), 12, 39–41, 44, 58, 109, 120, 132, 170, 171, 210–212, 246, 259, 288, 396, 403–406
Graphene oxide (GO), 3, 5, 20, 61, 116, 117, 132, 170, 259, 343
Green, 8, 57, 84, 108, 175, 194, 207, 242, 275, 315, 333, 372, 399
Green chemistry, 27–33, 45, 195–196, 251, 282, 284, 292
Green chemistry principles, 27–37, 61, 193, 278, 280
Greenhouse gas emissions, 17, 130, 139, 147, 202, 303–305, 311, 314, 324, 336–338, 358, 401
Green nanoparticles, 11, 31, 32, 44, 209–213, 216, 218, 223, 277, 288–289, 323, 403–406
Green nanoparticles in agriculture, 406
Green nanotechnology, 80, 108, 147, 193, 209, 242, 250–252, 283, 370, 372–374, 379, 382–384, 406, 407
Green solvents, 30, 31, 278
Green synthesis, 8, 27–46, 57, 84, 109, 129, 175, 193–194, 207, 273, 382, 393
Green synthesized nanomaterials, 58, 66–71, 108–109, 207–230, 403

H
Heavy metal remediation, 17, 60, 103, 116, 118
Heavy metals, 5, 17, 60, 61, 101, 103, 105–107, 110, 115–118, 138, 171, 242, 286, 341, 351, 356
High surface area nanoparticles, 224
Hormesis, 87, 122
Hormones, 59, 88, 136, 137
Horticultural applications, 346
Hybrid nanomaterials, 64–66, 70, 134–136
Hydroponics, 88, 130, 315, 316

I
Incineration, 250, 264, 289–290, 349, 351
Inherently safer design, 32
International standards, 265–266
Intracellular nanoparticle synthesis, 36, 42
Ionic liquids, 31
Iron oxide nanoparticles, 9, 58–60, 133, 138, 201, 224, 246

L
Life cycle assessment (LCA), 285–286, 344, 345, 349–356

Life-cycle NPs, 285
Life cycles, 101, 103, 229, 278, 285, 292, 306, 338, 345, 347, 349, 351, 352, 354, 356, 358, 382
Ligand chemistry, 87, 93
Limitations of green nanoparticles, 281–283

M
Machine learning, 20, 245–246, 250, 311
Manufacturing, 29, 57, 67, 228, 229, 243, 274, 281, 285–288, 292, 304, 307, 342, 381
Market adoption, 228
Mechanistic models, 245
Metabolic pathways, 33, 35, 91, 92
Metal-based nanomaterials, 58–60
Metal nanoparticles, 4, 12, 16, 30, 36, 37, 41, 57, 59, 60, 64–66, 70, 80–84, 108, 119, 122, 170, 171, 212, 382
Microbial biomass, 117
Microbial synthesis, 34, 196
Microbiomes, 314
Multimedia models, 245
Mycorrhizal fungi, 121, 130

N
Nanocarriers, 4, 20, 63, 150, 151, 156, 157, 162–164, 172, 175, 198
Nanocarrier systems, 59, 150, 157, 163
Nano-coatings, 214–216, 218, 224, 260
Nanocomposites (NCs), 2, 5, 131, 136, 174, 215–220, 224, 242, 282
Nano-enabled medical devices, 243
Nanoencapsulation, 157, 224
Nano-enhanced fertilizers, 199–200
Nanofertilizers, 114, 129–140, 148, 190, 225, 275, 393
Nano-formulated fertilizers, 195
Nanoformulations, 154, 157–159
Nanomaterial lifecycle, 263
Nanomaterials (NMs), 1–21, 33, 55–71, 89, 105, 113–118, 129, 148, 192, 207, 241, 257, 273, 339, 371, 393
Nanomaterials in agriculture, 58–71, 173, 380–382
Nanoparticle concentrations, 228
Nanoparticle morphology, 35, 37, 39, 42, 81, 199, 200
Nanoparticle reactivity, 226
Nanoparticles (NPs), 2, 33, 57, 91, 105, 113, 130, 148, 193, 207, 242, 257, 273, 343, 372, 394
Nanoparticle shape, 6, 12, 36, 42, 45, 84–86, 196, 212
Nanoparticle size, 14, 42, 45, 86, 208
Nanoparticle stability, 36, 37, 210, 274
Nanoparticle synthesis, 30–35, 37–39, 41, 42, 44, 45, 59, 81, 84, 175, 194–197, 281, 287, 403–404
Nanoparticle transport, 85, 87
Nano-pesticides, 16, 55, 116, 148–165, 175, 225, 393, 395, 404, 406, 407
Nanoregulation, 261–263, 267
Nanoscale nutrients, 130
Nano-sensors, 17–20, 376, 378, 382
Nanostructures, 57, 220, 225, 242, 258, 339
Nanotechnology, 1, 33, 56, 79, 106, 113, 138, 149, 189, 209, 242, 258, 273, 339, 375, 393
Nanotechnology in agriculture, 2, 13–20, 55–56, 106, 114, 166, 175, 377, 382–383
Nanotoxicity, 280, 288–289
Nano-waste, 292
Non-target species, 242, 246–248
Nutrient leaching, 69
Nutrient management, 68–69, 71, 199
Nutrient-use efficiency, 6, 194, 196, 198, 200, 372, 382, 393
Nutritional value of crops, 201

O
Organic acids (in biosynthesis), 42
Organic farming, 320, 321
Organic fertilizers, 16, 307, 308, 313, 314, 323, 325
Oxidation, 7, 28, 35, 116, 122, 219, 243, 245, 246, 348, 349, 351, 352, 354, 356, 358, 399
Oxidative stress, 90, 91, 104, 105, 107, 122, 132, 151, 163, 164, 167, 168, 170, 171, 197, 201, 242, 246–249, 257, 259, 405

P
Pathogen inhibition, 153
Pest management, 55, 60, 62, 65, 320, 394
pH effect in green synthesis, 405
Phenolic acid, 35, 276
Photooxidation, 246
Photosynthesis, 3, 5, 90–93, 102, 103, 105, 107, 110, 129, 132, 135, 168, 246, 281, 400
Photosynthesis enhancement, 168
Physiological processes, 90, 110

Phytochemicals, 19, 30, 34, 35, 42, 80, 208, 274, 281
Phytochemicals in synthesis, 35
Phytopathogens, 119, 121, 165, 166, 172, 174
Plant extracts, 11, 12, 14, 31, 34, 41–43, 57, 59, 66, 108, 110, 132, 136, 138, 193–195, 197–199, 208, 210, 213, 218, 219, 228, 276–278, 282, 287, 291, 393, 397, 403–406
Plant growth, 3, 4, 15, 55, 59, 69, 70, 80, 90, 101–103, 106, 109, 110, 117, 132–139, 148, 157, 163, 192, 198, 199, 247, 274, 286, 314, 343, 372, 395, 398
Plant growth and development, 79, 103, 122, 138, 139, 372
Plant growth promoters, 129–140
Plant growth promotion, 90, 122, 133, 137, 138, 314, 372
Plant-mediated synthesis, 32, 57
Plant-nanomaterial interaction, 6, 80
Plant nutrition, 14, 189–203, 274
Plant production, 80, 106
Plants, 2, 28, 55, 92, 101–110, 113, 130, 148, 190, 208, 242, 274, 307, 334, 372, 394
Policies, 147, 175, 250, 262, 266, 267, 317, 318, 326, 329, 338, 348, 357, 370, 374–377, 382–385
Policy frameworks, 357, 374–375, 384
Polymer-based nanomaterials, 3, 62–64, 70, 134–136, 167
Precautionary principles, 226, 263
Precision agriculture, 17–18, 20, 65, 70, 225, 311, 313, 314, 329, 357, 375, 378, 379, 382, 407
Precision farming, 328, 356, 374, 377, 383, 384
Precision nutrient delivery, 20, 68, 140
Principles, 1, 2, 20, 27–46, 57, 63, 84, 242, 250–252, 273, 283, 303–329, 338, 340, 358, 369, 370, 375, 378, 382, 384
Protein-mediated reduction, 35
Public perception, 225, 267, 381

R
Reactive oxygen species (ROS), 59, 62, 80, 91–93, 103–105, 107, 109, 110, 117–119, 122, 155, 163, 167, 169–171, 175, 248, 257, 288, 289, 343, 373
Reactive transport, 245
Recycling of nanoparticle, 289, 291
Recycling process, 291
Regulations, 21, 86, 103, 110, 163, 170, 175, 225–228, 249, 250, 257, 258, 260–265, 267, 287, 313, 318, 319, 348, 374, 375, 377, 394, 404, 407
Regulatory challenges, 165, 203, 267, 379, 394
Regulatory frameworks, 140, 175, 203, 227, 229, 242, 250, 252, 262, 263, 266–268, 357, 380
Renewable feedstocks, 31
Renewable resources, 27, 29, 31, 33, 62, 213, 225, 280, 369
Residue reduction, 277–278, 308
Resource efficiency, 46, 116, 140, 147, 242, 304, 310, 314, 317, 324, 328, 369, 372, 375, 398–399
Resource management, 56, 66, 71, 357–358
Responsible innovation, 209
Reuse, 202, 304, 308, 310, 312, 317, 323, 348, 351, 358, 401
Risk assessment, 67, 165, 226, 227, 242, 246, 248–252, 261–264, 266–268
Root uptake, 90

S
Saccharides in NP synthesis, 35
Safety issues, 71, 225
Safety protocols, 258, 263–265, 267
Sanitary landfilling, 290–291
Secondary metabolism, 80, 90
Secondary metabolites, 35, 42, 80, 88, 90–92, 104, 108, 110, 137, 276, 278, 287
Sedimentation, 243, 245, 246, 249, 291, 292
Silver nanoparticles, 1, 5, 12, 13, 16, 18, 19, 35, 39, 58, 59, 69, 130, 132, 136, 155, 163, 164, 166, 170, 208–212, 219, 221, 222, 225, 243, 245–247, 258–260, 396, 404
Slow-release fertilizers, 194, 373, 378, 382, 399
Smart delivery systems, 16, 70
Smart delivery technologies, 55
Smart farming, 17–18, 20, 65
Smart nutrient systems, 68–69, 71
Smart packaging, 216, 220
Soil biochemical properties, 117
Soil enzymatic activity (SEA), 115, 117
Soil fertility, 15, 60, 117, 122, 129, 134, 136, 198, 247, 308, 310, 314, 320, 342, 356, 375
Soil health, 17, 19, 64, 65, 69, 117, 118, 139, 175, 194, 202, 225, 247, 292, 308, 309, 314, 318, 320, 329, 338, 339, 341, 342, 357, 372, 376–379, 383
Soil microbial diversity, 119
Soil microbiome, 314
Soil mobility, 17, 114, 191, 244

Soil pH, 17, 115, 118
Soil physical properties, 114
Soil remediation, 5, 17, 20, 58, 60, 64, 116–118, 123
Soil structure, 17, 64, 65, 114, 194, 202, 308, 309, 314, 315
Soil toxicity, 115, 118, 136
Sonochemistry, 31, 278, 292
Stakeholder engagement, 252
Strategic planning, 376
Stress, 3, 56, 79, 106, 113, 138, 193, 227, 274, 338, 402
Stress tolerance, 2, 102, 110, 133, 139, 190, 292, 373
Sugarcane industry, 325–328
Surface charge, 85, 87–89, 93, 122, 259, 275
Surface functionalization, 42, 120, 243, 246, 258
Surface passivation, 246
Sustainability, 13, 44, 57, 79, 110, 139, 147, 190, 209, 242, 263, 274, 304, 335, 369, 399
Sustainability indicators, 17, 318, 370
Sustainable agriculture, 2, 4, 17, 60, 71, 106, 133, 139, 147–149, 151, 175, 189, 190, 193, 195, 198, 273, 280, 315, 326, 348, 371, 374–376, 378, 384, 385, 393–407
Sustainable chemistry, 27, 29, 33
Sustainable crop production, 198
Sustainable development, 29, 33, 57, 140, 266, 324, 336, 337, 346, 406
Sustainable Development Goals (SDGs), 209, 279, 306, 307, 336, 338, 375, 384
Sustainable farming, 21, 106, 147, 166, 309, 314, 407
Sustainable farming practices, 199, 371, 380, 382, 394–396, 402–406
Sustainable nanomaterial, 38, 62
Sustainable practices, 17, 29, 57, 229, 241, 319–321, 356, 357, 374, 375, 379, 384, 398
Systemic change, 318

T

Targeted delivery, 16, 58, 63, 172, 175, 189, 274, 374
Targeted nutrient delivery, 5, 16, 20
Technological transition, 350
Temperature effect in biosynthesis, 283
TiO_2 nanoparticles (TiO_2 NPs), 4, 81, 91, 107, 109, 117, 119, 120, 172, 174, 215, 216, 219, 242, 246, 248, 277, 285, 290, 291, 373
Top-down synthesis, 1, 80, 131
Toxicity, 5, 12–14, 29, 30, 34, 41, 57, 66, 88, 107, 109, 115–118, 121, 122, 134, 136, 148, 154, 157, 164, 167, 169–171, 175, 191, 208, 210, 225, 226, 228, 242–244, 246–249, 251, 252, 259, 263, 264, 273, 274, 280, 281, 284, 288–289, 349, 404, 405
Toxicity concerns, 292
Toxicological effects, 246–248
Transparency, 18, 216, 217, 227, 313, 318, 319, 357, 378
Transport behaviors, 244
12 Principles of green chemistry, 29

U

United Nations, 335, 336, 375

V

Value chains, 310, 318

W

Waste management, 17, 264, 305, 306, 308, 317, 323, 349
Waste minimization, 28, 276, 336, 372
Waste prevention, 27
Waste valorization, 44, 324, 358
Water conservation, 134, 336
Water management, 329, 348, 357
Water purification, 13, 61–62
Water transport, 244–245
Water use, 62, 317, 320, 347–349, 374, 376, 394

Y

Yeast, 12, 36, 37, 41, 84, 155, 208, 210, 211, 258, 315
Yeast-mediated synthesis, 12, 211

Z

Zinc oxide (ZnO), 4, 9, 16, 30, 37, 38, 58, 59, 106, 107, 130, 132, 163, 164, 167, 198, 213–215, 222, 247, 259, 277, 291, 382
Zinc oxide nanoparticle (ZnO NPs), 133, 163, 167, 170, 173, 174, 215, 217, 218, 246, 248, 259, 275, 405

MIX
Papier aus verantwortungsvollen Quellen
Paper from responsible sources
FSC® C105338
FSC
www.fsc.org

If you have any concerns about our products,
you can contact us on
ProductSafety@springernature.com

In case Publisher is established outside the EU,
the EU authorized representative is:
Springer Nature Customer Service Center GmbH
Europaplatz 3, 69115 Heidelberg, Germany

Printed by Libri Plureos GmbH
in Hamburg, Germany